LINEAR ALGEBRA

Second Edition

KENNETH HOFFMAN
Professor of Mathematics
Massachusetts Institute of Technology

RAY KUNZE
Professor of Mathematics
University of California, Irvine

Prentice Hall

PRENTICE HALL, Upper Saddle River, New Jersey 07458

Original edition, entitled LINEAR ALGEBRA, 2nd Edition, by HOFFMAN, KENNETH M; KUNZE, RAY, published by Pearson Education, Inc, publishing as Prentice Hall, Copyright © 1971.

KOREA edition published by PEARSON EDUCATION KOREA LTD, Copyright © 2013.

ISBN 978-89-450-0634-9

Preface

Our original purpose in writing this book was to provide a text for the undergraduate linear algebra course at the Massachusetts Institute of Technology. This course was designed for mathematics majors at the junior level, although three-fourths of the students were drawn from other scientific and technological disciplines and ranged from freshmen through graduate students. This description of the M.I.T. audience for the text remains generally accurate today. The ten years since the first edition have seen the proliferation of linear algebra courses throughout the country and have afforded one of the authors the opportunity to teach the basic material to a variety of groups at Brandeis University, Washington University (St. Louis), and the University of California (Irvine).

Our principal aim in revising *Linear Algebra* has been to increase the variety of courses which can easily be taught from it. On one hand, we have structured the chapters, especially the more difficult ones, so that there are several natural stopping points along the way, allowing the instructor in a one-quarter or one-semester course to exercise a considerable amount of choice in the subject matter. On the other hand, we have increased the amount of material in the text, so that it can be used for a rather comprehensive one-year course in linear algebra and even as a reference book for mathematicians.

The major changes have been in our treatments of canonical forms and inner product spaces. In Chapter 6 we no longer begin with the general spatial theory which underlies the theory of canonical forms. We first handle characteristic values in relation to triangulation and diagonalization theorems and then build our way up to the general theory. We have split Chapter 8 so that the basic material on inner product spaces and unitary diagonalization is followed by a Chapter 9 which treats sesqui-linear forms and the more sophisticated properties of normal operators, including normal operators on real inner product spaces.

We have also made a number of small changes and improvements from the first edition. But the basic philosophy behind the text is unchanged.

We have made no particular concession to the fact that the majority of the students may not be primarily interested in mathematics. For we believe a mathematics course should not give science, engineering, or social science students a hodgepodge of techniques, but should provide them with an understanding of basic mathematical concepts.

On the other hand, we have been keenly aware of the wide range of backgrounds which the students may possess and, in particular, of the fact that the students have had very little experience with abstract mathematical reasoning. For this reason, we have avoided the introduction of too many abstract ideas at the very beginning of the book. In addition, we have included an Appendix which presents such basic ideas as set, function, and equivalence relation. We have found it most profitable not to dwell on these ideas independently, but to advise the students to read the Appendix when these ideas arise.

Throughout the book we have included a great variety of examples of the important concepts which occur. The study of such examples is of fundamental importance and tends to minimize the number of students who can repeat definition, theorem, proof in logical order without grasping the meaning of the abstract concepts. The book also contains a wide variety of graded exercises (about six hundred), ranging from routine applications to ones which will extend the very best students. These exercises are intended to be an important part of the text.

Chapter 1 deals with systems of linear equations and their solution by means of elementary row operations on matrices. It has been our practice to spend about six lectures on this material. It provides the student with some picture of the origins of linear algebra and with the computational technique necessary to understand examples of the more abstract ideas occurring in the later chapters. Chapter 2 deals with vector spaces, subspaces, bases, and dimension. Chapter 3 treats linear transformations, their algebra, their representation by matrices, as well as isomorphism, linear functionals, and dual spaces. Chapter 4 defines the algebra of polynomials over a field, the ideals in that algebra, and the prime factorization of a polynomial. It also deals with roots, Taylor's formula, and the Lagrange interpolation formula. Chapter 5 develops determinants of square matrices, the determinant being viewed as an alternating n-linear function of the rows of a matrix, and then proceeds to multilinear functions on modules as well as the Grassman ring. The material on modules places the concept of determinant in a wider and more comprehensive setting than is usually found in elementary textbooks. Chapters 6 and 7 contain a discussion of the concepts which are basic to the analysis of a single linear transformation on a finite-dimensional vector space; the analysis of characteristic (eigen) values, triangulable and diagonalizable transformations; the concepts of the diagonalizable and nilpotent parts of a more general transformation, and the rational and Jordan canonical forms. The primary and cyclic decomposition theorems play a central role, the latter being arrived at through the study of admissible subspaces. Chapter 7 includes a discussion of matrices over a polynomial domain, the computation of invariant factors and elementary divisors of a matrix, and the development of the Smith canonical form. The chapter ends with a discussion of semi-simple operators, to round out the analysis of a single operator. Chapter 8 treats finite-dimensional inner product spaces in some detail. It covers the basic geometry, relating orthogonalization to the idea of 'best approximation to a vector' and leading to the concepts of the orthogonal projection of a vector onto a subspace and the orthogonal complement of a subspace. The chapter treats unitary operators and culminates in the diagonalization of self-adjoint and normal operators. Chapter 9 introduces sesqui-linear forms, relates them to positive and self-adjoint operators on an inner product space, moves on to the spectral theory of normal operators and then to more sophisticated results concerning normal operators on real or complex inner product spaces. Chapter 10 discusses bilinear forms, emphasizing canonical forms for symmetric and skew-symmetric forms, as well as groups preserving non-degenerate forms, especially the orthogonal, unitary, pseudo-orthogonal and Lorentz groups.

We feel that any course which uses this text should cover Chapters 1, 2, and 3 thoroughly, possibly excluding Sections 3.6 and 3.7 which deal with the double dual and the transpose of a linear transformation. Chapters 4 and 5, on polynomials and determinants, may be treated with varying degrees of thoroughness. In fact, polynomial ideals and basic properties of determinants may be covered quite sketchily without serious damage to the flow of the logic in the text; however, our inclination is to deal with these chapters carefully (except the results on modules), because the material illustrates so well the basic ideas of linear algebra. An elementary course may now be concluded nicely with the first four sections of Chapter 6, together with (the new) Chapter 8. If the rational and Jordan forms are to be included, a more extensive coverage of Chapter 6 is necessary.

Our indebtedness remains to those who contributed to the first edition, especially to Professors Harry Furstenberg, Louis Howard, Daniel Kan, Edward Thorp, to Mrs. Judith Bowers, Mrs. Betty Ann (Sargent) Rose and Miss Phyllis Ruby. In addition, we would like to thank the many students and colleagues whose perceptive comments led to this revision, and the staff of Prentice-Hall for their patience in dealing with two authors caught in the throes of academic administration. Lastly, special thanks are due to Mrs. Sophia Koulouras for both her skill and her tireless efforts in typing the revised manuscript.

K. M. H. / R. A. K.

Contents

1. Linear Equations

1.1. Fields

We assume that the reader is familiar with the elementary algebra of real and complex numbers. For a large portion of this book the algebraic properties of numbers which we shall use are easily deduced from the following brief list of properties of addition and multiplication. We let F denote either the set of real numbers or the set of complex numbers.

1. Addition is commutative,

$$x + y = y + x$$

for all x and y in F.

2. Addition is associative,

$$x + (y + z) = (x + y) + z$$

for all x, y, and z in F.

3. There is a unique element 0 (zero) in F such that $x + 0 = x$, for every x in F.

4. To each x in F there corresponds a unique element $(-x)$ in F such that $x + (-x) = 0$.

5. Multiplication is commutative,

$$xy = yx$$

for all x and y in F.

6. Multiplication is associative,

$$x(yz) = (xy)z$$

for all x, y, and z in F.

7. There is a unique non-zero element 1 (one) in F such that $x1 = x$, for every x in F.

8. To each non-zero x in F there corresponds a unique element x^{-1} (or $1/x$) in F such that $xx^{-1} = 1$.

9. Multiplication distributes over addition; that is, $x(y + z) = xy + xz$, for all x, y, and z in F.

Suppose one has a set F of objects $x, y, z, \ldots$ and two operations on the elements of F as follows. The first operation, called addition, associates with each pair of elements x, y in F an element $(x + y)$ in F; the second operation, called multiplication, associates with each pair x, y an element xy in F; and these two operations satisfy conditions (1)–(9) above. The set F, together with these two operations, is then called a **field.** Roughly speaking, a field is a set together with some operations on the objects in that set which behave like ordinary addition, subtraction, multiplication, and division of numbers in the sense that they obey the nine rules of algebra listed above. With the usual operations of addition and multiplication, the set C of complex numbers is a field, as is the set R of real numbers.

For most of this book the 'numbers' we use may as well be the elements from any field F. To allow for this generality, we shall use the word 'scalar' rather than 'number.' Not much will be lost to the reader if he always assumes that the field of scalars is a subfield of the field of complex numbers. A **subfield** of the field C is a set F of complex numbers which is itself a field under the usual operations of addition and multiplication of complex numbers. This means that 0 and 1 are in the set F, and that if x and y are elements of F, so are $(x + y)$, $-x$, xy, and x^{-1} (if $x \neq 0$). An example of such a subfield is the field R of real numbers; for, if we identify the real numbers with the complex numbers $(a + ib)$ for which $b = 0$, the 0 and 1 of the complex field are real numbers, and if x and y are real, so are $(x + y)$, $-x$, xy, and x^{-1} (if $x \neq 0$). We shall give other examples below. The point of our discussing subfields is essentially this: If we are working with scalars from a certain subfield of C, then the performance of the operations of addition, subtraction, multiplication, or division on these scalars does not take us out of the given subfield.

EXAMPLE 1. The set of **positive integers:** $1, 2, 3, \ldots$, is not a subfield of C, for a variety of reasons. For example, 0 is not a positive integer; for no positive integer n is $-n$ a positive integer; for no positive integer n except 1 is $1/n$ a positive integer.

EXAMPLE 2. The set of **integers:** $\ldots, -2, -1, 0, 1, 2, \ldots$, is not a subfield of C, because for an integer n, $1/n$ is not an integer unless n is 1 or

-1. With the usual operations of addition and multiplication, the set of integers satisfies all of the conditions (1)–(9) except condition (8).

EXAMPLE 3. The set of **rational numbers,** that is, numbers of the form p/q, where p and q are integers and $q \neq 0$, is a subfield of the field of complex numbers. The division which is not possible within the set of integers is possible within the set of rational numbers. The interested reader should verify that any subfield of C must contain every rational number.

EXAMPLE 4. The set of all complex numbers of the form $x + y\sqrt{2}$, where x and y are rational, is a subfield of C. We leave it to the reader to verify this.

In the examples and exercises of this book, the reader should assume that the field involved is a subfield of the complex numbers, unless it is expressly stated that the field is more general. We do not want to dwell on this point; however, we should indicate why we adopt such a convention. If F is a field, it may be possible to add the unit 1 to itself a finite number of times and obtain 0 (see Exercise 5 following Section 1.2):

$$1 + 1 + \cdots + 1 = 0.$$

That does not happen in the complex number field (or in any subfield thereof). If it does happen in F, then the least n such that the sum of n 1's is 0 is called the **characteristic** of the field F. If it does not happen in F, then (for some strange reason) F is called a field of **characteristic zero.** Often, when we assume F is a subfield of C, what we want to guarantee is that F is a field of characteristic zero; but, in a first exposure to linear algebra, it is usually better not to worry too much about characteristics of fields.

1.2. Systems of Linear Equations

Suppose F is a field. We consider the problem of finding n scalars (elements of F) $x_1, \ldots, x_n$ which satisfy the conditions

$$\begin{array}{c} A_{11}x_1 + A_{12}x_2 + \cdots + A_{1n}x_n = y_1 \\ A_{21}x_1 + A_{22}x_2 + \cdots + A_{2n}x_n = y_2 \\ \vdots \qquad\quad \vdots \qquad\qquad\quad \vdots \qquad\ \ \vdots \\ A_{m1}x_1 + A_{m2}x_2 + \cdots + A_{mn}x_n = y_m \end{array} \tag{1-1}$$

where $y_1, \ldots, y_m$ and A_{ij}, $1 \leq i \leq m$, $1 \leq j \leq n$, are given elements of F. We call (1-1) a **system of m linear equations in n unknowns.** Any n-tuple $(x_1, \ldots, x_n)$ of elements of F which satisfies each of the equations in (1-1) is called a **solution** of the system. If $y_1 = y_2 = \cdots = y_m = 0$, we say that the system is **homogeneous**, or that each of the equations is homogeneous.

Perhaps the most fundamental technique for finding the solutions of a system of linear equations is the technique of elimination. We can illustrate this technique on the homogeneous system

$$\begin{aligned} 2x_1 - x_2 + x_3 &= 0 \\ x_1 + 3x_2 + 4x_3 &= 0. \end{aligned}$$

If we add (-2) times the second equation to the first equation, we obtain

$$-7x_2 - 7x_3 = 0$$

or, $x_2 = -x_3$. If we add 3 times the first equation to the second equation, we obtain

$$7x_1 + 7x_3 = 0$$

or, $x_1 = -x_3$. So we conclude that if (x_1, x_2, x_3) is a solution then $x_1 = x_2 = -x_3$. Conversely, one can readily verify that any such triple is a solution. Thus the set of solutions consists of all triples $(-a, -a, a)$.

We found the solutions to this system of equations by 'eliminating unknowns,' that is, by multiplying equations by scalars and then adding to produce equations in which some of the x_j were not present. We wish to formalize this process slightly so that we may understand why it works, and so that we may carry out the computations necessary to solve a system in an organized manner.

For the general system (1-1), suppose we select m scalars $c_1, \ldots, c_m$, multiply the jth equation by c_j and then add. We obtain the equation

$$(c_1A_{11} + \cdots + c_mA_{m1})x_1 + \cdots + (c_1A_{1n} + \cdots + c_mA_{mn})x_n = c_1y_1 + \cdots + c_my_m.$$

Such an equation we shall call a **linear combination** of the equations in (1-1). Evidently, any solution of the entire system of equations (1-1) will also be a solution of this new equation. This is the fundamental idea of the elimination process. If we have another system of linear equations

$$\begin{aligned} B_{11}x_1 + \cdots + B_{1n}x_n &= z_1 \\ \vdots \qquad\qquad\quad \vdots \quad &\quad\ \vdots \\ B_{k1}x_1 + \cdots + B_{kn}x_n &= z_k \end{aligned} \tag{1-2}$$

in which each of the k equations is a linear combination of the equations in (1-1), then every solution of (1-1) is a solution of this new system. Of course it may happen that some solutions of (1-2) are not solutions of (1-1). This clearly does not happen if each equation in the original system is a linear combination of the equations in the new system. Let us say that two systems of linear equations are **equivalent** if each equation in each system is a linear combination of the equations in the other system. We can then formally state our observations as follows.

Theorem 1. *Equivalent systems of linear equations have exactly the same solutions.*

If the elimination process is to be effective in finding the solutions of a system like (1-1), then one must see how, by forming linear combinations of the given equations, to produce an equivalent system of equations which is easier to solve. In the next section we shall discuss one method of doing this.

Exercises

1. Verify that the set of complex numbers described in Example 4 is a subfield of C.

2. Let F be the field of complex numbers. Are the following two systems of linear equations equivalent? If so, express each equation in each system as a linear combination of the equations in the other system.

$$\begin{aligned} x_1 - x_2 &= 0 \\ 2x_1 + x_2 &= 0 \end{aligned} \qquad \begin{aligned} 3x_1 + x_2 &= 0 \\ x_1 + x_2 &= 0 \end{aligned}$$

3. Test the following systems of equations as in Exercise 2.

$$\begin{aligned} -x_1 + x_2 + 4x_3 &= 0 \\ x_1 + 3x_2 + 8x_3 &= 0 \\ \tfrac{1}{2}x_1 + x_2 + \tfrac{5}{2}x_3 &= 0 \end{aligned} \qquad \begin{aligned} x_1 \qquad - x_3 &= 0 \\ x_2 + 3x_3 &= 0 \end{aligned}$$

4. Test the following systems as in Exercise 2.

$$\begin{aligned} 2x_1 + (-1 + i)x_2 \qquad + x_4 &= 0 \\ 3x_2 - 2ix_3 + 5x_4 &= 0 \end{aligned} \qquad \begin{aligned} \left(1 + \frac{i}{2}\right)x_1 + 8x_2 - ix_3 - x_4 &= 0 \\ \tfrac{2}{3}x_1 - \tfrac{1}{2}x_2 + x_3 + 7x_4 &= 0 \end{aligned}$$

5. Let F be a set which contains exactly two elements, 0 and 1. Define an addition and multiplication by the tables:

+	0	1
0	0	1
1	1	0

·	0	1
0	0	0
1	0	1

Verify that the set F, together with these two operations, is a field.

6. Prove that if two homogeneous systems of linear equations in two unknowns have the same solutions, then they are equivalent.

7. Prove that each subfield of the field of complex numbers contains every rational number.

8. Prove that each field of characteristic zero contains a copy of the rational number field.

1.3. Matrices and Elementary Row Operations

One cannot fail to notice that in forming linear combinations of linear equations there is no need to continue writing the 'unknowns' $x_1, \ldots, x_n$, since one actually computes only with the coefficients A_{ij} and the scalars y_i. We shall now abbreviate the system (1-1) by

$$AX = Y$$

where

$$A = \begin{bmatrix} A_{11} & \cdots & A_{1n} \\ \vdots & & \vdots \\ A_{m1} & \cdots & A_{mn} \end{bmatrix}$$

$$X = \begin{bmatrix} x_1 \\ \vdots \\ x_n \end{bmatrix} \quad \text{and} \quad Y = \begin{bmatrix} y_1 \\ \vdots \\ y_m \end{bmatrix}.$$

We call A the **matrix of coefficients** of the system. Strictly speaking, the rectangular array displayed above is not a matrix, but is a representation of a matrix. An $m \times n$ **matrix over the field** F is a function A from the set of pairs of integers (i, j), $1 \leq i \leq m$, $1 \leq j \leq n$, into the field F. The **entries** of the matrix A are the scalars $A(i, j) = A_{ij}$, and quite often it is most convenient to describe the matrix by displaying its entries in a rectangular array having m rows and n columns, as above. Thus X (above) is, or defines, an $n \times 1$ matrix and Y is an $m \times 1$ matrix. For the time being, $AX = Y$ is nothing more than a shorthand notation for our system of linear equations. Later, when we have defined a multiplication for matrices, it will mean that Y is the product of A and X.

We wish now to consider operations on the rows of the matrix A which correspond to forming linear combinations of the equations in the system $AX = Y$. We restrict our attention to three **elementary row operations** on an $m \times n$ matrix A over the field F:

1. multiplication of one row of A by a non-zero scalar c;
2. replacement of the rth row of A by row r plus c times row s, c any scalar and $r \neq s$;
3. interchange of two rows of A.

An elementary row operation is thus a special type of function (rule) e which associated with each $m \times n$ matrix A an $m \times n$ matrix $e(A)$. One can precisely describe e in the three cases as follows:

1. $e(A)_{ij} = A_{ij}$ if $i \neq r$, $e(A)_{rj} = cA_{rj}$.
2. $e(A)_{ij} = A_{ij}$ if $i \neq r$, $e(A)_{rj} = A_{rj} + cA_{sj}$.
3. $e(A)_{ij} = A_{ij}$ if i is different from both r and s, $e(A)_{rj} = A_{sj}$, $e(A)_{sj} = A_{rj}$.

In defining $e(A)$, it is not really important how many columns A has, but the number of rows of A is crucial. For example, one must worry a little to decide what is meant by interchanging rows 5 and 6 of a 5×5 matrix. To avoid any such complications, we shall agree that an elementary row operation e is defined on the class of all $m \times n$ matrices over F, for some fixed m but any n. In other words, a particular e is defined on the class of all m-rowed matrices over F.

One reason that we restrict ourselves to these three simple types of row operations is that, having performed such an operation e on a matrix A, we can recapture A by performing a similar operation on $e(A)$.

Theorem 2. *To each elementary row operation* e *there corresponds an elementary row operation* e_1, *of the same type as* e, *such that* $e_1(e(A)) = e(e_1(A)) = A$ *for each* A. *In other words, the inverse operation (function) of an elementary row operation exists and is an elementary row operation of the same type.*

Proof. (1) Suppose e is the operation which multiplies the rth row of a matrix by the non-zero scalar c. Let e_1 be the operation which multiplies row r by c^{-1}. (2) Suppose e is the operation which replaces row r by row r plus c times row s, $r \neq s$. Let e_1 be the operation which replaces row r by row r plus $(-c)$ times row s. (3) If e interchanges rows r and s, let $e_1 = e$. In each of these three cases we clearly have $e_1(e(A)) = e(e_1(A)) = A$ for each A. ∎

Definition. *If* A *and* B *are* m × n *matrices over the field* F, *we say that* B **is row-equivalent to** A *if* B *can be obtained from* A *by a finite sequence of elementary row operations.*

Using Theorem 2, the reader should find it easy to verify the following. Each matrix is row-equivalent to itself; if B is row-equivalent to A, then A is row-equivalent to B; if B is row-equivalent to A and C is row-equivalent to B, then C is row-equivalent to A. In other words, row-equivalence is an equivalence relation (see Appendix).

Theorem 3. *If* A *and* B *are row-equivalent* m × n *matrices, the homogeneous systems of linear equations* $AX = 0$ *and* $BX = 0$ *have exactly the same solutions.*

Proof. Suppose we pass from A to B by a finite sequence of elementary row operations:

$$A = A_0 \rightarrow A_1 \rightarrow \cdots \rightarrow A_k = B.$$

It is enough to prove that the systems $A_jX = 0$ and $A_{j+1}X = 0$ have the same solutions, i.e., that one elementary row operation does not disturb the set of solutions.

So suppose that B is obtained from A by a single elementary row operation. No matter which of the three types the operation is, (1), (2), or (3), each equation in the system $BX = 0$ will be a linear combination of the equations in the system $AX = 0$. Since the inverse of an elementary row operation is an elementary row operation, each equation in $AX = 0$ will also be a linear combination of the equations in $BX = 0$. Hence these two systems are equivalent, and by Theorem 1 they have the same solutions. ∎

EXAMPLE 5. Suppose F is the field of rational numbers, and

$$A = \begin{bmatrix} 2 & -1 & 3 & 2 \\ 1 & 4 & 0 & -1 \\ 2 & 6 & -1 & 5 \end{bmatrix}.$$

We shall perform a finite sequence of elementary row operations on A, indicating by numbers in parentheses the type of operation performed.

$$\begin{bmatrix} 2 & -1 & 3 & 2 \\ 1 & 4 & 0 & -1 \\ 2 & 6 & -1 & 5 \end{bmatrix} \xrightarrow{(2)} \begin{bmatrix} 0 & -9 & 3 & 4 \\ 1 & 4 & 0 & -1 \\ 2 & 6 & -1 & 5 \end{bmatrix} \xrightarrow{(2)}$$

$$\begin{bmatrix} 0 & -9 & 3 & 4 \\ 1 & 4 & 0 & -1 \\ 0 & -2 & -1 & 7 \end{bmatrix} \xrightarrow{(1)} \begin{bmatrix} 0 & -9 & 3 & 4 \\ 1 & 4 & 0 & -1 \\ 0 & 1 & \frac{1}{2} & -\frac{7}{2} \end{bmatrix} \xrightarrow{(2)}$$

$$\begin{bmatrix} 0 & -9 & 3 & 4 \\ 1 & 0 & -2 & 13 \\ 0 & 1 & \frac{1}{2} & -\frac{7}{2} \end{bmatrix} \xrightarrow{(2)} \begin{bmatrix} 0 & 0 & \frac{15}{2} & -\frac{55}{2} \\ 1 & 0 & -2 & 13 \\ 0 & 1 & \frac{1}{2} & -\frac{7}{2} \end{bmatrix} \xrightarrow{(1)}$$

$$\begin{bmatrix} 0 & 0 & 1 & -\frac{11}{3} \\ 1 & 0 & -2 & 13 \\ 0 & 1 & \frac{1}{2} & -\frac{7}{2} \end{bmatrix} \xrightarrow{(2)} \begin{bmatrix} 0 & 0 & 1 & -\frac{11}{3} \\ 1 & 0 & 0 & \frac{17}{3} \\ 0 & 1 & \frac{1}{2} & -\frac{7}{2} \end{bmatrix} \xrightarrow{(2)}$$

$$\begin{bmatrix} 0 & 0 & 1 & -\frac{11}{3} \\ 1 & 0 & 0 & \frac{17}{3} \\ 0 & 1 & 0 & -\frac{5}{3} \end{bmatrix}$$

The row-equivalence of A with the final matrix in the above sequence tells us in particular that the solutions of

$$\begin{aligned} 2x_1 - x_2 + 3x_3 + 2x_4 &= 0 \\ x_1 + 4x_2 - x_4 &= 0 \\ 2x_1 + 6x_2 - x_3 + 5x_4 &= 0 \end{aligned}$$

and

$$\begin{aligned} x_3 - \tfrac{11}{3}x_4 &= 0 \\ x_1 \qquad + \tfrac{17}{3}x_4 &= 0 \\ x_2 \qquad - \tfrac{5}{3}x_4 &= 0 \end{aligned}$$

are exactly the same. In the second system it is apparent that if we assign any rational value c to x_4 we obtain a solution $(-\frac{17}{3}c, \frac{5}{3}, \frac{11}{3}c, c)$, and also that every solution is of this form.

EXAMPLE 6. Suppose F is the field of complex numbers and

$$A = \begin{bmatrix} -1 & i \\ -i & 3 \\ 1 & 2 \end{bmatrix}.$$

In performing row operations it is often convenient to combine several operations of type (2). With this in mind

$$\begin{bmatrix} -1 & i \\ -i & 3 \\ 1 & 2 \end{bmatrix} \xrightarrow{(2)} \begin{bmatrix} 0 & 2+i \\ 0 & 3+2i \\ 1 & 2 \end{bmatrix} \xrightarrow{(1)} \begin{bmatrix} 0 & 1 \\ 0 & 3+2i \\ 1 & 2 \end{bmatrix} \xrightarrow{(2)} \begin{bmatrix} 0 & 1 \\ 0 & 0 \\ 1 & 0 \end{bmatrix}.$$

Thus the system of equations

$$\begin{aligned} -x_1 + ix_2 &= 0 \\ -ix_1 + 3x_2 &= 0 \\ x_1 + 2x_2 &= 0 \end{aligned}$$

has only the trivial solution $x_1 = x_2 = 0$.

In Examples 5 and 6 we were obviously not performing row operations at random. Our choice of row operations was motivated by a desire to simplify the coefficient matrix in a manner analogous to 'eliminating unknowns' in the system of linear equations. Let us now make a formal definition of the type of matrix at which we were attempting to arrive.

Definition. *An* $m \times n$ *matrix* R *is called* **row-reduced** *if:*

(a) *the first non-zero entry in each non-zero row of* R *is equal to* 1;

(b) *each column of* R *which contains the leading non-zero entry of some row has all its other entries* 0.

EXAMPLE 7. One example of a row-reduced matrix is the $n \times n$ (square) **identity matrix** I. This is the $n \times n$ matrix defined by

$$I_{ij} = \delta_{ij} = \begin{cases} 1, & \text{if } i = j \\ 0, & \text{if } i \neq j. \end{cases}$$

This is the first of many occasions on which we shall use the **Kronecker delta** (δ).

In Examples 5 and 6, the final matrices in the sequences exhibited there are row-reduced matrices. Two examples of matrices which are *not* row-reduced are:

$$\begin{bmatrix} 1 & 0 & 0 & 0 \\ 0 & 1 & -1 & 0 \\ 0 & 0 & 1 & 0 \end{bmatrix}, \qquad \begin{bmatrix} 0 & 2 & 1 \\ 1 & 0 & -3 \\ 0 & 0 & 0 \end{bmatrix}.$$

The second matrix fails to satisfy condition (a), because the leading non-zero entry of the first row is not 1. The first matrix does satisfy condition (a), but fails to satisfy condition (b) in column 3.

We shall now prove that we can pass from any given matrix to a row-reduced matrix, by means of a finite number of elementary row opertions. In combination with Theorem 3, this will provide us with an effective tool for solving systems of linear equations.

Theorem 4. *Every* m × n *matrix over the field* F *is row-equivalent to a row-reduced matrix.*

Proof. Let A be an $m \times n$ matrix over F. If every entry in the first row of A is 0, then condition (a) is satisfied in so far as row 1 is concerned. If row 1 has a non-zero entry, let k be the smallest positive integer j for which $A_{1j} \neq 0$. Multiply row 1 by A_{1k}^{-1}, and then condition (a) is satisfied with regard to row 1. Now for each $i \geq 2$, add $(-A_{ik})$ times row 1 to row i. Now the leading non-zero entry of row 1 occurs in column k, that entry is 1, and every other entry in column k is 0.

Now consider the matrix which has resulted from above. If every entry in row 2 is 0, we do nothing to row 2. If some entry in row 2 is different from 0, we multiply row 2 by a scalar so that the leading non-zero entry is 1. In the event that row 1 had a leading non-zero entry in column k, this leading non-zero entry of row 2 cannot occur in column k; say it occurs in column $k_r \neq k$. By adding suitable multiples of row 2 to the various rows, we can arrange that all entries in column k' are 0, except the 1 in row 2. The important thing to notice is this: In carrying out these last operations, we will not change the entries of row 1 in columns $1, \ldots, k$; nor will we change any entry of column k. Of course, if row 1 was identically 0, the operations with row 2 will not affect row 1.

Working with one row at a time in the above manner, it is clear that in a finite number of steps we will arrive at a row-reduced matrix. ∎

Exercises

1. Find all solutions to the system of equations

$$\begin{aligned} (1 - i)x_1 - ix_2 &= 0 \\ 2x_1 + (1 - i)x_2 &= 0. \end{aligned}$$

2. If

$$A = \begin{bmatrix} 3 & -1 & 2 \\ 2 & 1 & 1 \\ 1 & -3 & 0 \end{bmatrix}$$

find all solutions of $AX = 0$ by row-reducing A.

3. If

$$A = \begin{bmatrix} 6 & -4 & 0 \\ 4 & -2 & 0 \\ -1 & 0 & 3 \end{bmatrix}$$

find all solutions of $AX = 2X$ and all solutions of $AX = 3X$. (The symbol cX denotes the matrix each entry of which is c times the corresponding entry of X.)

4. Find a row-reduced matrix which is row-equivalent to

$$A = \begin{bmatrix} i & -(1+i) & 0 \\ 1 & -2 & 1 \\ 1 & 2i & -1 \end{bmatrix}.$$

5. Prove that the following two matrices are *not* row-equivalent:

$$\begin{bmatrix} 2 & 0 & 0 \\ a & -1 & 0 \\ b & c & 3 \end{bmatrix}, \quad \begin{bmatrix} 1 & 1 & 2 \\ -2 & 0 & -1 \\ 1 & 3 & 5 \end{bmatrix}.$$

6. Let

$$A = \begin{bmatrix} a & b \\ c & d \end{bmatrix}$$

be a 2×2 matrix with complex entries. Suppose that A is row-reduced and also that $a + b + c + d = 0$. Prove that there are exactly three such matrices.

7. Prove that the interchange of two rows of a matrix can be accomplished by a finite sequence of elementary row operations of the other two types.

8. Consider the system of equations $AX = 0$ where

$$A = \begin{bmatrix} a & b \\ c & d \end{bmatrix}$$

is a 2×2 matrix over the field F. Prove the following.

(a) If every entry of A is 0, then every pair (x_1, x_2) is a solution of $AX = 0$.

(b) If $ad - bc \neq 0$, the system $AX = 0$ has only the trivial solution $x_1 = x_2 = 0$.

(c) If $ad - bc = 0$ and some entry of A is different from 0, then there is a solution (x_1^0, x_2^0) such that (x_1, x_2) is a solution if and only if there is a scalar y such that $x_1 = yx_1^0$, $x_2 = yx_2^0$.

1.4. Row-Reduced Echelon Matrices

Until now, our work with systems of linear equations was motivated by an attempt to find the solutions of such a system. In Section 1.3 we established a standardized technique for finding these solutions. We wish now to acquire some information which is slightly more theoretical, and for that purpose it is convenient to go a little beyond row-reduced matrices.

Definition. *An* m $\times$ n *matrix* R *is called a* **row-reduced echelon matrix** *if:*

(a) *R is row-reduced;*

(b) *every row of R which has all its entries 0 occurs below every row which has a non-zero entry;*

(c) *if rows 1, . . . , r are the non-zero rows of R, and if the leading non-zero entry of row i occurs in column k_i, i = 1, . . . , r, then $k_1 < k_2 < \cdots < k_r$.*

One can also describe an $m \times n$ row-reduced echelon matrix R as follows. Either every entry in R is 0, or there exists a positive integer r, $1 \leq r \leq m$, and r positive integers $k_1, \ldots, k_r$ with $1 \leq k_i \leq n$ and

(a) $R_{ij} = 0$ for $i > r$, and $R_{ij} = 0$ if $j < k_i$.

(b) $R_{ik_j} = \delta_{ij}$, $1 \leq i \leq r$, $1 \leq j \leq r$.

(c) $k_1 < \cdots < k_r$.

EXAMPLE 8. Two examples of row-reduced echelon matrices are the $n \times n$ identity matrix, and the $m \times n$ **zero matrix** $0^{m,n}$, in which all entries are 0. The reader should have no difficulty in making other examples, but we should like to give one non-trivial one:

$$\begin{bmatrix} 0 & 1 & -3 & 0 & \frac{1}{2} \\ 0 & 0 & 0 & 1 & 2 \\ 0 & 0 & 0 & 0 & 0 \end{bmatrix}.$$

Theorem 5. *Every $m \times n$ matrix A is row-equivalent to a row-reduced echelon matrix.*

Proof. We know that A is row-equivalent to a row-reduced matrix. All that we need observe is that by performing a finite number of row interchanges on a row-reduced matrix we can bring it to row-reduced echelon form. ∎

In Examples 5 and 6, we saw the significance of row-reduced matrices in solving homogeneous systems of linear equations. Let us now discuss briefly the system $RX = 0$, when R is a row-reduced echelon matrix. Let rows $1, \ldots, r$ be the non-zero rows of R, and suppose that the leading non-zero entry of row i occurs in column k_i. The system $RX = 0$ then consists of r non-trivial equations. Also the unknown x_{k_i} will occur (with non-zero coefficient) only in the ith equation. If we let $u_1, \ldots, u_{n-r}$ denote the $(n - r)$ unknowns which are different from $x_{k_1}, \ldots, x_{k_r}$, then the r non-trivial equations in $RX = 0$ are of the form

$$\begin{aligned} x_{k_1} + \sum_{j=1}^{n-r} C_{1j}u_j &= 0 \\ \vdots \qquad & \quad \vdots \\ x_{k_r} + \sum_{j=1}^{n-r} C_{rj}u_j &= 0. \end{aligned} \tag{1-3}$$

All the solutions to the system of equations $RX = 0$ are obtained by assigning any values whatsoever to $u_1, \ldots, u_{n-r}$ and then computing the corresponding values of $x_{k_1}, \ldots, x_{k_r}$ from (1-3). For example, if R is the matrix displayed in Example 8, then $r = 2$, $k_1 = 2$, $k_2 = 4$, and the two non-trivial equations in the system $RX = 0$ are

$$\begin{aligned} x_2 - 3x_3 \quad + \tfrac{1}{2}x_5 &= 0 \quad \text{or} \quad x_2 = 3x_3 - \tfrac{1}{2}x_5 \\ x_4 + 2x_5 &= 0 \quad \text{or} \quad x_4 = -2x_5. \end{aligned}$$

So we may assign any values to x_1, x_3, and x_5, say $x_1 = a$, $x_3 = b$, $x_5 = c$, and obtain the solution $(a, 3b - \frac{1}{2}c, b, -2c, c)$.

Let us observe one thing more in connection with the system of equations $RX = 0$. If the number r of non-zero rows in R is less than n, then the system $RX = 0$ has a non-trivial solution, that is, a solution $(x_1, \ldots, x_n)$ in which not every x_j is 0. For, since $r < n$, we can choose some x_j which is not among the r unknowns $x_{k_1}, \ldots, x_{k_r}$, and we can then construct a solution as above in which this x_j is 1. This observation leads us to one of the most fundamental facts concerning systems of homogeneous linear equations.

Theorem 6. *If* A *is an* m × n *matrix and* m < n, *then the homogeneous system of linear equations* AX = 0 *has a non-trivial solution.*

Proof. Let R be a row-reduced echelon matrix which is row-equivalent to A. Then the systems $AX = 0$ and $RX = 0$ have the same solutions by Theorem 3. If r is the number of non-zero rows in R, then certainly $r \leq m$, and since $m < n$, we have $r < n$. It follows immediately from our remarks above that $AX = 0$ has a non-trivial solution. ▌

Theorem 7. *If* A *is an* n × n *(square) matrix, then* A *is row-equivalent to the* n × n *identity matrix if and only if the system of equations* AX = 0 *has only the trivial solution.*

Proof. If A is row-equivalent to I, then $AX = 0$ and $IX = 0$ have the same solutions. Conversely, suppose $AX = 0$ has only the trivial solution $X = 0$. Let R be an $n \times n$ row-reduced echelon matrix which is row-equivalent to A, and let r be the number of non-zero rows of R. Then $RX = 0$ has no non-trivial solution. Thus $r \geq n$. But since R has n rows, certainly $r \leq n$, and we have $r = n$. Since this means that R actually has a leading non-zero entry of 1 in each of its n rows, and since these 1's occur each in a different one of the n columns, R must be the $n \times n$ identity matrix. ▌

Let us now ask what elementary row operations do toward solving a system of linear equations $AX = Y$ which is not homogeneous. At the outset, one must observe one basic difference between this and the homogeneous case, namely, that while the homogeneous system always has the trivial solution $x_1 = \cdots = x_n = 0$, an inhomogeneous system need have no solution at all.

We form the **augmented matrix** A' of the system $AX = Y$. This is the $m \times (n + 1)$ matrix whose first n columns are the columns of A and whose last column is Y. More precisely,

$$\begin{aligned} A'_{ij} &= A_{ij}, \quad \text{if} \quad j \leq n \\ A'_{i(n+1)} &= y_i. \end{aligned}$$

Suppose we perform a sequence of elementary row operations on A, arriving at a row-reduced echelon matrix R. If we perform this same sequence of row operations on the augmented matrix A', we will arrive at a matrix R' whose first n columns are the columns of R and whose last column contains certain scalars $z_1, \ldots, z_m$. The scalars z_i are the entries of the $m \times 1$ matrix

$$Z = \begin{bmatrix} z_1 \\ \vdots \\ z_m \end{bmatrix}$$

which results from applying the sequence of row operations to the matrix Y. It should be clear to the reader that, just as in the proof of Theorem 3, the systems $AX = Y$ and $RX = Z$ are equivalent and hence have the same solutions. It is very easy to determine whether the system $RX = Z$ has any solutions and to determine all the solutions if any exist. For, if R has r non-zero rows, with the leading non-zero entry of row i occurring in column k_i, $i = 1, \ldots, r$, then the first r equations of $RX = Z$ effectively express $x_{k_1}, \ldots, x_{k_r}$ in terms of the $(n - r)$ remaining x_j and the scalars $z_1, \ldots, z_r$. The last $(m - r)$ equations are

$$\begin{aligned} 0 &= z_{r+1} \\ \vdots\ & \quad\ \vdots \\ 0 &= z_m \end{aligned}$$

and accordingly the condition for the system to have a solution is $z_i = 0$ for $i > r$. If this condition is satisfied, all solutions to the system are found just as in the homogeneous case, by assigning arbitrary values to $(n - r)$ of the x_j and then computing x_{k_i} from the ith equation.

EXAMPLE 9. Let F be the field of rational numbers and

$$A = \begin{bmatrix} 1 & -2 & 1 \\ 2 & 1 & 1 \\ 0 & 5 & -1 \end{bmatrix}$$

and suppose that we wish to solve the system $AX = Y$ for some y_1, y_2, and y_3. Let us perform a sequence of row operations on the augmented matrix A' which row-reduces A:

$$\begin{bmatrix} 1 & -2 & 1 & y_1 \\ 2 & 1 & 1 & y_2 \\ 0 & 5 & -1 & y_3 \end{bmatrix} \xrightarrow{(2)} \begin{bmatrix} 1 & -2 & 1 & y_1 \\ 0 & 5 & -1 & (y_2 - 2y_1) \\ 0 & 5 & -1 & y_3 \end{bmatrix} \xrightarrow{(2)}$$

$$\begin{bmatrix} 1 & -2 & 1 & y_1 \\ 0 & 5 & -1 & (y_2 - 2y_1) \\ 0 & 0 & 0 & (y_3 - y_2 + 2y_1) \end{bmatrix} \xrightarrow{(1)} \begin{bmatrix} 1 & -2 & 1 & y_1 \\ 0 & 1 & -\frac{1}{5} & \frac{1}{5}(y_2 - 2y_1) \\ 0 & 0 & 0 & (y_3 - y_2 + 2y_1) \end{bmatrix} \xrightarrow{(2)}$$

$$\begin{bmatrix} 1 & 0 & \frac{3}{5} & \frac{1}{5}(y_1 + 2y_2) \\ 0 & 1 & -\frac{1}{5} & \frac{1}{5}(y_2 - 2y_1) \\ 0 & 0 & 0 & (y_3 - y_2 + 2y_1) \end{bmatrix}.$$

The condition that the system $AX = Y$ have a solution is thus

$$2y_1 - y_2 + y_3 = 0$$

and if the given scalars y_i satisfy this condition, all solutions are obtained by assigning a value c to x_3 and then computing

$$\begin{aligned} x_1 &= -\tfrac{3}{5}c + \tfrac{1}{5}(y_1 + 2y_2) \\ x_2 &= \tfrac{1}{5}c + \tfrac{1}{5}(y_2 - 2y_1). \end{aligned}$$

Let us observe one final thing about the system $AX = Y$. Suppose the entries of the matrix A and the scalars $y_1, \ldots, y_m$ happen to lie in a subfield F_1 of the field F. If the system of equations $AX = Y$ has a solution with $x_1, \ldots, x_n$ in F, it has a solution with $x_1, \ldots, x_n$ in F_1. For, over either field, the condition for the system to have a solution is that certain relations hold between $y_1, \ldots, y_m$ in F_1 (the relations $z_i = 0$ for $i > r$, above). For example, if $AX = Y$ is a system of linear equations in which the scalars y_k and A_{ij} are real numbers, and if there is a solution in which $x_1, \ldots, x_n$ are complex numbers, then there is a solution with $x_1, \ldots, x_n$ real numbers.

Exercises

1. Find all solutions to the following system of equations by row-reducing the coefficient matrix:

$$\begin{aligned} \tfrac{1}{3}x_1 + 2x_2 - 6x_3 &= 0 \\ -4x_1 + 5x_3 &= 0 \\ -3x_1 + 6x_2 - 13x_3 &= 0 \\ -\tfrac{7}{3}x_1 + 2x_2 - \tfrac{8}{3}x_3 &= 0 \end{aligned}$$

2. Find a row-reduced echelon matrix which is row-equivalent to

$$A = \begin{bmatrix} 1 & -i \\ 2 & 2 \\ i & 1+i \end{bmatrix}.$$

What are the solutions of $AX = 0$?

3. Describe explicitly all 2×2 row-reduced echelon matrices.

4. Consider the system of equations

$$\begin{aligned} x_1 - x_2 + 2x_3 &= 1 \\ 2x_1 + 2x_3 &= 1 \\ x_1 - 3x_2 + 4x_3 &= 2. \end{aligned}$$

Does this system have a solution? If so, describe explicitly all solutions.

5. Give an example of a system of two linear equations in two unknowns which has no solution.

6. Show that the system

$$\begin{aligned} x_1 - 2x_2 + x_3 + 2x_4 &= 1 \\ x_1 + x_2 - x_3 + x_4 &= 2 \\ x_1 + 7x_2 - 5x_3 - x_4 &= 3 \end{aligned}$$

has no solution.

7. Find all solutions of

$$\begin{aligned} 2x_1 - 3x_2 - 7x_3 + 5x_4 + 2x_5 &= -2 \\ x_1 - 2x_2 - 4x_3 + 3x_4 + x_5 &= -2 \\ 2x_1 - 4x_3 + 2x_4 + x_5 &= 3 \\ x_1 - 5x_2 - 7x_3 + 6x_4 + 2x_5 &= -7. \end{aligned}$$

8. Let

$$A = \begin{bmatrix} 3 & -1 & 2 \\ 2 & 1 & 1 \\ 1 & -3 & 0 \end{bmatrix}.$$

For which triples (y_1, y_2, y_3) does the system $AX = Y$ have a solution?

9. Let

$$A = \begin{bmatrix} 3 & -6 & 2 & -1 \\ -2 & 4 & 1 & 3 \\ 0 & 0 & 1 & 1 \\ 1 & -2 & 1 & 0 \end{bmatrix}.$$

For which (y_1, y_2, y_3, y_4) does the system of equations $AX = Y$ have a solution?

10. Suppose R and R' are 2×3 row-reduced echelon matrices and that the systems $RX = 0$ and $R'X = 0$ have exactly the same solutions. Prove that $R = R'$.

1.5. Matrix Multiplication

It is apparent (or should be, at any rate) that the process of forming linear combinations of the rows of a matrix is a fundamental one. For this reason it is advantageous to introduce a systematic scheme for indicating just what operations are to be performed. More specifically, suppose B is an $n \times p$ matrix over a field F with rows $\beta_1, \ldots, \beta_n$ and that from B we construct a matrix C with rows $\gamma_1, \ldots, \gamma_m$ by forming certain linear combinations

$$\gamma_i = A_{i1}\beta_1 + A_{i2}\beta_2 + \cdots + A_{in}\beta_n. \tag{1-4}$$

The rows of C are determined by the mn scalars A_{ij} which are themselves the entries of an $m \times n$ matrix A. If (1-4) is expanded to

$$(C_{i1} \cdots C_{ip}) = \sum_{r=1}^{n} (A_{ir}B_{r1} \cdots A_{ir}B_{rp})$$

we see that the entries of C are given by

$$C_{ij} = \sum_{r=1}^{n} A_{ir}B_{rj}.$$

Definition. *Let* A *be an* m $\times$ n *matrix over the field* F *and let* B *be an* n $\times$ p *matrix over* F. *The* **product** AB *is the* m $\times$ p *matrix* C *whose* i, j *entry is*

$$C_{ij} = \sum_{r=1}^{n} A_{ir}B_{rj}.$$

EXAMPLE 10. Here are some products of matrices with rational entries.

(a) $$\begin{bmatrix} 5 & -1 & 2 \\ 0 & 7 & 2 \end{bmatrix} = \begin{bmatrix} 1 & 0 \\ -3 & 1 \end{bmatrix} \begin{bmatrix} 5 & -1 & 2 \\ 15 & 4 & 8 \end{bmatrix}$$

Here

$$\begin{aligned} \gamma_1 &= (5 \quad -1 \quad 2) = 1 \cdot (5 \quad -1 \quad 2) + 0 \cdot (15 \quad 4 \quad 8) \\ \gamma_2 &= (0 \quad 7 \quad 2) = -3(5 \quad -1 \quad 2) + 1 \cdot (15 \quad 4 \quad 8) \end{aligned}$$

(b) $$\begin{bmatrix} 0 & 6 & 1 \\ 9 & 12 & -8 \\ 12 & 62 & -3 \\ 3 & 8 & -2 \end{bmatrix} = \begin{bmatrix} 1 & 0 \\ -2 & 3 \\ 5 & 4 \\ 0 & 1 \end{bmatrix} \begin{bmatrix} 0 & 6 & 1 \\ 3 & 8 & -2 \end{bmatrix}$$

Here

$$\begin{aligned} \gamma_2 &= (9 \quad 12 \quad -8) = -2(0 \quad 6 \quad 1) + 3(3 \quad 8 \quad -2) \\ \gamma_3 &= (12 \quad 62 \quad -3) = 5(0 \quad 6 \quad 1) + 4(3 \quad 8 \quad -2) \end{aligned}$$

(c) $$\begin{bmatrix} 8 \\ 29 \end{bmatrix} = \begin{bmatrix} 2 & 1 \\ 5 & 4 \end{bmatrix} \begin{bmatrix} 1 \\ 6 \end{bmatrix}$$

(d) $$\begin{bmatrix} -2 & -4 \\ 6 & 12 \end{bmatrix} = \begin{bmatrix} -1 \\ 3 \end{bmatrix} [2 \quad 4]$$

Here

$$\gamma_2 = (6 \quad 12) = 3(2 \quad 4)$$

(e) $$[2 \quad 4] \begin{bmatrix} -1 \\ 3 \end{bmatrix} = [10]$$

(f) $$\begin{bmatrix} 0 & 1 & 0 \\ 0 & 0 & 0 \\ 0 & 0 & 0 \end{bmatrix} \begin{bmatrix} 1 & -5 & 2 \\ 2 & 3 & 4 \\ 9 & -1 & 3 \end{bmatrix} = \begin{bmatrix} 2 & 3 & 4 \\ 0 & 0 & 0 \\ 0 & 0 & 0 \end{bmatrix}$$

(g) $$\begin{bmatrix} 1 & -5 & 2 \\ 2 & 3 & 4 \\ 9 & -1 & 3 \end{bmatrix} \begin{bmatrix} 0 & 1 & 0 \\ 0 & 0 & 0 \\ 0 & 0 & 0 \end{bmatrix} = \begin{bmatrix} 0 & 1 & 0 \\ 0 & 2 & 0 \\ 0 & 9 & 0 \end{bmatrix}$$

It is important to observe that the product of two matrices need not be defined; the product is defined if and only if the number of columns in the first matrix coincides with the number of rows in the second matrix. Thus it is meaningless to interchange the order of the factors in (a), (b), and (c) above. Frequently we shall write products such as AB without explicitly mentioning the sizes of the factors and in such cases it will be understood that the product is defined. From (d), (e), (f), (g) we find that even when the products AB and BA are both defined it need not be true that $AB = BA$; in other words, matrix multiplication is *not commutative.*

EXAMPLE 11.

(a) If I is the $m \times m$ identity matrix and A is an $m \times n$ matrix, $IA = A$.

(b) If I is the $n \times n$ identity matrix and A is an $m \times n$ matrix, $AI = A$.

(c) If $0^{k,m}$ is the $k \times m$ zero matrix, $0^{k,n} = 0^{k,m}A$. Similarly, $A0^{n,p} = 0^{m,p}$.

EXAMPLE 12. Let A be an $m \times n$ matrix over F. Our earlier short-hand notation, $AX = Y$, for systems of linear equations is consistent with our definition of matrix products. For if

$$X = \begin{bmatrix} x_1 \\ x_2 \\ \vdots \\ x_n \end{bmatrix}$$

with x_i in F, then AX is the $m \times 1$ matrix

$$Y = \begin{bmatrix} y_1 \\ y_2 \\ \vdots \\ y_m \end{bmatrix}$$

such that $y_i = A_{i1}x_1 + A_{i2}x_2 + \cdots + A_{in}x_n$.

The use of column matrices suggests a notation which is frequently useful. If B is an $n \times p$ matrix, the columns of B are the $1 \times n$ matrices $B_1, \ldots, B_p$ defined by

$$B_j = \begin{bmatrix} B_{1j} \\ \vdots \\ B_{nj} \end{bmatrix}, \qquad 1 \leq j \leq p.$$

The matrix B is the succession of these columns:

$$B = [B_1, \ldots, B_p].$$

The i, j entry of the product matrix AB is formed from the ith row of A

and the jth column of B. The reader should verify that the jth column of AB is AB_j:

$$AB = [AB_1, \ldots, AB_p].$$

In spite of the fact that a product of matrices depends upon the order in which the factors are written, it is independent of the way in which they are associated, as the next theorem shows.

Theorem 8. *If* A, B, C *are matrices over the field* F *such that the products* BC *and* A(BC) *are defined, then so are the products* AB, (AB)C *and*

$$A(BC) = (AB)C.$$

Proof. Suppose B is an $n \times p$ matrix. Since BC is defined, C is a matrix with p rows, and BC has n rows. Because $A(BC)$ is defined we may assume A is an $m \times n$ matrix. Thus the product AB exists and is an $m \times p$ matrix, from which it follows that the product $(AB)C$ exists. To show that $A(BC) = (AB)C$ means to show that

$$[A(BC)]_{ij} = [(AB)C]_{ij}$$

for each i, j. By definition

$$\begin{aligned}
[A(BC)]_{ij} &= \sum_r A_{ir}(BC)_{rj} \\
&= \sum_r A_{ir} \sum_s B_{rs}C_{sj} \\
&= \sum_r \sum_s A_{ir}B_{rs}C_{sj} \\
&= \sum_s \sum_r A_{ir}B_{rs}C_{sj} \\
&= \sum_s \Big(\sum_r A_{ir}B_{rs}\Big)C_{sj} \\
&= \sum_s (AB)_{is}C_{sj} \\
&= [(AB)C]_{ij}. \quad \blacksquare
\end{aligned}$$

When A is an $n \times n$ (square) matrix, the product AA is defined. We shall denote this matrix by A^2. By Theorem 8, $(AA)A = A(AA)$ or $A^2A = AA^2$, so that the product AAA is unambiguously defined. This product we denote by A^3. In general, the product $AA \cdots A$ (k times) is unambiguously defined, and we shall denote this product by A^k.

Note that the relation $A(BC) = (AB)C$ implies among other things that linear combinations of linear combinations of the rows of C are again linear combinations of the rows of C.

If B is a given matrix and C is obtained from B by means of an elementary row operation, then each row of C is a linear combination of the rows of B, and hence there is a matrix A such that $AB = C$. In general there are many such matrices A, and among all such it is convenient and possible to choose one having a number of special properties. Before going into this we need to introduce a class of matrices.

Definition. *An* m × n *matrix is said to be an* **elementary matrix** *if it can be obtained from the* m × m *identity matrix by means of a single elementary row operation.*

EXAMPLE 13. A 2 × 2 elementary matrix is necessarily one of the following:

$$\begin{bmatrix} 0 & 1 \\ 1 & 0 \end{bmatrix}, \quad \begin{bmatrix} 1 & c \\ 0 & 1 \end{bmatrix}, \quad \begin{bmatrix} 1 & 0 \\ c & 1 \end{bmatrix}$$

$$\begin{bmatrix} c & 0 \\ 0 & 1 \end{bmatrix}, \quad c \neq 0, \quad \begin{bmatrix} 1 & 0 \\ 0 & c \end{bmatrix}, \quad c \neq 0.$$

Theorem 9. *Let* e *be an elementary row operation and let* E *be the* m × m *elementary matrix* E = e(I). *Then, for every* m × n *matrix* A,

$$e(A) = EA.$$

Proof. The point of the proof is that the entry in the ith row and jth column of the product matrix EA is obtained from the ith row of E and the jth column of A. The three types of elementary row operations should be taken up separately. We shall give a detailed proof for an operation of type (ii). The other two cases are even easier to handle than this one and will be left as exercises. Suppose $r \neq s$ and e is the operation 'replacement of row r by row r plus c times row s.' Then

$$E_{ik} = \begin{cases} \delta_{ik}, & i \neq r \\ \delta_{rk} + c\delta_{sk}, & i = r. \end{cases}$$

Therefore,

$$(EA)_{ij} = \sum_{k=1}^{m} E_{ik}A_{kj} = \begin{cases} A_{ij}, & i \neq r \\ A_{rj} + cA_{sj}, & i = r. \end{cases}$$

In other words $EA = e(A)$. ∎

Corollary. *Let* A *and* B *be* m × n *matrices over the field* F. *Then* B *is row-equivalent to* A *if and only if* B = PA, *where* P *is a product of* m × m *elementary matrices.*

Proof. Suppose $B = PA$ where $P = E_s \cdots E_2E_1$ and the E_i are $m \times m$ elementary matrices. Then E_1A is row-equivalent to A, and $E_2(E_1A)$ is row-equivalent to E_1A. So E_2E_1A is row-equivalent to A; and continuing in this way we see that $(E_s \cdots E_1)A$ is row-equivalent to A.

Now suppose that B is row-equivalent to A. Let $E_1, E_2, \ldots, E_s$ be the elementary matrices corresponding to some sequence of elementary row operations which carries A into B. Then $B = (E_s \cdots E_1)A$. ∎

Exercises

1. Let

$$A = \begin{bmatrix} 2 & -1 & 1 \\ 1 & 2 & 1 \end{bmatrix}, \quad B = \begin{bmatrix} 3 \\ 1 \\ -1 \end{bmatrix}, \quad C = [1 \quad -1].$$

Compute ABC and CAB.

2. Let

$$A = \begin{bmatrix} 1 & -1 & 1 \\ 2 & 0 & 1 \\ 3 & 0 & 1 \end{bmatrix}, \quad B = \begin{bmatrix} 2 & -2 \\ 1 & 3 \\ 4 & 4 \end{bmatrix}.$$

Verify directly that $A(AB) = A^2B$.

3. Find two different 2×2 matrices A such that $A^2 = 0$ but $A \neq 0$.

4. For the matrix A of Exercise 2, find elementary matrices $E_1, E_2, \ldots, E_k$ such that

$$E_k \cdots E_2E_1A = I.$$

5. Let

$$A = \begin{bmatrix} 1 & -1 \\ 2 & 2 \\ 1 & 0 \end{bmatrix}, \quad B = \begin{bmatrix} 3 & 1 \\ -4 & 4 \end{bmatrix}.$$

Is there a matrix C such that $CA = B$?

6. Let A be an $m \times n$ matrix and B an $n \times k$ matrix. Show that the columns of $C = AB$ are linear combinations of the columns of A. If $\alpha_1, \ldots, \alpha_n$ are the columns of A and $\gamma_1, \ldots, \gamma_k$ are the columns of C, then

$$\gamma_j = \sum_{r=1}^{n} B_{rj}\alpha_r.$$

7. Let A and B be 2×2 matrices such that $AB = I$. Prove that $BA = I$.

8. Let

$$C = \begin{bmatrix} C_{11} & C_{12} \\ C_{21} & C_{22} \end{bmatrix}$$

be a 2×2 matrix. We inquire when it is possible to find 2×2 matrices A and B such that $C = AB - BA$. Prove that such matrices can be found if and only if $C_{11} + C_{22} = 0$.

1.6. Invertible Matrices

Suppose P is an $m \times m$ matrix which is a product of elementary matrices. For each $m \times n$ matrix A, the matrix $B = PA$ is row-equivalent to A; hence A is row-equivalent to B and there is a product Q of elementary matrices such that $A = QB$. In particular this is true when A is the

$m \times m$ identity matrix. In other words, there is an $m \times m$ matrix Q, which is itself a product of elementary matrices, such that $QP = I$. As we shall soon see, the existence of a Q with $QP = I$ is equivalent to the fact that P is a product of elementary matrices.

Definition. *Let* A *be an* $n \times n$ *(square) matrix over the field* F. *An* $n \times n$ *matrix* B *such that* $BA = I$ *is called a* **left inverse** *of* A; *an* $n \times n$ *matrix* B *such that* $AB = I$ *is called a* **right inverse** *of* A. *If* $AB = BA = I$, *then* B *is called a* **two-sided inverse** *of* A *and* A *is said to be* **invertible.**

Lemma. *If* A *has a left inverse* B *and a right inverse* C, *then* $B = C$.

Proof. Suppose $BA = I$ and $AC = I$. Then

$$B = BI = B(AC) = (BA)C = IC = C. \quad \blacksquare$$

Thus if A has a left and a right inverse, A is invertible and has a unique two-sided inverse, which we shall denote by A^{-1} and simply call **the inverse** of A.

Theorem 10. *Let* A *and* B *be* $n \times n$ *matrices over* F.

(i) *If* A *is invertible, so is* A^{-1} *and* $(A^{-1})^{-1} = A$.
(ii) *If both* A *and* B *are invertible, so is* AB, *and* $(AB)^{-1} = B^{-1}A^{-1}$.

Proof. The first statement is evident from the symmetry of the definition. The second follows upon verification of the relations

$$(AB)(B^{-1}A^{-1}) = (B^{-1}A^{-1})(AB) = I. \quad \blacksquare$$

Corollary. *A product of invertible matrices is invertible.*

Theorem 11. *An elementary matrix is invertible.*

Proof. Let E be an elementary matrix corresponding to the elementary row operation e. If e_1 is the inverse operation of e (Theorem 2) and $E_1 = e_1(I)$, then

$$EE_1 = e(E_1) = e(e_1(I)) = I$$

and

$$E_1E = e_1(E) = e_1(e(I)) = I$$

so that E is invertible and $E_1 = E^{-1}$. $\blacksquare$

EXAMPLE 14.

(a)
$$\begin{bmatrix} 0 & 1 \\ 1 & 0 \end{bmatrix}^{-1} = \begin{bmatrix} 0 & 1 \\ 1 & 0 \end{bmatrix}$$

(b)
$$\begin{bmatrix} 1 & c \\ 0 & 1 \end{bmatrix}^{-1} = \begin{bmatrix} 1 & -c \\ 0 & 1 \end{bmatrix}$$

(c)
$$\begin{bmatrix} 1 & 0 \\ c & 1 \end{bmatrix}^{-1} = \begin{bmatrix} 1 & 0 \\ -c & 1 \end{bmatrix}$$

(d) When $c \neq 0$,
$$\begin{bmatrix} c & 0 \\ 0 & 1 \end{bmatrix}^{-1} = \begin{bmatrix} c^{-1} & 0 \\ 0 & 1 \end{bmatrix} \text{ and } \begin{bmatrix} 1 & 0 \\ 0 & c \end{bmatrix}^{-1} = \begin{bmatrix} 1 & 0 \\ 0 & c^{-1} \end{bmatrix}.$$

Theorem 12. *If A is an n $\times$ n matrix, the following are equivalent.*

(i) A *is invertible.*
(ii) A *is row-equivalent to the* n $\times$ n *identity matrix.*
(iii) A *is a product of elementary matrices.*

Proof. Let R be a row-reduced echelon matrix which is row-equivalent to A. By Theorem 9 (or its corollary),
$$R = E_k \cdots E_2E_1A$$
where $E_1, \ldots, E_k$ are elementary matrices. Each E_j is invertible, and so
$$A = E_1^{-1} \cdots E_k^{-1}R.$$
Since products of invertible matrices are invertible, we see that A is invertible if and only if R is invertible. Since R is a (square) row-reduced echelon matrix, R is invertible if and only if each row of R contains a non-zero entry, that is, if and only if $R = I$. We have now shown that A is invertible if and only if $R = I$, and if $R = I$ then $A = E_k^{-1} \cdots E_1^{-1}$. It should now be apparent that (i), (ii), and (iii) are equivalent statements about A. ∎

Corollary. *If* A *is an invertible* n $\times$ n *matrix and if a sequence of elementary row operations reduces* A *to the identity, then that same sequence of operations when applied to* I *yields* A^{-1}.

Corollary. *Let* A *and* B *be* m $\times$ n *matrices. Then* B *is row-equivalent to* A *if and only if* B = PA *where* P *is an invertible* m $\times$ m *matrix.*

Theorem 13. *For an* n $\times$ n *matrix* A, *the following are equivalent.*

(i) A *is invertible.*
(ii) *The homogeneous system* AX = 0 *has only the trivial solution* X = 0.
(iii) *The system of equations* AX = Y *has a solution* X *for each* n $\times$ 1 *matrix* Y.

Proof. According to Theorem 7, condition (ii) is equivalent to the fact that A is row-equivalent to the identity matrix. By Theorem 12, (i) and (ii) are therefore equivalent. If A is invertible, the solution of $AX = Y$ is $X = A^{-1}Y$. Conversely, suppose $AX = Y$ has a solution for each given Y. Let R be a row-reduced echelon matrix which is row-equivalent to A. We wish to show that $R = I$. That amounts to showing that the last row of R is not (identically) 0. Let

$$E = \begin{bmatrix} 0 \\ 0 \\ \vdots \\ 0 \\ 1 \end{bmatrix}.$$

If the system $RX = E$ can be solved for X, the last row of R cannot be 0. We know that $R = PA$, where P is invertible. Thus $RX = E$ if and only if $AX = P^{-1}E$. According to (iii), the latter system has a solution. ∎

Corollary. *A square matrix with either a left or right inverse is invertible.*

Proof. Let A be an $n \times n$ matrix. Suppose A has a left inverse, i.e., a matrix B such that $BA = I$. Then $AX = 0$ has only the trivial solution, because $X = IX = B(AX)$. Therefore A is invertible. On the other hand, suppose A has a right inverse, i.e., a matrix C such that $AC = I$. Then C has a left inverse and is therefore invertible. It then follows that $A = C^{-1}$ and so A is invertible with inverse C. ∎

Corollary. *Let* $A = A_1A_2 \cdots A_k$, *where* $A_1 \ldots, A_k$ *are* $n \times n$ *(square) matrices. Then* A *is invertible if and only if each* A_j *is invertible.*

Proof. We have already shown that the product of two invertible matrices is invertible. From this one sees easily that if each A_j is invertible then A is invertible.

Suppose now that A is invertible. We first prove that A_k is invertible. Suppose X is an $n \times 1$ matrix and $A_kX = 0$. Then $AX = (A_1 \cdots A_{k-1})A_kX = 0$. Since A is invertible we must have $X = 0$. The system of equations $A_kX = 0$ thus has no non-trivial solution, so A_k is invertible. But now $A_1 \cdots A_{k-1} = AA_k^{-1}$ is invertible. By the preceding argument, A_{k-1} is invertible. Continuing in this way, we conclude that each A_j is invertible. ∎

We should like to make one final comment about the solution of linear equations. Suppose A is an $m \times n$ matrix and we wish to solve the system of equations $AX = Y$. If R is a row-reduced echelon matrix which is row-equivalent to A, then $R = PA$ where P is an $m \times m$ invertible matrix. The solutions of the system $AX = Y$ are exactly the same as the solutions of the system $RX = PY\ (= Z)$. In practice, it is not much more difficult to find the matrix P than it is to row-reduce A to R. For, suppose we form the augmented matrix A' of the system $AX = Y$, with arbitrary scalars $y_1, \ldots, y_m$ occurring in the last column. If we then perform on A' a sequence of elementary row operations which leads from A to R, it will become evident what the matrix P is. (The reader should refer to Example 9 where we essentially carried out this process.) In particular, if A is a square matrix, this process will make it clear whether or not A is invertible and if A is invertible what the inverse P is. Since we have already given the nucleus of one example of such a computation, we shall content ourselves with a 2×2 example.

EXAMPLE 15. Suppose F is the field of rational numbers and

$$A = \begin{bmatrix} 2 & -1 \\ 1 & 3 \end{bmatrix}.$$

Then

$$\begin{bmatrix} 2 & -1 & y_1 \\ 1 & 3 & y_2 \end{bmatrix} \xrightarrow{(3)} \begin{bmatrix} 1 & 3 & y_2 \\ 2 & -1 & y_1 \end{bmatrix} \xrightarrow{(2)} \begin{bmatrix} 1 & 3 & y_2 \\ 0 & -7 & y_1 - 2y_2 \end{bmatrix} \xrightarrow{(1)}$$

$$\begin{bmatrix} 1 & 3 & y_2 \\ 0 & 1 & \frac{1}{7}(2y_2 - y_1) \end{bmatrix} \xrightarrow{(2)} \begin{bmatrix} 1 & 0 & \frac{1}{7}(y_2 + 3y_1) \\ 0 & 1 & \frac{1}{7}(2y_2 - y_1) \end{bmatrix}$$

from which it is clear that A is invertible and

$$A^{-1} = \begin{bmatrix} \frac{3}{7} & \frac{1}{7} \\ -\frac{1}{7} & \frac{2}{7} \end{bmatrix}.$$

It may seem cumbersome to continue writing the arbitrary scalars $y_1, y_2, \ldots$ in the computation of inverses. Some people find it less awkward to carry along two sequences of matrices, one describing the reduction of A to the identity and the other recording the effect of the same sequence of operations starting from the identity. The reader may judge for himself which is a neater form of bookkeeping.

EXAMPLE 16. Let us find the inverse of

$$A = \begin{bmatrix} 1 & \frac{1}{2} & \frac{1}{3} \\ \frac{1}{2} & \frac{1}{3} & \frac{1}{4} \\ \frac{1}{3} & \frac{1}{4} & \frac{1}{5} \end{bmatrix}.$$

$$\begin{bmatrix} 1 & \frac{1}{2} & \frac{1}{3} \\ \frac{1}{2} & \frac{1}{3} & \frac{1}{4} \\ \frac{1}{3} & \frac{1}{4} & \frac{1}{5} \end{bmatrix}, \quad \begin{bmatrix} 1 & 0 & 0 \\ 0 & 1 & 0 \\ 0 & 0 & 1 \end{bmatrix}$$

$$\begin{bmatrix} 1 & \frac{1}{2} & \frac{1}{3} \\ 0 & \frac{1}{12} & \frac{1}{12} \\ 0 & \frac{1}{12} & \frac{4}{45} \end{bmatrix}, \quad \begin{bmatrix} 1 & 0 & 0 \\ -\frac{1}{2} & 1 & 0 \\ -\frac{1}{3} & 0 & 1 \end{bmatrix}$$

$$\begin{bmatrix} 1 & \frac{1}{2} & \frac{1}{3} \\ 0 & \frac{1}{12} & \frac{1}{12} \\ 0 & 0 & \frac{1}{180} \end{bmatrix}, \quad \begin{bmatrix} 1 & 0 & 0 \\ -\frac{1}{2} & 1 & 0 \\ \frac{1}{6} & -1 & 1 \end{bmatrix}$$

$$\begin{bmatrix} 1 & \frac{1}{2} & \frac{1}{3} \\ 0 & 1 & 1 \\ 0 & 0 & 1 \end{bmatrix}, \quad \begin{bmatrix} 1 & 0 & 0 \\ -6 & 12 & 0 \\ 30 & -180 & 180 \end{bmatrix}$$

$$\begin{bmatrix} 1 & \frac{1}{2} & 0 \\ 0 & 1 & 0 \\ 0 & 0 & 1 \end{bmatrix}, \quad \begin{bmatrix} -9 & 60 & -60 \\ -36 & 192 & -180 \\ 30 & -180 & 180 \end{bmatrix}$$

$$\begin{bmatrix} 1 & 0 & 0 \\ 0 & 1 & 0 \\ 0 & 0 & 1 \end{bmatrix}, \quad \begin{bmatrix} 9 & -36 & 30 \\ -36 & 192 & -180 \\ 30 & -180 & 180 \end{bmatrix}.$$

It must have occurred to the reader that we have carried on a lengthy discussion of the rows of matrices and have said little about the columns. We focused our attention on the rows because this seemed more natural from the point of view of linear equations. Since there is obviously nothing sacred about rows, the discussion in the last sections could have been carried on using columns rather than rows. If one defines an elementary column operation and column-equivalence in a manner analogous to that of elementary row operation and row-equivalence, it is clear that each $m \times n$ matrix will be column-equivalent to a 'column-reduced echelon' matrix. Also each elementary column operation will be of the form $A \to AE$, where E is an $n \times n$ elementary matrix—and so on.

Exercises

1. Let

$$A = \begin{bmatrix} 1 & 2 & 1 & 0 \\ -1 & 0 & 3 & 5 \\ 1 & -2 & 1 & 1 \end{bmatrix}.$$

Find a row-reduced echelon matrix R which is row-equivalent to A and an invertible 3×3 matrix P such that $R = PA$.

2. Do Exercise 1, but with

$$A = \begin{bmatrix} 2 & 0 & i \\ 1 & -3 & -i \\ i & 1 & 1 \end{bmatrix}.$$

3. For each of the two matrices

$$\begin{bmatrix} 2 & 5 & -1 \\ 4 & -1 & 2 \\ 6 & 4 & 1 \end{bmatrix}, \quad \begin{bmatrix} 1 & -1 & 2 \\ 3 & 2 & 4 \\ 0 & 1 & -2 \end{bmatrix}$$

use elementary row operations to discover whether it is invertible, and to find the inverse in case it is.

4. Let

$$A = \begin{bmatrix} 5 & 0 & 0 \\ 1 & 5 & 0 \\ 0 & 1 & 5 \end{bmatrix}.$$

For which X does there exist a scalar c such that $AX = cX$?

5. Discover whether

$$A = \begin{bmatrix} 1 & 2 & 3 & 4 \\ 0 & 2 & 3 & 4 \\ 0 & 0 & 3 & 4 \\ 0 & 0 & 0 & 4 \end{bmatrix}$$

is invertible, and find A^{-1} if it exists.

6. Suppose A is a 2×1 matrix and that B is a 1×2 matrix. Prove that $C = AB$ is not invertible.

7. Let A be an $n \times n$ (square) matrix. Prove the following two statements:

(a) If A is invertible and $AB = 0$ for some $n \times n$ matrix B, then $B = 0$.

(b) If A is not invertible, then there exists an $n \times n$ matrix B such that $AB = 0$ but $B \neq 0$.

8. Let

$$A = \begin{bmatrix} a & b \\ c & d \end{bmatrix}.$$

Prove, using elementary row operations, that A is invertible if and only if $(ad - bc) \neq 0$.

9. An $n \times n$ matrix A is called **upper-triangular** if $A_{ij} = 0$ for $i > j$, that is, if every entry below the main diagonal is 0. Prove that an upper-triangular (square) matrix is invertible if and only if every entry on its main diagonal is different from 0.

10. Prove the following generalization of Exercise 6. If A is an $m \times n$ matrix, B is an $n \times m$ matrix and $n < m$, then AB is not invertible.

11. Let A be an $m \times n$ matrix. Show that by means of a finite number of elementary row and/or column operations one can pass from A to a matrix R which is both 'row-reduced echelon' and 'column-reduced echelon,' i.e., $R_{ij} = 0$ if $i \neq j$, $R_{ii} = 1$, $1 \leq i \leq r$, $R_{ii} = 0$ if $i > r$. Show that $R = PAQ$, where P is an invertible $m \times m$ matrix and Q is an invertible $n \times n$ matrix.

12. The result of Example 16 suggests that perhaps the matrix

$$A = \begin{bmatrix} 1 & \frac{1}{2} & \cdots & \frac{1}{n} \\ \frac{1}{2} & \frac{1}{3} & \cdots & \frac{1}{n+1} \\ \vdots & \vdots & & \vdots \\ \frac{1}{n} & \frac{1}{n+1} & \cdots & \frac{1}{2n-1} \end{bmatrix}$$

is invertible and A^{-1} has integer entries. Can you prove that?

2. Vector Spaces

2.1. Vector Spaces

In various parts of mathematics, one is confronted with a set, such that it is both meaningful and interesting to deal with 'linear combinations' of the objects in that set. For example, in our study of linear equations we found it quite natural to consider linear combinations of the rows of a matrix. It is likely that the reader has studied calculus and has dealt there with linear combinations of functions; certainly this is so if he has studied differential equations. Perhaps the reader has had some experience with vectors in three-dimensional Euclidean space, and in particular, with linear combinations of such vectors.

Loosely speaking, linear algebra is that branch of mathematics which treats the common properties of algebraic systems which consist of a set, together with a reasonable notion of a 'linear combination' of elements in the set. In this section we shall define the mathematical object which experience has shown to be the most useful abstraction of this type of algebraic system.

Definition. *A* **vector space** *(or linear space) consists of the following:*

1. *a field* F *of scalars;*
2. *a set* V *of objects, called vectors;*
3. *a rule (or operation), called vector addition, which associates with each pair of vectors* α, β *in* V *a vector* $\alpha + \beta$ *in* V, *called the sum of* α *and* β, *in such a way that*
   (a) *addition is commutative,* $\alpha + \beta = \beta + \alpha$;
   (b) *addition is associative,* $\alpha + (\beta + \gamma) = (\alpha + \beta) + \gamma$;

(c) *there is a unique vector* 0 *in* V, *called the zero vector, such that* $\alpha + 0 = \alpha$ *for all* α *in* V;

(d) *for each vector* α *in* V *there is a unique vector* $-\alpha$ *in* V *such that* $\alpha + (-\alpha) = 0$;

4. *a rule (or operation), called scalar multiplication, which associates with each scalar* c *in* F *and vector* α *in* V *a vector* cα *in* V, *called the product of* c *and* α, *in such a way that*

(a) $1\alpha = \alpha$ *for every* α *in* V;

(b) $(c_1c_2)\alpha = c_1(c_2\alpha)$;

(c) $c(\alpha + \beta) = c\alpha + c\beta$;

(d) $(c_1 + c_2)\alpha = c_1\alpha + c_2\alpha$.

It is important to observe, as the definition states, that a vector space is a composite object consisting of a field, a set of 'vectors,' and two operations with certain special properties. The same set of vectors may be part of a number of distinct vector spaces (see Example 5 below). When there is no chance of confusion, we may simply refer to the vector space as V, or when it is desirable to specify the field, we shall say V is a **vector space over the field** F. The name 'vector' is applied to the elements of the set V largely as a matter of convenience. The origin of the name is to be found in Example 1 below, but one should not attach too much significance to the name, since the variety of objects occurring as the vectors in V may not bear much resemblance to any preassigned concept of vector which the reader has. We shall try to indicate this variety by a list of examples; our list will be enlarged considerably as we begin to study vector spaces.

EXAMPLE 1. **The n-tuple space, F^n.** Let F be any field, and let V be the set of all n-tuples $\alpha = (x_1, x_2, \ldots, x_n)$ of scalars x_i in F. If $\beta = (y_1, y_2, \ldots, y_n)$ with y_i in F, the sum of α and β is defined by

$$\alpha + \beta = (x_1 + y_1, x_2 + y_2, \ldots, x_n + y_n). \tag{2-1}$$

The product of a scalar c and vector α is defined by

$$c\alpha = (cx_1, cx_2, \ldots, cx_n). \tag{2-2}$$

The fact that this vector addition and scalar multiplication satisfy conditions (3) and (4) is easy to verify, using the similar properties of addition and multiplication of elements of F.

EXAMPLE 2. **The space of $m \times n$ matrices, $F^{m\times n}$.** Let F be any field and let m and n be positive integers. Let $F^{m\times n}$ be the set of all $m \times n$ matrices over the field F. The sum of two vectors A and B in $F^{m\times n}$ is defined by

$$(A + B)_{ij} = A_{ij} + B_{ij}. \tag{2-3}$$

The product of a scalar c and the matrix A is defined by

$$(cA)_{ij} = cA_{ij}. \tag{2-4}$$

Note that $F^{1\times n} = F^n$.

EXAMPLE 3. **The space of functions from a set to a field.** Let F be any field and let S be any non-empty set. Let V be the set of all functions from the set S into F. The sum of two vectors f and g in V is the vector $f + g$, i.e., the function from S into F, defined by

$$(f + g)(s) = f(s) + g(s). \tag{2-5}$$

The product of the scalar c and the function f is the function cf defined by

$$(cf)(s) = cf(s). \tag{2-6}$$

The preceding examples are special cases of this one. For an n-tuple of elements of F may be regarded as a function from the set S of integers $1, \ldots, n$ into F. Similarly, an $m \times n$ matrix over the field F is a function from the set S of pairs of integers, (i, j), $1 \leq i \leq m$, $1 \leq j \leq n$, into the field F. For this third example we shall indicate how one verifies that the operations we have defined satisfy conditions (3) and (4). For vector addition:

(a) Since addition in F is commutative,

$$f(s) + g(s) = g(s) + f(s)$$

for each s in S, so the functions $f + g$ and $g + f$ are identical.

(b) Since addition in F is associative,

$$f(s) + [g(s) + h(s)] = [f(s) + g(s)] + h(s)$$

for each s, so $f + (g + h)$ is the same function as $(f + g) + h$.

(c) The unique zero vector is the zero function which assigns to each element of S the scalar 0 in F.

(d) For each f in V, $(-f)$ is the function which is given by

$$(-f)(s) = -f(s).$$

The reader should find it easy to verify that scalar multiplication satisfies the conditions of (4), by arguing as we did with the vector addition.

EXAMPLE 4. **The space of polynomial functions over a field F.** Let F be a field and let V be the set of all functions f from F into F which have a rule of the form

$$f(x) = c_0 + c_1x + \cdots + c_nx^n \tag{2-7}$$

where $c_0, c_1, \ldots, c_n$ are fixed scalars in F (independent of x). A function of this type is called a **polynomial function on F.** Let addition and scalar multiplication be defined as in Example 3. One must observe here that if f and g are polynomial functions and c is in F, then $f + g$ and cf are again polynomial functions.

EXAMPLE 5. The field C of complex numbers may be regarded as a vector space over the field R of real numbers. More generally, let F be the field of real numbers and let V be the set of n-tuples $\alpha = (x_1, \ldots, x_n)$ where $x_1, \ldots, x_n$ are *complex* numbers. Define addition of vectors and scalar multiplication by (2-1) and (2-2), as in Example 1. In this way we obtain a vector space over the field R which is quite different from the space C^n and the space R^n.

There are a few simple facts which follow almost immediately from the definition of a vector space, and we proceed to derive these. If c is a scalar and 0 is the zero vector, then by 3(c) and 4(c)

$$c0 = c(0 + 0) = c0 + c0.$$

Adding $-(c0)$ and using 3(d), we obtain

$$c0 = 0. \tag{2-8}$$

Similarly, for the scalar 0 and any vector α we find that

$$0\alpha = 0. \tag{2-9}$$

If c is a non-zero scalar and α is a vector such that $c\alpha = 0$, then by (2-8), $c^{-1}(c\alpha) = 0$. But

$$c^{-1}(c\alpha) = (c^{-1}c)\alpha = 1\alpha = \alpha$$

hence, $\alpha = 0$. Thus we see that if c is a scalar and α a vector such that $c\alpha = 0$, then either c is the zero scalar or α is the zero vector.

If α is any vector in V, then

$$0 = 0\alpha = (1 - 1)\alpha = 1\alpha + (-1)\alpha = \alpha + (-1)\alpha$$

from which it follows that

$$(-1)\alpha = -\alpha. \tag{2-10}$$

Finally, the associative and commutative properties of vector addition imply that a sum involving a number of vectors is independent of the way in which these vectors are combined and associated. For example, if $\alpha_1, \alpha_2, \alpha_3, \alpha_4$ are vectors in V, then

$$(\alpha_1 + \alpha_2) + (\alpha_3 + \alpha_4) = [\alpha_2 + (\alpha_1 + \alpha_3)] + \alpha_4$$

and such a sum may be written without confusion as

$$\alpha_1 + \alpha_2 + \alpha_3 + \alpha_4.$$

Definition. *A vector β in V is said to be a* **linear combination** *of the vectors $\alpha_1, \ldots, \alpha_n$ in V provided there exist scalars $c_1, \ldots, c_n$ in F such that*

$$\begin{aligned} \beta &= c_1\alpha_1 + \cdots + c_n\alpha_n \\ &= \sum_{i=1}^{n} c_i\alpha_i. \end{aligned}$$

Other extensions of the associative property of vector addition and the distributive properties 4(c) and 4(d) of scalar multiplication apply to linear combinations:

$$\sum_{i=1}^{n} c_i\alpha_i + \sum_{i=1}^{n} d_i\alpha_i = \sum_{i=1}^{n} (c_i + d_i)\alpha_i$$

$$c \sum_{i=1}^{n} c_i\alpha_i = \sum_{i=1}^{n} (cc_i)\alpha_i.$$

Certain parts of linear algebra are intimately related to geometry. The very word 'space' suggests something geometrical, as does the word 'vector' to most people. As we proceed with our study of vector spaces, the reader will observe that much of the terminology has a geometrical connotation. Before concluding this introductory section on vector spaces, we shall consider the relation of vector spaces to geometry to an extent which will at least indicate the origin of the name 'vector space.' This will be a brief intuitive discussion.

Let us consider the vector space R^3. In analytic geometry, one identifies triples (x_1, x_2, x_3) of real numbers with the points in three-dimensional Euclidean space. In that context, a vector is usually defined as a directed line segment PQ, from a point P in the space to another point Q. This amounts to a careful formulation of the idea of the 'arrow' from P to Q. As vectors are used, it is intended that they should be determined by their length and direction. Thus one must identify two directed line segments if they have the same length and the same direction.

The directed line segment PQ, from the point $P = (x_1, x_2, x_3)$ to the point $Q = (y_1, y_2, y_3)$, has the same length and direction as the directed line segment from the origin $O = (0, 0, 0)$ to the point $(y_1 - x_1, y_2 - x_2, y_3 - x_3)$. Furthermore, this is the only segment emanating from the origin which has the same length and direction as PQ. Thus, if one agrees to treat only vectors which emanate from the origin, there is exactly one vector associated with each given length and direction.

The vector OP, from the origin to $P = (x_1, x_2, x_3)$, is completely determined by P, and it is therefore possible to identify this vector with the point P. In our definition of the vector space R^3, the vectors are simply defined to be the triples (x_1, x_2, x_3).

Given points $P = (x_1, x_2, x_3)$ and $Q = (y_1, y_2, y_3)$, the definition of the sum of the vectors OP and OQ can be given geometrically. If the vectors are not parallel, then the segments OP and OQ determine a plane and these segments are two of the edges of a parallelogram in that plane (see Figure 1). One diagonal of this parallelogram extends from O to a point S, and the sum of OP and OQ is defined to be the vector OS. The coordinates of the point S are $(x_1 + y_1, x_2 + y_2, x_3 + y_3)$ and hence this geometrical definition of vector addition is equivalent to the algebraic definition of Example 1.

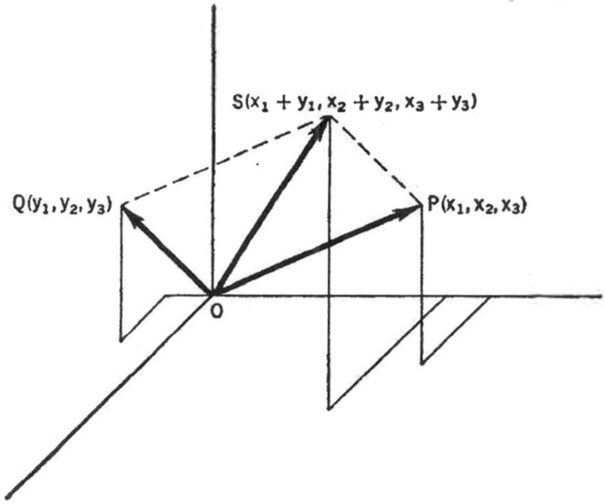

FIGURE 1

Scalar multiplication has a simpler geometric interpretation. If c is a real number, then the product of c and the vector OP is the vector from the origin with length $|c|$ times the length of OP and a direction which agrees with the direction of OP if $c > 0$, and which is opposite to the direction of OP if $c < 0$. This scalar multiplication just yields the vector OT where $T = (cx_1, cx_2, cx_3)$, and is therefore consistent with the algebraic definition given for R^3.

From time to time, the reader will probably find it helpful to 'think geometrically' about vector spaces, that is, to draw pictures for his own benefit to illustrate and motivate some of the ideas. Indeed, he should do this. However, in forming such illustrations he must bear in mind that, because we are dealing with vector spaces as algebraic systems, all proofs we give will be of an algebraic nature.

Exercises

1. If F is a field, verify that F^n (as defined in Example 1) is a vector space over the field F.

2. If V is a vector space over the field F, verify that

$$(\alpha_1 + \alpha_2) + (\alpha_3 + \alpha_4) = [\alpha_2 + (\alpha_3 + \alpha_1)] + \alpha_4$$

for all vectors α_1, α_2, α_3, and α_4 in V.

3. If C is the field of complex numbers, which vectors in C^3 are linear combinations of $(1, 0, -1)$, $(0, 1, 1)$, and $(1, 1, 1)$?

4. Let V be the set of all pairs (x, y) of real numbers, and let F be the field of real numbers. Define

$$(x, y) + (x_1, y_1) = (x + x_1, y + y_1)$$
$$c(x, y) = (cx, y).$$

Is V, with these operations, a vector space over the field of real numbers?

5. On R^n, define two operations

$$\alpha \oplus \beta = \alpha - \beta$$
$$c \cdot \alpha = -c\alpha.$$

The operations on the right are the usual ones. Which of the axioms for a vector space are satisfied by $(R^n, \oplus, \cdot)$?

6. Let V be the set of all complex-valued functions f on the real line such that (for all t in R)

$$f(-t) = \overline{f(t)}.$$

The bar denotes complex conjugation. Show that V, with the operations

$$(f + g)(t) = f(t) + g(t)$$
$$(cf)(t) = cf(t)$$

is a vector space over the field of *real* numbers. Give an example of a function in V which is not real-valued.

7. Let V be the set of pairs (x, y) of real numbers and let F be the field of real numbers. Define

$$(x, y) + (x_1, y_1) = (x + x_1, 0)$$
$$c(x, y) = (cx, 0).$$

Is V, with these operations, a vector space?

2.2. Subspaces

In this section we shall introduce some of the basic concepts in the study of vector spaces.

Definition. *Let* V *be a vector space over the field* F. *A* **subspace** *of* V *is a subset* W *of* V *which is itself a vector space over* F *with the operations of vector addition and scalar multiplication on* V.

A direct check of the axioms for a vector space shows that the subset W of V is a subspace if for each α and β in W the vector $\alpha + \beta$ is again in W; the 0 vector is in W; for each α in W the vector $(-\alpha)$ is in W; for each α in W and each scalar c the vector $c\alpha$ is in W. The commutativity and associativity of vector addition, and the properties (4)(a), (b), (c), and (d) of scalar multiplication do not need to be checked, since these are properties of the operations on V. One can simplify things still further.

Theorem 1. *A non-empty subset W of V is a subspace of V if and only if for each pair of vectors α, β in W and each scalar c in F the vector $c\alpha + \beta$ is again in W.*

Proof. Suppose that W is a non-empty subset of V such that $c\alpha + \beta$ belongs to W for all vectors α, β in W and all scalars c in F. Since W is non-empty, there is a vector ρ in W, and hence $(-1)\rho + \rho = 0$ is in W. Then if α is any vector in W and c any scalar, the vector $c\alpha = c\alpha + 0$ is in W. In particular, $(-1)\alpha = -\alpha$ is in W. Finally, if α and β are in W, then $\alpha + \beta = 1\alpha + \beta$ is in W. Thus W is a subspace of V.

Conversely, if W is a subspace of V, α and β are in W, and c is a scalar, certainly $c\alpha + \beta$ is in W. ∎

Some people prefer to use the $c\alpha + \beta$ property in Theorem 1 as the definition of a subspace. It makes little difference. The important point is that, if W is a non-empty subset of V such that $c\alpha + \beta$ is in V for all α, β in W and all c in F, then (with the operations inherited from V) W is a vector space. This provides us with many new examples of vector spaces.

EXAMPLE 6.

(a) If V is any vector space, V is a subspace of V; the subset consisting of the zero vector alone is a subspace of V, called the **zero subspace** of V.

(b) In F^n, the set of n-tuples $(x_1, \ldots, x_n)$ with $x_1 = 0$ is a subspace; however, the set of n-tuples with $x_1 = 1 + x_2$ is not a subspace $(n \geq 2)$.

(c) The space of polynomial functions over the field F is a subspace of the space of all functions from F into F.

(d) An $n \times n$ **(square) matrix** A over the field F is **symmetric** if $A_{ij} = A_{ji}$ for each i and j. The symmetric matrices form a subspace of the space of all $n \times n$ matrices over F.

(e) An $n \times n$ (square) matrix A over the field C of complex numbers is **Hermitian** (or **self-adjoint**) if

$$A_{jk} = \overline{A_{kj}}$$

for each j, k, the bar denoting complex conjugation. A 2×2 matrix is Hermitian if and only if it has the form

$$\begin{bmatrix} z & x + iy \\ x - iy & w \end{bmatrix}$$

where x, y, z, and w are real numbers. The set of all Hermitian matrices is *not* a subspace of the space of all $n \times n$ matrices over C. For if A is Hermitian, its diagonal entries A_{11}, A_{22}, . . . , are all real numbers, but the diagonal entries of iA are in general not real. On the other hand, it is easily verified that the set of $n \times n$ complex Hermitian matrices is a vector space over the field R of real numbers (with the usual operations).

EXAMPLE 7. **The solution space of a system of homogeneous linear equations.** Let A be an $m \times n$ matrix over F. Then the set of all $n \times 1$ (column) matrices X over F such that $AX = 0$ is a subspace of the space of all $n \times 1$ matrices over F. To prove this we must show that $A(cX + Y) = 0$ when $AX = 0$, $AY = 0$, and c is an arbitrary scalar in F. This follows immediately from the following general fact.

Lemma. *If* A *is an* m $\times$ n *matrix over* F *and* B, C *are* n $\times$ p *matrices over* F *then*

$$A(dB + C) = d(AB) + AC \tag{2-11}$$

for each scalar d *in* F.

Proof.

$$\begin{aligned}[A(dB + C)]_{ij} &= \sum_k A_{ik}(dB + C)_{kj} \\ &= \sum_k (dA_{ik}B_{kj} + A_{ik}C_{kj}) \\ &= d\sum_k A_{ik}B_{kj} + \sum_k A_{ik}C_{kj} \\ &= d(AB)_{ij} + (AC)_{ij} \\ &= [d(AB) + AC]_{ij}. \quad \blacksquare\end{aligned}$$

Similarly one can show that $(dB + C)A = d(BA) + CA$, if the matrix sums and products are defined.

Theorem 2. *Let* V *be a vector space over the field* F. *The intersection of any collection of subspaces of* V *is a subspace of* V.

Proof. Let $\{W_a\}$ be a collection of subspaces of V, and let $W = \bigcap_a W_a$ be their intersection. Recall that W is defined as the set of all elements belonging to every W_a (see Appendix). Since each W_a is a subspace, each contains the zero vector. Thus the zero vector is in the intersection W, and W is non-empty. Let α and β be vectors in W and let c be a scalar. By definition of W, both α and β belong to each W_a, and because each W_a is a subspace, the vector $(c\alpha + \beta)$ is in every W_a. Thus $(c\alpha + \beta)$ is again in W. By Theorem 1, W is a subspace of V. $\blacksquare$

From Theorem 2 it follows that if S is any collection of vectors in V, then there is a smallest subspace of V which contains S, that is, a subspace which contains S and which is contained in every other subspace containing S.

Definition. *Let* S *be a set of vectors in a vector space* V. *The* **subspace spanned** *by* S *is defined to be the intersection* W *of all subspaces of* V *which contain* S. *When* S *is a finite set of vectors,* S = $\{\alpha_1, \alpha_2, \ldots, \alpha_n\}$, *we shall simply call* W *the* **subspace spanned by the vectors** $\alpha_1, \alpha_2, \ldots, \alpha_n$.

Theorem 3. *The subspace spanned by a non-empty subset* S *of a vector space* V *is the set of all linear combinations of vectors in* S.

Proof. Let W be the subspace spanned by S. Then each linear combination

$$\alpha = x_1\alpha_1 + x_2\alpha_2 + \cdots + x_m\alpha_m$$

of vectors $\alpha_1, \alpha_2, \ldots, \alpha_m$ in S is clearly in W. Thus W contains the set L of all linear combinations of vectors in S. The set L, on the other hand, contains S and is non-empty. If α, β belong to L then α is a linear combination,

$$\alpha = x_1\alpha_1 + x_2\alpha_2 + \cdots + x_m\alpha_m$$

of vectors α_i in S, and β is a linear combination,

$$\beta = y_1\beta_1 + y_2\beta_2 + \cdots + y_n\beta_n$$

of vectors β_j in S. For each scalar c,

$$c\alpha + \beta = \sum_{i=1}^{m} (cx_i)\alpha_i + \sum_{j=1}^{n} y_j\beta_j.$$

Hence $c\alpha + \beta$ belongs to L. Thus L is a subspace of V.

Now we have shown that L is a subspace of V which contains S, and also that any subspace which contains S contains L. It follows that L is the intersection of all subspaces containing S, i.e., that L is the subspace spanned by the set S. ∎

Definition. *If* $S_1, S_2, \ldots, S_k$ *are subsets of a vector space* V, *the set of all sums*

$$\alpha_1 + \alpha_2 + \cdots + \alpha_k$$

of vectors α_i *in* S_i *is called the* **sum** *of the subsets* $S_1, S_2, \ldots, S_k$ *and is denoted by*

$$S_1 + S_2 + \cdots + S_k$$

or by

$$\sum_{i=1}^{k} S_i.$$

If $W_1, W_2, \ldots, W_k$ are subspaces of V, then the sum

$$W = W_1 + W_2 + \cdots + W_k$$

is easily seen to be a subspace of V which contains each of the subspaces W_i. From this it follows, as in the proof of Theorem 3, that W is the subspace spanned by the union of $W_1, W_2, \ldots, W_k$.

EXAMPLE 8. Let F be a subfield of the field C of complex numbers. Suppose

$$\begin{aligned}\alpha_1 &= (1, 2, 0, 3, 0)\\ \alpha_2 &= (0, 0, 1, 4, 0)\\ \alpha_3 &= (0, 0, 0, 0, 1).\end{aligned}$$

By Theorem 3, a vector α is in the subspace W of F^5 spanned by α_1, α_2, α_3 if and only if there exist scalars c_1, c_2, c_3 in F such that

$$\alpha = c_1\alpha_1 + c_2\alpha_2 + c_3\alpha_3.$$

Thus W consists of all vectors of the form

$$\alpha = (c_1, 2c_1, c_2, 3c_1 + 4c_2, c_3)$$

where c_1, c_2, c_3 are arbitrary scalars in F. Alternatively, W can be described as the set of all 5-tuples

$$\alpha = (x_1, x_2, x_3, x_4, x_5)$$

with x_i in F such that

$$\begin{aligned}x_2 &= 2x_1\\ x_4 &= 3x_1 + 4x_3.\end{aligned}$$

Thus $(-3, -6, 1, -5, 2)$ is in W, whereas $(2, 4, 6, 7, 8)$ is not.

Example 9. Let F be a subfield of the field C of complex numbers, and let V be the vector space of all 2×2 matrices over F. Let W_1 be the subset of V consisting of all matrices of the form

$$\begin{bmatrix} x & y \\ z & 0 \end{bmatrix}$$

where x, y, z are arbitrary scalars in F. Finally, let W_2 be the subset of V consisting of all matrices of the form

$$\begin{bmatrix} x & 0 \\ 0 & y \end{bmatrix}$$

where x and y are arbitrary scalars in F. Then W_1 and W_2 are subspaces of V. Also

$$V = W_1 + W_2$$

because

$$\begin{bmatrix} a & b \\ c & d \end{bmatrix} = \begin{bmatrix} a & b \\ c & 0 \end{bmatrix} + \begin{bmatrix} 0 & 0 \\ 0 & d \end{bmatrix}.$$

The subspace $W_1 \cap W_2$ consists of all matrices of the form

$$\begin{bmatrix} x & 0 \\ 0 & 0 \end{bmatrix}.$$

Example 10. Let A be an $m \times n$ matrix over a field F. The **row vectors** of A are the vectors in F^n given by $\alpha_i = (A_{i1}, \dots, A_{in})$, $i = 1, \dots, m$. The subspace of F^n spanned by the row vectors of A is called the **row**

space of A. The subspace considered in Example 8 is the row space of the matrix

$$A = \begin{bmatrix} 1 & 2 & 0 & 3 & 0 \\ 0 & 0 & 1 & 4 & 0 \\ 0 & 0 & 0 & 0 & 1 \end{bmatrix}.$$

It is also the row space of the matrix

$$B = \begin{bmatrix} 1 & 2 & 0 & 3 & 0 \\ 0 & 0 & 1 & 4 & 0 \\ 0 & 0 & 0 & 0 & 1 \\ -4 & -8 & 1 & -8 & 0 \end{bmatrix}.$$

EXAMPLE 11. Let V be the space of all polynomial functions over F. Let S be the subset of V consisting of the polynomial functions $f_0, f_1, f_2, \ldots$ defined by

$$f_n(x) = x^n, \qquad n = 0, 1, 2, \ldots.$$

Then V is the subspace spanned by the set S.

Exercises

1. Which of the following sets of vectors $\alpha = (a_1, \ldots, a_n)$ in R^n are subspaces of R^n $(n \geq 3)$?

(a) all α such that $a_1 \geq 0$;
(b) all α such that $a_1 + 3a_2 = a_3$;
(c) all α such that $a_2 = a_1^2$;
(d) all α such that $a_1a_2 = 0$;
(e) all α such that a_2 is rational.

2. Let V be the (real) vector space of all functions f from R into R. Which of the following sets of functions are subspaces of V?

(a) all f such that $f(x^2) = f(x)^2$;
(b) all f such that $f(0) = f(1)$;
(c) all f such that $f(3) = 1 + f(-5)$;
(d) all f such that $f(-1) = 0$;
(e) all f which are continuous.

3. Is the vector $(3, -1, 0, -1)$ in the subspace of R^5 spanned by the vectors $(2, -1, 3, 2)$, $(-1, 1, 1, -3)$, and $(1, 1, 9, -5)$?

4. Let W be the set of all $(x_1, x_2, x_3, x_4, x_5)$ in R^5 which satisfy

$$\begin{aligned} 2x_1 - x_2 + \tfrac{4}{3}x_3 - x_4 \qquad &= 0 \\ x_1 \qquad + \tfrac{2}{3}x_3 \qquad - x_5 &= 0 \\ 9x_1 - 3x_2 + 6x_3 - 3x_4 - 3x_5 &= 0. \end{aligned}$$

Find a finite set of vectors which spans W.

5. Let F be a field and let n be a positive integer ($n \geq 2$). Let V be the vector space of all $n \times n$ matrices over F. Which of the following sets of matrices A in V are subspaces of V?

(a) all invertible A;
(b) all non-invertible A;
(c) all A such that $AB = BA$, where B is some fixed matrix in V;
(d) all A such that $A^2 = A$.

6. (a) Prove that the only subspaces of R^1 are R^1 and the zero subspace.

(b) Prove that a subspace of R^2 is R^2, or the zero subspace, or consists of all scalar multiples of some fixed vector in R^2. (The last type of subspace is, intuitively, a straight line through the origin.)

(c) Can you describe the subspaces of R^3?

7. Let W_1 and W_2 be subspaces of a vector space V such that the set-theoretic union of W_1 and W_2 is also a subspace. Prove that one of the spaces W_i is contained in the other.

8. Let V be the vector space of all functions from R into R; let V_e be the subset of even functions, $f(-x) = f(x)$; let V_o be the subset of odd functions, $f(-x) = -f(x)$.

(a) Prove that V_e and V_o are subspaces of V.
(b) Prove that $V_e + V_o = V$.
(c) Prove that $V_e \cap V_o = \{0\}$.

9. Let W_1 and W_2 be subspaces of a vector space V such that $W_1 + W_2 = V$ and $W_1 \cap W_2 = \{0\}$. Prove that for each vector α in V there are *unique* vectors α_1 in W_1 and α_2 in W_2 such that $\alpha = \alpha_1 + \alpha_2$.

2.3. Bases and Dimension

We turn now to the task of assigning a dimension to certain vector spaces. Although we usually associate 'dimension' with something geometrical, we must find a suitable algebraic definition of the dimension of a vector space. This will be done through the concept of a basis for the space.

Definition. *Let V be a vector space over F. A subset S of V is said to be* **linearly dependent** *(or simply,* **dependent***) if there exist distinct vectors $\alpha_1, \alpha_2, \ldots, \alpha_n$ in S and scalars $c_1, c_2, \ldots, c_n$ in F, not all of which are 0, such that*

$$c_1\alpha_1 + c_2\alpha_2 + \cdots + c_n\alpha_n = 0.$$

A set which is not linearly dependent is called **linearly independent.** *If the set S contains only finitely many vectors $\alpha_1, \alpha_2, \ldots, \alpha_n$, we sometimes say that $\alpha_1, \alpha_2, \ldots, \alpha_n$ are dependent (or independent) instead of saying S is dependent (or independent).*

The following are easy consequences of the definition.

1. Any set which contains a linearly dependent set is linearly dependent.
2. Any subset of a linearly independent set is linearly independent.
3. Any set which contains the 0 vector is linearly dependent; for $1 \cdot 0 = 0$.
4. A set S of vectors is linearly independent if and only if each finite subset of S is linearly independent, i.e., if and only if for any distinct vectors $\alpha_1, \ldots, \alpha_n$ of S, $c_1\alpha_1 + \cdots + c_n\alpha_n = 0$ implies each $c_i = 0$.

Definition. *Let* V *be a vector space. A* **basis** *for* V *is a linearly independent set of vectors in* V *which spans the space* V. *The space* V *is* **finite-dimensional** *if it has a finite basis.*

EXAMPLE 12. Let F be a subfield of the complex numbers. In F^3 the vectors

$$\begin{aligned} \alpha_1 &= (\ \ 3, 0, -3) \\ \alpha_2 &= (-1, 1, \ \ 2) \\ \alpha_3 &= (\ \ 4, 2, -2) \\ \alpha_4 &= (\ \ 2, 1, \ \ 1) \end{aligned}$$

are linearly dependent, since

$$2\alpha_1 + 2\alpha_2 - \alpha_3 + 0 \cdot \alpha_4 = 0.$$

The vectors

$$\begin{aligned} \epsilon_1 &= (1, 0, 0) \\ \epsilon_2 &= (0, 1, 0) \\ \epsilon_3 &= (0, 0, 1) \end{aligned}$$

are linearly independent

EXAMPLE 13. Let F be a field and in F^n let S be the subset consisting of the vectors $\epsilon_1, \epsilon_2, \ldots, \epsilon_n$ defined by

$$\begin{aligned} \epsilon_1 &= (1, 0, 0, \ldots, 0) \\ \epsilon_2 &= (0, 1, 0, \ldots, 0) \\ &\cdots\cdots\cdots \\ \epsilon_n &= (0, 0, 0, \ldots, 1). \end{aligned}$$

Let $x_1, x_2, \ldots, x_n$ be scalars in F and put $\alpha = x_1\epsilon_1 + x_2\epsilon_2 + \cdots + x_n\epsilon_n$. Then

$$\alpha = (x_1, x_2, \ldots, x_n). \tag{2-12}$$

This shows that $\epsilon_1, \ldots, \epsilon_n$ span F^n. Since $\alpha = 0$ if and only if $x_1 = x_2 = \cdots = x_n = 0$, the vectors $\epsilon_1, \ldots, \epsilon_n$ are linearly independent. The set $S = \{\epsilon_1, \ldots, \epsilon_n\}$ is accordingly a basis for F^n. We shall call this particular basis the **standard basis** of F^n.

EXAMPLE 14. Let P be an invertible $n \times n$ matrix with entries in the field F. Then $P_1, \ldots, P_n$, the columns of P, form a basis for the space of column matrices, $F^{n\times 1}$. We see that as follows. If X is a column matrix, then

$$PX = x_1P_1 + \cdots + x_nP_n.$$

Since $PX = 0$ has only the trivial solution $X = 0$, it follows that $\{P_1, \ldots, P_n\}$ is a linearly independent set. Why does it span $F^{n\times 1}$? Let Y be any column matrix. If $X = P^{-1}Y$, then $Y = PX$, that is,

$$Y = x_1P_1 + \cdots + x_nP_n.$$

So $\{P_1, \ldots, P_n\}$ is a basis for $F^{n\times 1}$.

EXAMPLE 15. Let A be an $m \times n$ matrix and let S be the solution space for the homogeneous system $AX = 0$ (Example 7). Let R be a row-reduced echelon matrix which is row-equivalent to A. Then S is also the solution space for the system $RX = 0$. If R has r non-zero rows, then the system of equations $RX = 0$ simply expresses r of the unknowns $x_1, \ldots, x_n$ in terms of the remaining $(n - r)$ unknowns x_j. Suppose that the leading non-zero entries of the non-zero rows occur in columns $k_1, \ldots, k_r$. Let J be the set consisting of the $n - r$ indices different from $k_1, \ldots, k_r$:

$$J = \{1, \ldots, n\} - \{k_1, \ldots, k_r\}.$$

The system $RX = 0$ has the form

$$\begin{aligned} x_{k_1} + \sum_J c_{1j}x_j &= 0 \\ \vdots \qquad \vdots \quad &\quad \vdots \\ x_{k_r} + \sum_J c_{rj}x_j &= 0 \end{aligned}$$

where the c_{ij} are certain scalars. All solutions are obtained by assigning (arbitrary) values to those x_j's with j in J and computing the corresponding values of $x_{k_1}, \ldots, x_{k_r}$. For each j in J, let E_j be the solution obtained by setting $x_j = 1$ and $x_i = 0$ for all other i in J. We assert that the $(n - r)$ vectors E_j, j in J, form a basis for the solution space.

Since the column matrix E_j has a 1 in row j and zeros in the rows indexed by other elements of J, the reasoning of Example 13 shows us that the set of these vectors is linearly independent. That set spans the solution space, for this reason. If the column matrix T, with entries $t_1, \ldots, t_n$, is in the solution space, the matrix

$$N = \sum_J t_jE_j$$

is also in the solution space *and* is a solution such that $x_j = t_j$ for each j in J. The solution with that property is unique; hence, $N = T$ and T is in the span of the vectors E_j.

EXAMPLE 16. We shall now give an example of an infinite basis. Let F be a subfield of the complex numbers and let V be the space of polynomial functions over F. Recall that these functions are the functions from F into F which have a rule of the form

$$f(x) = c_0 + c_1x + \cdots + c_nx^n.$$

Let $f_k(x) = x_k$, $k = 0, 1, 2, \ldots$. The (infinite) set $\{f_0, f_1, f_2, \ldots\}$ is a basis for V. Clearly the set spans V, because the function f (above) is

$$f = c_0f_0 + c_1f_1 + \cdots + c_nf_n.$$

The reader should see that this is virtually a repetition of the definition of polynomial function, that is, a function f from F into F is a polynomial function if and only if there exists an integer n and scalars $c_0, \ldots, c_n$ such that $f = c_0f_0 + \cdots + c_nf_n$. Why are the functions independent? To show that the set $\{f_0, f_1, f_2, \ldots\}$ is independent means to show that each finite subset of it is independent. It will suffice to show that, for each n, the set $\{f_0, \ldots, f_n\}$ is independent. Suppose that

$$c_0f_0 + \cdots + c_nf_n = 0.$$

This says that

$$c_0 + c_1x + \cdots + c_nx^n = 0$$

for every x in F; in other words, every x in F is a root of the polynomial $f(x) = c_0 + c_1x + \cdots + c_nx^n$. We assume that the reader knows that a polynomial of degree n with complex coefficients cannot have more than n distinct roots. It follows that $c_0 = c_1 = \cdots = c_n = 0$.

We have exhibited an infinite basis for V. Does that mean that V is not finite-dimensional? As a matter of fact it does; however, that is not immediate from the definition, because for all we know V might also have a finite basis. That possibility is easily eliminated. (We shall eliminate it in general in the next theorem.) Suppose that we have a finite number of polynomial functions $g_1, \ldots, g_r$. There will be a largest power of x which appears (with non-zero coefficient) in $g_1(x), \ldots, g_r(x)$. If that power is k, clearly $f_{k+1}(x) = x^{k+1}$ is not in the linear span of $g_1, \ldots, g_r$. So V is not finite-dimensional.

A final remark about this example is in order. Infinite bases have nothing to do with 'infinite linear combinations.' The reader who feels an irresistible urge to inject power series

$$\sum_{k=0}^{\infty} c_kx^k$$

into this example should study the example carefully again. If that does not effect a cure, he should consider restricting his attention to finite-dimensional spaces from now on.

Theorem 4. *Let V be a vector space which is spanned by a finite set of vectors $\beta_1, \beta_2, \ldots, \beta_m$. Then any independent set of vectors in V is finite and contains no more than m elements.*

Proof. To prove the theorem it suffices to show that every subset S of V which contains more than m vectors is linearly dependent. Let S be such a set. In S there are distinct vectors $\alpha_1, \alpha_2, \ldots, \alpha_n$ where $n > m$. Since $\beta_1, \ldots, \beta_m$ span V, there exist scalars A_{ij} in F such that

$$\alpha_j = \sum_{i=1}^{m} A_{ij}\beta_i.$$

For any n scalars $x_1, x_2, \ldots, x_n$ we have

$$\begin{aligned} x_1\alpha_1 + \cdots + x_n\alpha_n &= \sum_{j=1}^{n} x_j\alpha_j \\ &= \sum_{j=1}^{n} x_j \sum_{i=1}^{m} A_{ij}\beta_i \\ &= \sum_{j=1}^{n} \sum_{i=1}^{m} (A_{ij}x_j)\beta_i \\ &= \sum_{i=1}^{m} \left(\sum_{j=1}^{n} A_{ij}x_j \right)\beta_i. \end{aligned}$$

Since $n > m$, Theorem 6 of Chapter 1 implies that there exist scalars $x_1, x_2, \ldots, x_n$ not all 0 such that

$$\sum_{j=1}^{n} A_{ij}x_j = 0, \qquad 1 \leq i \leq m.$$

Hence $x_1\alpha_1 + x_2\alpha_2 + \cdots + x_n\alpha_n = 0$. This shows that S is a linearly dependent set. ∎

Corollary 1. *If V is a finite-dimensional vector space, then any two bases of V have the same (finite) number of elements.*

Proof. Since V is finite-dimensional, it has a finite basis

$$\{\beta_1, \beta_2, \ldots, \beta_m\}.$$

By Theorem 4 every basis of V is finite and contains no more than m elements. Thus if $\{\alpha_1, \alpha_2, \ldots, \alpha_n\}$ is a basis, $n \leq m$. By the same argument, $m \leq n$. Hence $m = n$. ∎

This corollary allows us to define the **dimension** of a finite-dimensional vector space as the number of elements in a basis for V. We shall denote the dimension of a finite-dimensional space V by dim V. This allows us to reformulate Theorem 4 as follows.

Corollary 2. *Let V be a finite-dimensional vector space and let n = dim V. Then*

(a) *any subset of* V *which contains more than* n *vectors is linearly dependent;*

(b) *no subset of* V *which contains fewer than* n *vectors can span* V.

EXAMPLE 17. If F is a field, the dimension of F^n is n, because the standard basis for F^n contains n vectors. The matrix space $F^{m\times n}$ has dimension mn. That should be clear by analogy with the case of F^n, because the mn matrices which have a 1 in the i, j place with zeros elsewhere form a basis for $F^{m\times n}$. If A is an $m \times n$ matrix, then the solution space for A has dimension $n - r$, where r is the number of non-zero rows in a row-reduced echelon matrix which is row-equivalent to A. See Example 15.

If V is any vector space over F, the zero subspace of V is spanned by the vector 0, but $\{0\}$ is a linearly dependent set and not a basis. For this reason, we shall agree that the zero subspace has dimension 0. Alternatively, we could reach the same conclusion by arguing that the empty set is a basis for the zero subspace. The empty set spans $\{0\}$, because the intersection of all subspaces containing the empty set is $\{0\}$, and the empty set is linearly independent because it contains no vectors.

Lemma. *Let* S *be a linearly independent subset of a vector space* V. *Suppose* β *is a vector in* V *which is not in the subspace spanned by* S. *Then the set obtained by adjoining* β *to* S *is linearly independent.*

Proof. Suppose $\alpha_1, \ldots, \alpha_m$ are distinct vectors in S and that

$$c_1\alpha_1 + \cdots + c_m\alpha_m + b\beta = 0.$$

Then $b = 0$; for otherwise,

$$\beta = \left(-\frac{c_1}{b}\right)\alpha_1 + \cdots + \left(-\frac{c_m}{b}\right)\alpha_m$$

and β is in the subspace spanned by S. Thus $c_1\alpha_1 + \cdots + c_m\alpha_m = 0$, and since S is a linearly independent set each $c_i = 0$. ∎

Theorem 5. *If* W *is a subspace of a finite-dimensional vector space* V, *every linearly independent subset of* W *is finite and is part of a (finite) basis for* W.

Proof. Suppose S_0 is a linearly independent subset of W. If S is a linearly independent subset of W containing S_0, then S is also a linearly independent subset of V; since V is finite-dimensional, S contains no more than dim V elements.

We extend S_0 to a basis for W, as follows. If S_0 spans W, then S_0 is a basis for W and we are done. If S_0 does not span W, we use the preceding lemma to find a vector β_1 in W such that the set $S_1 = S_0 \cup \{\beta_1\}$ is independent. If S_1 spans W, fine. If not, apply the lemma to obtain a vector β_2

in W such that $S_2 = S_1 \cup \{\beta_2\}$ is independent. If we continue in this way, then (in not more than dim V steps) we reach a set

$$S_m = S_0 \cup \{\beta_1, \ldots, \beta_m\}$$

which is a basis for W. ∎

Corollary 1. *If* W *is a proper subspace of a finite-dimensional vector space* V, *then* W *is finite-dimensional and dim* W < *dim* V.

Proof. We may suppose W contains a vector $\alpha \neq 0$. By Theorem 5 and its proof, there is a basis of W containing α which contains no more than dim V elements. Hence W is finite-dimensional, and dim $W \leq$ dim V. Since W is a proper subspace, there is a vector β in V which is not in W. Adjoining β to any basis of W, we obtain a linearly independent subset of V. Thus dim $W <$ dim V. ∎

Corollary 2. *In a finite-dimensional vector space* V *every non-empty linearly independent set of vectors is part of a basis.*

Corollary 3. *Let* A *be an* n × n *matrix over a field* F, *and suppose the row vectors of* A *form a linearly independent set of vectors in* F^n. *Then* A *is invertible.*

Proof. Let $\alpha_1, \alpha_2, \ldots, \alpha_n$ be the row vectors of A, and suppose W is the subspace of F^n spanned by $\alpha_1, \alpha_2, \ldots, \alpha_n$. Since $\alpha_1, \alpha_2, \ldots, \alpha_n$ are linearly independent, the dimension of W is n. Corollary 1 now shows that $W = F^n$. Hence there exist scalars B_{ij} in F such that

$$\epsilon_i = \sum_{j=1}^{n} B_{ij}\alpha_j, \qquad 1 \leq i \leq n$$

where $\{\epsilon_1, \epsilon_2, \ldots, \epsilon_n\}$ is the standard basis of F^n. Thus for the matrix B with entries B_{ij} we have

$$BA = I. \quad ∎$$

Theorem 6. *If* W_1 *and* W_2 *are finite-dimensional subspaces of a vector space* V, *then* $\mathrm{W}_1 + \mathrm{W}_2$ *is finite-dimensional and*

$$\dim \mathrm{W}_1 + \dim \mathrm{W}_2 = \dim (\mathrm{W}_1 \cap \mathrm{W}_2) + \dim (\mathrm{W}_1 + \mathrm{W}_2).$$

Proof. By Theorem 5 and its corollaries, $W_1 \cap W_2$ has a finite basis $\{\alpha_1, \ldots, \alpha_k\}$ which is part of a basis

$$\{\alpha_1, \ldots, \alpha_k, \quad \beta_1, \ldots, \beta_m\} \quad \text{for} \quad W_1$$

and part of a basis

$$\{\alpha_1, \ldots, \alpha_k, \quad \gamma_1, \ldots, \gamma_n\} \quad \text{for} \quad W_2.$$

The subspace $W_1 + W_2$ is spanned by the vectors

$$\alpha_1, \ldots, \alpha_k, \qquad \beta_1, \ldots, \beta_m, \qquad \gamma_1, \ldots, \gamma_n$$

and these vectors form an independent set. For suppose

$$\Sigma\, x_i\alpha_i + \Sigma\, y_j\beta_j + \Sigma\, z_r\gamma_r = 0.$$

Then

$$-\Sigma\, z_r\gamma_r = \Sigma\, x_i\alpha_i + \Sigma\, y_j\beta_j$$

which shows that $\Sigma\, z_r\gamma_r$ belongs to W_1. As $\Sigma\, z_r\gamma_r$ also belongs to W_2 it follows that

$$\Sigma\, z_r\gamma_r = \Sigma\, c_i\alpha_i$$

for certain scalars $c_1, \ldots, c_k$. Because the set

$$\{\alpha_1, \ldots, \alpha_k, \quad \gamma_1, \ldots, \gamma_n\}$$

is independent, each of the scalars $z_r = 0$. Thus

$$\Sigma\, x_i\alpha_i + \Sigma\, y_j\beta_j = 0$$

and since

$$\{\alpha_1, \ldots, \alpha_k, \quad \beta_1, \ldots, \beta_m\}$$

is also an independent set, each $x_i = 0$ and each $y_j = 0$. Thus,

$$\{\alpha_1, \ldots, \alpha_k, \quad \beta_1, \ldots, \beta_m, \quad \gamma_1, \ldots, \gamma_n\}$$

is a basis for $W_1 + W_2$. Finally

$$\begin{aligned}\dim W_1 + \dim W_2 &= (k + m) + (k + n)\\ &= k + (m + k + n)\\ &= \dim (W_1 \cap W_2) + \dim (W_1 + W_2).\end{aligned}$$ ∎

Let us close this section with a remark about linear independence and dependence. We defined these concepts for sets of vectors. It is useful to have them defined for finite sequences (ordered n-tuples) of vectors: $\alpha_1, \ldots, \alpha_n$. We say that the vectors $\alpha_1, \ldots, \alpha_n$ are **linearly dependent** if there exist scalars $c_1, \ldots, c_n$, not all 0, such that $c_1\alpha_1 + \cdots + c_n\alpha_n = 0$. This is all so natural that the reader may find that he has been using this terminology already. What is the difference between a finite sequence $\alpha_1, \ldots, \alpha_n$ and a set $\{\alpha_1, \ldots, \alpha_n\}$? There are two differences, identity and order.

If we discuss the set $\{\alpha_1, \ldots, \alpha_n\}$, usually it is presumed that no two of the vectors $\alpha_1, \ldots, \alpha_n$ are identical. In a sequence $\alpha_1, \ldots, \alpha_n$ all the α_i's may be the same vector. If $\alpha_i = \alpha_j$ for some $i \neq j$, then the sequence $\alpha_1, \ldots, \alpha_n$ is linearly dependent:

$$\alpha_i + (-1)\alpha_j = 0.$$

Thus, if $\alpha_1, \ldots, \alpha_n$ are linearly independent, they are distinct and we may talk about the set $\{\alpha_1, \ldots, \alpha_n\}$ and know that it has n vectors in it. So, clearly, no confusion will arise in discussing bases and dimension. The dimension of a finite-dimensional space V is the largest n such that some n-tuple of vectors in V is linearly independent—and so on. The reader who feels that this paragraph is much ado about nothing might ask himself whether the vectors

$$\alpha_1 = (e^{\pi/2}, 1)$$
$$\alpha_2 = (\sqrt[3]{110}, 1)$$

are linearly independent in R^2.

The elements of a sequence are enumerated in a specific order. A set is a collection of objects, with no specified arrangement or order. Of course, to describe the set we may list its members, and that requires choosing an order. But, the order is not part of the set. The sets $\{1, 2, 3, 4\}$ and $\{4, 3, 2, 1\}$ are identical, whereas $1, 2, 3, 4$ is quite a different sequence from $4, 3, 2, 1$. The order aspect of sequences has no bearing on questions of independence, dependence, etc., because dependence (as defined) is not affected by the order. The sequence $\alpha_n, \ldots, \alpha_1$ is dependent if and only if the sequence $\alpha_1, \ldots, \alpha_n$ is dependent. In the next section, order will be important.

Exercises

1. Prove that if two vectors are linearly dependent, one of them is a scalar multiple of the other.

2. Are the vectors

$$\alpha_1 = (1, 1, 2, 4), \qquad \alpha_2 = (2, -1, -5, 2)$$
$$\alpha_3 = (1, -1, -4, 0), \qquad \alpha_4 = (2, 1, 1, 6)$$

linearly independent in R^4?

3. Find a basis for the subspace of R^4 spanned by the four vectors of Exercise 2.

4. Show that the vectors

$$\alpha_1 = (1, 0, -1), \qquad \alpha_2 = (1, 2, 1), \qquad \alpha_3 = (0, -3, 2)$$

form a basis for R^3. Express each of the standard basis vectors as linear combinations of α_1, α_2, and α_3.

5. Find three vectors in R^3 which are linearly dependent, and are such that any two of them are linearly independent.

6. Let V be the vector space of all 2×2 matrices over the field F. Prove that V has dimension 4 by exhibiting a basis for V which has four elements.

7. Let V be the vector space of Exercise 6. Let W_1 be the set of matrices of the form

$$\begin{bmatrix} x & -x \\ y & z \end{bmatrix}$$

and let W_2 be the set of matrices of the form

$$\begin{bmatrix} a & b \\ -a & c \end{bmatrix}.$$

(a) Prove that W_1 and W_2 are subspaces of V.

(b) Find the dimensions of W_1, W_2, $W_1 + W_2$, and $W_1 \cap W_2$.

8. Again let V be the space of 2×2 matrices over F. Find a basis $\{A_1, A_2, A_3, A_4\}$ for V such that $A_j^2 = A_j$ for each j.

9. Let V be a vector space over a subfield F of the complex numbers. Suppose α, β, and γ are linearly independent vectors in V. Prove that $(\alpha + \beta)$, $(\beta + \gamma)$, and $(\gamma + \alpha)$ are linearly independent.

10. Let V be a vector space over the field F. Suppose there are a finite number of vectors $\alpha_1, \ldots, \alpha_r$ in V which span V. Prove that V is finite-dimensional.

11. Let V be the set of all 2×2 matrices A with *complex* entries which satisfy $A_{11} + A_{22} = 0$.

(a) Show that V is a vector space over the field of *real* numbers, with the usual operations of matrix addition and multiplication of a matrix by a scalar.

(b) Find a basis for this vector space.

(c) Let W be the set of all matrices A in V such that $A_{21} = -\bar{A}_{12}$ (the bar denotes complex conjugation). Prove that W is a subspace of V and find a basis for W.

12. Prove that the space of all $m \times n$ matrices over the field F has dimension mn, by exhibiting a basis for this space.

13. Discuss Exercise 9, when V is a vector space over the field with two elements described in Exercise 5, Section 1.1.

14. Let V be the set of real numbers. Regard V as a vector space over the field of *rational* numbers, with the usual operations. Prove that this vector space is *not* finite-dimensional.

2.4. Coordinates

One of the useful features of a basis $\mathcal{B}$ in an n-dimensional space V is that it essentially enables one to introduce coordinates in V analogous to the 'natural coordinates' x_i of a vector $\alpha = (x_1, \ldots, x_n)$ in the space F^n. In this scheme, the coordinates of a vector α in V relative to the basis $\mathcal{B}$ will be the scalars which serve to express α as a linear combination of the vectors in the basis. Thus, we should like to regard the natural coordinates of a vector α in F^n as being defined by α and the standard basis for F^n; however, in adopting this point of view we must exercise a certain amount of care. If

$$\alpha = (x_1, \ldots, x_n) = \Sigma\, x_i \epsilon_i$$

and $\mathcal{B}$ is the standard basis for F^n, just how are the coordinates of α determined by $\mathcal{B}$ and α? One way to phrase the answer is this. A given vector α has a unique expression as a linear combination of the standard basis vectors, and the ith coordinate x_i of α is the coefficient of ϵ_i in this expression. From this point of view we are able to say which is the ith coordinate because we have a 'natural' ordering of the vectors in the standard basis, that is, we have a rule for determining which is the 'first' vector in the basis, which is the 'second,' and so on. If $\mathcal{B}$ is an arbitrary basis of the n-dimensional space V, we shall probably have no natural ordering of the vectors in $\mathcal{B}$, and it will therefore be necessary for us to impose some order on these vectors before we can define 'the ith coordinate of α relative to $\mathcal{B}$.' To put it another way, coordinates will be defined relative to sequences of vectors rather than sets of vectors.

Definition. *If* V *is a finite-dimensional vector space, an* **ordered basis** *for* V *is a finite sequence of vectors which is linearly independent and spans* V.

If the sequence $\alpha_1, \ldots, \alpha_n$ is an ordered basis for V, then the set $\{\alpha_1, \ldots, \alpha_n\}$ is a basis for V. The ordered basis is the set, together with the specified ordering. We shall engage in a slight abuse of notation and describe all that by saying that

$$\mathcal{B} = \{\alpha_1, \ldots, \alpha_n\}$$

is an ordered basis for V.

Now suppose V is a finite-dimensional vector space over the field F and that

$$\mathcal{B} = \{\alpha_1, \ldots, \alpha_n\}$$

is an ordered basis for V. Given α in V, there is a unique n-tuple $(x_1, \ldots, x_n)$ of scalars such that

$$\alpha = \sum_{i=1}^{n} x_i\alpha_i.$$

The n-tuple is unique, because if we also have

$$\alpha = \sum_{i=1}^{n} z_i\alpha_i$$

then

$$\sum_{i=1}^{n} (x_i - z_i)\alpha_i = 0$$

and the linear independence of the α_i tells us that $x_i - z_i = 0$ for each i. We shall call x_i the ith **coordinate of α relative to the ordered basis**

$$\mathcal{B} = \{\alpha_1, \ldots, \alpha_n\}.$$

If

$$\beta = \sum_{i=1}^{n} y_i\alpha_i$$

then

$$\alpha + \beta = \sum_{i=1}^{n} (x_i + y_i)\alpha_i$$

so that the ith coordinate of $(\alpha + \beta)$ in this ordered basis is $(x_i + y_i)$.

Similarly, the ith coordinate of $(c\alpha)$ is cx_i. One should also note that every n-tuple $(x_1, \ldots, x_n)$ in F^n is the n-tuple of coordinates of some vector in V, namely the vector

$$\sum_{i=1}^{n} x_i\alpha_i.$$

To summarize, each ordered basis for V determines a one-one correspondence

$$\alpha \to (x_1, \ldots, x_n)$$

between the set of all vectors in V and the set of all n-tuples in F^n. This correspondence has the property that the correspondent of $(\alpha + \beta)$ is the sum in F^n of the correspondents of α and β, and that the correspondent of $(c\alpha)$ is the product in F^n of the scalar c and the correspondent of α.

One might wonder at this point why we do not simply select some ordered basis for V and describe each vector in V by its corresponding n-tuple of coordinates, since we would then have the convenience of operating only with n-tuples. This would defeat our purpose, for two reasons. First, as our axiomatic definition of vector space indicates, we are attempting to learn to reason with vector spaces as abstract algebraic systems. Second, even in those situations in which we use coordinates, the significant results follow from our ability to change the coordinate system, i.e., to change the ordered basis.

Frequently, it will be more convenient for us to use the **coordinate matrix of α relative to the ordered basis $\mathcal{B}$**:

$$X = \begin{bmatrix} x_1 \\ \vdots \\ x_n \end{bmatrix}$$

rather than the n-tuple $(x_1, \ldots, x_n)$ of coordinates. To indicate the dependence of this coordinate matrix on the basis, we shall use the symbol

$$[\alpha]_{\mathcal{B}}$$

for the coordinate matrix of the vector α relative to the ordered basis $\mathcal{B}$. This notation will be particularly useful as we now proceed to describe what happens to the coordinates of a vector α as we change from one ordered basis to another.

Suppose then that V is n-dimensional and that

$$\mathcal{B} = \{\alpha_1, \ldots, \alpha_n\} \quad \text{and} \quad \mathcal{B}' = \{\alpha_1', \ldots, \alpha_n'\}$$

are two ordered bases for V. There are unique scalars P_{ij} such that

$$\alpha_j' = \sum_{i=1}^{n} P_{ij}\alpha_i, \qquad 1 \leq j \leq n. \tag{2-13}$$

Let $x_1', \ldots, x_n'$ be the coordinates of a given vector α in the ordered basis $\mathcal{B}'$. Then

$$\begin{aligned}\alpha &= x_1'\alpha_1' + \cdots + x_n'\alpha_n' \\ &= \sum_{j=1}^{n} x_j'\alpha_j' \\ &= \sum_{j=1}^{n} x_j' \sum_{i=1}^{n} P_{ij}\alpha_i \\ &= \sum_{j=1}^{n} \sum_{i=1}^{n} (P_{ij}x_j')\alpha_i \\ &= \sum_{i=1}^{n} \left(\sum_{j=1}^{n} P_{ij}x_j'\right) \alpha_i.\end{aligned}$$

Thus we obtain the relation

$$\alpha = \sum_{i=1}^{n} \left(\sum_{j=1}^{n} P_{ij}x_j'\right) \alpha_i. \tag{2-14}$$

Since the coordinates $x_1, x_2, \ldots, x_n$ of α in the ordered basis $\mathcal{B}$ are uniquely determined, it follows from (2-14) that

$$x_i = \sum_{j=1}^{n} P_{ij}x_j', \qquad 1 \leq i \leq n. \tag{2-15}$$

Let P be the $n \times n$ matrix whose i, j entry is the scalar P_{ij}, and let X and X' be the coordinate matrices of the vector α in the ordered bases $\mathcal{B}$ and $\mathcal{B}'$. Then we may reformulate (2-15) as

$$X = PX'. \tag{2-16}$$

Since $\mathcal{B}$ and $\mathcal{B}'$ are linearly independent sets, $X = 0$ if and only if $X' = 0$. Thus from (2-16) and Theorem 7 of Chapter 1, it follows that P is invertible. Hence

$$X' = P^{-1}X. \tag{2-17}$$

If we use the notation introduced above for the coordinate matrix of a vector relative to an ordered basis, then (2-16) and (2-17) say

$$\begin{aligned}[\alpha]_{\mathcal{B}} &= P[\alpha]_{\mathcal{B}'} \\ [\alpha]_{\mathcal{B}'} &= P^{-1}[\alpha]_{\mathcal{B}}.\end{aligned}$$

Thus the preceding discussion may be summarized as follows.

Theorem 7. *Let* V *be an* n*-dimensional vector space over the field* F, *and let* $\mathcal{B}$ *and* $\mathcal{B}'$ *be two ordered bases of* V. *Then there is a unique, necessarily invertible,* $n \times n$ *matrix* P *with entries in* F *such that*

$$\begin{aligned}&\text{(i)} \qquad [\alpha]_{\mathcal{B}} = \mathrm{P}[\alpha]_{\mathcal{B}'} \\ &\text{(ii)} \qquad [\alpha]_{\mathcal{B}'} = \mathrm{P}^{-1}[\alpha]_{\mathcal{B}}\end{aligned}$$

for every vector α *in* V. *The columns of* P *are given by*

$$\mathrm{P}_j = [\alpha_j']_{\mathcal{B}}, \qquad j = 1, \ldots, n.$$

To complete the above analysis we shall also prove the following result.

Theorem 8. *Suppose* P *is an* n × n *invertible matrix over* F. *Let* V *be an* n-*dimensional vector space over* F, *and let* $\mathcal{B}$ *be an ordered basis of* V. *Then there is a unique ordered basis* $\mathcal{B}'$ *of* V *such that*

$$\text{(i)} \qquad [\alpha]_{\mathcal{B}} = P[\alpha]_{\mathcal{B}'}$$
$$\text{(ii)} \qquad [\alpha]_{\mathcal{B}'} = P^{-1}[\alpha]_{\mathcal{B}}$$

for every vector α *in* V.

Proof. Let $\mathcal{B}$ consist of the vectors $\alpha_1, \ldots, \alpha_n$. If $\mathcal{B}' = \{\alpha'_1, \ldots, \alpha'_n\}$ is an ordered basis of V for which (i) is valid, it is clear that

$$\alpha'_j = \sum_{i=1}^{n} P_{ij}\alpha_i.$$

Thus we need only show that the vectors α'_j, defined by these equations, form a basis. Let $Q = P^{-1}$. Then

$$\begin{aligned}\sum_j Q_{jk}\alpha'_j &= \sum_j Q_{jk} \sum_i P_{ij}\alpha_i \\ &= \sum_j \sum_i P_{ij}Q_{jk}\, \alpha_i \\ &= \sum_i \Big(\sum_j P_{ij}Q_{jk}\Big)\, \alpha_i \\ &= \alpha_k.\end{aligned}$$

Thus the subspace spanned by the set

$$\mathcal{B}' = \{\alpha'_1, \ldots, \alpha'_n\}$$

contains $\mathcal{B}$ and hence equals V. Thus $\mathcal{B}'$ is a basis, and from its definition and Theorem 7, it is clear that (i) is valid and hence also (ii). ∎

EXAMPLE 18. Let F be a field and let

$$\alpha = (x_1, x_2, \ldots, x_n)$$

be a vector in F^n. If $\mathcal{B}$ is the standard ordered basis of F^n,

$$\mathcal{B} = \{\epsilon_1, \ldots, \epsilon_n\}$$

the coordinate matrix of the vector α in the basis $\mathcal{B}$ is given by

$$[\alpha]_{\mathcal{B}} = \begin{bmatrix} x_1 \\ x_2 \\ \vdots \\ x_n \end{bmatrix}.$$

EXAMPLE 19. Let R be the field of the real numbers and let θ be a fixed real number. The matrix

$$P = \begin{bmatrix} \cos\theta & -\sin\theta \\ \sin\theta & \cos\theta \end{bmatrix}$$

is invertible with inverse,

$$P^{-1} = \begin{bmatrix} \cos\theta & \sin\theta \\ -\sin\theta & \cos\theta \end{bmatrix}.$$

Thus for each θ the set $\mathcal{B}'$ consisting of the vectors $(\cos\theta, \sin\theta)$, $(-\sin\theta, \cos\theta)$ is a basis for R^2; intuitively this basis may be described as the one obtained by rotating the standard basis through the angle θ. If α is the vector (x_1, x_2), then

$$[\alpha]_{\mathcal{B}'} = \begin{bmatrix} \cos\theta & \sin\theta \\ -\sin\theta & \cos\theta \end{bmatrix} \begin{bmatrix} x_1 \\ x_2 \end{bmatrix}$$

or

$$\begin{aligned} x_1' &= \quad x_1 \cos\theta + x_2 \sin\theta \\ x_2' &= -x_1 \sin\theta + x_2 \cos\theta. \end{aligned}$$

EXAMPLE 20. Let F be a subfield of the complex numbers. The matrix

$$P = \begin{bmatrix} -1 & 4 & 5 \\ 0 & 2 & -3 \\ 0 & 0 & 8 \end{bmatrix}$$

is invertible with inverse

$$P^{-1} = \begin{bmatrix} -1 & 2 & \frac{11}{8} \\ 0 & \frac{1}{2} & \frac{3}{16} \\ 0 & 0 & \frac{1}{8} \end{bmatrix}.$$

Thus the vectors

$$\begin{aligned} \alpha_1' &= (-1, \quad 0, 0) \\ \alpha_2' &= (\ \ 4, \quad 2, 0) \\ \alpha_3' &= (\ \ 5, -3, 8) \end{aligned}$$

form a basis $\mathcal{B}'$ of F^3. The coordinates x_1', x_2', x_3' of the vector $\alpha = (x_1, x_2, x_3)$ in the basis $\mathcal{B}'$ are given by

$$\begin{bmatrix} x_1' \\ x_2' \\ x_3' \end{bmatrix} = \begin{bmatrix} -x_1 + 2x_2 + \frac{11}{8}x_3 \\ \frac{1}{2}x_2 + \frac{3}{16}x_3 \\ \frac{1}{8}x_3 \end{bmatrix} = \begin{bmatrix} -1 & 2 & \frac{11}{8} \\ 0 & \frac{1}{2} & \frac{3}{16} \\ 0 & 0 & \frac{1}{8} \end{bmatrix} \begin{bmatrix} x_1 \\ x_2 \\ x_3 \end{bmatrix}.$$

In particular,

$$(3, 2, -8) = -10\alpha_1' - \tfrac{1}{2}\alpha_2' - \alpha_3'.$$

Exercises

1. Show that the vectors

$$\begin{aligned} \alpha_1 &= (1, 1, 0, 0), & \alpha_2 &= (0, 0, 1, 1) \\ \alpha_3 &= (1, 0, 0, 4), & \alpha_4 &= (0, 0, 0, 2) \end{aligned}$$

form a basis for R^4. Find the coordinates of each of the standard basis vectors in the ordered basis $\{\alpha_1, \alpha_2, \alpha_3, \alpha_4\}$.

2. Find the coordinate matrix of the vector (1, 0, 1) in the basis of C^3 consisting of the vectors $(2i, 1, 0)$, $(2, -1, 1)$, $(0, 1 + i, 1 - i)$, in that order.

3. Let $\mathcal{B} = \{\alpha_1, \alpha_2, \alpha_3\}$ be the ordered basis for R^3 consisting of

$$\alpha_1 = (1, 0, -1), \qquad \alpha_2 = (1, 1, 1), \qquad \alpha_3 = (1, 0, 0).$$

What are the coordinates of the vector (a, b, c) in the ordered basis $\mathcal{B}$?

4. Let W be the subspace of C^3 spanned by $\alpha_1 = (1, 0, i)$ and $\alpha_2 = (1 + i, 1, -1)$.
(a) Show that α_1 and α_2 form a basis for W.
(b) Show that the vectors $\beta_1 = (1, 1, 0)$ and $\beta_2 = (1, i, 1 + i)$ are in W and form another basis for W.
(c) What are the coordinates of α_1 and α_2 in the ordered basis $\{\beta_1, \beta_2\}$ for W?

5. Let $\alpha = (x_1, x_2)$ and $\beta = (y_1, y_2)$ be vectors in R^2 such that

$$x_1y_1 + x_2y_2 = 0, \qquad x_1^2 + x_2^2 = y_1^2 + y_2^2 = 1.$$

Prove that $\mathcal{B} = \{\alpha, \beta\}$ is a basis for R^2. Find the coordinates of the vector (a, b) in the ordered basis $\mathcal{B} = \{\alpha, \beta\}$. (The conditions on α and β say, geometrically, that α and β are perpendicular and each has length 1.)

6. Let V be the vector space over the complex numbers of all functions from R into C, i.e., the space of all complex-valued functions on the real line. Let $f_1(x) = 1$, $f_2(x) = e^{ix}$, $f_3(x) = e^{-ix}$.
(a) Prove that f_1, f_2, and f_3 are linearly independent.
(b) Let $g_1(x) = 1$, $g_2(x) = \cos x$, $g_3(x) = \sin x$. Find an invertible 3×3 matrix P such that

$$g_j = \sum_{i=1}^{3} P_{ij} f_i.$$

7. Let V be the (real) vector space of all polynomial functions from R into R of degree 2 or less, i.e., the space of all functions f of the form

$$f(x) = c_0 + c_1x + c_2x^2.$$

Let t be a fixed real number and define

$$g_1(x) = 1, \qquad g_2(x) = x + t, \qquad g_3(x) = (x + t)^2.$$

Prove that $\mathcal{B} = \{g_1, g_2, g_3\}$ is a basis for V. If

$$f(x) = c_0 + c_1x + c_2x^2$$

what are the coordinates of f in this ordered basis $\mathcal{B}$?

2.5. Summary of Row-Equivalence

In this section we shall utilize some elementary facts on bases and dimension in finite-dimensional vector spaces to complete our discussion of row-equivalence of matrices. We recall that if A is an $m \times n$ matrix over the field F the row vectors of A are the vectors $\alpha_1, \ldots, \alpha_m$ in F^n defined by

$$\alpha_i = (A_{i1}, \ldots, A_{in})$$

and that the row space of A is the subspace of F^n spanned by these vectors. The **row rank** of A is the dimension of the row space of A.

If P is a $k \times m$ matrix over F, then the product $B = PA$ is a $k \times n$ matrix whose row vectors $\beta_1, \ldots, \beta_k$ are linear combinations

$$\beta_i = P_{i1}\alpha_1 + \cdots + P_{im}\alpha_m$$

of the row vectors of A. Thus the row space of B is a subspace of the row space of A. If P is an $m \times m$ invertible matrix, then B is row-equivalent to A so that the symmetry of row-equivalence, or the equation $A = P^{-1}B$, implies that the row space of A is also a subspace of the row space of B.

Theorem 9. *Row-equivalent matrices have the same row space.*

Thus we see that to study the row space of A we may as well study the row space of a row-reduced echelon matrix which is row-equivalent to A. This we proceed to do.

Theorem 10. *Let R be a non-zero row-reduced echelon matrix. Then the non-zero row vectors of R form a basis for the row space of R.*

Proof. Let $\rho_1, \ldots, \rho_r$ be the non-zero row vectors of R:

$$\rho_i = (R_{i1}, \ldots, R_{in}).$$

Certainly these vectors span the row space of R; we need only prove they are linearly independent. Since R is a row-reduced echelon matrix, there are positive integers $k_1, \ldots, k_r$ such that, for $i \leq r$

$$\text{(2-18)} \qquad \begin{aligned} &\text{(a)}\ R(i, j) = 0 \quad \text{if} \quad j < k_i \\ &\text{(b)}\ R(i, k_j) = \delta_{ij} \\ &\text{(c)}\ k_1 < \cdots < k_r. \end{aligned}$$

Suppose $\beta = (b_1, \ldots, b_n)$ is a vector in the row space of R:

$$\text{(2-19)} \qquad \beta = c_1\rho_1 + \cdots + c_r\rho_r.$$

Then we claim that $c_j = b_{k_j}$. For, by (2-18)

$$\text{(2-20)} \qquad \begin{aligned} b_{k_j} &= \sum_{i=1}^{r} c_i R(i, k_j) \\ &= \sum_{i=1}^{r} c_i \delta_{ij} \\ &= c_j. \end{aligned}$$

In particular, if $\beta = 0$, i.e., if $c_1\rho_1 + \cdots + c_r\rho_r = 0$, then c_j must be the k_jth coordinate of the zero vector so that $c_j = 0$, $j = 1, \ldots, r$. Thus $\rho_1, \ldots, \rho_r$ are linearly independent. ∎

Theorem 11. *Let m and n be positive integers and let F be a field. Suppose W is a subspace of F^n and dim W $\leq$ m. Then there is precisely one m $\times$ n row-reduced echelon matrix over F which has W as its row space.*

Proof. There is at least one $m \times n$ row-reduced echelon matrix with row space W. Since $\dim W \leq m$, we can select some m vectors $\alpha_1, \ldots, \alpha_m$ in W which span W. Let A be the $m \times n$ matrix with row vectors $\alpha_1, \ldots, \alpha_m$ and let R be a row-reduced echelon matrix which is row-equivalent to A. Then the row space of R is W.

Now let R be any row-reduced echelon matrix which has W as its row space. Let $\rho_1, \ldots, \rho_r$ be the non-zero row vectors of R and suppose that the leading non-zero entry of ρ_i occurs in column k_i, $i = 1, \ldots, r$. The vectors $\rho_1, \ldots, \rho_r$ form a basis for W. In the proof of Theorem 10, we observed that if $\beta = (b_1, \ldots, b_n)$ is in W, then

$$\beta = c_1\rho_1 + \cdots + c_r\rho_r,$$

and $c_i = b_{k_i}$; in other words, the unique expression for β as a linear combination of $\rho_1, \ldots, \rho_r$ is

$$\beta = \sum_{i=1}^{r} b_{k_i}\rho_i. \tag{2-21}$$

Thus any vector β is determined if one knows the coordinates b_{k_i}, $i = 1, \ldots, r$. For example, ρ_s is the unique vector in W which has k_sth coordinate 1 and k_ith coordinate 0 for $i \neq s$.

Suppose β is in W and $\beta \neq 0$. We claim the first non-zero coordinate of β occurs in one of the columns k_s. Since

$$\beta = \sum_{i=1}^{r} b_{k_i}\rho_i$$

and $\beta \neq 0$, we can write

$$\beta = \sum_{i=s}^{r} b_{k_i}\rho_i, \qquad b_{k_s} \neq 0. \tag{2-22}$$

From the conditions (2-18) one has $R_{ij} = 0$ if $i > s$ and $j \leq k_s$. Thus

$$\beta = (0, \ldots, 0, \quad b_{k_s}, \ldots, b_n), \qquad b_{k_s} \neq 0$$

and the first non-zero coordinate of β occurs in column k_s. Note also that for each k_s, $s = 1, \ldots, r$, there exists a vector in W which has a non-zero k_sth coordinate, namely ρ_s.

It is now clear that R is uniquely determined by W. The description of R in terms of W is as follows. We consider all vectors $\beta = (b_1, \ldots, b_n)$ in W. If $\beta \neq 0$, then the first non-zero coordinate of β must occur in some column t:

$$\beta = (0, \ldots, 0, \quad b_t, \ldots, b_n), \qquad b_t \neq 0.$$

Let $k_1, \ldots, k_r$ be those positive integers t such that there is some $\beta \neq 0$ in W, the first non-zero coordinate of which occurs in column t. Arrange $k_1, \ldots, k_r$ in the order $k_1 < k_2 < \cdots < k_r$. For each of the positive integers k_s there will be one and only one vector ρ_s in W such that the k_sth coordinate of ρ_s is 1 and the k_ith coordinate of ρ_s is 0 for $i \neq s$. Then R is the $m \times n$ matrix which has row vectors $\rho_1, \ldots, \rho_r, 0, \ldots, 0$. ∎

Corollary. *Each* m × n *matrix* A *is row-equivalent to one and only one row-reduced echelon matrix.*

Proof. We know that A is row-equivalent to at least one row-reduced echelon matrix R. If A is row-equivalent to another such matrix R', then R is row-equivalent to R'; hence, R and R' have the same row space and must be identical. ∎

Corollary. *Let* A *and* B *be* m × n *matrices over the field* F. *Then* A *and* B *are row-equivalent if and only if they have the same row space.*

Proof. We know that if A and B are row-equivalent, then they have the same row space. So suppose that A and B have the same row space. Now A is row-equivalent to a row-reduced echelon matrix R and B is row-equivalent to a row-reduced echelon matrix R'. Since A and B have the same row space, R and R' have the same row space. Thus $R = R'$ and A is row-equivalent to B. ∎

To summarize—if A and B are $m \times n$ matrices over the field F, the following statements are equivalent:

1. A and B are row-equivalent.
2. A and B have the same row space.
3. $B = PA$, where P is an invertible $m \times m$ matrix.

A fourth equivalent statement is that the homogeneous systems $AX = 0$ and $BX = 0$ have the same solutions; however, although we know that the row-equivalence of A and B implies that these systems have the same solutions, it seems best to leave the proof of the converse until later.

2.6. Computations Concerning Subspaces

We should like now to show how elementary row operations provide a standardized method of answering certain concrete questions concerning subspaces of F^n. We have already derived the facts we shall need. They are gathered here for the convenience of the reader. The discussion applies to any n-dimensional vector space over the field F, if one selects a fixed ordered basis $\mathcal{B}$ and describes each vector α in V by the n-tuple $(x_1, \ldots, x_n)$ which gives the coordinates of α in the ordered basis $\mathcal{B}$.

Suppose we are given m vectors $\alpha_1, \ldots, \alpha_m$ in F^n. We consider the following questions.

1. How does one determine if the vectors $\alpha_1, \ldots, \alpha_m$ are linearly independent? More generally, how does one find the dimension of the subspace W spanned by these vectors?

2. Given β in F^n, how does one determine whether β is a linear combination of $\alpha_1, \ldots, \alpha_m$, i.e., whether β is in the subspace W?

3. How can one give an explicit description of the subspace W?

The third question is a little vague, since it does not specify what is meant by an 'explicit description'; however, we shall clear up this point by giving the sort of description we have in mind. With this description, questions (1) and (2) can be answered immediately.

Let A be the $m \times n$ matrix with row vectors α_i:

$$\alpha_i = (A_{i1}, \ldots, A_{in}).$$

Perform a sequence of elementary row operations, starting with A and terminating with a row-reduced echelon matrix R. We have previously described how to do this. At this point, the dimension of W (the row space of A) is apparent, since this dimension is simply the number of non-zero row vectors of R. If $\rho_1, \ldots, \rho_r$ are the non-zero row vectors of R, then $\mathcal{B} = \{\rho_1, \ldots, \rho_r\}$ is a basis for W. If the first non-zero coordinate of ρ_i is the k_ith one, then we have for $i \leq r$

(a) $$R(i, j) = 0, \quad \text{if} \quad j < k_i$$

(b) $$R(i, k_j) = \delta_{ij}$$

(c) $$k_1 < \cdots < k_r.$$

The subspace W consists of all vectors

$$\beta = c_1\rho_1 + \cdots + c_r\rho_r$$

$$= \sum_{i=1}^{r} c_i(R_{i1}, \ldots, R_{in}).$$

The coordinates $b_1, \ldots, b_n$ of such a vector β are then

$$b_j = \sum_{i=1}^{r} c_i R_{ij}. \tag{2-23}$$

In particular, $b_{k_j} = c_j$, and so if $\beta = (b_1, \ldots, b_n)$ is a linear combination of the ρ_i, it must be the particular linear combination

$$\beta = \sum_{i=1}^{r} b_{k_i}\rho_i. \tag{2-24}$$

The conditions on β that (2-24) should hold are

$$b_j = \sum_{i=1}^{r} b_{k_i} R_{ij}, \qquad j = 1, \ldots, n. \tag{2-25}$$

Now (2-25) is the explicit description of the subspace W spanned by $\alpha_1, \ldots, \alpha_m$, that is, the subspace consists of all vectors β in F^n whose coordinates satisfy (2-25). What kind of description is (2-25)? In the first place it describes W as all solutions $\beta = (b_1, \ldots, b_n)$ of the system of homogeneous linear equations (2-25). This system of equations is of a very special nature, because it expresses $(n - r)$ of the coordinates as linear combinations of the r distinguished coordinates $b_{k_1}, \ldots, b_{k_r}$. One has complete freedom of choice in the coordinates b_{k_i}, that is, if $c_1, \ldots, c_r$ are *any* r scalars, there is one and only one vector β in W which has c_i as its k_ith coordinate.

The significant point here is this: Given the vectors α_i, row-reduction is a straightforward method of determining the integers $r, k_1, \ldots, k_r$ and the scalars R_{ij} which give the description (2-25) of the subspace spanned by $\alpha_1, \ldots, \alpha_m$. One should observe as we did in Theorem 11 that every subspace W of F^n has a description of the type (2-25). We should also point out some things about question (2). We have already stated how one can find an invertible $m \times m$ matrix P such that $R = PA$, in Section 1.4. The knowledge of P enables one to find the scalars $x_1, \ldots, x_m$ such that

$$\beta = x_1\alpha_1 + \cdots + x_m\alpha_m$$

when this is possible. For the row vectors of R are given by

$$\rho_i = \sum_{j=1}^{m} P_{ij}\alpha_j$$

so that if β is a linear combination of the α_j, we have

$$\begin{aligned}\beta &= \sum_{i=1}^{r} b_{k_i}\rho_i \\ &= \sum_{i=1}^{r} b_{k_i} \sum_{j=1}^{m} P_{ij}\alpha_j \\ &= \sum_{j=1}^{m} \sum_{i=1}^{r} b_{k_i}P_{ij}\alpha_j\end{aligned}$$

and thus

$$x_j = \sum_{i=1}^{r} b_{k_i}P_{ij}$$

is one possible choice for the x_j (there may be many).

The question of whether $\beta = (b_1, \ldots, b_n)$ is a linear combination of the α_i, and if so, what the scalars x_i are, can also be looked at by asking whether the system of equations

$$\sum_{i=1}^{m} A_{ij}x_i = b_j, \qquad j = 1, \ldots, n$$

has a solution and what the solutions are. The coefficient matrix of this system of equations is the $n \times m$ matrix B with *column* vectors $\alpha_1, \ldots, \alpha_m$. In Chapter 1 we discussed the use of elementary row operations in solving a system of equations $BX = Y$. Let us consider one example in which we adopt both points of view in answering questions about subspaces of F^n.

EXAMPLE 21. Let us pose the following problem. Let W be the subspace of R^4 spanned by the vectors

$$\begin{aligned}\alpha_1 &= (1, 2, 2, 1)\\ \alpha_2 &= (0, 2, 0, 1)\\ \alpha_3 &= (-2, 0, -4, 3).\end{aligned}$$

(a) Prove that α_1, α_2, α_3 form a basis for W, i.e., that these vectors are linearly independent.

(b) Let $\beta = (b_1, b_2, b_3, b_4)$ be a vector in W. What are the coordinates of β relative to the ordered basis $\{\alpha_1, \alpha_2, \alpha_3\}$?

(c) Let

$$\begin{aligned}\alpha_1' &= (1, 0, 2, 0)\\ \alpha_2' &= (0, 2, 0, 1)\\ \alpha_3' &= (0, 0, 0, 3).\end{aligned}$$

Show that α_1', α_2', α_3' form a basis for W.

(d) If β is in W, let X denote the coordinate matrix of β relative to the α-basis and X' the coordinate matrix of β relative to the α'-basis. Find the 3×3 matrix P such that $X = PX'$ for every such β.

To answer these questions by the first method we form the matrix A with row vectors α_1, α_2, α_3, find the row-reduced echelon matrix R which is row-equivalent to A and simultaneously perform the same operations on the identity to obtain the invertible matrix Q such that $R = QA$:

$$\begin{bmatrix} 1 & 2 & 2 & 1\\ 0 & 2 & 0 & 1\\ -2 & 0 & -4 & 3\end{bmatrix} \to R = \begin{bmatrix} 1 & 0 & 2 & 0\\ 0 & 1 & 0 & 0\\ 0 & 0 & 0 & 1\end{bmatrix}$$

$$\begin{bmatrix} 1 & 0 & 0\\ 0 & 1 & 0\\ 0 & 0 & 1\end{bmatrix} \to Q = \tfrac{1}{6}\begin{bmatrix} 6 & -6 & 0\\ -2 & 5 & -1\\ 4 & -4 & 2\end{bmatrix}$$

(a) Clearly R has rank 3, so α_1, α_2 and α_3 are independent.

(b) Which vectors $\beta = (b_1, b_2, b_3, b_4)$ are in W? We have the basis for W given by ρ_1, ρ_2, ρ_3, the row vectors of R. One can see at a glance that the span of ρ_1, ρ_2, ρ_3 consists of the vectors β for which $b_3 = 2b_1$. For such a β we have

$$\begin{aligned}\beta &= b_1\rho_1 + b_2\rho_2 + b_4\rho_3\\ &= [b_1, b_2, b_4]R\\ &= [b_1 \quad b_2 \quad b_4]QA\\ &= x_1\alpha_1 + x_2\alpha_2 + x_3\alpha_3\end{aligned}$$

where $x_i = [b_1 \quad b_2 \quad b_4]Q_i$:

$$\begin{aligned}x_1 &= b_1 - \tfrac{1}{3}b_2 + \tfrac{2}{3}b_4\\ x_2 &= -b_1 + \tfrac{5}{6}b_2 - \tfrac{2}{3}b_4\\ x_3 &= \quad -\tfrac{1}{6}b_2 + \tfrac{1}{3}b_4.\end{aligned} \tag{2-26}$$

(c) The vectors $\alpha_1', \alpha_2', \alpha_3'$ are all of the form (y_1, y_2, y_3, y_4) with $y_3 = 2y_1$ and thus they are in W. One can see at a glance that they are independent.

(d) The matrix P has for its columns

$$P_j = [\alpha_j']_{\mathcal{B}}$$

where $\mathcal{B} = \{\alpha_1, \alpha_2, \alpha_3\}$. The equations (2-26) tell us how to find the coordinate matrices for $\alpha_1', \alpha_2', \alpha_3'$. For example with $\beta = \alpha_1'$ we have $b_1 = 1$, $b_2 = 0$, $b_3 = 2$, $b_4 = 0$, and

$$\begin{aligned} x_1 &= 1 - \tfrac{1}{3}(0) + \tfrac{2}{3}(0) = 1 \\ x_2 &= -1 + \tfrac{5}{6}(0) - \tfrac{2}{3}(0) = -1 \\ x_3 &= \; - \tfrac{1}{6}(0) + \tfrac{1}{3}(0) = 0. \end{aligned}$$

Thus $\alpha_1' = \alpha_1 - \alpha_2$. Similarly we obtain $\alpha_2' = \alpha_2$ and $\alpha_3' = 2\alpha_1 - 2\alpha_2 + \alpha_3$. Hence

$$P = \begin{bmatrix} 1 & 0 & 2 \\ -1 & 1 & -2 \\ 0 & 0 & 1 \end{bmatrix}$$

Now let us see how we would answer the questions by the second method which we described. We form the 4×3 matrix B with column vectors $\alpha_1, \alpha_2, \alpha_3$:

$$B = \begin{bmatrix} 1 & 0 & -2 \\ 2 & 2 & 0 \\ 2 & 0 & -4 \\ 1 & 1 & 3 \end{bmatrix}$$

We inquire for which y_1, y_2, y_3, y_4 the system $BX = Y$ has a solution.

$$\begin{bmatrix} 1 & 0 & -2 & y_1 \\ 2 & 2 & 0 & y_2 \\ 2 & 0 & -4 & y_3 \\ 1 & 1 & 3 & y_4 \end{bmatrix} \longrightarrow \begin{bmatrix} 1 & 0 & -2 & y_1 \\ 0 & 2 & 4 & y_2 - 2y_1 \\ 0 & 0 & 0 & y_3 - 2y_1 \\ 0 & 1 & 5 & y_4 - y_1 \end{bmatrix} \longrightarrow$$

$$\begin{bmatrix} 1 & 0 & -2 & y_1 \\ 0 & 0 & -6 & y_2 - 2y_4 \\ 0 & 1 & 5 & y_4 - y_1 \\ 0 & 0 & 0 & y_3 - 2y_1 \end{bmatrix} \longrightarrow \begin{bmatrix} 1 & 0 & 0 & y_1 - \tfrac{1}{3}y_2 + \tfrac{2}{3}y_4 \\ 0 & 0 & 1 & \tfrac{1}{6}(2y_4 - y_2) \\ 0 & 1 & 0 & -y_1 + \tfrac{5}{6}y_2 - \tfrac{2}{3}y_4 \\ 0 & 0 & 0 & y_3 - 2y_1 \end{bmatrix}$$

Thus the condition that the system $BX = Y$ have a solution is $y_3 = 2y_1$. So $\beta = (b_1, b_2, b_3, b_4)$ is in W if and only if $b_3 = 2b_1$. If β is in W, then the coordinates (x_1, x_2, x_3) in the ordered basis $\{\alpha_1, \alpha_2, \alpha_3\}$ can be read off from the last matrix above. We obtain once again the formulas (2-26) for those coordinates.

The questions (c) and (d) are now answered as before.

EXAMPLE 22. We consider the 5×5 matrix

$$A = \begin{bmatrix} 1 & 2 & 0 & 3 & 0 \\ 1 & 2 & -1 & -1 & 0 \\ 0 & 0 & 1 & 4 & 0 \\ 2 & 4 & 1 & 10 & 1 \\ 0 & 0 & 0 & 0 & 1 \end{bmatrix}$$

and the following problems concerning A

(a) Find an invertible matrix P such that PA is a row-reduced echelon matrix R.

(b) Find a basis for the row space W of A.

(c) Say which vectors $(b_1, b_2, b_3, b_4, b_5)$ are in W.

(d) Find the coordinate matrix of each vector $(b_1, b_2, b_3, b_4, b_5)$ in W in the ordered basis chosen in (b).

(e) Write each vector $(b_1, b_2, b_3, b_4, b_5)$ in W as a linear combination of the rows of A.

(f) Give an explicit description of the vector space V of all 5×1 column matrices X such that $AX = 0$.

(g) Find a basis for V.

(h) For what 5×1 column matrices Y does the equation $AX = Y$ have solutions X?

To solve these problems we form the augmented matrix A' of the system $AX = Y$ and apply an appropriate sequence of row operations to A'.

$$\begin{bmatrix} 1 & 2 & 0 & 3 & 0 & y_1 \\ 1 & 2 & -1 & -1 & 0 & y_2 \\ 0 & 0 & 1 & 4 & 0 & y_3 \\ 2 & 4 & 1 & 10 & 1 & y_4 \\ 0 & 0 & 0 & 0 & 1 & y_5 \end{bmatrix} \rightarrow \begin{bmatrix} 1 & 2 & 0 & 3 & 0 & y_1 \\ 0 & 0 & -1 & -4 & 0 & -y_1 + y_2 \\ 0 & 0 & 1 & 4 & 0 & y_3 \\ 0 & 0 & 1 & 4 & 1 & -2y_1 + y_4 \\ 0 & 0 & 0 & 0 & 1 & y_5 \end{bmatrix} \rightarrow$$

$$\begin{bmatrix} 1 & 2 & 0 & 3 & 0 & y_1 \\ 0 & 0 & 1 & 4 & 0 & y_1 - y_2 \\ 0 & 0 & 0 & 0 & 0 & -y_1 + y_2 + y_3 \\ 0 & 0 & 0 & 0 & 1 & -3y_1 + y_2 + y_4 \\ 0 & 0 & 0 & 0 & 1 & y_5 \end{bmatrix} \rightarrow$$

$$\begin{bmatrix} 1 & 2 & 0 & 3 & 0 & y_1 \\ 0 & 0 & 1 & 4 & 0 & y_1 - y_2 \\ 0 & 0 & 0 & 0 & 1 & y_5 \\ 0 & 0 & 0 & 0 & 0 & -y_1 + y_2 + y_3 \\ 0 & 0 & 0 & 0 & 0 & -3y_1 + y_2 + y_4 - y_5 \end{bmatrix}$$

(a) If

$$PY = \begin{bmatrix} y_1 \\ y_1 - y_2 \\ y_5 \\ -y_1 + y_2 + y_3 \\ -3y_1 + y_2 + y_4 - y_5 \end{bmatrix}$$

for all Y, then

$$P = \begin{bmatrix} 1 & 0 & 0 & 0 & 0 \\ 1 & -1 & 0 & 0 & 0 \\ 0 & 0 & 0 & 0 & 1 \\ -1 & 1 & 1 & 0 & 0 \\ -3 & 1 & 0 & 1 & -1 \end{bmatrix}$$

hence PA is the row-reduced echelon matrix

$$R = \begin{bmatrix} 1 & 2 & 0 & 3 & 0 \\ 0 & 0 & 1 & 4 & 0 \\ 0 & 0 & 0 & 0 & 1 \\ 0 & 0 & 0 & 0 & 0 \\ 0 & 0 & 0 & 0 & 0 \end{bmatrix}.$$

It should be stressed that the matrix P is not unique. There are, in fact, many invertible matrices P (which arise from different choices for the operations used to reduce A') such that $PA = R$.

(b) As a basis for W we may take the non-zero rows

$$\begin{aligned} \rho_1 &= (1 \quad 2 \quad 0 \quad 3 \quad 0) \\ \rho_2 &= (0 \quad 0 \quad 1 \quad 4 \quad 0) \\ \rho_3 &= (0 \quad 0 \quad 0 \quad 0 \quad 1) \end{aligned}$$

of R.

(c) The row-space W consists of all vectors of the form

$$\begin{aligned} \beta &= c_1\rho_1 + c_2\rho_2 + c_3\rho_3 \\ &= (c_1, 2c_1, c_2, 3c_1 + 4c_2, c_3) \end{aligned}$$

where c_1, c_2, c_3 are arbitrary scalars. Thus $(b_1, b_2, b_3, b_4, b_5)$ is in W if and only if

$$(b_1, b_2, b_3, b_4, b_5) = b_1\rho_1 + b_3\rho_2 + b_5\rho_3$$

which is true if and only if

$$\begin{aligned} b_2 &= 2b_1 \\ b_4 &= 3b_1 + 4b_3. \end{aligned}$$

These equations are instances of the general system (2-25), and using them we may tell at a glance whether a given vector lies in W. Thus $(-5, -10, 1, -11, 20)$ is a linear combination of the rows of A, but $(1, 2, 3, 4, 5)$ is not.

(d) The coordinate matrix of the vector $(b_1, 2b_1, b_3, 3b_1 + 4b_3, b_5)$ in the basis $\{\rho_1, \rho_2, \rho_3\}$ is evidently

$$\begin{bmatrix} b_1 \\ b_3 \\ b_5 \end{bmatrix}.$$

(e) There are many ways to write the vectors in W as linear combinations of the rows of A. Perhaps the easiest method is to follow the first procedure indicated before Example 21:

$$\begin{aligned} \beta &= (b_1, 2b_1, b_3, 3b_1 + 4b_3, b_5) \\ &= [b_1, b_3, b_5, 0, 0] \cdot R \\ &= [b_1, b_3, b_5, 0, 0] \cdot PA \\ &= [b_1, b_3, b_5, 0, 0] \begin{bmatrix} 1 & 0 & 0 & 0 & 0 \\ 1 & -1 & 0 & 0 & 0 \\ 0 & 0 & 0 & 0 & 1 \\ -1 & 1 & 1 & 0 & 0 \\ -3 & 1 & 0 & 1 & -1 \end{bmatrix} \cdot A \\ &= [b_1 + b_3, -b_3, 0, 0, b_5] \cdot A. \end{aligned}$$

In particular, with $\beta = (-5, -10, 1, -11, 20)$ we have

$$\beta = (-4, -1, 0, 0, 20) \begin{bmatrix} 1 & 2 & 0 & 3 & 0 \\ 1 & 2 & -1 & -1 & 0 \\ 0 & 0 & 1 & 4 & 0 \\ 2 & 4 & 1 & 10 & 1 \\ 0 & 0 & 0 & 0 & 1 \end{bmatrix}.$$

(f) The equations in the system $RX = 0$ are

$$\begin{aligned} x_1 + 2x_2 + 3x_4 &= 0 \\ x_3 + 4x_4 &= 0 \\ x_5 &= 0. \end{aligned}$$

Thus V consists of all columns of the form

$$X = \begin{bmatrix} -2x_2 - 3x_4 \\ x_2 \\ -4x_4 \\ x_4 \\ 0 \end{bmatrix}$$

where x_2 and x_4 are arbitrary.

(g) The columns

$$\begin{bmatrix} -2 \\ 1 \\ 0 \\ 0 \\ 0 \end{bmatrix} \quad \begin{bmatrix} -3 \\ 0 \\ -4 \\ 1 \\ 0 \end{bmatrix}$$

form a basis of V. This is an example of the basis described in Example 15.

(h) The equation $AX = Y$ has solutions X if and only if

$$\begin{aligned} -y_1 + y_2 + y_3 \qquad &= 0 \\ -3y_1 + y_2 + y_4 - y_5 &= 0. \end{aligned}$$

Exercises

1. Let $s < n$ and A an $s \times n$ matrix with entries in the field F. Use Theorem 4 (not its proof) to show that there is a non-zero X in $F^{n\times 1}$ such that $AX = 0$.

2. Let

$$\alpha_1 = (1, 1, -2, 1), \qquad \alpha_2 = (3, 0, 4, -1), \qquad \alpha_3 = (-1, 2, 5, 2).$$

Let

$$\alpha = (4, -5, 9, -7), \qquad \beta = (3, 1, -4, 4), \qquad \gamma = (-1, 1, 0, 1).$$

(a) Which of the vectors α, β, γ are in the subspace of R^4 spanned by the α_i?
(b) Which of the vectors α, β, γ are in the subspace of C^4 spanned by the α_i?
(c) Does this suggest a theorem?

3. Consider the vectors in R^4 defined by

$$\alpha_1 = (-1, 0, 1, 2), \qquad \alpha_2 = (3, 4, -2, 5), \qquad \alpha_3 = (1, 4, 0, 9).$$

Find a system of homogeneous linear equations for which the space of solutions is exactly the subspace of R^4 spanned by the three given vectors.

4. In C^3, let

$$\alpha_1 = (1, 0, -i), \qquad \alpha_2 = (1 + i, 1 - i, 1), \qquad \alpha_3 = (i, i, i).$$

Prove that these vectors form a basis for C^3. What are the coordinates of the vector (a, b, c) in this basis?

5. Give an explicit description of the type (2-25) for the vectors

$$\beta = (b_1, b_2, b_3, b_4, b_5)$$

in R^5 which are linear combinations of the vectors

$$\begin{aligned} \alpha_1 &= (1, 0, 2, 1, -1), \qquad & \alpha_2 &= (-1, 2, -4, 2, 0) \\ \alpha_3 &= (2, -1, 5, 2, 1), \qquad & \alpha_4 &= (2, 1, 3, 5, 2). \end{aligned}$$

6. Let V be the real vector space spanned by the rows of the matrix

$$A = \begin{bmatrix} 3 & 21 & 0 & 9 & 0 \\ 1 & 7 & -1 & -2 & -1 \\ 2 & 14 & 0 & 6 & 1 \\ 6 & 42 & -1 & 13 & 0 \end{bmatrix}.$$

(a) Find a basis for V.
(b) Tell which vectors $(x_1, x_2, x_3, x_4, x_5)$ are elements of V.
(c) If $(x_1, x_2, x_3, x_4, x_5)$ is in V what are its coordinates in the basis chosen in part (a)?

7. Let A be an $m \times n$ matrix over the field F, and consider the system of equations $AX = Y$. Prove that this system of equations has a solution if and only if the row rank of A is equal to the row rank of the augmented matrix of the system.

3. Linear Transformations

3.1. Linear Transformations

We shall now introduce linear transformations, the objects which we shall study in most of the remainder of this book. The reader may find it helpful to read (or reread) the discussion of functions in the Appendix, since we shall freely use the terminology of that discussion.

Definition. *Let* V *and* W *be vector spaces over the field* F. *A* **linear transformation from** V **into** W *is a function* T *from* V *into* W *such that*

$$T(c\alpha + \beta) = c(T\alpha) + T\beta$$

for all α *and* β *in* V *and all scalars* c *in* F.

EXAMPLE 1. If V is any vector space, the identity transformation I, defined by $I\alpha = \alpha$, is a linear transformation from V into V. The **zero transformation** 0, defined by $0\alpha = 0$, is a linear transformation from V into V.

EXAMPLE 2. Let F be a field and let V be the space of polynomial functions f from F into F, given by

$$f(x) = c_0 + c_1x + \cdots + c_kx^k.$$

Let

$$(Df)(x) = c_1 + 2c_2x + \cdots + kc_kx^{k-1}.$$

Then D is a linear transformation from V into V—the differentiation transformation.

EXAMPLE 3. Let A be a fixed $m \times n$ matrix with entries in the field F. The function T defined by $T(X) = AX$ is a linear transformation from $F^{n\times 1}$ into $F^{m\times 1}$. The function U defined by $U(\alpha) = \alpha A$ is a linear transformation from F^m into F^n.

EXAMPLE 4. Let P be a fixed $m \times m$ matrix with entries in the field F and let Q be a fixed $n \times n$ matrix over F. Define a function T from the space $F^{m\times n}$ into itself by $T(A) = PAQ$. Then T is a linear transformation from $F^{m\times n}$ into $F^{m\times n}$, because

$$\begin{aligned} T(cA + B) &= P(cA + B)Q \\ &= (cPA + PB)Q \\ &= cPAQ + PBQ \\ &= cT(A) + T(B). \end{aligned}$$

EXAMPLE 5. Let R be the field of real numbers and let V be the space of all functions from R into R which are *continuous*. Define T by

$$(Tf)(x) = \int_0^x f(t)\,dt.$$

Then T is a linear transformation from V into V. The function Tf is not only continuous but has a continuous first derivative. The linearity of integration is one of its fundamental properties.

The reader should have no difficulty in verifying that the transformations defined in Examples 1, 2, 3, and 5 are linear transformations. We shall expand our list of examples considerably as we learn more about linear transformations.

It is important to note that if T is a linear transformation from V into W, then $T(0) = 0$; one can see this from the definition because

$$T(0) = T(0 + 0) = T(0) + T(0).$$

This point is often confusing to the person who is studying linear algebra for the first time, since he probably has been exposed to a slightly different use of the term 'linear function.' A brief comment should clear up the confusion. Suppose V is the vector space R^1. A linear transformation from V into V is then a particular type of real-valued function on the real line R. In a calculus course, one would probably call such a function linear if its graph is a straight line. A linear transformation from R^1 into R^1, according to our definition, will be a function from R into R, the graph of which is a straight line *passing through the origin*.

In addition to the property $T(0) = 0$, let us point out another property of the general linear transformation T. Such a transformation 'preserves' linear combinations; that is, if $\alpha_1, \ldots, \alpha_n$ are vectors in V and $c_1, \ldots, c_n$ are scalars, then

$$T(c_1\alpha_1 + \cdots + c_n\alpha_n) = c_1(T\alpha_1) + \cdots + c_n(T\alpha_n).$$

This follows readily from the definition. For example,

$$\begin{aligned} T(c_1\alpha_1 + c_2\alpha_2) &= c_1(T\alpha_1) + T(c_2\alpha_2) \\ &= c_1(T\alpha_1) + c_2(T\alpha_2). \end{aligned}$$

Theorem 1. *Let V be a finite-dimensional vector space over the field F and let $\{\alpha_1, \ldots, \alpha_n\}$ be an ordered basis for V. Let W be a vector space over the same field F and let $\beta_1, \ldots, \beta_n$ be any vectors in W. Then there is precisely one linear transformation T from V into W such that*

$$T\alpha_j = \beta_j, \qquad j = 1, \ldots, n.$$

Proof. To prove there is some linear transformation T with $T\alpha_j = \beta_j$ we proceed as follows. Given α in V, there is a unique n-tuple $(x_1, \ldots, x_n)$ such that

$$\alpha = x_1\alpha_1 + \cdots + x_n\alpha_n.$$

For this vector α we define

$$T\alpha = x_1\beta_1 + \cdots + x_n\beta_n.$$

Then T is a well-defined rule for associating with each vector α in V a vector $T\alpha$ in W. From the definition it is clear that $T\alpha_j = \beta_j$ for each j. To see that T is linear, let

$$\beta = y_1\alpha_1 + \cdots + y_n\alpha_n$$

be in V and let c be any scalar. Now

$$c\alpha + \beta = (cx_1 + y_1)\alpha_1 + \cdots + (cx_n + y_n)\alpha_n$$

and so by definition

$$T(c\alpha + \beta) = (cx_1 + y_1)\beta_1 + \cdots + (cx_n + y_n)\beta_n.$$

On the other hand,

$$\begin{aligned} c(T\alpha) + T\beta &= c\sum_{i=1}^{n} x_i\beta_i + \sum_{i=1}^{n} y_i\beta_i \\ &= \sum_{i=1}^{n} (cx_i + y_i)\beta_i \end{aligned}$$

and thus

$$T(c\alpha + \beta) = c(T\alpha) + T\beta.$$

If U is a linear transformation from V into W with $U\alpha_j = \beta_j$, $j = 1, \ldots, n$, then for the vector $\alpha = \sum_{i=1}^{n} x_i\alpha_i$ we have

$$\begin{aligned} U\alpha &= U\left(\sum_{i=1}^{n} x_i\alpha_i\right) \\ &= \sum_{i=1}^{n} x_i(U\alpha_i) \\ &= \sum_{i=1}^{n} x_i\beta_i \end{aligned}$$

so that U is exactly the rule T which we defined above. This shows that the linear transformation T with $T\alpha_j = \beta_j$ is unique. ∎

Theorem 1 is quite elementary; however, it is so basic that we have stated it formally. The concept of function is very general. If V and W are (non-zero) vector spaces, there is a multitude of functions from V into W. Theorem 1 helps to underscore the fact that the functions which are linear are extremely special.

EXAMPLE 6. The vectors

$$\alpha_1 = (1, 2)$$
$$\alpha_2 = (3, 4)$$

are linearly independent and therefore form a basis for R^2. According to Theorem 1, there is a unique linear transformation from R^2 into R^3 such that

$$T\alpha_1 = (3, 2, 1)$$
$$T\alpha_2 = (6, 5, 4).$$

If so, we must be able to find $T(\epsilon_1)$. We find scalars c_1, c_2 such that $\epsilon_1 = c_1\alpha_1 + c_2\alpha_2$ and then we know that $T\epsilon_1 = c_1T\alpha_1 + c_2T\alpha_2$. If $(1, 0) = c_1(1, 2) + c_2(3, 4)$ then $c_1 = -2$ and $c_2 = 1$. Thus

$$\begin{aligned} T(1, 0) &= -2(3, 2, 1) + (6, 5, 4) \\ &= (0, 1, 2). \end{aligned}$$

EXAMPLE 7. Let T be a linear transformation from the m-tuple space F^m into the n-tuple space F^n. Theorem 1 tells us that T is uniquely determined by the sequence of vectors $\beta_1, \ldots, \beta_m$ where

$$\beta_i = T\epsilon_i, \qquad i = 1, \ldots, m.$$

In short, T is uniquely determined by the images of the standard basis vectors. The determination is

$$\alpha = (x_1, \ldots, x_m)$$
$$T\alpha = x_1\beta_1 + \cdots + x_m\beta_m.$$

If B is the $m \times n$ matrix which has row vectors $\beta_1, \ldots, \beta_m$, this says that

$$T\alpha = \alpha B.$$

In other words, if $\beta_i = (B_{i1}, \ldots, B_{in})$, then

$$T(x_1, \ldots, x_n) = [x_1 \cdots x_m] \begin{bmatrix} B_{11} & \cdots & B_{1n} \\ \vdots & & \vdots \\ B_{m1} & \cdots & B_{mn} \end{bmatrix}.$$

This is a very explicit description of the linear transformation. In Section 3.4 we shall make a serious study of the relationship between linear transformations and matrices. We shall not pursue the particular description $T\alpha = \alpha B$ because it has the matrix B on the right of the vector α, and that can lead to some confusion. The point of this example is to show that we can give an explicit and reasonably simple description of all linear transformations from F^m into F^n.

If T is a linear transformation from V into W, then the range of T is not only a subset of W; it is a subspace of W. Let R_T be the range of T, that is, the set of all vectors β in W such that $\beta = T\alpha$ for some α in V. Let β_1 and β_2 be in R_T and let c be a scalar. There are vectors α_1 and α_2 in V such that $T\alpha_1 = \beta_1$ and $T\alpha_2 = \beta_2$. Since T is linear

$$\begin{aligned} T(c\alpha_1 + \alpha_2) &= cT\alpha_1 + T\alpha_2 \\ &= c\beta_1 + \beta_2, \end{aligned}$$

which shows that $c\beta_1 + \beta_2$ is also in R_T.

Another interesting subspace associated with the linear transformation T is the set N consisting of the vectors α in V such that $T\alpha = 0$. It is a subspace of V because

(a) $T(0) = 0$, so that N is non-empty;
(b) if $T\alpha_1 = T\alpha_2 = 0$, then

$$\begin{aligned} T(c\alpha_1 + \alpha_2) &= cT\alpha_1 + T\alpha_2 \\ &= c0 + 0 \\ &= 0 \end{aligned}$$

so that $c\alpha_1 + \alpha_2$ is in N.

Definition. *Let* V *and* W *be vector spaces over the field* F *and let* T *be a linear transformation from* V *into* W. *The* **null space** *of* T *is the set of all vectors* α *in* V *such that* $T\alpha = 0$.

If V *is finite-dimensional, the* **rank** *of* T *is the dimension of the range of* T *and the* **nullity** *of* T *is the dimension of the null space of* T.

The following is one of the most important results in linear algebra.

Theorem 2. *Let* V *and* W *be vector spaces over the field* F *and let* T *be a linear transformation from* V *into* W. *Suppose that* V *is finite-dimensional. Then*

$$\text{rank (T)} + \text{nullity (T)} = \dim \text{V}.$$

Proof. Let $\{\alpha_1, \ldots, \alpha_k\}$ be a basis for N, the null space of T. There are vectors $\alpha_{k+1}, \ldots, \alpha_n$ in V such that $\{\alpha_1, \ldots, \alpha_n\}$ is a basis for V. We shall now prove that $\{T\alpha_{k+1}, \ldots, T\alpha_n\}$ is a basis for the range of T. The vectors $T\alpha_1, \ldots, T\alpha_n$ certainly span the range of T, and since $T\alpha_j = 0$, for $j \leq k$, we see that $T\alpha_{k+1}, \ldots, T\alpha_n$ span the range. To see that these vectors are independent, suppose we have scalars c_i such that

$$\sum_{i=k+1}^{n} c_i(T\alpha_i) = 0.$$

This says that

$$T\left(\sum_{i=k+1}^{n} c_i\alpha_i\right) = 0$$

and accordingly the vector $\alpha = \sum_{i=k+1}^{n} c_i\alpha_i$ is in the null space of T. Since $\alpha_1, \ldots, \alpha_k$ form a basis for N, there must be scalars $b_1, \ldots, b_k$ such that

$$\alpha = \sum_{i=1}^{k} b_i\alpha_i.$$

Thus

$$\sum_{i=1}^{k} b_i\alpha_i - \sum_{j=k+1}^{n} c_j\alpha_j = 0$$

and since $\alpha_1, \ldots, \alpha_n$ are linearly independent we must have

$$b_1 = \cdots = b_k = c_{k+1} = \cdots = c_n = 0.$$

If r is the rank of T, the fact that $T\alpha_{k+1}, \ldots, T\alpha_n$ form a basis for the range of T tells us that $r = n - k$. Since k is the nullity of T and n is the dimension of V, we are done. ∎

Theorem 3. *If* A *is an* m × n *matrix with entries in the field* F, *then*

$$\textit{row rank}\ (A) = \textit{column rank}\ (A).$$

Proof. Let T be the linear transformation from $F^{n\times 1}$ into $F^{m\times 1}$ defined by $T(X) = AX$. The null space of T is the solution space for the system $AX = 0$, i.e., the set of all column matrices X such that $AX = 0$. The range of T is the set of all $m \times 1$ column matrices Y such that $AX = Y$ has a solution for X. If $A_1, \ldots, A_n$ are the columns of A, then

$$AX = x_1A_1 + \cdots + x_nA_n$$

so that the range of T is the subspace spanned by the columns of A. In other words, the range of T is the column space of A. Therefore,

$$\text{rank}\ (T) = \text{column rank}\ (A).$$

Theorem 2 tells us that if S is the solution space for the system $AX = 0$, then

$$\dim S + \text{column rank}\ (A) = n.$$

We now refer to Example 15 of Chapter 2. Our deliberations there showed that, if r is the dimension of the row space of A, then the solution space S has a basis consisting of $n - r$ vectors:

$$\dim S = n - \text{row rank}\ (A).$$

It is now apparent that

$$\text{row rank}\ (A) = \text{column rank}\ (A). \quad ∎$$

The proof of Theorem 3 which we have just given depends upon explicit calculations concerning systems of linear equations. There is a more conceptual proof which does not rely on such calculations. We shall give such a proof in Section 3.7.

Exercises

1. Which of the following functions T from R^2 into R^2 are linear transformations?

(a) $T(x_1, x_2) = (1 + x_1, x_2)$;
(b) $T(x_1, x_2) = (x_2, x_1)$;
(c) $T(x_1, x_2) = (x_1^2, x_2)$;
(d) $T(x_1, x_2) = (\sin x_1, x_2)$;
(e) $T(x_1, x_2) = (x_1 - x_2, 0)$.

2. Find the range, rank, null space, and nullity for the zero transformation and the identity transformation on a finite-dimensional space V.

3. Describe the range and the null space for the differentiation transformation of Example 2. Do the same for the integration transformation of Example 5.

4. Is there a linear transformation T from R^3 into R^2 such that $T(1, -1, 1) = (1, 0)$ and $T(1, 1, 1) = (0, 1)$?

5. If

$$\begin{aligned} \alpha_1 &= (1, -1), & \beta_1 &= (1, 0) \\ \alpha_2 &= (2, -1), & \beta_2 &= (0, 1) \\ \alpha_3 &= (-3, 2), & \beta_3 &= (1, 1) \end{aligned}$$

is there a linear transformation T from R^2 into R^2 such that $T\alpha_i = \beta_i$ for $i = 1, 2$ and 3?

6. Describe explicitly (as in Exercises 1 and 2) the linear transformation T from F^2 into F^2 such that $T\epsilon_1 = (a, b)$, $T\epsilon_2 = (c, d)$.

7. Let F be a subfield of the complex numbers and let T be the function from F^3 into F^3 defined by

$$T(x_1, x_2, x_3) = (x_1 - x_2 + 2x_3, 2x_1 + x_2, -x_1 - 2x_2 + 2x_3).$$

(a) Verify that T is a linear transformation.

(b) If (a, b, c) is a vector in F^3, what are the conditions on a, b, and c that the vector be in the range of T? What is the rank of T?

(c) What are the conditions on a, b, and c that (a, b, c) be in the null space of T? What is the nullity of T?

8. Describe explicitly a linear transformation from R^3 into R^3 which has as its range the subspace spanned by $(1, 0, -1)$ and $(1, 2, 2)$.

9. Let V be the vector space of all $n \times n$ matrices over the field F, and let B be a fixed $n \times n$ matrix. If

$$T(A) = AB - BA$$

verify that T is a linear transformation from V into V.

10. Let V be the set of all complex numbers regarded as a vector space over the field of *real* numbers (usual operations). Find a function from V into V which is a linear transformation on the above vector space, but which is not a linear transformation on C^1, i.e., which is not complex linear.

11. Let V be the space of $n \times 1$ matrices over F and let W be the space of $m \times 1$ matrices over F. Let A be a fixed $m \times n$ matrix over F and let T be the linear transformation from V into W defined by $T(X) = AX$. Prove that T is the zero transformation if and only if A is the zero matrix.

12. Let V be an n-dimensional vector space over the field F and let T be a linear transformation from V into V such that the range and null space of T are identical. Prove that n is even. (Can you give an example of such a linear transformation T?)

13. Let V be a vector space and T a linear transformation from V into V. Prove that the following two statements about T are equivalent.

(a) The intersection of the range of T and the null space of T is the zero subspace of V.

(b) If $T(T\alpha) = 0$, then $T\alpha = 0$.

3.2. The Algebra of Linear Transformations

In the study of linear transformations from V into W, it is of fundamental importance that the set of these transformations inherits a natural vector space structure. The set of linear transformations from a space V into itself has even more algebraic structure, because ordinary composition of functions provides a 'multiplication' of such transformations. We shall explore these ideas in this section.

Theorem 4. *Let* V *and* W *be vector spaces over the field* F. *Let* T *and* U *be linear transformations from* V *into* W. *The function* (T + U) *defined by*

$$(T + U)(\alpha) = T\alpha + U\alpha$$

is a linear transformation from V *into* W. *If* c *is any element of* F, *the function* (cT) *defined by*

$$(cT)(\alpha) = c(T\alpha)$$

is a linear transformation from V *into* W. *The set of all linear transformations from* V *into* W, *together with the addition and scalar multiplication defined above, is a vector space over the field* F.

Proof. Suppose T and U are linear transformations from V into W and that we define $(T + U)$ as above. Then

$$\begin{aligned}(T + U)(c\alpha + \beta) &= T(c\alpha + \beta) + U(c\alpha + \beta)\\ &= c(T\alpha) + T\beta + c(U\alpha) + U\beta\\ &= c(T\alpha + U\alpha) + (T\beta + U\beta)\\ &= c(T + U)(\alpha) + (T + U)(\beta)\end{aligned}$$

which shows that $(T + U)$ is a linear transformation. Similarly,

$$\begin{aligned}(cT)(d\alpha + \beta) &= c[T(d\alpha + \beta)]\\ &= c[d(T\alpha) + T\beta]\\ &= cd(T\alpha) + c(T\beta)\\ &= d[c(T\alpha)] + c(T\beta)\\ &= d[(cT)\alpha] + (cT)\beta\end{aligned}$$

which shows that (cT) is a linear transformation.

To verify that the set of linear transformations of V into W (together with these operations) is a vector space, one must directly check each of the conditions on the vector addition and scalar multiplication. We leave the bulk of this to the reader, and content ourselves with this comment: The zero vector in this space will be the zero transformation, which sends every vector of V into the zero vector in W; each of the properties of the two operations follows from the corresponding property of the operations in the space W. ∎

We should perhaps mention another way of looking at this theorem. If one defines sum and scalar multiple as we did above, then the set of *all* functions from V into W becomes a vector space over the field F. This has nothing to do with the fact that V is a vector space, only that V is a non-empty set. When V is a vector space we can define a linear transformation from V into W, and Theorem 4 says that the linear transformations are a subspace of the space of all functions from V into W.

We shall denote the space of linear transformations from V into W by $L(V, W)$. We remind the reader that $L(V, W)$ is defined only when V and W are vector spaces over the same field.

Theorem 5. *Let* V *be an* n*-dimensional vector space over the field* F, *and let* W *be an* m*-dimensional vector space over* F. *Then the space* L(V, W) *is finite-dimensional and has dimension* mn.

Proof. Let

$$\mathcal{B} = \{\alpha_1, \ldots, \alpha_n\} \quad \text{and} \quad \mathcal{B}' = \{\beta_1, \ldots, \beta_m\}$$

be ordered bases for V and W, respectively. For each pair of integers (p, q) with $1 \leq p \leq m$ and $1 \leq q \leq n$, we define a linear transformation $E^{p,q}$ from V into W by

$$\begin{aligned}E^{p,q}(\alpha_i) &= \begin{cases}0, & \text{if } i \neq q\\ \beta_p, & \text{if } i = q\end{cases}\\ &= \delta_{iq}\beta_p.\end{aligned}$$

According to Theorem 1, there is a unique linear transformation from V into W satisfying these conditions. The claim is that the mn transformations $E^{p,q}$ form a basis for $L(V, W)$.

Let T be a linear transformation from V into W. For each j, $1 \leq j \leq n$, let $A_{ij}, \ldots, A_{mj}$ be the coordinates of the vector $T\alpha_j$ in the ordered basis $\mathcal{B}'$, i.e.,

$$T\alpha_j = \sum_{p=1}^{m} A_{pj}\beta_p. \tag{3-1}$$

We wish to show that

$$T = \sum_{p=1}^{m} \sum_{q=1}^{n} A_{pq}E^{p,q}. \tag{3-2}$$

Let U be the linear transformation in the right-hand member of (3-2). Then for each j

$$\begin{aligned} U\alpha_j &= \sum_p \sum_q A_{pq}E^{p,q}(\alpha_j) \\ &= \sum_p \sum_q A_{pq}\delta_{jq}\beta_p \\ &= \sum_{p=1}^{m} A_{pj}\beta_p \\ &= T\alpha_j \end{aligned}$$

and consequently $U = T$. Now (3-2) shows that the $E^{p,q}$ span $L(V, W)$; we must prove that they are independent. But this is clear from what we did above; for, if the transformation

$$U = \sum_p \sum_q A_{pq}E^{p,q}$$

is the zero transformation, then $U\alpha_j = 0$ for each j, so

$$\sum_{p=1}^{m} A_{pj}\beta_p = 0$$

and the independence of the β_p implies that $A_{pj} = 0$ for every p and j. ∎

Theorem 6. *Let* V, W, *and* Z *be vector spaces over the field* F. *Let* T *be a linear transformation from* V *into* W *and* U *a linear transformation from* W *into* Z. *Then the composed function* UT *defined by* $(UT)(\alpha) = U(T(\alpha))$ *is a linear transformation from* V *into* Z.

Proof.

$$\begin{aligned} (UT)(c\alpha + \beta) &= U[T(c\alpha + \beta)] \\ &= U(cT\alpha + T\beta) \\ &= c[U(T\alpha)] + U(T\beta) \\ &= c(UT)(\alpha) + (UT)(\beta). \end{aligned}$$ ∎

In what follows, we shall be primarily concerned with linear transformation of a vector space into itself. Since we would so often have to write 'T is a linear transformation from V into V,' we shall replace this with 'T is a linear operator on V.'

Definition. *If* V *is a vector space over the field* F, *a* **linear operator on** V *is a linear transformation from* V *into* V.

In the case of Theorem 6 when $V = W = Z$, so that U and T are linear operators on the space V, we see that the composition UT is again a linear operator on V. Thus the space $L(V, V)$ has a 'multiplication' defined on it by composition. In this case the operator TU is also defined, and one should note that in general $UT \neq TU$, i.e., $UT - TU \neq 0$. We should take special note of the fact that if T is a linear operator on V then we can compose T with T. We shall use the notation $T^2 = TT$, and in general $T^n = T \cdots T$ (n times) for $n = 1, 2, 3, \ldots$. We define $T^0 = I$ if $T \neq 0$.

Lemma. *Let* V *be a vector space over the field* F; *let* U, T_1 *and* T_2 *be linear operators on* V; *let* c *be an element of* F.

(a) $IU = UI = U$;

(b) $U(T_1 + T_2) = UT_1 + UT_2$; $(T_1 + T_2)U = T_1U + T_2U$;

(c) $c(UT_1) = (cU)T_1 = U(cT_1)$.

Proof. (a) This property of the identity function is obvious. We have stated it here merely for emphasis.

(b)
$$\begin{aligned}[U(T_1 + T_2)](\alpha) &= U[(T_1 + T_2)(\alpha)] \\ &= U(T_1\alpha + T_2\alpha) \\ &= U(T_1\alpha) + U(T_2\alpha) \\ &= (UT_1)(\alpha) + (UT_2)(\alpha)\end{aligned}$$

so that $U(T_1 + T_2) = UT_1 + UT_2$. Also

$$\begin{aligned}[(T_1 + T_2)U](\alpha) &= (T_1 + T_2)(U\alpha) \\ &= T_1(U\alpha) + T_2(U\alpha) \\ &= (T_1U)(\alpha) + (T_2U)(\alpha)\end{aligned}$$

so that $(T_1 + T_2)U = T_1U + T_2U$. (The reader may note that the proofs of these two distributive laws do not use the fact that T_1 and T_2 are linear, and the proof of the second one does not use the fact that U is linear either.)

(c) We leave the proof of part (c) to the reader. ∎

The contents of this lemma and a portion of Theorem 5 tell us that the vector space $L(V, V)$, together with the composition operation, is what is known as a linear algebra with identity. We shall discuss this in Chapter 4.

EXAMPLE 8. If A is an $m \times n$ matrix with entries in F, we have the linear transformation T defined by $T(X) = AX$, from $F^{n\times 1}$ into $F^{m\times 1}$. If B is a $p \times m$ matrix, we have the linear transformation U from $F^{m\times 1}$ into $F^{p\times 1}$ defined by $U(Y) = BY$. The composition UT is easily described:

$$\begin{aligned}(UT)(X) &= U(T(X)) \\ &= U(AX) \\ &= B(AX) \\ &= (BA)X.\end{aligned}$$

Thus UT is 'left multiplication by the product matrix BA.'

Example 9. Let F be a field and V the vector space of all polynomial functions from F into F. Let D be the differentiation operator defined in Example 2, and let T be the linear operator 'multiplication by x':

$$(Tf)(x) = xf(x).$$

Then $DT \neq TD$. In fact, the reader should find it easy to verify that $DT - TD = I$, the identity operator.

Even though the 'multiplication' we have on $L(V, V)$ is not commutative, it is nicely related to the vector space operations of $L(V, V)$.

Example 10. Let $\mathcal{B} = \{\alpha_1, \ldots, \alpha_n\}$ be an ordered basis for a vector space V. Consider the linear operators $E^{p,q}$ which arose in the proof of Theorem 5:

$$E^{p,q}(\alpha_i) = \delta_{iq}\alpha_p.$$

These n^2 linear operators form a basis for the space of linear operators on V. What is $E^{p,q}E^{r,s}$? We have

$$\begin{aligned} (E^{p,q}E^{r,s})(\alpha_i) &= E^{p,q}(\delta_{is}\alpha_r) \\ &= \delta_{is}E^{p,q}(\alpha_r) \\ &= \delta_{is}\delta_{rq}\alpha_p. \end{aligned}$$

Therefore,

$$E^{p,q}E^{r,s} = \begin{cases} 0, & \text{if } r \neq q \\ E^{p,s}, & \text{if } q = r. \end{cases}$$

Let T be a linear operator on V. We showed in the proof of Theorem 5 that if

$$\begin{aligned} A_j &= [T\alpha_j]_{\mathcal{B}} \\ A &= [A_1, \ldots, A_n] \end{aligned}$$

then

$$T = \sum_p \sum_q A_{pq}E^{p,q}.$$

If

$$U = \sum_r \sum_s B_{rs}E^{r,s}$$

is another linear operator on V, then the last lemma tells us that

$$\begin{aligned} TU &= (\sum_p \sum_q A_{pq}E^{p,q})(\sum_r \sum_s B_{rs}E^{r,s}) \\ &= \sum_p \sum_q \sum_r \sum_s A_{pq}B_{rs}E^{p,q}E^{r,s}. \end{aligned}$$

As we have noted, the only terms which survive in this huge sum are the terms where $q = r$, and since $E^{p,r}E^{r,s} = E^{p,s}$, we have

$$\begin{aligned} TU &= \sum_p \sum_s (\sum_r A_{pr}B_{rs})E^{p,s} \\ &= \sum_p \sum_s (AB)_{ps}E^{p,s}. \end{aligned}$$

Thus, the effect of composing T and U is to multiply the matrices A and B.

In our discussion of algebraic operations with linear transformations we have not yet said anything about invertibility. One specific question of interest is this. For which linear operators T on the space V does there exist a linear operator T^{-1} such that $TT^{-1} = T^{-1}T = I$?

The function T from V into W is called **invertible** if there exists a function U from W into V such that UT is the identity function on V and TU is the identity function on W. If T is invertible, the function U is unique and is denoted by T^{-1}. (See Appendix.) Furthermore, T is invertible if and only if

1. T is 1:1, that is, $T\alpha = T\beta$ implies $\alpha = \beta$;
2. T is onto, that is, the range of T is (all of) W.

Theorem 7. *Let* V *and* W *be vector spaces over the field* F *and let* T *be a linear transformation from* V *into* W. *If* T *is invertible, then the inverse function* T^{-1} *is a linear transformation from* W *onto* V.

Proof. We repeat ourselves in order to underscore a point. When T is one-one and onto, there is a uniquely determined inverse function T^{-1} which maps W onto V such that $T^{-1}T$ is the identity function on V, and TT^{-1} is the identity function on W. What we are proving here is that if a linear function T is invertible, then the inverse T^{-1} is also linear.

Let β_1 and β_2 be vectors in W and let c be a scalar. We wish to show that

$$T^{-1}(c\beta_1 + \beta_2) = cT^{-1}\beta_1 + T^{-1}\beta_2.$$

Let $\alpha_i = T^{-1}\beta_i$, $i = 1, 2$, that is, let α_i be the unique vector in V such that $T\alpha_i = \beta_i$. Since T is linear,

$$\begin{aligned} T(c\alpha_1 + \alpha_2) &= cT\alpha_1 + T\alpha_2 \\ &= c\beta_1 + \beta_2. \end{aligned}$$

Thus $c\alpha_1 + \alpha_2$ is the unique vector in V which is sent by T into $c\beta_1 + \beta_2$, and so

$$\begin{aligned} T^{-1}(c\beta_1 + \beta_2) &= c\alpha_1 + \alpha_2 \\ &= c(T^{-1}\beta_1) + T^{-1}\beta_2 \end{aligned}$$

and T^{-1} is linear. ∎

Suppose that we have an invertible linear transformation T from V onto W and an invertible linear transformation U from W onto Z. Then UT is invertible and $(UT)^{-1} = T^{-1}U^{-1}$. That conclusion does not require the linearity nor does it involve checking separately that UT is 1:1 and onto. All it involves is verifying that $T^{-1}U^{-1}$ is both a left and a right inverse for UT.

If T is linear, then $T(\alpha - \beta) = T\alpha - T\beta$; hence, $T\alpha = T\beta$ if and only if $T(\alpha - \beta) = 0$. This simplifies enormously the verification that T is 1:1. Let us call a linear transformation T **non-singular** if $T\gamma = 0$ implies

$\gamma = 0$, i.e., if the null space of T is $\{0\}$. Evidently, T is 1:1 if and only if T is non-singular. The extension of this remark is that non-singular linear transformations are those which preserve linear independence.

Theorem 8. *Let* T *be a linear transformation from* V *into* W. *Then* T *is non-singular if and only if* T *carries each linearly independent subset of* V *onto a linearly independent subset of* W.

Proof. First suppose that T is non-singular. Let S be a linearly independent subset of V. If $\alpha_1, \ldots, \alpha_k$ are vectors in S, then the vectors $T\alpha_1, \ldots, T\alpha_k$ are linearly independent; for if

$$c_1(T\alpha_1) + \cdots + c_k(T\alpha_k) = 0$$

then

$$T(c_1\alpha_1 + \cdots + c_k\alpha_k) = 0$$

and since T is non-singular

$$c_1\alpha_1 + \cdots + c_k\alpha_k = 0$$

from which it follows that each $c_i = 0$ because S is an independent set. This argument shows that the image of S under T is independent.

Suppose that T carries independent subsets onto independent subsets. Let α be a non-zero vector in V. Then the set S consisting of the one vector α is independent. The image of S is the set consisting of the one vector $T\alpha$, and this set is independent. Therefore $T\alpha \neq 0$, because the set consisting of the zero vector alone is dependent. This shows that the null space of T is the zero subspace, i.e., T is non-singular. ∎

EXAMPLE 11. Let F be a subfield of the complex numbers (or a field of characteristic zero) and let V be the space of polynomial functions over F. Consider the differentiation operator D and the 'multiplication by x' operator T, from Example 9. Since D sends all constants into 0, D is singular; however, V is not finite dimensional, the range of D is all of V, and it is possible to define a right inverse for D. For example, if E is the indefinite integral operator:

$$E(c_0 + c_1x + \cdots + c_nx^n) = c_0x + \frac{1}{2}c_1x^2 + \cdots + \frac{1}{n+1}c_nx^{n+1}$$

then E is a linear operator on V and $DE = I$. On the other hand, $ED \neq I$ because ED sends the constants into 0. The operator T is in what we might call the reverse situation. If $xf(x) = 0$ for all x, then $f = 0$. Thus T is non-singular and it is possible to find a left inverse for T. For example if U is the operation 'remove the constant term and divide by x':

$$U(c_0 + c_1x + \cdots + c_nx^n) = c_1 + c_2x + \cdots + c_nx^{n-1}$$

then U is a linear operator on V and $UT = I$. But $TU \neq I$ since every function in the range of TU is in the range of T, which is the space of polynomial functions f such that $f(0) = 0$.

EXAMPLE 12. Let F be a field and let T be the linear operator on F^2 defined by

$$T(x_1, x_2) = (x_1 + x_2, x_1).$$

Then T is non-singular, because if $T(x_1, x_2) = 0$ we have

$$\begin{aligned} x_1 + x_2 &= 0 \\ x_1 &= 0 \end{aligned}$$

so that $x_1 = x_2 = 0$. We also see that T is onto; for, let (z_1, z_2) be any vector in F^2. To show that (z_1, z_2) is in the range of T we must find scalars x_1 and x_2 such that

$$\begin{aligned} x_1 + x_2 &= z_1 \\ x_1 &= z_2 \end{aligned}$$

and the obvious solution is $x_1 = z_2$, $x_2 = z_1 - z_2$. This last computation gives us an explicit formula for T^{-1}, namely,

$$T^{-1}(z_1, z_2) = (z_2, z_1 - z_2).$$

We have seen in Example 11 that a linear transformation may be non-singular without being onto and may be onto without being non-singular. The present example illustrates an important case in which that cannot happen.

Theorem 9. *Let* V *and* W *be finite-dimensional vector spaces over the field* F *such that dim* V = *dim* W. *If* T *is a linear transformation from* V *into* W, *the following are equivalent:*

(i) T *is invertible.*
(ii) T *is non-singular.*
(iii) T *is onto, that is, the range of* T *is* W.

Proof. Let $n = \dim V = \dim W$. From Theorem 2 we know that

$$\text{rank } (T) + \text{nullity } (T) = n.$$

Now T is non-singular if and only if nullity $(T) = 0$, and (since $n = \dim W$) the range of T is W if and only if rank $(T) = n$. Since the rank plus the nullity is n, the nullity is 0 precisely when the rank is n. Therefore T is non-singular if and only if $T(V) = W$. So, if either condition (ii) or (iii) holds, the other is satisfied as well and T is invertible. ∎

We caution the reader not to apply Theorem 9 except in the presence of finite-dimensionality and with $\dim V = \dim W$. Under the hypotheses of Theorem 9, the conditions (i), (ii), and (iii) are also equivalent to these.

(iv) *If* $\{\alpha_1, \ldots, \alpha_n\}$ *is basis for* V, *then* $\{T\alpha_1, \ldots, T\alpha_n\}$ *is a basis for* W.

(v) *There is some basis* $\{\alpha_1, \ldots, \alpha_n\}$ *for* V *such that* $\{T\alpha_1, \ldots, T\alpha_n\}$ *is a basis for* W.

We shall give a proof of the equivalence of the five conditions which contains a different proof that (i), (ii), and (iii) are equivalent.

(i) → (ii). If T is invertible, T is non-singular. (ii) → (iii). Suppose T is non-singular. Let $\{\alpha_1, \ldots, \alpha_n\}$ be a basis for V. By Theorem 8, $\{T\alpha_1, \ldots, T\alpha_n\}$ is a linearly independent set of vectors in W, and since the dimension of W is also n, this set of vectors is a basis for W. Now let β be any vector in W. There are scalars $c_1, \ldots, c_n$ such that

$$\begin{aligned}\beta &= c_1(T\alpha_1) + \cdots + c_n(T\alpha_n)\\ &= T(c_1\alpha_1 + \cdots + c_n\alpha_n)\end{aligned}$$

which shows that β is in the range of T. (iii) → (iv). We now assume that T is onto. If $\{\alpha_1, \ldots, \alpha_n\}$ is any basis for V, the vectors $T\alpha_1, \ldots, T\alpha_n$ span the range of T, which is all of W by assumption. Since the dimension of W is n, these n vectors must be linearly independent, that is, must comprise a basis for W. (iv) → (v). This requires no comment. (v) → (i). Suppose there is some basis $\{\alpha_1, \ldots, \alpha_n\}$ for V such that $\{T\alpha_1, \ldots, T\alpha_n\}$ is a basis for W. Since the $T\alpha_i$ span W, it is clear that the range of T is all of W. If $\alpha = c_1\alpha_1 + \cdots + c_n\alpha_n$ is in the null space of T, then

$$T(c_1\alpha_1 + \cdots + c_n\alpha_n) = 0$$

or

$$c_1(T\alpha_1) + \cdots + c_n(T\alpha_n) = 0$$

and since the $T\alpha_i$ are independent each $c_i = 0$, and thus $\alpha = 0$. We have shown that the range of T is W, and that T is non-singular, hence T is invertible.

The set of invertible linear operators on a space V, with the operation of composition, provides a nice example of what is known in algebra as a 'group.' Although we shall not have time to discuss groups in any detail, we shall at least give the definition.

Definition. *A* **group** *consists of the following.*

1. *A set* G;

2. *A rule (or operation) which associates with each pair of elements* x, y *in* G *an element* xy *in* G *in such a way that*

(a) x(yz) = (xy)z, *for all* x, y, *and* z *in* G *(associativity);*

(b) *there is an element* e *in* G *such that* ex = xe = x, *for every* x *in* G;

(c) *to each element* x *in* G *there corresponds an element* x^{-1} *in* G *such that* $xx^{-1} = x^{-1}x = e$.

We have seen that composition $(U, T) \rightarrow UT$ associates with each pair of invertible linear operators on a space V another invertible operator on V. Composition is an associative operation. The identity operator I

satisfies $IT = TI$ for each T, and for an invertible T there is (by Theorem 7) an invertible linear operator T^{-1} such that $TT^{-1} = T^{-1}T = I$. Thus the set of invertible linear operators on V, together with this operation, is a group. The set of invertible $n \times n$ matrices with matrix multiplication as the operation is another example of a group. A group is called **commutative** if it satisfies the condition $xy = yx$ for each x and y. The two examples we gave above are not commutative groups, in general. One often writes the operation in a commutative group as $(x, y) \to x + y$, rather than $(x, y) \to xy$, and then uses the symbol 0 for the 'identity' element e. The set of vectors in a vector space, together with the operation of vector addition, is a commutative group. A field can be described as a set with two operations, called addition and multiplication, which is a commutative group under addition, and in which the non-zero elements form a commutative group under multiplication, with the distributive law $x(y + z) = xy + xz$ holding.

Exercises

1. Let T and U be the linear operators on R^2 defined by

$$T(x_1, x_2) = (x_2, x_1) \quad \text{and} \quad U(x_1, x_2) = (x_1, 0).$$

(a) How would you describe T and U geometrically?

(b) Give rules like the ones defining T and U for each of the transformations $(U + T)$, UT, TU, T^2, U^2.

2. Let T be the (unique) linear operator on C^3 for which

$$T\epsilon_1 = (1, 0, i), \qquad T\epsilon_2 = (0, 1, 1), \qquad T\epsilon_3 = (i, 1, 0).$$

Is T invertible?

3. Let T be the linear operator on R^3 defined by

$$T(x_1, x_2, x_3) = (3x_1, x_1 - x_2, 2x_1 + x_2 + x_3).$$

Is T invertible? If so, find a rule for T^{-1} like the one which defines T.

4. For the linear operator T of Exercise 3, prove that

$$(T^2 - I)(T - 3I) = 0.$$

5. Let $C^{2\times 2}$ be the complex vector space of 2×2 matrices with complex entries. Let

$$B = \begin{bmatrix} 1 & -1 \\ -4 & 4 \end{bmatrix}$$

and let T be the linear operator on $C^{2\times 2}$ defined by $T(A) = BA$. What is the rank of T? Can you describe T^2?

6. Let T be a linear transformation from R^3 into R^2, and let U be a linear transformation from R^2 into R^3. Prove that the transformation UT is not invertible. Generalize the theorem.

7. Find two linear operators T and U on R^2 such that $TU = 0$ but $UT \neq 0$.

8. Let V be a vector space over the field F and T a linear operator on V. If $T^2 = 0$, what can you say about the relation of the range of T to the null space of T? Give an example of a linear operator T on R^2 such that $T^2 = 0$ but $T \neq 0$.

9. Let T be a linear operator on the finite-dimensional space V. Suppose there is a linear operator U on V such that $TU = I$. Prove that T is invertible and $U = T^{-1}$. Give an example which shows that this is false when V is not finite-dimensional. (*Hint:* Let $T = D$, the differentiation operator on the space of polynomial functions.)

10. Let A be an $m \times n$ matrix with entries in F and let T be the linear transformation from $F^{n \times 1}$ into $F^{m \times 1}$ defined by $T(X) = AX$. Show that if $m < n$ it may happen that T is onto without being non-singular. Similarly, show that if $m > n$ we may have T non-singular but not onto.

11. Let V be a finite-dimensional vector space and let T be a linear operator on V. Suppose that rank (T^2) = rank (T). Prove that the range and null space of T are disjoint, i.e., have only the zero vector in common.

12. Let p, m, and n be positive integers and F a field. Let V be the space of $m \times n$ matrices over F and W the space of $p \times n$ matrices over F. Let B be a fixed $p \times m$ matrix and let T be the linear transformation from V into W defined by $T(A) = BA$. Prove that T is invertible if and only if $p = m$ and B is an invertible $m \times m$ matrix.

3.3. Isomorphism

If V and W are vector spaces over the field F, any one-one linear transformation T of V onto W is called an **isomorphism of V onto W**. If there exists an isomorphism of V onto W, we say that V is **isomorphic** to W.

Note that V is trivially isomorphic to V, the identity operator being an isomorphism of V onto V. Also, if V is isomorphic to W via an isomorphism T, then W is isomorphic to V, because T^{-1} is an isomorphism of W onto V. The reader should find it easy to verify that if V is isomorphic to W and W is isomorphic to Z, then V is isomorphic to Z. Briefly, isomorphism is an equivalence relation on the class of vector spaces. If there exists an isomorphism of V onto W, we may sometimes say that V and W are isomorphic, rather than V is isomorphic to W. This will cause no confusion because V is isomorphic to W if and only if W is isomorphic to V.

Theorem 10. *Every n-dimensional vector space over the field F is isomorphic to the space F^n.*

Proof. Let V be an n-dimensional space over the field F and let $\mathcal{B} = \{\alpha_1, \ldots, \alpha_n\}$ be an ordered basis for V. We define a function T

from V into F^n, as follows: If α is in V, let $T\alpha$ be the n-tuple $(x_1, \ldots, x_n)$ of coordinates of α relative to the ordered basis $\mathcal{B}$, i.e., the n-tuple such that

$$\alpha = x_1\alpha_1 + \cdots + x_n\alpha_n.$$

In our discussion of coordinates in Chapter 2, we verified that this T is linear, one-one, and maps V onto F^n. ∎

For many purposes one often regards isomorphic vector spaces as being 'the same,' although the vectors and operations in the spaces may be quite different, that is, one often identifies isomorphic spaces. We shall not attempt a lengthy discussion of this idea at present but shall let the understanding of isomorphism and the sense in which isomorphic spaces are 'the same' grow as we continue our study of vector spaces.

We shall make a few brief comments. Suppose T is an isomorphism of V onto W. If S is a subset of V, then Theorem 8 tells us that S is linearly independent if and only if the set $T(S)$ in W is independent. Thus in deciding whether S is independent it doesn't matter whether we look at S or $T(S)$. From this one sees that an isomorphism is 'dimension preserving,' that is, any finite-dimensional subspace of V has the same dimension as its image under T. Here is a very simple illustration of this idea. Suppose A is an $m \times n$ matrix over the field F. We have really given two definitions of the solution space of the matrix A. The first is the set of all n-tuples $(x_1, \ldots, x_n)$ in F^n which satisfy each of the equations in the system $AX = 0$. The second is the set of all $n \times 1$ column matrices X such that $AX = 0$. The first solution space is thus a subspace of F^n and the second is a subspace of the space of all $n \times 1$ matrices over F. Now there is a completely obvious isomorphism between F^n and $F^{n\times 1}$, namely,

$$(x_1, \ldots, x_n) \rightarrow \begin{bmatrix} x_1 \\ \vdots \\ x_n \end{bmatrix}.$$

Under this isomorphism, the first solution space of A is carried onto the second solution space. These spaces have the same dimension, and so if we want to prove a theorem about the dimension of the solution space, it is immaterial which space we choose to discuss. In fact, the reader would probably not balk if we chose to identify F^n and the space of $n \times 1$ matrices. We may do this when it is convenient, and when it is not convenient we shall not.

Exercises

1. Let V be the set of complex numbers and let F be the field of real numbers. With the usual operations, V is a vector space over F. Describe explicitly an isomorphism of this space onto R^2.

2. Let V be a vector space over the field of complex numbers, and suppose there is an isomorphism T of V onto C^3. Let $\alpha_1, \alpha_2, \alpha_3, \alpha_4$ be vectors in V such that

$$T\alpha_1 = (1, 0, i), \qquad T\alpha_2 = (-2, 1 + i, 0),$$
$$T\alpha_3 = (-1, 1, 1), \qquad T\alpha_4 = (\sqrt{2}, i, 3).$$

(a) Is α_1 in the subspace spanned by α_2 and α_3?

(b) Let W_1 be the subspace spanned by α_1 and α_2, and let W_2 be the subspace spanned by α_3 and α_4. What is the intersection of W_1 and W_2?

(c) Find a basis for the subspace of V spanned by the four vectors α_j.

3. Let W be the set of all 2×2 complex Hermitian matrices, that is, the set of 2×2 complex matrices A such that $A_{ij} = \overline{A_{ji}}$ (the bar denoting complex conjugation). As we pointed out in Example 6 of Chapter 2, W is a vector space over the field of *real* numbers, under the usual operations. Verify that

$$(x, y, z, t) \rightarrow \begin{bmatrix} t + x & y + iz \\ y - iz & t - x \end{bmatrix}$$

is an isomorphism of R^4 onto W.

4. Show that $F^{m \times n}$ is isomorphic to F^{mn}.

5. Let V be the set of complex numbers regarded as a vector space over the field of real numbers (Exercise 1). We define a function T from V into the space of 2×2 real matrices, as follows. If $z = x + iy$ with x and y real numbers, then

$$T(z) = \begin{bmatrix} x + 7y & 5y \\ -10y & x - 7y \end{bmatrix}.$$

(a) Verify that T is a one-one (real) linear transformation of V into the space of 2×2 real matrices.

(b) Verify that $T(z_1z_2) = T(z_1)T(z_2)$.

(c) How would you describe the range of T?

6. Let V and W be finite-dimensional vector spaces over the field F. Prove that V and W are isomorphic if and only if $\dim V = \dim W$.

7. Let V and W be vector spaces over the field F and let U be an isomorphism of V onto W. Prove that $T \rightarrow UTU^{-1}$ is an isomorphism of $L(V, V)$ onto $L(W, W)$.

3.4. Representation of Transformations by Matrices

Let V be an n-dimensional vector space over the field F and let W be an m-dimensional vector space over F. Let $\mathcal{B} = \{\alpha_1, \ldots, \alpha_n\}$ be an ordered basis for V and $\mathcal{B}' = \{\beta_1, \ldots, \beta_m\}$ an ordered basis for W. If T is any linear transformation from V into W, then T is determined by its action on the vectors α_j. Each of the n vectors $T\alpha_j$ is uniquely expressible as a linear combination

$$T\alpha_j = \sum_{i=1}^{m} A_{ij}\beta_i \tag{3-3}$$

of the β_i, the scalars $A_{1j}, \ldots, A_{mj}$ being the coordinates of $T\alpha_j$ in the ordered basis $\mathcal{B}'$. Accordingly, the transformation T is determined by the mn scalars A_{ij} via the formulas (3-3). The $m \times n$ matrix A defined by $A(i, j) = A_{ij}$ is called **the matrix of T relative to the pair of ordered bases $\mathcal{B}$ and $\mathcal{B}'$**. Our immediate task is to understand explicitly how the matrix A determines the linear transformation T.

If $\alpha = x_1\alpha_1 + \cdots + x_n\alpha_n$ is a vector in V, then

$$\begin{aligned} T\alpha &= T\left(\sum_{j=1}^{n} x_j\alpha_j\right) \\ &= \sum_{j=1}^{n} x_j(T\alpha_j) \\ &= \sum_{j=1}^{n} x_j \sum_{i=1}^{m} A_{ij}\beta_i \\ &= \sum_{i=1}^{m} \left(\sum_{j=1}^{n} A_{ij}x_j\right) \beta_i. \end{aligned}$$

If X is the coordinate matrix of α in the ordered basis $\mathcal{B}$, then the computation above shows that AX is the coordinate matrix of the vector $T\alpha$ in the ordered basis $\mathcal{B}'$, because the scalar

$$\sum_{j=1}^{n} A_{ij}x_j$$

is the entry in the ith row of the column matrix AX. Let us also observe that if A is any $m \times n$ matrix over the field F, then

$$T\left(\sum_{j=1}^{n} x_j\alpha_j\right) = \sum_{i=1}^{m} \left(\sum_{j=1}^{n} A_{ij}x_j\right) \beta_i \tag{3-4}$$

defines a linear transformation T from V into W, the matrix of which is A, relative to $\mathcal{B}$, $\mathcal{B}'$. We summarize formally:

Theorem 11. *Let V be an n-dimensional vector space over the field F and W an m-dimensional vector space over F. Let $\mathcal{B}$ be an ordered basis for V and $\mathcal{B}'$ an ordered basis for W. For each linear transformation T from V into W, there is an m $\times$ n matrix A with entries in F such that*

$$[T\alpha]_{\mathcal{B}'} = A[\alpha]_{\mathcal{B}}$$

for every vector α in V. Furthermore, $T \to A$ is a one-one correspondence between the set of all linear transformations from V into W and the set of all m $\times$ n matrices over the field F.

The matrix A which is associated with T in Theorem 11 is called the **matrix of T relative to the ordered bases $\mathcal{B}$, $\mathcal{B}'$**. Note that Equation (3-3) says that A is the matrix whose columns $A_1, \ldots, A_n$ are given by

$$A_j = [T\alpha_j]_{\mathcal{B}'}, \qquad j = 1, \ldots, n.$$

If U is another linear transformation from V into W and $B = [B_1, \ldots, B_n]$ is the matrix of U relative to the ordered bases $\mathcal{B}$, $\mathcal{B}'$ then $cA + B$ is the matrix of $cT + U$ relative to $\mathcal{B}$, $\mathcal{B}'$. That is clear because

$$\begin{aligned} cA_j + B_j &= c[T\alpha_j]_{\mathcal{B}'} + [U\alpha_j]_{\mathcal{B}'} \\ &= [cT\alpha_j + U\alpha_j]_{\mathcal{B}'} \\ &= [(cT + U)\alpha_j]_{\mathcal{B}'}. \end{aligned}$$

Theorem 12. *Let V be an n-dimensional vector space over the field F and let W be an m-dimensional vector space over F. For each pair of ordered bases $\mathcal{B}$, $\mathcal{B}'$ for V and W respectively, the function which assigns to a linear transformation T its matrix relative to $\mathcal{B}$, $\mathcal{B}'$ is an isomorphism between the space L(V, W) and the space of all m × n matrices over the field F.*

Proof. We observed above that the function in question is linear, and as stated in Theorem 11, this function is one-one and maps $L(V, W)$ onto the set of $m \times n$ matrices. ∎

We shall be particularly interested in the representation by matrices of linear transformations of a space into itself, i.e., linear operators on a space V. In this case it is most convenient to use the same ordered basis in each case, that is, to take $\mathcal{B} = \mathcal{B}'$. We shall then call the representing matrix simply the **matrix of T relative to the ordered basis** $\mathcal{B}$. Since this concept will be so important to us, we shall review its definition. If T is a linear operator on the finite-dimensional vector space V and $\mathcal{B} = \{\alpha_1, \ldots, \alpha_n\}$ is an ordered basis for V, the matrix of T relative to $\mathcal{B}$ (or, the matrix of T in the ordered basis $\mathcal{B}$) is the $n \times n$ matrix A whose entries A_{ij} are defined by the equations

$$T\alpha_j = \sum_{i=1}^{n} A_{ij}\alpha_i, \qquad j = 1, \ldots, n. \tag{3-5}$$

One must always remember that this matrix representing T depends upon the ordered basis $\mathcal{B}$, and that there is a representing matrix for T in each ordered basis for V. (For transformations of one space into another the matrix depends upon two ordered bases, one for V and one for W.) In order that we shall not forget this dependence, we shall use the notation

$$[T]_{\mathcal{B}}$$

for the matrix of the linear operator T in the ordered basis $\mathcal{B}$. The manner in which this matrix and the ordered basis describe T is that for each α in V

$$[T\alpha]_{\mathcal{B}} = [T]_{\mathcal{B}}[\alpha]_{\mathcal{B}}.$$

EXAMPLE 13. Let V be the space of $n \times 1$ column matrices over the field F; let W be the space of $m \times 1$ matrices over F; and let A be a fixed $m \times n$ matrix over F. Let T be the linear transformation of V into W defined by $T(X) = AX$. Let $\mathcal{B}$ be the ordered basis for V analogous to the standard basis in F^n, i.e., the ith vector in $\mathcal{B}$ in the $n \times 1$ matrix X_i with a 1 in row i and all other entries 0. Let $\mathcal{B}'$ be the corresponding ordered basis for W, i.e., the jth vector in $\mathcal{B}'$ is the $m \times 1$ matrix Y_j with a 1 in row j and all other entries 0. Then the matrix of T relative to the pair $\mathcal{B}$, $\mathcal{B}'$ is the matrix A itself. This is clear because the matrix AX_j is the jth column of A.

Example 14. Let F be a field and let T be the operator on F^2 defined by

$$T(x_1, x_2) = (x_1, 0).$$

It is easy to see that T is a linear operator on F^2. Let $\mathcal{B}$ be the standard ordered basis for F^2, $\mathcal{B} = \{\epsilon_1, \epsilon_2\}$. Now

$$T\epsilon_1 = T(1, 0) = (1, 0) = 1\epsilon_1 + 0\epsilon_2$$
$$T\epsilon_2 = T(0, 1) = (0, 0) = 0\epsilon_1 + 0\epsilon_2$$

so the matrix of T in the ordered basis $\mathcal{B}$ is

$$[T]_{\mathcal{B}} = \begin{bmatrix} 1 & 0 \\ 0 & 0 \end{bmatrix}.$$

Example 15. Let V be the space of all polynomial functions from R into R of the form

$$f(x) = c_0 + c_1x + c_2x^2 + c_3x^3$$

that is, the space of polynomial functions of degree three or less. The differentiation operator D of Example 2 maps V into V, since D is 'degree decreasing.' Let $\mathcal{B}$ be the ordered basis for V consisting of the four functions f_1, f_2, f_3, f_4 defined by $f_j(x) = x^{j-1}$. Then

$$\begin{aligned} (Df_1)(x) &= 0, & Df_1 &= 0f_1 + 0f_2 + 0f_3 + 0f_4 \\ (Df_2)(x) &= 1, & Df_2 &= 1f_1 + 0f_2 + 0f_3 + 0f_4 \\ (Df_3)(x) &= 2x, & Df_3 &= 0f_1 + 2f_2 + 0f_3 + 0f_4 \\ (Df_4)(x) &= 3x^2, & Df_4 &= 0f_1 + 0f_2 + 3f_3 + 0f_4 \end{aligned}$$

so that the matrix of D in the ordered basis $\mathcal{B}$ is

$$[D]_{\mathcal{B}} = \begin{bmatrix} 0 & 1 & 0 & 0 \\ 0 & 0 & 2 & 0 \\ 0 & 0 & 0 & 3 \\ 0 & 0 & 0 & 0 \end{bmatrix}.$$

We have seen what happens to representing matrices when transformations are added, namely, that the matrices add. We should now like to ask what happens when we compose transformations. More specifically, let V, W, and Z be vector spaces over the field F of respective dimensions n, m, and p. Let T be a linear transformation from V into W and U a linear transformation from W into Z. Suppose we have ordered bases

$$\mathcal{B} = \{\alpha_1, \ldots, \alpha_n\}, \qquad \mathcal{B}' = \{\beta_1, \ldots, \beta_m\}, \qquad \mathcal{B}'' = \{\gamma_1, \ldots, \gamma_p\}$$

for the respective spaces V, W, and Z. Let A be the matrix of T relative to the pair $\mathcal{B}$, $\mathcal{B}'$ and let B be the matrix of U relative to the pair $\mathcal{B}'$, $\mathcal{B}''$. It is then easy to see that the matrix C of the transformation UT relative to the pair $\mathcal{B}$, $\mathcal{B}''$ is the product of B and A; for, if α is any vector in V

$$[T\alpha]_{\mathcal{B}'} = A[\alpha]_{\mathcal{B}}$$
$$[U(T\alpha)]_{\mathcal{B}''} = B[T\alpha]_{\mathcal{B}'}$$

and so

$$[(UT)(\alpha)]_{\mathcal{B}''} = BA[\alpha]_{\mathcal{B}}$$

and hence, by the definition and uniqueness of the representing matrix, we must have $C = BA$. One can also see this by carrying out the computation

$$\begin{aligned}(UT)(\alpha_j) &= U(T\alpha_j)\\ &= U\left(\sum_{k=1}^{m} A_{kj}\beta_k\right)\\ &= \sum_{k=1}^{m} A_{kj}(U\beta_k)\\ &= \sum_{k=1}^{m} A_{kj}\sum_{i=1}^{p} B_{ik}\gamma_i\\ &= \sum_{i=1}^{p}\left(\sum_{k=1}^{m} B_{ik}A_{kj}\right)\gamma_i\end{aligned}$$

so that we must have

$$C_{ij} = \sum_{k=1}^{m} B_{ik}A_{kj}. \tag{3-6}$$

We motivated the definition (3-6) of matrix multiplication via operations on the rows of a matrix. One sees here that a very strong motivation for the definition is to be found in composing linear transformations. Let us summarize formally.

Theorem 13. *Let* V, W, *and* Z *be finite-dimensional vector spaces over the field* F; *let* T *be a linear transformation from* V *into* W *and* U *a linear transformation from* W *into* Z. *If* $\mathcal{B}$, $\mathcal{B}'$, *and* $\mathcal{B}''$ *are ordered bases for the spaces* V, W, *and* Z, *respectively, if* A *is the matrix of* T *relative to the pair* $\mathcal{B}$, $\mathcal{B}'$, *and* B *is the matrix of* U *relative to the pair* $\mathcal{B}'$, $\mathcal{B}''$, *then the matrix of the composition* UT *relative to the pair* $\mathcal{B}$, $\mathcal{B}''$ *is the product matrix* C = BA.

We remark that Theorem 13 gives a proof that matrix multiplication is associative—a proof which requires no calculations and is independent of the proof we gave in Chapter 1. We should also point out that we proved a special case of Theorem 13 in Example 12.

It is important to note that if T and U are linear operators on a space V and we are representing by a single ordered basis $\mathcal{B}$, then Theorem 13 assumes the simple form $[UT]_{\mathcal{B}} = [U]_{\mathcal{B}}[T]_{\mathcal{B}}$. Thus in this case, the correspondence which $\mathcal{B}$ determines between operators and matrices is not only a vector space isomorphism but also preserves products. A simple consequence of this is that the linear operator T is invertible if and only if $[T]_{\mathcal{B}}$ is an invertible matrix. For, the identity operator I is represented by the identity matrix in any ordered basis, and thus

$$UT = TU = I$$

is equivalent to

$$[U]_{\mathcal{B}}[T]_{\mathcal{B}} = [T]_{\mathcal{B}}[U]_{\mathcal{B}} = I.$$

Of course, when T is invertible

$$[T^{-1}]_{\mathcal{B}} = [T]_{\mathcal{B}}^{-1}.$$

Now we should like to inquire what happens to representing matrices when the ordered basis is changed. For the sake of simplicity, we shall consider this question only for linear operators on a space V, so that we can use a single ordered basis. The specific question is this. Let T be a linear operator on the finite-dimensional space V, and let

$$\mathcal{B} = \{\alpha_1, \ldots, \alpha_n\} \quad \text{and} \quad \mathcal{B}' = \{\alpha'_1, \ldots, \alpha'_n\}$$

be two ordered bases for V. How are the matrices $[T]_{\mathcal{B}}$ and $[T]_{\mathcal{B}'}$ related? As we observed in Chapter 2, there is a unique (invertible) $n \times n$ matrix P such that

$$[\alpha]_{\mathcal{B}} = P[\alpha]_{\mathcal{B}'} \tag{3-7}$$

for every vector α in V. It is the matrix $P = [P_1, \ldots, P_n]$ where $P_j = [\alpha'_j]_{\mathcal{B}}$. By definition

$$[T\alpha]_{\mathcal{B}} = [T]_{\mathcal{B}}[\alpha]_{\mathcal{B}}. \tag{3-8}$$

Applying (3-7) to the vector $T\alpha$, we have

$$[T\alpha]_{\mathcal{B}} = P[T\alpha]_{\mathcal{B}'}. \tag{3-9}$$

Combining (3-7), (3-8), and (3-9), we obtain

$$[T]_{\mathcal{B}}P[\alpha]_{\mathcal{B}'} = P[T\alpha]_{\mathcal{B}'}$$

or

$$P^{-1}[T]_{\mathcal{B}}P[\alpha]_{\mathcal{B}'} = [T\alpha]_{\mathcal{B}'}$$

and so it must be that

$$[T]_{\mathcal{B}'} = P^{-1}[T]_{\mathcal{B}}P. \tag{3-10}$$

This answers our question.

Before stating this result formally, let us observe the following. There is a unique linear operator U which carries $\mathcal{B}$ onto $\mathcal{B}'$, defined by

$$U\alpha_j = \alpha'_j, \qquad j = 1, \ldots, n.$$

This operator U is invertible since it carries a basis for V onto a basis for

V. The matrix P (above) is precisely the matrix of the operator U in the ordered basis $\mathcal{B}$. For, P is defined by

$$\alpha'_j = \sum_{i=1}^{n} P_{ij}\alpha_i$$

and since $U\alpha_j = \alpha'_j$, this equation can be written

$$U\alpha_j = \sum_{i=1}^{n} P_{ij}\alpha_i.$$

So $P = [U]_{\mathcal{B}}$, by definition.

Theorem 14. *Let* V *be a finite-dimensional vector space over the field* F, *and let*

$$\mathcal{B} = \{\alpha_1, \ldots, \alpha_n\} \quad and \quad \mathcal{B}' = \{\alpha'_1, \ldots, \alpha'_n\}$$

be ordered bases for V. *Suppose* T *is a linear operator on* V. *If* $P = [P_1, \ldots, P_n]$ *is the* $n \times n$ *matrix with columns* $P_j = [\alpha'_j]_{\mathcal{B}}$, *then*

$$[T]_{\mathcal{B}'} = P^{-1}[T]_{\mathcal{B}}P.$$

Alternatively, if U *is the invertible operator on* V *defined by* $U\alpha_j = \alpha'_j$, $j = 1, \ldots, n$, *then*

$$[T]_{\mathcal{B}'} = [U]_{\mathcal{B}}^{-1}[T]_{\mathcal{B}}[U]_{\mathcal{B}}.$$

EXAMPLE 16. Let T be the linear operator on R^2 defined by $T(x_1, x_2) = (x_1, 0)$. In Example 14 we showed that the matrix of T in the standard ordered basis $\mathcal{B} = \{\epsilon_1, \epsilon_2\}$ is

$$[T]_{\mathcal{B}} = \begin{bmatrix} 1 & 0 \\ 0 & 0 \end{bmatrix}.$$

Suppose $\mathcal{B}'$ is the ordered basis for R^2 consisting of the vectors $\epsilon'_1 = (1, 1)$, $\epsilon'_2 = (2, 1)$. Then

$$\begin{aligned} \epsilon'_1 &= \epsilon_1 + \epsilon_2 \\ \epsilon'_2 &= 2\epsilon_1 + \epsilon_2 \end{aligned}$$

so that P is the matrix

$$P = \begin{bmatrix} 1 & 2 \\ 1 & 1 \end{bmatrix}.$$

By a short computation

$$P^{-1} = \begin{bmatrix} -1 & 2 \\ 1 & -1 \end{bmatrix}.$$

Thus

$$\begin{aligned} [T]_{\mathcal{B}'} &= P^{-1}[T]_{\mathcal{B}}P \\ &= \begin{bmatrix} -1 & 2 \\ 1 & -1 \end{bmatrix}\begin{bmatrix} 1 & 0 \\ 0 & 0 \end{bmatrix}\begin{bmatrix} 1 & 2 \\ 1 & 1 \end{bmatrix} \\ &= \begin{bmatrix} -1 & 2 \\ 1 & -1 \end{bmatrix}\begin{bmatrix} 1 & 2 \\ 0 & 0 \end{bmatrix} \\ &= \begin{bmatrix} -1 & -2 \\ 1 & 2 \end{bmatrix}. \end{aligned}$$

We can easily check that this is correct because

$$\begin{aligned} T\epsilon_1' &= (1, 0) = -\epsilon_1' + \epsilon_2' \\ T\epsilon_2' &= (2, 0) = -2\epsilon_1' + 2\epsilon_2'. \end{aligned}$$

EXAMPLE 17. Let V be the space of polynomial functions from R into R which have 'degree' less than or equal to 3. As in Example 15, let D be the differentiation operator on V, and let

$$\mathfrak{B} = \{f_1, f_2, f_3, f_4\}$$

be the ordered basis for V defined by $f_i(x) = x^{i-1}$. Let t be a real number and define $g_i(x) = (x + t)^{i-1}$, that is

$$\begin{aligned} g_1 &= f_1 \\ g_2 &= tf_1 + f_2 \\ g_3 &= t^2f_1 + 2tf_2 + f_3 \\ g_4 &= t^3f_1 + 3t^2f_2 + 3tf_3 + f_4. \end{aligned}$$

Since the matrix

$$P = \begin{bmatrix} 1 & t & t^2 & t^3 \\ 0 & 1 & 2t & 3t^2 \\ 0 & 0 & 1 & 3t \\ 0 & 0 & 0 & 1 \end{bmatrix}$$

is easily seen to be invertible with

$$P^{-1} = \begin{bmatrix} 1 & -t & t^2 & -t^3 \\ 0 & 1 & -2t & 3t^2 \\ 0 & 0 & 1 & -3t \\ 0 & 0 & 0 & 1 \end{bmatrix}$$

it follows that $\mathfrak{B}' = \{g_1, g_2, g_3, g_4\}$ is an ordered basis for V. In Example 15, we found that the matrix of D in the ordered basis $\mathfrak{B}$ is

$$[D]_{\mathfrak{B}} = \begin{bmatrix} 0 & 1 & 0 & 0 \\ 0 & 0 & 2 & 0 \\ 0 & 0 & 0 & 3 \\ 0 & 0 & 0 & 0 \end{bmatrix}.$$

The matrix of D in the ordered basis $\mathfrak{B}'$ is thus

$$\begin{aligned} P^{-1}[D]_{\mathfrak{B}}P &= \begin{bmatrix} 1 & -t & t^2 & t^3 \\ 0 & 1 & -2t & 3t^2 \\ 0 & 0 & 1 & -3t \\ 0 & 0 & 0 & 1 \end{bmatrix} \begin{bmatrix} 0 & 1 & 0 & 0 \\ 0 & 0 & 2 & 0 \\ 0 & 0 & 0 & 3 \\ 0 & 0 & 0 & 0 \end{bmatrix} \begin{bmatrix} 1 & t & t^2 & t^3 \\ 0 & 1 & 2t & 3t^2 \\ 0 & 0 & 1 & 3t \\ 0 & 0 & 0 & 1 \end{bmatrix} \\ &= \begin{bmatrix} 1 & -t & t^2 & t^3 \\ 0 & 1 & -2t & 3t^2 \\ 0 & 0 & 1 & -3t \\ 0 & 0 & 0 & 1 \end{bmatrix} \begin{bmatrix} 0 & 1 & 2t & 3t^2 \\ 0 & 0 & 2 & 6t \\ 0 & 0 & 0 & 3 \\ 0 & 0 & 0 & 0 \end{bmatrix} \\ &= \begin{bmatrix} 0 & 1 & 0 & 0 \\ 0 & 0 & 2 & 0 \\ 0 & 0 & 0 & 3 \\ 0 & 0 & 0 & 0 \end{bmatrix}. \end{aligned}$$

Thus D is represented by the same matrix in the ordered bases $\mathcal{B}$ and $\mathcal{B}'$. Of course, one can see this somewhat more directly since

$$\begin{aligned} Dg_1 &= 0 \\ Dg_2 &= g_1 \\ Dg_3 &= 2g_2 \\ Dg_4 &= 3g_3. \end{aligned}$$

This example illustrates a good point. If one knows the matrix of a linear operator in some ordered basis $\mathcal{B}$ and wishes to find the matrix in another ordered basis $\mathcal{B}'$, it is often most convenient to perform the coordinate change using the invertible matrix P; however, it may be a much simpler task to find the representing matrix by a direct appeal to its definition.

Definition. *Let* A *and* B *be* n $\times$ n *(square) matrices over the field* F. *We say that* B **is similar to** A **over** F *if there is an invertible* n $\times$ n *matrix* P *over* F *such that* B = P^{-1}AP.

According to Theorem 14, we have the following: If V is an n-dimensional vector space over F and $\mathcal{B}$ and $\mathcal{B}'$ are two ordered bases for V, then for each linear operator T on V the matrix $B = [T]_{\mathcal{B}'}$ is similar to the matrix $A = [T]_{\mathcal{B}}$. The argument also goes in the other direction. Suppose A and B are $n \times n$ matrices and that B is similar to A. Let V be any n-dimensional space over F and let $\mathcal{B}$ be an ordered basis for V. Let T be the linear operator on V which is represented in the basis $\mathcal{B}$ by A. If $B = P^{-1}AP$, let $\mathcal{B}'$ be the ordered basis for V obtained from $\mathcal{B}$ by P, i.e.,

$$\alpha'_j = \sum_{i=1}^{n} P_{ij}\alpha_i.$$

Then the matrix of T in the ordered basis $\mathcal{B}'$ will be B.

Thus the statement that B is similar to A means that on each n-dimensional space over F the matrices A and B represent the same linear transformation in two (possibly) different ordered bases.

Note that each $n \times n$ matrix A is similar to itself, using $P = I$; if B is similar to A, then A is similar to B, for $B = P^{-1}AP$ implies that $A = (P^{-1})^{-1}BP^{-1}$; if B is similar to A and C is similar to B, then C is similar to A, for $B = P^{-1}AP$ and $C = Q^{-1}BQ$ imply that $C = (PQ)^{-1}A(PQ)$. Thus, similarity is an equivalence relation on the set of $n \times n$ matrices over the field F. Also note that the only matrix similar to the identity matrix I is I itself, and that the only matrix similar to the zero matrix is the zero matrix itself.

Exercises

1. Let T be the linear operator on C^2 defined by $T(x_1, x_2) = (x_1, 0)$. Let $\mathcal{B}$ be the standard ordered basis for C^2 and let $\mathcal{B}' = \{\alpha_1, \alpha_2\}$ be the ordered basis defined by $\alpha_1 = (1, i)$, $\alpha_2 = (-i, 2)$.

(a) What is the matrix of T relative to the pair $\mathcal{B}, \mathcal{B}'$?

(b) What is the matrix of T relative to the pair $\mathcal{B}', \mathcal{B}$?

(c) What is the matrix of T in the ordered basis $\mathcal{B}'$?

(d) What is the matrix of T in the ordered basis $\{\alpha_2, \alpha_1\}$?

2. Let T be the linear transformation from R^3 into R^2 defined by

$$T(x_1, x_2, x_3) = (x_1 + x_2, 2x_3 - x_1).$$

(a) If $\mathcal{B}$ is the standard ordered basis for R^3 and $\mathcal{B}'$ is the standard ordered basis for R^2, what is the matrix of T relative to the pair $\mathcal{B}, \mathcal{B}'$?

(b) If $\mathcal{B} = \{\alpha_1, \alpha_2, \alpha_3\}$ and $\mathcal{B}' = \{\beta_1, \beta_2\}$, where

$$\alpha_1 = (1, 0, -1), \quad \alpha_2 = (1, 1, 1), \quad \alpha_3 = (1, 0, 0), \quad \beta_1 = (0, 1), \quad \beta_2 = (1, 0)$$

what is the matrix of T relative to the pair $\mathcal{B}, \mathcal{B}'$?

3. Let T be a linear operator on F^n, let A be the matrix of T in the standard ordered basis for F^n, and let W be the subspace of F^n spanned by the column vectors of A. What does W have to do with T?

4. Let V be a two-dimensional vector space over the field F, and let $\mathcal{B}$ be an ordered basis for V. If T is a linear operator on V and

$$[T]_{\mathcal{B}} = \begin{bmatrix} a & b \\ c & d \end{bmatrix}$$

prove that $T^2 - (a + d)T + (ad - bc)I = 0$.

5. Let T be the linear operator on R^3, the matrix of which in the standard ordered basis is

$$A = \begin{bmatrix} 1 & 2 & 1 \\ 0 & 1 & 1 \\ -1 & 3 & 4 \end{bmatrix}.$$

Find a basis for the range of T and a basis for the null space of T.

6. Let T be the linear operator on R^2 defined by

$$T(x_1, x_2) = (-x_2, x_1).$$

(a) What is the matrix of T in the standard ordered basis for R^2?

(b) What is the matrix of T in the ordered basis $\mathcal{B} = \{\alpha_1, \alpha_2\}$, where $\alpha_1 = (1, 2)$ and $\alpha_2 = (1, -1)$?

(c) Prove that for every real number c the operator $(T - cI)$ is invertible.

(d) Prove that if $\mathcal{B}$ is any ordered basis for R^2 and $[T]_{\mathcal{B}} = A$, then $A_{12}A_{21} \neq 0$.

7. Let T be the linear operator on R^3 defined by

$$T(x_1, x_2, x_3) = (3x_1 + x_3, -2x_1 + x_2, -x_1 + 2x_2 + 4x_3).$$

(a) What is the matrix of T in the standard ordered basis for R^3?

(b) What is the matrix of T in the ordered basis

$$\{\alpha_1, \alpha_2, \alpha_3\}$$

where $\alpha_1 = (1, 0, 1)$, $\alpha_2 = (-1, 2, 1)$, and $\alpha_3 = (2, 1, 1)$?

(c) Prove that T is invertible and give a rule for T^{-1} like the one which defines T.

8. Let θ be a real number. Prove that the following two matrices are similar over the field of complex numbers:

$$\begin{bmatrix} \cos\theta & -\sin\theta \\ \sin\theta & \cos\theta \end{bmatrix}, \qquad \begin{bmatrix} e^{i\theta} & 0 \\ 0 & e^{-i\theta} \end{bmatrix}$$

(*Hint:* Let T be the linear operator on C^2 which is represented by the first matrix in the standard ordered basis. Then find vectors α_1 and α_2 such that $T\alpha_1 = e^{i\theta}\alpha_1$, $T\alpha_2 = e^{-i\theta}\alpha_2$, and $\{\alpha_1, \alpha_2\}$ is a basis.)

9. Let V be a finite-dimensional vector space over the field F and let S and T be linear operators on V. We ask: When do there exist ordered bases $\mathcal{B}$ and $\mathcal{B}'$ for V such that $[S]_{\mathcal{B}} = [T]_{\mathcal{B}'}$? Prove that such bases exist if and only if there is an invertible linear operator U on V such that $T = USU^{-1}$. (*Outline of proof:* If $[S]_{\mathcal{B}} = [T]_{\mathcal{B}'}$, let U be the operator which carries $\mathcal{B}$ onto $\mathcal{B}'$ and show that $S = UTU^{-1}$. Conversely, if $T = USU^{-1}$ for some invertible U, let $\mathcal{B}$ be any ordered basis for V and let $\mathcal{B}'$ be its image under U. Then show that $[S]_{\mathcal{B}} = [T]_{\mathcal{B}'}$.)

10. We have seen that the linear operator T on R^2 defined by $T(x_1, x_2) = (x_1, 0)$ is represented in the standard ordered basis by the matrix

$$A = \begin{bmatrix} 1 & 0 \\ 0 & 0 \end{bmatrix}.$$

This operator satisfies $T^2 = T$. Prove that if S is a linear operator on R^2 such that $S^2 = S$, then $S = 0$, or $S = I$, or there is an ordered basis $\mathcal{B}$ for R^2 such that $[S]_{\mathcal{B}} = A$ (above).

11. Let W be the space of all $n \times 1$ column matrices over a field F. If A is an $n \times n$ matrix over F, then A defines a linear operator L_A on W through left multiplication: $L_A(X) = AX$. Prove that every linear operator on W is left multiplication by some $n \times n$ matrix, i.e., is L_A for some A.

Now suppose V is an n-dimensional vector space over the field F, and let $\mathcal{B}$ be an ordered basis for V. For each α in V, define $U\alpha = [\alpha]_{\mathcal{B}}$. Prove that U is an isomorphism of V onto W. If T is a linear operator on V, then UTU^{-1} is a linear operator on W. Accordingly, UTU^{-1} is left multiplication by some $n \times n$ matrix A. What is A?

12. Let V be an n-dimensional vector space over the field F, and let $\mathcal{B} = \{\alpha_1, \ldots, \alpha_n\}$ be an ordered basis for V.

(a) According to Theorem 1, there is a unique linear operator T on V such that

$$T\alpha_j = \alpha_{j+1}, \qquad j = 1, \ldots, n-1, \qquad T\alpha_n = 0.$$

What is the matrix A of T in the ordered basis $\mathcal{B}$?

(b) Prove that $T^n = 0$ but $T^{n-1} \neq 0$.

(c) Let S be any linear operator on V such that $S^n = 0$ but $S^{n-1} \neq 0$. Prove that there is an ordered basis $\mathcal{B}'$ for V such that the matrix of S in the ordered basis $\mathcal{B}'$ is the matrix A of part (a).

(d) Prove that if M and N are $n \times n$ matrices over F such that $M^n = N^n = 0$ but $M^{n-1} \neq 0 \neq N^{n-1}$, then M and N are similar.

13. Let V and W be finite-dimensional vector spaces over the field F and let T be a linear transformation from V into W. If

$$\mathfrak{B} = \{\alpha_1, \ldots, \alpha_n\} \quad \text{and} \quad \mathfrak{B}' = \{\beta_1, \ldots, \beta_m\}$$

are ordered bases for V and W, respectively, define the linear transformations $E^{p,q}$ as in the proof of Theorem 5: $E^{p,q}(\alpha_i) = \delta_{iq}\beta_p$. Then the $E^{p,q}$, $1 \leq p \leq m$, $1 \leq q \leq n$, form a basis for $L(V, W)$, and so

$$T = \sum_{p=1}^{m} \sum_{q=1}^{n} A_{pq}E^{p,q}$$

for certain scalars A_{pq} (the coordinates of T in this basis for $L(V, W)$). Show that the matrix A with entries $A(p, q) = A_{pq}$ is precisely the matrix of T relative to the pair $\mathfrak{B}$, $\mathfrak{B}'$.

3.5. Linear Functionals

If V is a vector space over the field F, a linear transformation f from V into the scalar field F is also called a **linear functional** on V. If we start from scratch, this means that f is a function from V into F such that

$$f(c\alpha + \beta) = cf(\alpha) + f(\beta)$$

for all vectors α and β in V and all scalars c in F. The concept of linear functional is important in the study of finite-dimensional spaces because it helps to organize and clarify the discussion of subspaces, linear equations, and coordinates.

Example 18. Let F be a field and let $a_1, \ldots, a_n$ be scalars in F. Define a function f on F^n by

$$f(x_1, \ldots, x_n) = a_1x_1 + \cdots + a_nx_n.$$

Then f is a linear functional on F^n. It is the linear functional which is represented by the matrix $[a_1 \cdots a_n]$ relative to the standard ordered basis for F^n and the basis $\{1\}$ for F:

$$a_j = f(\epsilon_j), \qquad j = 1, \ldots, n.$$

Every linear functional on F^n is of this form, for some scalars $a_1, \ldots, a_n$. That is immediate from the definition of linear functional because we define $a_j = f(\epsilon_j)$ and use the linearity

$$\begin{aligned} f(x_1, \ldots, x_n) &= f\left(\sum_j x_j\epsilon_j\right) \\ &= \sum_j x_jf(\epsilon_j) \\ &= \sum_j a_jx_j. \end{aligned}$$

EXAMPLE 19. Here is an important example of a linear functional. Let n be a positive integer and F a field. If A is an $n \times n$ matrix with entries in F, the **trace** of A is the scalar

$$\operatorname{tr} A = A_{11} + A_{22} + \cdots + A_{nn}.$$

The trace function is a linear functional on the matrix space $F^{n\times n}$ because

$$\begin{aligned}\operatorname{tr}(cA + B) &= \sum_{i=1}^{n} (cA_{ii} + B_{ii})\\ &= c\sum_{i=1}^{n} A_{ii} + \sum_{i=1}^{n} B_{ii}\\ &= c \operatorname{tr} A + \operatorname{tr} B.\end{aligned}$$

EXAMPLE 20. Let V be the space of all polynomial functions from the field F into itself. Let t be an element of F. If we define

$$L_t(p) = p(t)$$

then L_t is a linear functional on V. One usually describes this by saying that, for each t, 'evaluation at t' is a linear functional on the space of polynomial functions. Perhaps we should remark that the fact that the functions are polynomials plays no role in this example. Evaluation at t is a linear functional on the space of all functions from F into F.

EXAMPLE 21. This may be the most important linear functional in mathematics. Let $[a, b]$ be a closed interval on the real line and let $C([a, b])$ be the space of continuous real-valued functions on $[a, b]$. Then

$$L(g) = \int_a^b g(t)\,dt$$

defines a linear functional L on $C([a, b])$.

If V is a vector space, the collection of all linear functionals on V forms a vector space in a natural way. It is the space $L(V, F)$. We denote this space by V^* and call it the **dual space** of V:

$$V^* = L(V, F).$$

If V is finite-dimensional, we can obtain a rather explicit description of the dual space V^*. From Theorem 5 we know something about the space V^*, namely that

$$\dim V^* = \dim V.$$

Let $\mathcal{B} = \{\alpha_1, \ldots, \alpha_n\}$ be a basis for V. According to Theorem 1, there is (for each i) a unique linear functional f_i on V such that

$$f_i(\alpha_j) = \delta_{ij}. \tag{3-11}$$

In this way we obtain from $\mathcal{B}$ a set of n distinct linear functionals $f_1, \ldots, f_n$ on V. These functionals are also linearly independent. For, suppose

(3-12)
$$f = \sum_{i=1}^{n} c_i f_i.$$

Then

$$\begin{aligned} f(\alpha_j) &= \sum_{i=1}^{n} c_i f_i(\alpha_j) \\ &= \sum_{i=1}^{n} c_i \delta_{ij} \\ &= c_j. \end{aligned}$$

In particular, if f is the zero functional, $f(\alpha_j) = 0$ for each j and hence the scalars c_j are all 0. Now $f_1, \ldots, f_n$ are n linearly independent functionals, and since we know that V^* has dimension n, it must be that $\mathcal{B}^* = \{f_1, \ldots, f_n\}$ is a basis for V^*. This basis is called the **dual basis** of $\mathcal{B}$.

Theorem 15. *Let* V *be a finite-dimensional vector space over the field* F, *and let* $\mathcal{B} = \{\alpha_1, \ldots, \alpha_n\}$ *be a basis for* V. *Then there is a unique dual basis* $\mathcal{B}^* = \{f_1, \ldots, f_n\}$ *for* V^* *such that* $f_i(\alpha_j) = \delta_{ij}$. *For each linear functional* f *on* V *we have*

(3-13)
$$f = \sum_{i=1}^{n} f(\alpha_i) f_i$$

and for each vector α *in* V *we have*

(3-14)
$$\alpha = \sum_{i=1}^{n} f_i(\alpha)\alpha_i.$$

Proof. We have shown above that there is a unique basis which is 'dual' to $\mathcal{B}$. If f is a linear functional on V, then f is some linear combination (3-12) of the f_i, and as we observed after (3-12) the scalars c_j must be given by $c_j = f(\alpha_j)$. Similarly, if

$$\alpha = \sum_{i=1}^{n} x_i \alpha_i$$

is a vector in V, then

$$\begin{aligned} f_j(\alpha) &= \sum_{i=1}^{n} x_i f_j(\alpha_i) \\ &= \sum_{i=1}^{n} x_i \delta_{ij} \\ &= x_j \end{aligned}$$

so that the unique expression for α as a linear combination of the α_i is

$$\alpha = \sum_{i=1}^{n} f_i(\alpha)\alpha_i. \quad \blacksquare$$

Equation (3-14) provides us with a nice way of describing what the dual basis is. It says, if $\mathcal{B} = \{\alpha_1, \ldots, \alpha_n\}$ is an ordered basis for V and

$\mathfrak{B}^* = \{f_1, \ldots, f_n\}$ is the dual basis, then f_i is precisely the function which assigns to each vector α in V the ith coordinate of α relative to the ordered basis $\mathfrak{B}$. Thus we may also call the f_i the coordinate functions for $\mathfrak{B}$. The formula (3-13), when combined with (3-14) tells us the following: If f is in V^*, and we let $f(\alpha_i) = a_i$, then when

$$\alpha = x_1\alpha_1 + \cdots + x_n\alpha_n$$

we have

$$f(\alpha) = a_1x_1 + \cdots + a_nx_n. \tag{3-15}$$

In other words, if we choose an ordered basis $\mathfrak{B}$ for V and describe each vector in V by its n-tuple of coordinates $(x_1, \ldots, x_n)$ relative to $\mathfrak{B}$, then every linear functional on V has the form (3-15). This is the natural generalization of Example 18, which is the special case $V = F^n$ and $\mathfrak{B} = \{\epsilon_1, \ldots, \epsilon_n\}$.

Example 22. Let V be the vector space of all polynomial functions from R into R which have degree less than or equal to 2. Let t_1, t_2, and t_3 be any three *distinct* real numbers, and let

$$L_i(p) = p(t_i).$$

Then L_1, L_2, and L_3 are linear functionals on V. These functionals are linearly independent; for, suppose

$$L = c_1L_1 + c_2L_2 + c_3L_3.$$

If $L = 0$, i.e., if $L(p) = 0$ for each p in V, then applying L to the particular polynomial 'functions' 1, x, x^2, we obtain

$$\begin{aligned} c_1 + c_2 + c_3 &= 0 \\ t_1c_1 + t_2c_2 + t_3c_3 &= 0 \\ t_1^2c_1 + t_2^2c_2 + t_3^2c_3 &= 0 \end{aligned}$$

From this it follows that $c_1 = c_2 = c_3 = 0$, because (as a short computation shows) the matrix

$$\begin{bmatrix} 1 & 1 & 1 \\ t_1 & t_2 & t_3 \\ t_1^2 & t_2^2 & t_3^2 \end{bmatrix}$$

is invertible when t_1, t_2, and t_3 are distinct. Now the L_i are independent, and since V has dimension 3, these functionals form a basis for V^*. What is the basis for V, of which this is the dual? Such a basis $\{p_1, p_2, p_3\}$ for V must satisfy

$$L_i(p_j) = \delta_{ij}$$

or

$$p_j(t_i) = \delta_{ij}.$$

These polynomial functions are rather easily seen to be

$$p_1(x) = \frac{(x - t_2)(x - t_3)}{(t_1 - t_2)(t_1 - t_3)}$$

$$p_2(x) = \frac{(x - t_1)(x - t_3)}{(t_2 - t_1)(t_2 - t_3)}$$

$$p_3(x) = \frac{(x - t_1)(x - t_2)}{(t_3 - t_1)(t_3 - t_2)}.$$

The basis $\{p_1, p_2, p_3\}$ for V is interesting, because according to (3-14) we have for each p in V

$$p = p(t_1)p_1 + p(t_2)p_2 + p(t_3)p_3.$$

Thus, if c_1, c_2, and c_3 are any real numbers, there is exactly one polynomial function p over R which has degree at most 2 and satisfies $p(t_j) = c_j$, $j = 1, 2, 3$. This polynomial function is $p = c_1p_1 + c_2p_2 + c_3p_3$.

Now let us discuss the relationship between linear functionals and subspaces. If f is a non-zero linear functional, then the rank of f is 1 because the range of f is a non-zero subspace of the scalar field and must (therefore) be the scalar field. If the underlying space V is finite-dimensional, the rank plus nullity theorem (Theorem 2) tells us that the null space N_f has dimension

$$\dim N_f = \dim V - 1.$$

In a vector space of dimension n, a subspace of dimension $n - 1$ is called a **hyperspace.** Such spaces are sometimes called hyperplanes or subspaces of codimension 1. Is every hyperspace the null space of a linear functional? The answer is easily seen to be yes. It is not much more difficult to show that each d-dimensional subspace of an n-dimensional space is the intersection of the null spaces of $(n - d)$ linear functionals (Theorem 16 below).

Definition. *If* V *is a vector space over the field* F *and* S *is a subset of* V, *the* **annihilator** *of* S *is the set* S^0 *of linear functionals* f *on* V *such that* $f(\alpha) = 0$ *for every* α *in* S.

It should be clear to the reader that S^0 is a subspace of V^*, whether S is a subspace of V or not. If S is the set consisting of the zero vector alone, then $S^0 = V^*$. If $S = V$, then S^0 is the zero subspace of V^*. (This is easy to see when V is finite-dimensional.)

Theorem 16. *Let* V *be a finite-dimensional vector space over the field* F, *and let* W *be a subspace of* V. *Then*

$$\dim W + \dim W^0 = \dim V.$$

Proof. Let k be the dimension of W and $\{\alpha_1, \ldots, \alpha_k\}$ a basis for W. Choose vectors $\alpha_{k+1}, \ldots, \alpha_n$ in V such that $\{\alpha_1, \ldots, \alpha_n\}$ is a basis for V. Let $\{f_1, \ldots, f_n\}$ be the basis for V^* which is dual to this basis for V.

The claim is that $\{f_{k+1}, \ldots, f_n\}$ is a basis for the annihilator W^0. Certainly f_i belongs to W^0 for $i \geq k + 1$, because

$$f_i(\alpha_j) = \delta_{ij}$$

and $\delta_{ij} = 0$ if $i \geq k + 1$ and $j \leq k$; from this it follows that, for $i \geq k + 1$, $f_i(\alpha) = 0$ whenever α is a linear combination of $\alpha_1, \ldots, \alpha_k$. The functionals $f_{k+1}, \ldots, f_n$ are independent, so all we must show is that they span W^0. Suppose f is in V^*. Now

$$f = \sum_{i=1}^{n} f(\alpha_i)f_i$$

so that if f is in W^0 we have $f(\alpha_i) = 0$ for $i \leq k$ and

$$f = \sum_{i=k+1}^{n} f(\alpha_i)f_i.$$

We have shown that if $\dim W = k$ and $\dim V = n$ then $\dim W^0 = n - k$. ∎

Corollary. *If* W *is a* k*-dimensional subspace of an* n*-dimensional vector space* V, *then* W *is the intersection of* (n − k) *hyperspaces in* V.

Proof. This is a corollary of the proof of Theorem 16 rather than its statement. In the notation of the proof, W is exactly the set of vectors α such that $f_i(\alpha) = 0$, $i = k + 1, \ldots, n$. In case $k = n - 1$, W is the null space of f_n. ∎

Corollary. *If* W_1 *and* W_2 *are subspaces of a finite-dimensional vector space, then* $W_1 = W_2$ *if and only if* $W_1^0 = W_2^0$.

Proof. If $W_1 = W_2$, then of course $W_1^0 = W_2^0$. If $W_1 \neq W_2$, then one of the two subspaces contains a vector which is not in the other. Suppose there is a vector α which is in W_2 but not in W_1. By the previous corollaries (or the proof of Theorem 16) there is a linear functional f such that $f(\beta) = 0$ for all β in W, but $f(\alpha) \neq 0$. Then f is in W_1^0 but not in W_2^0 and $W_1^0 \neq W_2^0$. ∎

In the next section we shall give different proofs for these two corollaries. The first corollary says that, if we select some ordered basis for the space, each k-dimensional subspace can be described by specifying $(n - k)$ homogeneous linear conditions on the coordinates relative to that basis.

Let us look briefly at systems of homogeneous linear equations from the point of view of linear functionals. Suppose we have a system of linear equations,

$$\begin{aligned} A_{11}x_1 + \cdots + A_{1n}x_n &= 0 \\ \vdots \qquad\qquad & \;\vdots \\ A_{m1}x_1 + \cdots + A_{mn}x_n &= 0 \end{aligned}$$

for which we wish to find the solutions. If we let f_i, $i = 1, \ldots, m$, be the linear functional on F^n defined by

$$f_i(x_1, \ldots, x_n) = A_{i1}x_1 + \cdots + A_{in}x_n$$

then we are seeking the subspace of F^n of all α such that

$$f_i(\alpha) = 0, \qquad i = 1, \ldots, m.$$

In other words, we are seeking the subspace annihilated by $f_1, \ldots, f_m$. Row-reduction of the coefficient matrix provides us with a systematic method of finding this subspace. The n-tuple $(A_{i1}, \ldots, A_{in})$ gives the coordinates of the linear functional f_i relative to the basis which is dual to the standard basis for F^n. The row space of the coefficient matrix may thus be regarded as the space of linear functionals spanned by $f_1, \ldots, f_m$. The solution space is the subspace annihilated by this space of functionals.

Now one may look at the system of equations from the 'dual' point of view. That is, suppose that we are given m vectors in F^n

$$\alpha_i = (A_{i1}, \ldots, A_{in})$$

and we wish to find the annihilator of the subspace spanned by these vectors. Since a typical linear functional on F^n has the form

$$f(x_1, \ldots, x_n) = c_1x_1 + \cdots + c_nx_n$$

the condition that f be in this annihilator is that

$$\sum_{j=1}^{n} A_{ij}c_j = 0, \qquad i = 1, \ldots, m$$

that is, that $(c_1, \ldots, c_n)$ be a solution of the system $AX = 0$. From this point of view, row-reduction gives us a systematic method of finding the annihilator of the subspace spanned by a given finite set of vectors in F^n.

EXAMPLE 23. Here are three linear functionals on R^4:

$$\begin{aligned} f_1(x_1, x_2, x_3, x_4) &= x_1 + 2x_2 + 2x_3 + x_4 \\ f_2(x_1, x_2, x_3, x_4) &= 2x_2 + x_4 \\ f_3(x_1, x_2, x_3, x_4) &= -2x_1 - 4x_3 + 3x_4. \end{aligned}$$

The subspace which they annihilate may be found explicitly by finding the row-reduced echelon form of the matrix

$$A = \begin{bmatrix} 1 & 2 & 2 & 1 \\ 0 & 2 & 0 & 1 \\ -2 & 0 & -4 & 3 \end{bmatrix}.$$

A short calculation, or a peek at Example 21 of Chapter 2, shows that

$$R = \begin{bmatrix} 1 & 0 & 2 & 0 \\ 0 & 1 & 0 & 0 \\ 0 & 0 & 0 & 1 \end{bmatrix}.$$

Therefore, the linear functionals

$$\begin{aligned} g_1(x_1, x_2, x_3, x_4) &= x_1 + 2x_3 \\ g_2(x_1, x_2, x_3, x_4) &= x_2 \\ g_3(x_1, x_2, x_3, x_4) &= x_4 \end{aligned}$$

span the same subspace of $(R^4)^*$ and annihilate the same subspace of R^4 as do f_1, f_2, f_3. The subspace annihilated consists of the vectors with

$$\begin{aligned} x_1 &= -2x_3 \\ x_2 &= x_4 = 0. \end{aligned}$$

EXAMPLE 24. Let W be the subspace of R^5 which is spanned by the vectors

$$\begin{aligned} \alpha_1 &= (2, -2, 3, 4, -1), & \alpha_3 &= (0, 0, -1, -2, 3) \\ \alpha_2 &= (-1, 1, 2, 5, 2), & \alpha_4 &= (1, -1, 2, 3, 0). \end{aligned}$$

How does one describe W^0, the annihilator of W? Let us form the 4×5 matrix A with row vectors $\alpha_1, \alpha_2, \alpha_3, \alpha_4$, and find the row-reduced echelon matrix R which is row-equivalent to A:

$$A = \begin{bmatrix} 2 & -2 & 3 & 4 & -1 \\ -1 & 1 & 2 & 5 & 2 \\ 0 & 0 & -1 & -2 & 3 \\ 1 & -1 & 2 & 3 & 0 \end{bmatrix} \longrightarrow R = \begin{bmatrix} 1 & -1 & 0 & -1 & 0 \\ 0 & 0 & 1 & 2 & 0 \\ 0 & 0 & 0 & 0 & 1 \\ 0 & 0 & 0 & 0 & 0 \end{bmatrix}.$$

If f is a linear functional on R^5:

$$f(x_1, \ldots, x_5) = \sum_{j=1}^{5} c_j x_j$$

then f is in W^0 if and only if $f(\alpha_i) = 0$, $i = 1, 2, 3, 4$, i.e., if and only if

$$\sum_{j=1}^{5} A_{ij} c_j = 0, \qquad 1 \leq i \leq 4.$$

This is equivalent to

$$\sum_{j=1}^{5} R_{ij} c_j = 0, \qquad 1 \leq i \leq 3$$

or

$$\begin{aligned} c_1 - c_2 - c_4 &= 0 \\ c_3 + 2c_4 &= 0 \\ c_5 &= 0. \end{aligned}$$

We obtain all such linear functionals f by assigning arbitrary values to c_2 and c_4, say $c_2 = a$ and $c_4 = b$, and then finding the corresponding $c_1 = a + b$, $c_3 = -2b$, $c_5 = 0$. So W^0 consists of all linear functionals f of the form

$$f(x_1, x_2, x_3, x_4, x_5) = (a + b)x_1 + ax_2 - 2bx_3 + bx_4.$$

The dimension of W^0 is 2 and a basis $\{f_1, f_2\}$ for W^0 can be found by first taking $a = 1$, $b = 0$ and then $a = 0$, $b = 1$:

$$f_1(x_1, \ldots, x_5) = x_1 + x_2$$
$$f_2(x_1, \ldots, x_5) = x_1 - 2x_3 + x_4.$$

The above general f in W^0 is $f = af_1 + bf_2$.

Exercises

1. In R^3, let $\alpha_1 = (1, 0, 1)$, $\alpha_2 = (0, 1, -2)$, $\alpha_3 = (-1, -1, 0)$.
   (a) If f is a linear functional on R^3 such that

$$f(\alpha_1) = 1, \qquad f(\alpha_2) = -1, \qquad f(\alpha_3) = 3,$$

and if $\alpha = (a, b, c)$, find $f(\alpha)$.
   (b) Describe explicitly a linear functional f on R^3 such that

$$f(\alpha_1) = f(\alpha_2) = 0 \quad \text{but} \quad f(\alpha_3) \neq 0.$$

   (c) Let f be any linear functional such that

$$f(\alpha_1) = f(\alpha_2) = 0 \quad \text{and} \quad f(\alpha_3) \neq 0.$$

If $\alpha = (2, 3, -1)$, show that $f(\alpha) \neq 0$.

2. Let $\mathcal{B} = \{\alpha_1, \alpha_2, \alpha_3\}$ be the basis for C^3 defined by

$$\alpha_1 = (1, 0, -1), \qquad \alpha_2 = (1, 1, 1), \qquad \alpha_3 = (2, 2, 0).$$

Find the dual basis of $\mathcal{B}$.

3. If A and B are $n \times n$ matrices over the field F, show that trace (AB) = trace (BA). Now show that similar matrices have the same trace.

4. Let V be the vector space of all polynomial functions p from R into R which have degree 2 or less:

$$p(x) = c_0 + c_1x + c_2x^2.$$

Define three linear functionals on V by

$$f_1(p) = \int_0^1 p(x)\,dx, \qquad f_2(p) = \int_0^2 p(x)\,dx, \qquad f_3(p) = \int_0^{-1} p(x)\,dx.$$

Show that $\{f_1, f_2, f_3\}$ is a basis for V^* by exhibiting the basis for V of which it is the dual.

5. If A and B are $n \times n$ complex matrices, show that $AB - BA = I$ is impossible.

6. Let m and n be positive integers and F a field. Let $f_1, \ldots, f_m$ be linear functionals on F^n. For α in F^n define

$$T\alpha = (f_1(\alpha), \ldots, f_m(\alpha)).$$

Show that T is a linear transformation from F^n into F^m. Then show that every linear transformation from F^n into F^m is of the above form, for some $f_1, \ldots, f_m$.

7. Let $\alpha_1 = (1, 0, -1, 2)$ and $\alpha_2 = (2, 3, 1, 1)$, and let W be the subspace of R^4 spanned by α_1 and α_2. Which linear functionals f:

$$f(x_1, x_2, x_3, x_4) = c_1x_1 + c_2x_2 + c_3x_3 + c_4x_4$$

are in the annihilator of W?

8. Let W be the subspace of R^5 which is spanned by the vectors

$$\alpha_1 = \epsilon_1 + 2\epsilon_2 + \epsilon_3, \qquad \alpha_2 = \epsilon_2 + 3\epsilon_3 + 3\epsilon_4 + \epsilon_5$$
$$\alpha_3 = \epsilon_1 + 4\epsilon_2 + 6\epsilon_3 + 4\epsilon_4 + \epsilon_5.$$

Find a basis for W^0.

9. Let V be the vector space of all 2×2 matrices over the field of real numbers, and let

$$B = \begin{bmatrix} 2 & -2 \\ -1 & 1 \end{bmatrix}.$$

Let W be the subspace of V consisting of all A such that $AB = 0$. Let f be a linear functional on V which is in the annihilator of W. Suppose that $f(I) = 0$ and $f(C) = 3$, where I is the 2×2 identity matrix and

$$C = \begin{bmatrix} 0 & 0 \\ 0 & 1 \end{bmatrix}.$$

Find $f(B)$.

10. Let F be a subfield of the complex numbers. We define n linear functionals on F^n $(n \geq 2)$ by

$$f_k(x_1, \ldots, x_n) = \sum_{j=1}^{n} (k - j)x_j, \qquad 1 \leq k \leq n.$$

What is the dimension of the subspace annihilated by $f_1, \ldots, f_n$?

11. Let W_1 and W_2 be subspaces of a finite-dimensional vector space V.

(a) Prove that $(W_1 + W_2)^0 = W_1^0 \cap W_2^0$.

(b) Prove that $(W_1 \cap W_2)^0 = W_1^0 + W_2^0$.

12. Let V be a finite-dimensional vector space over the field F and let W be a subspace of V. If f is a linear functional on W, prove that there is a linear functional g on V such that $g(\alpha) = f(\alpha)$ for each α in the subspace W.

13. Let F be a subfield of the field of complex numbers and let V be any vector space over F. Suppose that f and g are linear functionals on V such that the function h defined by $h(\alpha) = f(\alpha)g(\alpha)$ is also a linear functional on V. Prove that either $f = 0$ or $g = 0$.

14. Let F be a field of characteristic zero and let V be a finite-dimensional vector space over F. If $\alpha_1, \ldots, \alpha_m$ are finitely many vectors in V, each different from the zero vector, prove that there is a linear functional f on V such that

$$f(\alpha_i) \neq 0, \qquad i = 1, \ldots, m.$$

15. According to Exercise 3, similar matrices have the same trace. Thus we can define the trace of a linear operator on a finite-dimensional space to be the trace of any matrix which represents the operator in an ordered basis. This is well-defined since all such representing matrices for one operator are similar.

Now let V be the space of all 2×2 matrices over the field F and let P be a fixed 2×2 matrix. Let T be the linear operator on V defined by $T(A) = PA$. Prove that trace $(T) = 2$ trace (P).

16. Show that the trace functional on $n \times n$ matrices is unique in the following sense. If W is the space of $n \times n$ matrices over the field F and if f is a linear functional on W such that $f(AB) = f(BA)$ for each A and B in W, then f is a scalar multiple of the trace function. If, in addition, $f(I) = n$, then f is the trace function.

17. Let W be the space of $n \times n$ matrices over the field F, and let W_0 be the subspace spanned by the matrices C of the form $C = AB - BA$. Prove that W_0 is exactly the subspace of matrices which have trace zero. (*Hint:* What is the dimension of the space of matrices of trace zero? Use the matrix 'units,' i.e., matrices with exactly one non-zero entry, to construct enough linearly independent matrices of the form $AB - BA$.)

3.6. The Double Dual

One question about dual bases which we did not answer in the last section was whether every basis for V^* is the dual of some basis for V. One way to answer that question is to consider V^{**}, the dual space of V^*.

If α is a vector in V, then α induces a linear functional L_α on V^* defined by

$$L_\alpha(f) = f(\alpha), \qquad f \text{ in } V^*.$$

The fact that L_α is linear is just a reformulation of the definition of linear operations in V^*:

$$\begin{aligned} L_\alpha(cf + g) &= (cf + g)(\alpha) \\ &= (cf)(\alpha) + g(\alpha) \\ &= cf(\alpha) + g(\alpha) \\ &= cL_\alpha(f) + L_\alpha(g). \end{aligned}$$

If V is finite-dimensional and $\alpha \neq 0$, then $L_\alpha \neq 0$; in other words, there exists a linear functional f such that $f(\alpha) \neq 0$. The proof is very simple and was given in Section 3.5: Choose an ordered basis $\mathcal{B} = \{\alpha_1, \ldots, \alpha_n\}$ for V such that $\alpha_1 = \alpha$ and let f be the linear functional which assigns to each vector in V its first coordinate in the ordered basis $\mathcal{B}$.

Theorem 17. *Let* V *be a finite-dimensional vector space over the field* F. *For each vector* α *in* V *define*

$$\mathrm{L}_\alpha(\mathrm{f}) = \mathrm{f}(\alpha), \qquad \mathrm{f} \textit{ in } \mathrm{V}^*.$$

The mapping $\alpha \rightarrow \mathrm{L}_\alpha$ *is then an isomorphism of* V *onto* V^{**}.

Proof. We showed that for each α the function L_α is linear. Suppose α and β are in V and c is in F, and let $\gamma = c\alpha + \beta$. Then for each f in V^*

$$\begin{aligned} L_\gamma(f) &= f(\gamma) \\ &= f(c\alpha + \beta) \\ &= cf(\alpha) + f(\beta) \\ &= cL_\alpha(f) + L_\beta(f) \end{aligned}$$

and so

$$L_\gamma = cL_\alpha + L_\beta.$$

This shows that the mapping $\alpha \to L_\alpha$ is a linear transformation from V into V^{**}. This transformation is non-singular; for, according to the remarks above $L_\alpha = 0$ if and only if $\alpha = 0$. Now $\alpha \to L_\alpha$ is a non-singular linear transformation from V into V^{**}, and since

$$\dim V^{**} = \dim V^* = \dim V$$

Theorem 9 tells us that this transformation is invertible, and is therefore an isomorphism of V onto V^{**}. ∎

Corollary. *Let V be a finite-dimensional vector space over the field F. If L is a linear functional on the dual space V* of V, then there is a unique vector α in V such that*

$$L(f) = f(\alpha)$$

for every f in V.*

Corollary. *Let V be a finite-dimensional vector space over the field F. Each basis for V* is the dual of some basis for V.*

Proof. Let $\mathcal{B}^* = \{f_1, \ldots, f_n\}$ be a basis for V^*. By Theorem 15, there is a basis $\{L_1, \ldots, L_n\}$ for V^{**} such that

$$L_i(f_j) = \delta_{ij}.$$

Using the corollary above, for each i there is a vector α_i in V such that

$$L_i(f) = f(\alpha_i)$$

for every f in V^*, i.e., such that $L_i = L_{\alpha_i}$. It follows immediately that $\{\alpha_1, \ldots, \alpha_n\}$ is a basis for V and that $\mathcal{B}^*$ is the dual of this basis. ∎

In view of Theorem 17, we usually identify α with L_α and say that V 'is' the dual space of V^* or that the spaces V, V^* are naturally in duality with one another. Each is the dual space of the other. In the last corollary we have an illustration of how that can be useful. Here is a further illustration.

If E is a subset of V^*, then the annihilator E^0 is (technically) a subset of V^{**}. If we choose to identify V and V^{**} as in Theorem 17, then E^0 is a subspace of V, namely, the set of all α in V such that $f(\alpha) = 0$ for all f in E. In a corollary of Theorem 16 we noted that each subspace W is determined by its annihilator W^0. How is it determined? The answer is that W is the subspace annihilated by all f in W^0, that is, the intersection of the null spaces of all f's in W^0. In our present notation for annihilators, the answer may be phrased very simply: $W = (W^0)^0$.

Theorem 18. *If S is any subset of a finite-dimensional vector space V, then $(S^0)^0$ is the subspace spanned by S.*

Proof. Let W be the subspace spanned by S. Clearly $W^0 = S^0$. Therefore, what we are to prove is that $W = W^{00}$. We have given one proof. Here is another. By Theorem 16

$$\dim W + \dim W^0 = \dim V$$
$$\dim W^0 + \dim W^{00} = \dim V^*$$

and since $\dim V = \dim V^*$ we have

$$\dim W = \dim W^{00}.$$

Since W is a subspace of W^{00}, we see that $W = W^{00}$. ∎

The results of this section hold for arbitrary vector spaces; however, the proofs require the use of the so-called Axiom of Choice. We want to avoid becoming embroiled in a lengthy discussion of that axiom, so we shall not tackle annihilators for general vector spaces. But, there are two results about linear functionals on arbitrary vector spaces which are so fundamental that we should include them.

Let V be a vector space. We want to define hyperspaces in V. Unless V is finite-dimensional, we cannot do that with the dimension of the hyperspace. But, we can express the idea that a space N falls just one dimension short of filling out V, in the following way:

1. N is a proper subspace of V;
2. if W is a subspace of V which contains N, then either $W = N$ or $W = V$.

Conditions (1) and (2) together say that N is a proper subspace and there is no larger proper subspace, in short, N is a maximal proper subspace.

Definition. *If* V *is a vector space, a* **hyperspace** *in* V *is a maximal proper subspace of* V.

Theorem 19. *If* f *is a non-zero linear functional on the vector space* V, *then the null space of* f *is a hyperspace in* V. *Conversely, every hyperspace in* V *is the null space of a (not unique) non-zero linear functional on* V.

Proof. Let f be a non-zero linear functional on V and N_f its null space. Let α be a vector in V which is not in N_f, i.e., a vector such that $f(\alpha) \neq 0$. We shall show that every vector in V is in the subspace spanned by N_f and α. That subspace consists of all vectors

$$\gamma + c\alpha, \qquad \gamma \text{ in } N_f,\ c \text{ in } F.$$

Let β be in V. Define

$$c = \frac{f(\beta)}{f(\alpha)}$$

which makes sense because $f(\alpha) \neq 0$. Then the vector $\gamma = \beta - c\alpha$ is in N_f since

$$\begin{aligned} f(\gamma) &= f(\beta - c\alpha) \\ &= f(\beta) - cf(\alpha) \\ &= 0. \end{aligned}$$

So β is in the subspace spanned by N_f and α.

Now let N be a hyperspace in V. Fix some vector α which is not in N. Since N is a maximal proper subspace, the subspace spanned by N and α is the entire space V. Therefore each vector β in V has the form

$$\beta = \gamma + c\alpha, \qquad \gamma \text{ in } N,\ c \text{ in } F.$$

The vector γ and the scalar c are uniquely determined by β. If we have also

$$\beta = \gamma' + c'\alpha, \qquad \gamma' \text{ in } N,\ c' \text{ in } F.$$

then

$$(c' - c)\alpha = \gamma - \gamma'.$$

If $c' - c \neq 0$, then α would be in N; hence, $c' = c$ and $\gamma' = \gamma$. Another way to phrase our conclusion is this: If β is in V, there is a unique scalar c such that $\beta - c\alpha$ is in N. Call that scalar $g(\beta)$. It is easy to see that g is a linear functional on V and that N is the null space of g. ∎

Lemma. *If f and g are linear functionals on a vector space V, then g is a scalar multiple of f if and only if the null space of g contains the null space of f, that is, if and only if* $f(\alpha) = 0$ *implies* $g(\alpha) = 0$.

Proof. If $f = 0$ then $g = 0$ as well and g is trivially a scalar multiple of f. Suppose $f \neq 0$ so that the null space N_f is a hyperspace in V. Choose some vector α in V with $f(\alpha) \neq 0$ and let

$$c = \frac{g(\alpha)}{f(\alpha)}.$$

The linear functional $h = g - cf$ is 0 on N_f, since both f and g are 0 there, and $h(\alpha) = g(\alpha) - cf(\alpha) = 0$. Thus h is 0 on the subspace spanned by N_f and α—and that subspace is V. We conclude that $h = 0$, i.e., that $g = cf$. ∎

Theorem 20. *Let* $g, f_1, \ldots, f_r$ *be linear functionals on a vector space V with respective null spaces* $N, N_1, \ldots, N_r$. *Then g is a linear combination of* $f_1, \ldots, f_r$ *if and only if N contains the intersection* $N_1 \cap \cdots \cap N_r$.

Proof. If $g = c_1f_1 + \cdots + c_rf_r$ and $f_i(\alpha) = 0$ for each i, then clearly $g(\alpha) = 0$. Therefore, N contains $N_1 \cap \cdots \cap N_r$.

We shall prove the converse (the 'if' half of the theorem) by induction on the number r. The preceding lemma handles the case $r = 1$. Suppose we know the result for $r = k - 1$, and let $f_1, \ldots, f_k$ be linear functionals with null spaces $N_1, \ldots, N_k$ such that $N_1 \cap \cdots \cap N_k$ is contained in N, the null space of g. Let $g', f'_1, \ldots, f'_{k-1}$ be the restrictions of $g, f_1, \ldots, f_{k-1}$ to the subspace N_k. Then $g', f'_1, \ldots, f'_{k-1}$ are linear functionals on the vector space N_k. Furthermore, if α is a vector in N_k and $f'_i(\alpha) = 0$, $i = 1, \ldots, k-1$, then α is in $N_1 \cap \cdots \cap N_k$ and so $g'(\alpha) = 0$. By the induction hypothesis (the case $r = k - 1$), there are scalars c_i such that

$$g' = c_1 f'_1 + \cdots + c_{k-1} f'_{k-1}.$$

Now let

$$h = g - \sum_{i=1}^{k-1} c_i f_i. \tag{3-16}$$

Then h is a linear functional on V and (3-16) tells us that $h(\alpha) = 0$ for every α in N_k. By the preceding lemma, h is a scalar multiple of f_k. If $h = c_k f_k$, then

$$g = \sum_{i=1}^{k} c_i f_i. \quad \blacksquare$$

Exercises

1. Let n be a positive integer and F a field. Let W be the set of all vectors $(x_1, \ldots, x_n)$ in F^n such that $x_1 + \cdots + x_n = 0$.

(a) Prove that W^0 consists of all linear functionals f of the form

$$f(x_1, \ldots, x_n) = c \sum_{j=1}^{n} x_j.$$

(b) Show that the dual space W^* of W can be 'naturally' identified with the linear functionals

$$f(x_1, \ldots, x_n) = c_1 x_1 + \cdots + c_n x_n$$

on F^n which satisfy $c_1 + \cdots + c_n = 0$.

2. Use Theorem 20 to prove the following. If W is a subspace of a finite-dimensional vector space V and if $\{g_1, \ldots, g_r\}$ is any basis for W^0, then

$$W = \bigcap_{i=1}^{r} N_{g_i}.$$

3. Let S be a set, F a field, and $V(S; F)$ the space of all functions from S into F:

$$(f + g)(x) = f(x) + g(x)$$
$$(cf)(x) = cf(x).$$

Let W be any n-dimensional subspace of $V(S; F)$. Show that there exist points $x_1, \ldots, x_n$ in S and functions $f_1, \ldots, f_n$ in W such that $f_i(x_j) = \delta_{ij}$.

3.7. The Transpose of a Linear Transformation

Suppose that we have two vector spaces over the field F, V, and W, and a linear transformation T from V into W. Then T induces a linear transformation from W^* into V^*, as follows. Suppose g is a linear functional on W, and let

$$f(\alpha) = g(T\alpha) \tag{3-17}$$

for each α in V. Then (3-17) defines a function f from V into F, namely, the composition of T, a function from V into W, with g, a function from W into F. Since both T and g are linear, Theorem 6 tells us that f is also linear, i.e., f is a linear functional on V. Thus T provides us with a rule T^t which associates with each linear functional g on W a linear functional $f = T^tg$ on V, defined by (3-17). Note also that T^t is actually a linear transformation from W^* into V^*; for, if g_1 and g_2 are in W^* and c is a scalar

$$\begin{aligned}[T^t(cg_1 + g_2)](\alpha) &= (cg_1 + g_2)(T\alpha)\\ &= cg_1(T\alpha) + g_2(T\alpha)\\ &= c(T^tg_1)(\alpha) + (T^tg_2)(\alpha)\end{aligned}$$

so that $T^t(cg_1 + g_2) = cT^tg_1 + T^tg_2$. Let us summarize.

Theorem 21. *Let* V *and* W *be vector spaces over the field* F. *For each linear transformation* T *from* V *into* W, *there is a unique linear transformation* T^t *from* W^* *into* V^* *such that*

$$(T^tg)(\alpha) = g(T\alpha)$$

for every g *in* W^* *and* α *in* V.

We shall call T^t the **transpose** of T. This transformation T^t is often called the adjoint of T; however, we shall not use this terminology.

Theorem 22. *Let* V *and* W *be vector spaces over the field* F, *and let* T *be a linear transformation from* V *into* W. *The null space of* T^t *is the annihilator of the range of* T. *If* V *and* W *are finite-dimensional, then*

(i) *rank* (T^t) = *rank* (T)
(ii) *the range of* T^t *is the annihilator of the null space of* T.

Proof. If g is in W^*, then by definition

$$(T^tg)(\alpha) = g(T\alpha)$$

for each α in V. The statement that g is in the null space of T^t means that $g(T\alpha) = 0$ for every α in V. Thus the null space of T^t is precisely the annihilator of the range of T.

Suppose that V and W are finite-dimensional, say $\dim V = n$ and $\dim W = m$. For (i): Let r be the rank of T, i.e., the dimension of the range of T. By Theorem 16, the annihilator of the range of T then has dimension $(m - r)$. By the first statement of this theorem, the nullity of T^t must be $(m - r)$. But then since T^t is a linear transformation on an m-dimensional space, the rank of T^t is $m - (m - r) = r$, and so T and T^t have the same rank. For (ii): Let N be the null space of T. Every functional in the range of T^t is in the annihilator of N; for, suppose $f = T^tg$ for some g in W^*; then, if α is in N

$$f(\alpha) = (T^tg)(\alpha) = g(T\alpha) = g(0) = 0.$$

Now the range of T^t is a subspace of the space N^0, and

$$\dim N^0 = n - \dim N = \text{rank } (T) = \text{rank } (T^t)$$

so that the range of T^t must be exactly N^0. ∎

Theorem 23. *Let V and W be finite-dimensional vector spaces over the field F. Let $\mathcal{B}$ be an ordered basis for V with dual basis $\mathcal{B}^*$, and let $\mathcal{B}'$ be an ordered basis for W with dual basis $\mathcal{B}'^*$. Let T be a linear transformation from V into W; let A be the matrix of T relative to $\mathcal{B}$, $\mathcal{B}'$ and let B be the matrix of T^t relative to $\mathcal{B}'^*$, $\mathcal{B}^*$. Then $B_{ij} = A_{ji}$.*

Proof. Let

$$\mathcal{B} = \{\alpha_1, \ldots, \alpha_n\}, \qquad \mathcal{B}' = \{\beta_1, \ldots, \beta_m\},$$
$$\mathcal{B}^* = \{f_1, \ldots, f_n\}, \qquad \mathcal{B}'^* = \{g_1, \ldots, g_m\}.$$

By definition,

$$T\alpha_j = \sum_{i=1}^{m} A_{ij}\beta_i, \qquad j = 1, \ldots, n$$

$$T^tg_j = \sum_{i=1}^{n} B_{ij}f_i, \qquad j = 1, \ldots, m.$$

On the other hand,

$$\begin{aligned}(T^tg_j)(\alpha_i) &= g_j(T\alpha_i)\\ &= g_j\left(\sum_{k=1}^{m} A_{ki}\beta_k\right)\\ &= \sum_{k=1}^{m} A_{ki}g_j(\beta_k)\\ &= \sum_{k=1}^{m} A_{ki}\delta_{jk}\\ &= A_{ji}.\end{aligned}$$

For any linear functional f on V

$$f = \sum_{i=1}^{m} f(\alpha_i)f_i.$$

If we apply this formula to the functional $f = T^tg_j$ and use the fact that $(T^tg_j)(\alpha_i) = A_{ji}$, we have

$$T^tg_j = \sum_{i=1}^{n} A_{ji}f_i$$

from which it immediately follows that $B_{ij} = A_{ji}$. ∎

Definition. *If* A *is an* m × n *matrix over the field* F, *the* **transpose** *of* A *is the* n × m *matrix* A^t *defined by* $A^t_{ij} = A_{ji}$.

Theorem 23 thus states that if T is a linear transformation from V into W, the matrix of which in some pair of bases is A, then the transpose transformation T^t is represented in the dual pair of bases by the transpose matrix A^t.

Theorem 24. *Let* A *be any* m × n *matrix over the field* F. *Then the row rank of* A *is equal to the column rank of* A.

Proof. Let $\mathcal{B}$ be the standard ordered basis for F^n and $\mathcal{B}'$ the standard ordered basis for F^m. Let T be the linear transformation from F^n into F^m such that the matrix of T relative to the pair $\mathcal{B}$, $\mathcal{B}'$ is A, i.e.,

$$T(x_1, \ldots, x_n) = (y_1, \ldots, y_m)$$

where

$$y_i = \sum_{j=1}^{n} A_{ij}x_j.$$

The column rank of A is the rank of the transformation T, because the range of T consists of all m-tuples which are linear combinations of the column vectors of A.

Relative to the dual bases $\mathcal{B}'^*$ and $\mathcal{B}^*$, the transpose mapping T^t is represented by the matrix A^t. Since the columns of A^t are the rows of A, we see by the same reasoning that the row rank of A (the column rank of A^t) is equal to the rank of T^t. By Theorem 22, T and T^t have the same rank, and hence the row rank of A is equal to the column rank of A. ∎

Now we see that if A is an $m \times n$ matrix over F and T is the linear transformation from F^n into F^m defined above, then

$$\text{rank } (T) = \text{row rank } (A) = \text{column rank } (A)$$

and we shall call this number simply the **rank** of A.

Example 25. This example will be of a general nature—more discussion than example. Let V be an n-dimensional vector space over the field F, and let T be a linear operator on V. Suppose $\mathcal{B} = \{\alpha_1, \ldots, \alpha_n\}$ is an ordered basis for V. The matrix of T in the ordered basis $\mathcal{B}$ is defined to be the $n \times n$ matrix A such that

$$T\alpha_j = \sum_{j=1}^{n} A_{ij}\alpha_i$$

in other words, A_{ij} is the ith coordinate of the vector $T\alpha_j$ in the ordered basis $\mathcal{B}$. If $\{f_1, \ldots, f_n\}$ is the dual basis of $\mathcal{B}$, this can be stated simply

$$A_{ij} = f_i(T\alpha_j).$$

Let us see what happens when we change basis. Suppose

$$\mathcal{B}' = \{\alpha'_1, \ldots, \alpha'_n\}$$

is another ordered basis for V, with dual basis $\{f'_1, \ldots, f'_n\}$. If B is the matrix of T in the ordered basis $\mathcal{B}'$, then

$$B_{ij} = f'_i(T\alpha'_j).$$

Let U be the invertible linear operator such that $U\alpha_j = \alpha'_j$. Then the transpose of U is given by $U^t f'_i = f_i$. It is easy to verify that since U is invertible, so is U^t and $(U^t)^{-1} = (U^{-1})^t$. Thus $f'_i = (U^{-1})^t f_i$, $i = 1, \ldots, n$. Therefore,

$$\begin{aligned} B_{ij} &= [(U^{-1})^t f_i](T\alpha'_j) \\ &= f_i(U^{-1}T\alpha'_j) \\ &= f_i(U^{-1}TU\alpha_j). \end{aligned}$$

Now what does this say? Well, $f_i(U^{-1}TU\alpha_j)$ is the i, j entry of the matrix of $U^{-1}TU$ in the ordered basis $\mathcal{B}$. Our computation above shows that this scalar is also the i, j entry of the matrix of T in the ordered basis $\mathcal{B}'$. In other words

$$\begin{aligned} [T]_{\mathcal{B}'} &= [U^{-1}TU]_{\mathcal{B}} \\ &= [U^{-1}]_{\mathcal{B}}[T]_{\mathcal{B}}[U]_{\mathcal{B}} \\ &= [U]_{\mathcal{B}}^{-1}[T]_{\mathcal{B}}[U]_{\mathcal{B}} \end{aligned}$$

and this is precisely the change-of-basis formula which we derived earlier.

Exercises

1. Let F be a field and let f be the linear functional on F^2 defined by $f(x_1, x_2) = ax_1 + bx_2$. For each of the following linear operators T, let $g = T^t f$, and find $g(x_1, x_2)$.

(a) $T(x_1, x_2) = (x_1, 0)$;

(b) $T(x_1, x_2) = (-x_2, x_1)$;

(c) $T(x_1, x_2) = (x_1 - x_2, x_1 + x_2)$.

2. Let V be the vector space of all polynomial functions over the field of real numbers. Let a and b be fixed real numbers and let f be the linear functional on V defined by

$$f(p) = \int_a^b p(x)\, dx.$$

If D is the differentiation operator on V, what is $D^t f$?

3. Let V be the space of all $n \times n$ matrices over a field F and let B be a fixed $n \times n$ matrix. If T is the linear operator on V defined by $T(A) = AB - BA$, and if f is the trace function, what is $T^t f$?

4. Let V be a finite-dimensional vector space over the field F and let T be a linear operator on V. Let c be a scalar and suppose there is a non-zero vector α in V such that $T\alpha = c\alpha$. Prove that there is a non-zero linear functional f on V such that $T^t f = cf$.

5. Let A be an $m \times n$ matrix with *real* entries. Prove that $A = 0$ if and only if trace $(A^tA) = 0$.

6. Let n be a positive integer and let V be the space of all polynomial functions over the field of real numbers which have degree at most n, i.e., functions of the form

$$f(x) = c_0 + c_1x + \cdots + c_nx^n.$$

Let D be the differentiation operator on V. Find a basis for the null space of the transpose operator D^t.

7. Let V be a finite-dimensional vector space over the field F. Show that $T \rightarrow T^t$ is an isomorphism of $L(V, V)$ onto $L(V^*, V^*)$.

8. Let V be the vector space of $n \times n$ matrices over the field F.

(a) If B is a fixed $n \times n$ matrix, define a function f_B on V by $f_B(A) =$ trace (B^tA). Show that f_B is a linear functional on V.

(b) Show that every linear functional on V is of the above form, i.e., is f_B for some B.

(c) Show that $B \rightarrow f_B$ is an isomorphism of V onto V^*.

4. Polynomials

4.1. Algebras

The purpose of this chapter is to establish a few of the basic properties of the algebra of polynomials over a field. The discussion will be facilitated if we first introduce the concept of a linear algebra over a field.

Definition. *Let* F *be a field. A* **linear algebra over the field** F *is a vector space* $\mathcal{A}$ *over* F *with an additional operation called* **multiplication of vectors** *which associates with each pair of vectors* α, β *in* $\mathcal{A}$ *a vector* $\alpha\beta$ *in* $\mathcal{A}$ *called the* **product** *of* α *and* β *in such a way that*

(a) *multiplication is associative,*

$$\alpha(\beta\gamma) = (\alpha\beta)\gamma$$

(b) *multiplication is distributive with respect to addition,*

$$\alpha(\beta + \gamma) = \alpha\beta + \alpha\gamma \quad and \quad (\alpha + \beta)\gamma = \alpha\gamma + \beta\gamma$$

(c) *for each scalar* c *in* F,

$$c(\alpha\beta) = (c\alpha)\beta = \alpha(c\beta).$$

If there is an element 1 *in* $\mathcal{A}$ *such that* $1\alpha = \alpha 1 = \alpha$ *for each* α *in* $\mathcal{A}$, *we call* $\mathcal{A}$ *a* **linear algebra with identity over** F, *and call* 1 *the* **identity** *of* $\mathcal{A}$. *The algebra* $\mathcal{A}$ *is called* **commutative** *if* $\alpha\beta = \beta\alpha$ *for all* α *and* β *in* $\mathcal{A}$.

EXAMPLE 1. The set of $n \times n$ matrices over a field, with the usual operations, is a linear algebra with identity; in particular the field itself is an algebra with identity. This algebra is not commutative if $n \geq 2$. The field itself is (of course) commutative.

EXAMPLE 2. The space of all linear operators on a vector space, with composition as the product, is a linear algebra with identity. It is commutative if and only if the space is one-dimensional.

The reader may have had some experience with the dot product and cross product of vectors in R^3. If so, he should observe that neither of these products is of the type described in the definition of a linear algebra. The dot product is a 'scalar product,' that is, it associates with a pair of vectors a scalar, and thus it is certainly not the type of product we are presently discussing. The cross product does associate a vector with each pair of vectors in R^3; however, this is not an associative multiplication.

The rest of this section will be devoted to the construction of an algebra which is significantly different from the algebras in either of the preceding examples. Let F be a field and S the set of non-negative integers. By Example 3 of Chapter 2, the set of all functions from S into F is a vector space over F. We shall denote this vector space by F^∞. The vectors in F^∞ are therefore infinite sequences $f = (f_0, f_1, f_2, \ldots)$ of scalars f_i in F. If $g = (g_0, g_1, g_2, \ldots)$, g_i in F, and a, b are scalars in F, $af + bg$ is the infinite sequence given by

$$af + bg = (af_0 + bg_0, af_1 + bg_1, af_2 + bg_2, \ldots). \tag{4-1}$$

We define a product in F^∞ by associating with each pair of vectors f and g in F^∞ the vector fg which is given by

$$(fg)_n = \sum_{i=0}^{n} f_i g_{n-i}, \qquad n = 0, 1, 2, \ldots. \tag{4-2}$$

Thus

$$fg = (f_0g_0, f_0g_1 + f_1g_0, f_0g_2 + f_1g_1 + f_2g_0, \ldots)$$

and as

$$(gf)_n = \sum_{i=0}^{n} g_i f_{n-i} = \sum_{i=0}^{n} f_i g_{n-i} = (fg)_n$$

for $n = 0, 1, 2, \ldots$, it follows that multiplication is commutative, $fg = gf$. If h also belongs to F^∞, then

$$\begin{aligned}
[(fg)h]_n &= \sum_{i=0}^{n} (fg)_i h_{n-i} \\
&= \sum_{i=0}^{n} \left(\sum_{j=0}^{i} f_j g_{i-j} \right) h_{n-i} \\
&= \sum_{i=0}^{n} \sum_{j=0}^{i} f_j g_{i-j} h_{n-i} \\
&= \sum_{j=0}^{n} f_j \sum_{i=0}^{n-j} g_i h_{n-i-j} \\
&= \sum_{j=0}^{n} f_j (gh)_{n-j} = [f(gh)]_n
\end{aligned}$$

for $n = 0, 1, 2, \ldots$, so that

$$(fg)h = f(gh). \tag{4-3}$$

We leave it to the reader to verify that the multiplication defined by (4-2) satisfies (b) and (c) in the definition of a linear algebra, and that the vector $1 = (1, 0, 0, \ldots)$ serves as an identity for F^∞. Then F^∞, with the operations defined above, is a commutative linear algebra with identity over the field F.

The vector $(0, 1, 0, \ldots, 0, \ldots)$ plays a distinguished role in what follows and we shall consistently denote it by x. Throughout this chapter x will never be used to denote an element of the field F. The product of x with itself n times will be denoted by x^n and we shall put $x^0 = 1$. Then

$$x^2 = (0, 0, 1, 0, \ldots), \qquad x^3 = (0, 0, 0, 1, 0, \ldots)$$

and in general for each integer $k \geq 0$, $(x^k)_k = 1$ and $(x^k)_n = 0$ for all non-negative integers $n \neq k$. In concluding this section we observe that the set consisting of $1, x, x^2, \ldots$ is both independent and infinite. Thus the algebra F^∞ is not finite-dimensional.

The algebra F^∞ is sometimes called the **algebra of formal power series** over F. The element $f = (f_0, f_1, f_2, \ldots)$ is frequently written

$$f = \sum_{n=0}^{\infty} f_n x^n. \tag{4-4}$$

This notation is very convenient for dealing with the algebraic operations. When used, it must be remembered that it is purely formal. There are no 'infinite sums' in algebra, and the power series notation (4-4) is not intended to suggest anything about convergence, if the reader knows what that is. By using sequences, we were able to define carefully an algebra in which the operations behave like addition and multiplication of formal power series, without running the risk of confusion over such things as infinite sums.

4.2. The Algebra of Polynomials

We are now in a position to define a polynomial over the field F.

Definition. *Let* $F[x]$ *be the subspace of* F^∞ *spanned by the vectors* $1, x, x^2, \ldots$. *An element of* $F[x]$ *is called a* **polynomial over** F.

Since $F[x]$ consists of all (finite) linear combinations of x and its powers, a non-zero vector f in F^∞ is a polynomial if and only if there is an integer $n \geq 0$ such that $f_n \neq 0$ and such that $f_k = 0$ for all integers $k > n$; this integer (when it exists) is obviously unique and is called the **degree** of f. We denote the degree of a polynomial f by deg f, and do not assign a degree to the 0-polynomial. If f is a non-zero polynomial of degree n it follows that

$$f = f_0x^0 + f_1x + f_2x^2 + \cdots + f_nx^n, \qquad f_n \neq 0. \tag{4-5}$$

The scalars $f_0, f_1, \ldots, f_n$ are sometimes called the **coefficients** of f, and we may say that f is a polynomial with coefficients in F. We shall call polynomials of the form cx^0 **scalar polynomials,** and frequently write c for cx^0. A non-zero polynomial f of degree n such that $f_n = 1$ is said to be a **monic** polynomial.

The reader should note that polynomials are not the same sort of objects as the polynomial functions on F which we have discussed on several occasions. If F contains an infinite number of elements, there is a natural isomorphism between $F[x]$ and the algebra of polynomial functions on F. We shall discuss that in the next section. Let us verify that $F[x]$ is an algebra.

Theorem 1. *Let* f *and* g *be non-zero polynomials over* F. *Then*

(i) fg *is a non-zero polynomial;*

(ii) *deg* (fg) = *deg* f + *deg* g;

(iii) fg *is a monic polynomial if both* f *and* g *are monic polynomials;*

(iv) fg *is a scalar polynomial if and only if both* f *and* g *are scalar polynomials;*

(v) *if* f + g ≠ 0,

$$\deg (f + g) \leq \max (\deg f, \deg g).$$

Proof. Suppose f has degree m and that g has degree n. If k is a non-negative integer,

$$(fg)_{m+n+k} = \sum_{i=0}^{m+n+k} f_ig_{m+n+k-i}.$$

In order that $f_ig_{m+n+k-i} \neq 0$, it is necessary that $i \leq m$ and $m + n + k - i \leq n$. Hence it is necessary that $m + k \leq i \leq m$, which implies $k = 0$ and $i = m$. Thus

$$(fg)_{m+n} = f_mg_n \tag{4-6}$$

and

$$(fg)_{m+n+k} = 0, \qquad k > 0. \tag{4-7}$$

The statements (i), (ii), (iii) follow immediately from (4-6) and (4-7), while (iv) is a consequence of (i) and (ii). We leave the verification of (v) to the reader. ∎

Corollary 1. *The set of all polynomials over a given field* F *equipped with the operations (4-1) and (4-2) is a commutative linear algebra with identity over* F.

Proof. Since the operations (4-1) and (4-2) are those defined in the algebra F^∞ and since $F[x]$ is a subspace of F^∞, it suffices to prove that the product of two polynomials is again a polynomial. This is trivial when one of the factors is 0 and otherwise follows from (i). ∎

Corollary 2. *Suppose* f, g, *and* h *are polynomials over the field* F *such that* f $\neq$ 0 *and* fg = fh. *Then* g = h.

Proof. Since $fg = fh$, $f(g - h) = 0$, and as $f \neq 0$ it follows at once from (i) that $g - h = 0$. ∎

Certain additional facts follow rather easily from the proof of Theorem 1, and we shall mention some of these.

Suppose

$$f = \sum_{i=0}^{m} f_i x^i \quad \text{and} \quad g = \sum_{j=0}^{n} g_j x^j.$$

Then from (4-7) we obtain,

$$fg = \sum_{s=0}^{m+n} \left(\sum_{r=0}^{s} f_r g_{s-r} \right) x^s. \tag{4-8}$$

The reader should verify, in the special case $f = cx^m$, $g = dx^n$ with c, d in F, that (4-8) reduces to

$$(cx^m)(dx^n) = cdx^{m+n}. \tag{4-9}$$

Now from (4-9) and the distributive laws in $F[x]$, it follows that the product in (4-8) is also given by

$$\sum_{i,j} f_i g_j x^{i+j} \tag{4-10}$$

where the sum is extended over all integer pairs i, j such that $0 \leq i \leq m$, and $0 \leq j \leq n$.

Definition. *Let $\mathcal{A}$ be a linear algebra with identity over the field* F. *We shall denote the identity of $\mathcal{A}$ by* 1 *and make the convention that* $\alpha^0 = 1$ *for each* α *in* $\mathcal{A}$. *Then to each polynomial* $f = \sum_{i=0}^{n} f_i x^i$ *over* F *and* α *in* $\mathcal{A}$ *we associate an element* $f(\alpha)$ *in* $\mathcal{A}$ *by the rule*

$$f(\alpha) = \sum_{i=0}^{n} f_i \alpha^i.$$

EXAMPLE 3. Let C be the field of complex numbers and let $f = x^2 + 2$.

(a) If $\mathcal{A} = C$ and z belongs to C, $f(z) = z^2 + 2$, in particular $f(2) = 6$ and

$$f\left(\frac{1+i}{1-i}\right) = 1.$$

(b) If $\mathfrak{A}$ is the algebra of all 2×2 matrices over C and if

$$B = \begin{bmatrix} 1 & 0 \\ -1 & 2 \end{bmatrix}$$

then

$$f(B) = 2\begin{bmatrix} 1 & 0 \\ 0 & 1 \end{bmatrix} + \begin{bmatrix} 1 & 0 \\ -1 & 2 \end{bmatrix}^2 = \begin{bmatrix} 3 & 0 \\ -3 & 6 \end{bmatrix}.$$

(c) If $\mathfrak{A}$ is the algebra of all linear operators on C^3 and T is the element of $\mathfrak{A}$ given by

$$T(c_1, c_2, c_3) = (i\sqrt{2}\, c_1, c_2, i\sqrt{2}\, c_3)$$

then $f(T)$ is the linear operator on C^3 defined by

$$f(T)(c_1, c_2, c_3) = (0, 3c_2, 0).$$

(d) If $\mathfrak{A}$ is the algebra of all polynomials over C and $g = x^4 + 3i$, then $f(g)$ is the polynomial in $\mathfrak{A}$ given by

$$f(g) = -7 + 6ix^4 + x^8.$$

The observant reader may notice in connection with this last example that if f is a polynomial over any field and x is the polynomial $(0, 1, 0, \ldots)$ then $f = f(x)$, but he is advised to forget this fact.

Theorem 2. *Let* F *be a field and* $\mathfrak{A}$ *be a linear algebra with identity over* F. *Suppose* f *and* g *are polynomials over* F, *that* α *is an element of* $\mathfrak{A}$, *and that* c *belongs to* F. *Then*

(i) $(\mathrm{cf} + \mathrm{g})(\alpha) = \mathrm{cf}(\alpha) + \mathrm{g}(\alpha)$;
(ii) $(\mathrm{fg})(\alpha) = \mathrm{f}(\alpha)\mathrm{g}(\alpha)$.

Proof. As (i) is quite easy to establish, we shall only prove (ii). Suppose

$$f = \sum_{i=0}^{m} f_i x^i \quad \text{and} \quad g = \sum_{j=0}^{n} g_j x^j.$$

By (4-10),

$$fg = \sum_{i,j} f_i g_j x^{i+j}$$

and hence by (i),

$$\begin{aligned} (fg)(\alpha) &= \sum_{i,j} f_i g_j \alpha^{i+j} \\ &= \left(\sum_{i=0}^{m} f_i \alpha^i\right)\left(\sum_{j=0}^{n} g_j \alpha^j\right) \\ &= f(\alpha)g(\alpha). \quad \blacksquare \end{aligned}$$

Exercises

1. Let F be a subfield of the complex numbers and let A be the following 2×2 matrix over F

$$A = \begin{bmatrix} 2 & 1 \\ -1 & 3 \end{bmatrix}.$$

For each of the following polynomials f over F, compute $f(A)$.

(a) $f = x^2 - x + 2$;
(b) $f = x^3 - 1$;
(c) $f = x^2 - 5x + 7$.

2. Let T be the linear operator on R^3 defined by

$$T(x_1, x_2, x_3) = (x_1, x_3, -2x_2 - x_3).$$

Let f be the polynomial over R defined by $f = -x^3 + 2$. Find $f(T)$.

3. Let A be an $n \times n$ diagonal matrix over the field F, i.e., a matrix satisfying $A_{ij} = 0$ for $i \neq j$. Let f be the polynomial over F defined by

$$f = (x - A_{11}) \cdots (x - A_{nn}).$$

What is the matrix $f(A)$?

4. If f and g are independent polynomials over a field F and h is a non-zero polynomial over F, show that fh and gh are independent.

5. If F is a field, show that the product of two non-zero elements of F^∞ is non-zero.

6. Let S be a set of non-zero polynomials over a field F. If no two elements of S have the same degree, show that S is an independent set in $F[x]$.

7. If a and b are elements of a field F and $a \neq 0$, show that the polynomials 1, $ax + b$, $(ax + b)^2$, $(ax + b)^3$, . . . form a basis of $F[x]$.

8. If F is a field and h is a polynomial over F of degree ≥ 1, show that the mapping $f \to f(h)$ is a one-one linear transformation of $F[x]$ into $F[x]$. Show that this transformation is an isomorphism of $F[x]$ onto $F[x]$ if and only if $\deg h = 1$.

9. Let F be a subfield of the complex numbers and let T, D be the transformations on $F[x]$ defined by

$$T\left(\sum_{i=0}^{n} c_i x^i\right) = \sum_{i=0}^{n} \frac{c_i}{1+i} x^{i+1}$$

and

$$D\left(\sum_{i=0}^{n} c_i x^i\right) = \sum_{i=1}^{n} i c_i x^{i-1}.$$

(a) Show that T is a non-singular linear operator on $F[x]$. Show also that T is not invertible.

(b) Show that D is a linear operator on $F[x]$ and find its null space.

(c) Show that $DT = I$, and $TD \neq I$.

(d) Show that $T[(Tf)g] = (Tf)(Tg) - T[f(Tg)]$ for all f, g in $F[x]$.

(e) State and prove a rule for D similar to the one given for T in (d).

(f) Suppose V is a non-zero subspace of $F[x]$ such that Tf belongs to V for each f in V. Show that V is not finite-dimensional.

(g) Suppose V is a finite-dimensional subspace of $F[x]$. Prove there is an integer $m \geq 0$ such that $D^m f = 0$ for each f in V.

4.3. Lagrange Interpolation

Throughout this section we shall assume F is a fixed field and that $t_0, t_1, \ldots, t_n$ are $n + 1$ *distinct* elements of F. Let V be the subspace of $F[x]$ consisting of all polynomials of degree less than or equal to n (together with the 0-polynomial), and let L_i be the function from V into F defined for f in V by

$$L_i(f) = f(t_i), \qquad 0 \le i \le n.$$

By part (i) of Theorem 2, each L_i is a linear functional on V, and one of the things we intend to show is that the set consisting of $L_0, L_1, \ldots, L_n$ is a basis for V^*, the dual space of V.

Of course in order that this be so, it is sufficient (cf. Theorem 15 of Chapter 3) that $\{L_0, L_1, \ldots, L_n\}$ be the dual of a basis $\{P_0, P_1, \ldots, P_n\}$ of V. There is at most one such basis, and if it exists it is characterized by

$$L_j(P_i) = P_i(t_j) = \delta_{ij}. \tag{4-11}$$

The polynomials

$$\begin{aligned} P_i &= \frac{(x - t_0) \cdots (x - t_{i-1})(x - t_{i+1}) \cdots (x - t_n)}{(t_i - t_0) \cdots (t_i - t_{i-1})(t_i - t_{i+1}) \cdots (t_i - t_n)} \\ &= \prod_{j \neq i} \left(\frac{x - t_j}{t_i - t_j} \right) \end{aligned} \tag{4-12}$$

are of degree n, hence belong to V, and by Theorem 2, they satisfy (4-11).

If $f = \sum\limits_i c_i P_i$, then for each j

$$f(t_j) = \sum_i c_i P_i(t_j) = c_j. \tag{4-13}$$

Since the 0-polynomial has the property that $0(t) = 0$ for each t in F, it follows from (4-13) that the polynomials $P_0, P_1, \ldots, P_n$ are linearly independent. The polynomials $1, x, \ldots, x^n$ form a basis of V and hence the dimension of V is $(n + 1)$. So, the independent set $\{P_0, P_1, \ldots, P_n\}$ must also be a basis for V. Thus for each f in V

$$f = \sum_{i=0}^{n} f(t_i) P_i. \tag{4-14}$$

The expression (4-14) is called **Lagrange's interpolation formula.** Setting $f = x^j$ in (4-14) we obtain

$$x^j = \sum_{i=0}^{n} (t_i)^j P_i.$$

Now from Theorem 7 of Chapter 2 it follows that the matrix

$$\begin{bmatrix} 1 & t_0 & t_0^2 & \cdots & t_0^n \\ 1 & t_1 & t_1^2 & \cdots & t_1^n \\ \vdots & \vdots & \vdots & & \vdots \\ 1 & t_n & t_n^2 & \cdots & t_n^n \end{bmatrix} \tag{4-15}$$

is invertible. The matrix in (4-15) is called a **Vandermonde matrix**; it is an interesting exercise to show directly that such a matrix is invertible, when $t_0, t_1, \ldots, t_n$ are $n + 1$ distinct elements of F.

If f is any polynomial over F we shall, in our present discussion, denote by $f^\sim$ the polynomial function from F into F taking each t in F into $f(t)$. By definition (cf. Example 4, Chapter 2) every polynomial function arises in this way; however, it may happen that $f^\sim = g^\sim$ for two polynomials f and g such that $f \neq g$. Fortunately, as we shall see, this unpleasant situation only occurs in the case where F is a field having only a finite number of distinct elements. In order to describe in a precise way the relation between polynomials and polynomial functions, we need to define the product of two polynomial functions. If f, g are polynomials over F, the product of $f^\sim$ and $g^\sim$ is the function $f^\sim g^\sim$ from F into F given by

$$(f^\sim g^\sim)(t) = f^\sim(t)g^\sim(t), \qquad t \text{ in } F. \tag{4-16}$$

By part (ii) of Theorem 2, $(fg)(t) = f(t)g(t)$, and hence

$$(fg)^\sim(t) = f^\sim(t)g^\sim(t)$$

for each t in F. Thus $f^\sim g^\sim = (fg)^\sim$, and is a polynomial function. At this point it is a straightforward matter, which we leave to the reader, to verify that the vector space of polynomial functions over F becomes a linear algebra with identity over F if multiplication is defined by (4-16).

Definition. *Let* F *be a field and let* $\mathcal{A}$ *and* $\mathcal{A}^\sim$ *be linear algebras over* F. *The algebras* $\mathcal{A}$ *and* $\mathcal{A}^\sim$ *are said to be* **isomorphic** *if there is a one-to-one mapping* $\alpha \to \alpha^\sim$ *of* $\mathcal{A}$ *onto* $\mathcal{A}^\sim$ *such that*

$$\text{(a)} \qquad (c\alpha + d\beta)^\sim = c\alpha^\sim + d\beta^\sim$$

$$\text{(b)} \qquad (\alpha\beta)^\sim = \alpha^\sim\beta^\sim$$

for all α, β *in* $\mathcal{A}$ *and all scalars* c, d *in* F. *The mapping* $\alpha \to \alpha^\sim$ *is called an* **isomorphism** *of* $\mathcal{A}$ *onto* $\mathcal{A}^\sim$. *An isomorphism of* $\mathcal{A}$ *onto* $\mathcal{A}^\sim$ *is thus a vector-space isomorphism of* $\mathcal{A}$ *onto* $\mathcal{A}^\sim$ *which has the additional property* (b) *of 'preserving' products.*

EXAMPLE 4. Let V be an n-dimensional vector space over the field F. By Theorem 13 of Chapter 3 and subsequent remarks, each ordered basis $\mathcal{B}$ of V determines an isomorphism $T \to [T]_\mathcal{B}$ of the algebra of linear operators on V onto the algebra of $n \times n$ matrices over F. Suppose now that U is a fixed linear operator on V and that we are given a polynomial

$$f = \sum_{i=0}^{n} c_i x^i$$

with coefficients c_i in F. Then

$$f(U) = \sum_{i=0}^{n} c_i U^i$$

and since $T \to [T]_{\mathcal{B}}$ is a linear mapping

$$[f(U)]_{\mathcal{B}} = \sum_{i=0}^{n} c_i [U^i]_{\mathcal{B}}.$$

Now from the additional fact that

$$[T_1 T_2]_{\mathcal{B}} = [T_1]_{\mathcal{B}} [T_2]_{\mathcal{B}}$$

for all T_1, T_2 in $L(V, V)$ it follows that

$$[U^i]_{\mathcal{B}} = ([U]_{\mathcal{B}})^i, \qquad 2 \leq i \leq n.$$

As this relation is also valid for $i = 0, 1$ we obtain the result that

$$[f(U)]_{\mathcal{B}} = f([U]_{\mathcal{B}}). \tag{4-17}$$

In words, if U is a linear operator on V, the matrix of a polynomial in U, in a given basis, is the same polynomial in the matrix of U.

Theorem 3. *If F is a field containing an infinite number of distinct elements, the mapping* $f \to f^{\sim}$ *is an isomorphism of the algebra of polynomials over F onto the algebra of polynomial functions over F.*

Proof. By definition, the mapping is onto, and if f, g belong to $F[x]$ it is evident that

$$(cf + dg)^{\sim} = df^{\sim} + dg^{\sim}$$

for all scalars c and d. Since we have already shown that $(fg)^{\sim} = f^{\sim}g^{\sim}$, we need only show that the mapping is one-to-one. To do this it suffices by linearity to show that $f^{\sim} = 0$ implies $f = 0$. Suppose then that f is a polynomial of degree n or less such that $f' = 0$. Let $t_0, t_1, \ldots, t_n$ be any $n + 1$ distinct elements of F. Since $f^{\sim} = 0$, $f(t_i) = 0$ for $i = 0, 1, \ldots, n$, and it is an immediate consequence of (4-14) that $f = 0$. ∎

From the results of the next section we shall obtain an altogether different proof of this theorem.

Exercises

1. Use the Lagrange interpolation formula to find a polynomial f with real coefficients such that f has degree ≤ 3 and $f(-1) = -6$, $f(0) = 2$, $f(1) = -2$, $f(2) = 6$.

2. Let $\alpha, \beta, \gamma, \delta$ be real numbers. We ask when it is possible to find a polynomial f over R, of *degree not more than* 2, such that $f(-1) = \alpha$, $f(1) = \beta$, $f(3) = \gamma$ and $f(0) = \delta$. Prove that this is possible if and only if

$$3\alpha + 6\beta - \gamma - 8\delta = 0.$$

3. Let F be the field of real numbers,

$$A = \begin{bmatrix} 2 & 0 & 0 & 0 \\ 0 & 2 & 0 & 0 \\ 0 & 0 & 3 & 0 \\ 0 & 0 & 0 & 1 \end{bmatrix}$$

$$p = (x - 2)(x - 3)(x - 1).$$

(a) Show that $p(A) = 0$.

(b) Let P_1, P_2, P_3 be the Lagrange polynomials for $t_1 = 2$, $t_2 = 3$, $t_3 = 1$. Compute $E_i = P_i(A)$, $i = 1, 2, 3$.

(c) Show that $E_1 + E_2 + E_3 = I$, $E_iE_j = 0$ if $i \neq j$, $E_i^2 = E_i$.

(d) Show that $A = 2E_1 + 3E_2 + E_3$.

4. Let $p = (x - 2)(x - 3)(x - 1)$ and let T be any linear operator on R^4 such that $p(T) = 0$. Let P_1, P_2, P_3 be the Lagrange polynomials of Exercise 3, and let $E_i = P_i(T)$, $i = 1, 2, 3$. Prove that

$$E_1 + E_2 + E_3 = I, \qquad E_iE_j = 0 \quad \text{if} \quad i \neq j,$$

$$E_i^2 = E_i, \quad \text{and} \quad T = 2E_1 + 3E_2 + E_3.$$

5. Let n be a positive integer and F a field. Suppose A is an $n \times n$ matrix over F and P is an invertible $n \times n$ matrix over F. If f is any polynomial over F, prove that

$$f(P^{-1}AP) = P^{-1}f(A)P.$$

6. Let F be a field. We have considered certain special linear functionals on $F[x]$ obtained via 'evaluation at t':

$$L(f) = f(t).$$

Such functionals are not only linear but also have the property that $L(fg) = L(f)L(g)$. Prove that if L is any linear functional on $F[x]$ such that

$$L(fg) = L(f)L(g)$$

for all f and g, then either $L = 0$ or there is a t in F such that $L(f) = f(t)$ for all f.

4.4. Polynomial Ideals

In this section we are concerned with results which depend primarily on the multiplicative structure of the algebra of polynomials over a field.

Lemma. *Suppose* f *and* d *are non-zero polynomials over a field* F *such that deg* d $\leq$ *deg* f. *Then there exists a polynomial* g *in* F[x] *such that either*

$$\mathrm{f - dg = 0} \quad \textit{or} \quad \textit{deg}\ \mathrm{(f - dg)} < \textit{deg}\ \mathrm{f}.$$

Proof. Suppose

$$f = a_mx^m + \sum_{i=0}^{m-1} a_ix^i, \qquad a_m \neq 0$$

and that

$$d = b_nx^n + \sum_{i=0}^{n-1} b_ix^i, \qquad b_n \neq 0.$$

Then $m \geq n$, and

$$f - \left(\frac{a_m}{b_n}\right)x^{m-n}d = 0 \quad \text{or} \quad \deg\left[f - \left(\frac{a_m}{b_n}\right)x^{m-n}d\right] < \deg f.$$

Thus we may take $g = \left(\frac{a_m}{b_n}\right) x^{m-n}$. ∎

Using this lemma we can show that the familiar process of 'long division' of polynomials with real or complex coefficients is possible over any field.

Theorem 4. *If* f, d *are polynomials over a field* F *and* d *is different from* 0 *then there exist polynomials* q, r *in* F[x] *such that*

(i) $f = dq + r$.
(ii) *either* $r = 0$ *or deg* $r <$ *deg* d.

The polynomials q, r *satisfying* (i) *and* (ii) *are unique.*

Proof. If f is 0 or $\deg f < \deg d$ we may take $q = 0$ and $r = f$. In case $f \neq 0$ and $\deg f \geq \deg d$, the preceding lemma shows we may choose a polynomial g such that $f - dg = 0$ or $\deg (f - dg) < \deg f$. If $f - dg \neq 0$ and $\deg (f - dg) \geq \deg d$ we choose a polynomial h such that $(f - dg) - dh = 0$ or

$$\deg [f - d(g + h)] < \deg (f - dg).$$

Continuing this process as long as necessary, we ultimately obtain polynomials q, r such that $r = 0$ or $\deg r < \deg d$, and $f = dq + r$. Now suppose we also have $f = dq_1 + r_1$ where $r_1 = 0$ or $\deg r_1 < \deg d$. Then $dq + r = dq_1 + r_1$, and $d(q - q_1) = r_1 - r$. If $q - q_1 \neq 0$ then $d(q - q_1) \neq 0$ and

$$\deg d + \deg (q - q_1) = \deg (r_1 - r).$$

But as the degree of $r_1 - r$ is less than the degree of d, this is impossible and $q - q_1 = 0$. Hence also $r_1 - r = 0$. ∎

Definition. *Let* d *be a non-zero polynomial over the field* F. *If* f *is in* F[x], *the preceding theorem shows there is at most one polynomial* q *in* F[x] *such that* f = dq. *If such a* q *exists we say that* d **divides** f, *that* f *is* **divisible** *by* d, *that* f *is a* **multiple** *of* d, *and call* q *the* **quotient** *of* f *and* d. *We also write* q = f/d.

Corollary 1. *Let* f *be a polynomial over the field* F, *and let* c *be an element of* F. *Then* f *is divisible by* x − c *if and only if* f(c) = 0.

Proof. By the theorem, $f = (x - c)q + r$ where r is a scalar polynomial. By Theorem 2,

$$f(c) = 0q(c) + r(c) = r(c).$$

Hence $r = 0$ if and only if $f(c) = 0$. ∎

Definition. *Let* F *be a field. An element* c *in* F *is said to be a* **root** *or a* **zero** *of a given polynomial* f *over* F *if* f(c) = 0.

Corollary 2. *A polynomial* f *of degree* n *over a field* F *has at most* n *roots in* F.

Proof. The result is obviously true for polynomials of degree 0 and degree 1. We assume it to be true for polynomials of degree $n - 1$. If a is a root of f, $f = (x - a)q$ where q has degree $n - 1$. Since $f(b) = 0$ if and only if $a = b$ or $q(b) = 0$, it follows by our inductive assumption that f has at most n roots. ∎

The reader should observe that the main step in the proof of Theorem 3 follows immediately from this corollary.

The formal derivatives of a polynomial are useful in discussing multiple roots. The **derivative** of the polynomial

$$f = c_0 + c_1x + \cdots + c_nx^n$$

is the polynomial

$$f' = c_1 + 2c_2x + \cdots + nc_nx^{n-1}.$$

We also use the notation $Df = f'$. Differentiation is linear, that is, D is a linear operator on $F[x]$. We have the higher order formal derivatives $f'' = D^2f$, $f^{(3)} = D^3f$, and so on.

Theorem 5 (Taylor's Formula). *Let* F *be a field of characteristic zero,* c *an element of* F, *and* n *a positive integer. If* f *is a polynomial over* f *with* deg f ≤ n, *then*

$$f = \sum_{k=0}^{n} \frac{(D^kf)}{k!}(c)(x - c)^k.$$

Proof. Taylor's formula is a consequence of the binomial theorem and the linearity of the operators $D, D^2, \ldots, D^n$. The binomial theorem is easily proved by induction and asserts that

$$(a + b)^m = \sum_{k=0}^{m} \binom{m}{k} a^{m-k}b^k$$

where

$$\binom{m}{k} = \frac{m!}{k!(m-k)!} = \frac{m(m-1)\cdots(m-k+1)}{1 \cdot 2 \cdots k}$$

is the familiar binomial coefficient giving the number of combinations of m objects taken k at a time. By the binomial theorem

$$\begin{aligned} x^m &= [c + (x - c)]^m \\ &= \sum_{k=0}^{m} \binom{m}{k} c^{m-k}(x - c)^k \\ &= c^m + mc^{m-1}(x - c) + \cdots + (x - c)^m \end{aligned}$$

and this is the statement of Taylor's formula for the case $f = x^m$. If

$$f = \sum_{m=0}^{n} a_m x^m$$

then

$$D^k f(c) = \sum_m a_m (D^k x^m)(c)$$

and

$$\begin{aligned}\sum_{k=0}^{n} \frac{D^k f(c)}{k!}(x-c)^k &= \sum_k \sum_m a_m \frac{(D^k x^m)}{k!}(c)(x-c)^k \\ &= \sum_m a_m \sum_k \frac{(D^k x^m)}{k!}(c)(x-c)^k \\ &= \sum_m a_m x^m \\ &= f. \quad \blacksquare\end{aligned}$$

It should be noted that because the polynomials $1, (x-c), \ldots, (x-c)^n$ are linearly independent (cf. Exercise 6, Section 4.2) Taylor's formula provides the unique method for writing f as a linear combination of the polynomials $(x-c)^k$ $(0 \leq k \leq n)$.

Although we shall not give any details, it is perhaps worth mentioning at this point that with the proper interpretation Taylor's formula is also valid for polynomials over fields of finite characteristic. If the field F has finite characteristic (the sum of some finite number of 1's in F is 0) then we may have $k! = 0$ in F, in which case the division of $(D^k f)(c)$ by $k!$ is meaningless. Nevertheless, sense can be made out of the division of $D^k f$ by $k!$, because every coefficient of $D^k f$ is an element of F multiplied by an integer divisible by $k!$ If all of this seems confusing, we advise the reader to restrict his attention to fields of characteristic 0 or to subfields of the complex numbers.

If c is a root of the polynomial f, the **multiplicity** of c as a root of f is the largest positive integer r such that $(x-c)^r$ divides f.

The multiplicity of a root is clearly less than or equal to the degree of f. For polynomials over fields of characteristic zero, the multiplicity of c as a root of f is related to the number of derivatives of f that are 0 at c.

Theorem 6. *Let F be a field of characteristic zero and f a polynomial over F with deg f $\leq$ n. Then the scalar c is a root of f of multiplicity r if and only if*

$$(D^k f)(c) = 0, \qquad 0 \leq k \leq r-1$$
$$(D^r f)(c) \neq 0.$$

Proof. Suppose that r is the multiplicity of c as a root of f. Then there is a polynomial g such that $f = (x-c)^r g$ and $g(c) \neq 0$. For otherwise f would be divisible by $(x - c)^{r+1}$, by Corollary 1 of Theorem 4. By Taylor's formula applied to g

$$f = (x - c)^r \left[\sum_{m=0}^{n-r} \frac{(D^m g)}{m!} (c) (x - c)^m \right]$$

$$= \sum_{m=0}^{n-r} \frac{(D^m g)}{m!} (x - c)^{r+m}$$

Since there is only one way to write f as a linear combination of the powers $(x - c)^k$ $(0 \leq k \leq n)$ it follows that

$$\frac{(D^k f)(c)}{k!} = \begin{cases} 0 \text{ if } 0 \leq k \leq r - 1 \\ \dfrac{D^{k-r}g(c)}{(k - r)!} \text{ if } r \leq k \leq n. \end{cases}$$

Therefore, $D^k f(c) = 0$ for $0 \leq k \leq r - 1$, and $D^r f(c) = g(c) \neq 0$. Conversely, if these conditions are satisfied, it follows at once from Taylor's formula that there is a polynomial g such that $f = (x - c)^r g$ and $g(c) \neq 0$. Now suppose that r is not the largest positive integer such that $(x - c)^r$ divides f. Then there is a polynomial h such that $f = (x - c)^{r+1}h$. But this implies $g = (x - c)h$, by Corollary 2 of Theorem 1; hence $g(c) = 0$, a contradiction. ∎

Definition. *Let* F *be a field. An* **ideal** *in* F[x] *is a subspace* M *of* F[x] *such that* fg *belongs to* M *whenever* f *is in* F[x] *and* g *is in* M.

Example 5. If F is a field and d is a polynomial over F, the set $M = dF[x]$, of all multiples df of d by arbitrary f in $F[x]$, is an ideal. For M is non-empty, M in fact contains d. If f, g belong to $F[x]$ and c is a scalar, then

$$c(df) - dg = d(cf - g)$$

belongs to M, so that M is a subspace. Finally M contains $(df)g = d(fg)$ as well. The ideal M is called the **principal ideal generated by** d.

Example 6. Let $d_1, \ldots, d_n$ be a finite number of polynomials over F. Then the sum M of the subspaces $d_iF[x]$ is a subspace and is also an ideal. For suppose p belongs to M. Then there exist polynomials $f_1, \ldots, f_n$ in $F[x]$ such that $p = d_1f_1 + \cdots + d_nf_n$. If g is an arbitrary polynomial over F, then

$$pg = d_1(f_1g) + \cdots + d_n(f_ng)$$

so that pg also belongs to M. Thus M is an ideal, and we say that M is the ideal **generated** by the polynomials, $d_1, \ldots, d_n$.

Example 7. Let F be a subfield of the complex numbers, and consider the ideal

$$M = (x + 2)F[x] + (x^2 + 8x + 16)F[x].$$

We assert that $M = F[x]$. For M contains

$$x^2 + 8x + 16 - x(x + 2) = 6x + 16$$

and hence M contains $6x + 16 - 6(x + 2) = 4$. Thus the scalar polynomial 1 belongs to M as well as all its multiples.

Theorem 7. *If* F *is a field, and* M *is any non-zero ideal in* F[x], *there is a unique monic polynomial* d *in* F[x] *such that* M *is the principal ideal generated by* d.

Proof. By assumption, M contains a non-zero polynomial; among all non-zero polynomials in M there is a polynomial d of minimal degree. We may assume d is monic, for otherwise we can multiply d by a scalar to make it monic. Now if f belongs to M, Theorem 4 shows that $f = dq + r$ where $r = 0$ or $\deg r < \deg d$. Since d is in M, dq and $f - dq = r$ also belong to M. Because d is an element of M of minimal degree we cannot have $\deg r < \deg d$, so $r = 0$. Thus $M = dF[x]$. If g is another monic polynomial such that $M = gF[x]$, then there exist non-zero polynomials p, q such that $d = gp$ and $g = dq$. Thus $d = dpq$ and

$$\deg d = \deg d + \deg p + \deg q.$$

Hence $\deg p = \deg q = 0$, and as d, g are monic, $p = q = 1$. Thus $d = g$. ▌

It is worth observing that in the proof just given we have used a special case of a more general and rather useful fact; namely, if p is a non-zero polynomial in an ideal M and if f is a polynomial in M which is not divisible by p, then $f = pq + r$ where the 'remainder' r belongs to M, is different from 0, and has smaller degree than p. We have already made use of this fact in Example 7 to show that the scalar polynomial 1 is the monic generator of the ideal considered there. In principle it is always possible to find the monic polynomial generating a given non-zero ideal. For one can ultimately obtain a polynomial in the ideal of minimal degree by a finite number of successive divisions.

Corollary. *If* $p_1, \ldots, p_n$ *are polynomials over a field* F, *not all of which are* 0, *there is a unique monic polynomial* d *in* F[x] *such that*

(a) d *is in the ideal generated by* $p_1, \ldots, p_n$;

(b) d *divides each of the polynomials* p_i.

Any polynomial satisfying (a) *and* (b) *necessarily satisfies*

(c) d *is divisible by every polynomial which divides each of the polynomials* $p_1, \ldots, p_n$.

Proof. Let d be the monic generator of the ideal

$$p_1F[x] + \cdots + p_nF[x].$$

Every member of this ideal is divisible by d; thus each of the polynomials p_i is divisible by d. Now suppose f is a polynomial which divides each of the polynomials $p_1, \ldots, p_n$. Then there exist polynomials $g_1, \ldots, g_n$ such that $p_i = fg_i$, $1 \leq i \leq n$. Also, since d is in the ideal

$$p_1F[x] + \cdots + p_nF[x],$$

there exist polynomials $q_1, \ldots, q_n$ in $F[x]$ such that

$$d = p_1q_1 + \cdots + p_nq_n.$$

Thus

$$d = f[g_1q_1 + \cdots + g_nq_n].$$

We have shown that d is a monic polynomial satisfying (a), (b), and (c). If d' is any polynomial satisfying (a) and (b) it follows, from (a) and the definition of d, that d' is a scalar multiple of d and satisfies (c) as well. Finally, in case d' is a monic polynomial, we have $d' = d$. ∎

Definition. *If* $p_1, \ldots, p_n$ *are polynomials over a field* F, *not all of which are* 0, *the monic generator* d *of the ideal*

$$p_1F[x] + \cdots + p_nF[x]$$

is called the **greatest common divisor** *(g.c.d.) of* $p_1, \ldots, p_n$. *This terminology is justified by the preceding corollary. We say that the polynomials* $p_1, \ldots, p_n$ *are* **relatively prime** *if their greatest common divisor is* 1, *or equivalently if the ideal they generate is all of* F[x].

Example 8. Let C be the field of complex numbers. Then

(a) g.c.d. $(x + 2, x^2 + 8x + 16) = 1$ (see Example 7);

(b) g.c.d. $((x - 2)^2(x + i), (x - 2)(x^2 + 1)) = (x - 2)(x + i)$. For, the ideal

$$(x - 2)^2(x + i)F[x] + (x - 2)(x^2 + 1)F[x]$$

contains

$$(x - 2)^2(x + i) - (x - 2)(x^2 + 1) = (x - 2)(x + i)(i - 2).$$

Hence it contains $(x - 2)(x + i)$, which is monic and divides both

$$(x - 2)^2(x + i) \quad \text{and} \quad (x - 2)(x^2 + 1).$$

Example 9. Let F be the field of rational numbers and in $F[x]$ let M be the ideal generated by

$$(x - 1)(x + 2)^2, \qquad (x + 2)^2(x - 3), \qquad \text{and} \qquad (x - 3).$$

Then M contains

$$\tfrac{1}{2}(x + 2)^2[(x - 1) - (x - 3)] = (x + 2)^2$$

and since

$$(x + 2)^2 = (x - 3)(x + 7) - 17$$

M contains the scalar polynomial 1. Thus $M = F[x]$ and the polynomials

$$(x-1)(x+2)^2, \quad (x+2)^2(x-3), \quad \text{and} \quad (x-3)$$

are relatively prime.

Exercises

1. Let Q be the field of rational numbers. Determine which of the following subsets of $Q[x]$ are ideals. When the set is an ideal, find its monic generator.
   (a) all f of even degree;
   (b) all f of degree ≥ 5;
   (c) all f such that $f(0) = 0$;
   (d) all f such that $f(2) = f(4) = 0$;
   (e) all f in the range of the linear operator T defined by

$$T\left(\sum_{i=0}^{n} c_i x^i\right) = \sum_{i=0}^{n} \frac{c_i}{i+1} x^{i+1}.$$

2. Find the g.c.d. of each of the following pairs of polynomials
   (a) $2x^5 - x^3 - 3x^2 - 6x + 4$, $x^4 + x^3 - x^2 - 2x - 2$;
   (b) $3x^4 + 8x^2 - 3$, $x^3 + 2x^2 + 3x + 6$;
   (c) $x^4 - 2x^3 - 2x^2 - 2x - 3$, $x^3 + 6x^2 + 7x + 1$.

3. Let A be an $n \times n$ matrix over a field F. Show that the set of all polynomials f in $F[x]$ such that $f(A) = 0$ is an ideal.

4. Let F be a subfield of the complex numbers, and let

$$A = \begin{bmatrix} 1 & -2 \\ 0 & 3 \end{bmatrix}.$$

Find the monic generator of the ideal of all polynomials f in $F[x]$ such that $f(A) = 0$.

5. Let F be a field. Show that the intersection of any number of ideals in $F[x]$ is an ideal.

6. Let F be a field. Show that the ideal generated by a finite number of polynomials $f_1, \ldots, f_n$ in $F[x]$ is the intersection of all ideals containing $f_1, \ldots, f_n$.

7. Let K be a subfield of a field F, and suppose f, g are polynomials in $K[x]$. Let M_K be the ideal generated by f and g in $K[x]$ and M_F be the ideal they generate in $F[x]$. Show that M_K and M_F have the same monic generator.

4.5. The Prime Factorization of a Polynomial

In this section we shall prove that each polynomial over the field F can be written as a product of 'prime' polynomials. This factorization provides us with an effective tool for finding the greatest common divisor of a finite number of polynomials, and in particular, provides an effective means for deciding when the polynomials are relatively prime.

Definition. *Let* F *be a field. A polynomial* f *in* F[x] *is said to be* **reducible over** F *if there exist polynomials* g, h *in* F[x] *of degree* ≥ 1 *such that* f = gh, *and if not,* f *is said to be* **irreducible over** F. *A non-scalar irreducible polynomial over* F *is called a* **prime polynomial over** F, *and we sometimes say it is a* **prime in** F[x].

EXAMPLE 10. The polynomial $x^2 + 1$ is reducible over the field C of complex numbers. For

$$x^2 + 1 = (x + i)(x - i)$$

and the polynomials $x + i$, $x - i$ belong to $C[x]$. On the other hand, $x^2 + 1$ is irreducible over the field R of real numbers. For if

$$x^2 + 1 = (ax + b)(a'x + b')$$

with a, a', b, b' in R, then

$$aa' = 1, \qquad ab' + ba' = 0, \qquad bb' = 1.$$

These relations imply $a^2 + b^2 = 0$, which is impossible with real numbers a and b, unless $a = b = 0$.

Theorem 8. *Let* p, f, *and* g *be polynomials over the field* F. *Suppose that* p *is a prime polynomial and that* p *divides the product* fg. *Then either* p *divides* f *or* p *divides* g.

Proof. It is no loss of generality to assume that p is a monic prime polynomial. The fact that p is prime then simply says that the only monic divisors of p are 1 and p. Let d be the g.c.d. of f and p. Then either $d = 1$ or $d = p$, since d is a monic polynomial which divides p. If $d = p$, then p divides f and we are done. So suppose $d = 1$, i.e., suppose f and p are relatively prime. We shall prove that p divides g. Since $(f, p) = 1$, there are polynomials f_0 and p_0 such that $1 = f_0 f + p_0 p$. Multiplying by g, we obtain

$$\begin{aligned} g &= f_0 f g + p_0 p g \\ &= (fg) f_0 + p(p_0 g). \end{aligned}$$

Since p divides fg it divides $(fg)f_0$, and certainly p divides $p(p_0 g)$. Thus p divides g. ∎

Corollary. *If* p *is a prime and divides a product* $f_1 \cdots f_n$, *then* p *divides one of the polynomials* $f_1, \ldots, f_n$.

Proof. The proof is by induction. When $n = 2$, the result is simply the statement of Theorem 6. Suppose we have proved the corollary for $n = k$, and that p divides the product $f_1 \cdots f_{k+1}$ of some $(k + 1)$ polynomials. Since p divides $(f_1 \cdots f_k)f_{k+1}$, either p divides f_{k+1} or p divides $f_1 \cdots f_k$. By the induction hypothesis, if p divides $f_1 \cdots f_k$, then p divides f_j for some j, $1 \leq j \leq k$. So we see that in any case p must divide some f_j, $1 \leq j \leq k + 1$. ∎

Theorem 9. *If* F *is a field, a non-scalar monic polynomial in* F[x] *can be factored as a product of monic primes in* F[x] *in one and, except for order, only one way.*

Proof. Suppose f is a non-scalar monic polynomial over F. As polynomials of degree one are irreducible, there is nothing to prove if $\deg f = 1$. Suppose f has degree $n > 1$. By induction we may assume the theorem is true for all non-scalar monic polynomials of degree less than n. If f is irreducible, it is already factored as a product of monic primes, and otherwise $f = gh$ where g and h are non-scalar monic polynomials of degree less than n. Thus g and h can be factored as products of monic primes in $F[x]$ and hence so can f. Now suppose

$$f = p_1 \cdots p_m = q_1 \cdots q_n$$

where $p_1, \ldots, p_m$ and $q_1, \ldots, q_n$ are monic primes in $F[x]$. Then p_m divides the product $q_1 \cdots q_n$. By the above corollary, p_m must divide some q_i. Since q_i and p_m are both monic primes, this means that

$$q_i = p_m. \tag{4-16}$$

From (4-16) we see that $m = n = 1$ if either $m = 1$ or $n = 1$. For

$$\deg f = \sum_{i=1}^{m} \deg p_i = \sum_{j=1}^{n} \deg q_j.$$

In this case there is nothing more to prove, so we may assume $m > 1$ and $n > 1$. By rearranging the q's we can then assume $p_m = q_n$, and that

$$p_1 \cdots p_{m-1}p_m = q_1 \cdots q_{n-1}p_m.$$

Now by Corollary 2 of Theorem 1 it follows that

$$p_1 \cdots p_{m-1} = q_1 \cdots q_{n-1}.$$

As the polynomial $p_1 \cdots p_{m-1}$ has degree less than n, our inductive assumption applies and shows that the sequence $q_1, \ldots, q_{n-1}$ is at most a rearrangement of the sequence $p_1, \ldots, p_{m-1}$. This together with (4-16) shows that the factorization of f as a product of monic primes is unique up to the order of the factors. ∎

In the above factorization of a given non-scalar monic polynomial f, some of the monic prime factors may be repeated. If $p_1, p_2, \ldots, p_r$ are the distinct monic primes occurring in this factorization of f, then

$$f = p_1^{n_1}p_2^{n_2} \cdots p_r^{n_r}, \tag{4-17}$$

the exponent n_i being the number of times the prime p_i occurs in the factorization. This decomposition is also clearly unique, and is called the **primary decomposition** of f. It is easily verified that every monic divisor of f has the form

$$p_1^{m_1}p_2^{m_2}\cdots p_r^{m_r}, \qquad 0 \leq m_i \leq n_i. \tag{4-18}$$

From (4-18) it follows that the g.c.d. of a finite number of non-scalar monic polynomials $f_1, \ldots, f_s$ is obtained by combining all those monic primes which occur simultaneously in the factorizations of $f_1, \ldots, f_s$. The exponent to which each prime is to be taken is the largest for which the corresponding prime power is a factor of each f_i. If no (non-trivial) prime power is a factor of each f_i, the polynomials are relatively prime.

EXAMPLE 11. Suppose F is a field, and let a, b, c be distinct elements of F. Then the polynomials $x - a$, $x - b$, $x - c$ are distinct monic primes in $F[x]$. If m, n, and s are positive integers, $(x - c)^s$ is the g.c.d. of the polynomials.

$$(x - b)^n(x - c)^s \quad \text{and} \quad (x - a)^m(x - c)^s$$

whereas the three polynomials

$$(x - b)^n(x - c)^s, \qquad (x - a)^m(x - c)^s, \qquad (x - a)^m(x - b)^n$$

are relatively prime.

Theorem 10. *Let* f *be a non-scalar monic polynomial over the field* F *and let*

$$f = p_1^{n_1} \cdots p_k^{n_k}$$

be the prime factorization of f. *For each* j, $1 \leq j \leq k$, *let*

$$f_j = f/p_j^{n_j} = \prod_{i \neq j} p_i^{n_i}.$$

Then $f_1, \ldots, f_k$ *are relatively prime.*

Proof. We leave the (easy) proof of this to the reader. We have stated this theorem largely because we wish to refer to it later. ∎

Theorem 11. *Let* f *be a polynomial over the field* F *with derivative* f'. *Then* f *is a product of distinct irreducible polynomials over* F *if and only if* f *and* f' *are relatively prime.*

Proof. Suppose in the prime factorization of f over the field F that some (non-scalar) prime polynomial p is repeated. Then $f = p^2h$ for some h in $F[x]$. Then

$$f' = p^2h' + 2pp'h$$

and p is also a divisor of f'. Hence f and f' are not relatively prime.

Now suppose $f = p_1 \cdots p_k$, where $p_1, \ldots, p_k$ are distinct non-scalar irreducible polynomials over F. Let $f_j = f/p_j$. Then

$$f' = p_1'f_1 + p_2'f_2 + \cdots + p_k'f_k.$$

Let p be a prime polynomial which divides both f and f'. Then $p = p_i$ for some i. Now p_i divides f_j for $j \neq i$, and since p_i also divides

$$f' = \sum_{j=1}^{k} p_j' f_j$$

we see that p_i must divide $p_i' f_i$. Therefore p_i divides either f_i or p_i'. But p_i does not divide f_i since $p_1, \ldots, p_k$ are distinct. So p_i divides p_i'. This is not possible, since p_i' has degree one less than the degree of p_i. We conclude that no prime divides both f and f', or that, f and f' are relatively prime. ∎

Definition. *The field* F *is called* **algebraically closed** *if every prime polynomial over* F *has degree* 1.

To say that F is algebraically closed means every non-scalar irreducible monic polynomial over F is of the form $(x - c)$. We have already observed that each such polynomial is irreducible for any F. Accordingly, an equivalent definition of an algebraically closed field is a field F such that each non-scalar polynomial f in $F[x]$ can be expressed in the form

$$f = c(x - c_1)^{n_1} \cdots (x - c_k)^{n_k}$$

where c is a scalar, $c_1, \ldots, c_k$ are distinct elements of F, and $n_1, \ldots, n_k$ are positive integers. Still another formulation is that if f is a non-scalar polynomial over F, then there is an element c in F such that $f(c) = 0$.

The field R of real numbers is not algebraically closed, since the polynomial $(x^2 + 1)$ is irreducible over R but not of degree 1, or, because there is no real number c such that $c^2 + 1 = 0$. The so-called Fundamental Theorem of Algebra states that the field C of complex numbers is algebraically closed. We shall not prove this theorem, although we shall use it somewhat later in this book. The proof is omitted partly because of the limitations of time and partly because the proof depends upon a 'non-algebraic' property of the system of real numbers. For one possible proof the interested reader may consult the book by Schreier and Sperner in the Bibliography.

The Fundamental Theorem of Algebra also makes it clear what the possibilities are for the prime factorization of a polynomial with real coefficients. If f is a polynomial with real coefficients and c is a complex root of f, then the complex conjugate $\bar{c}$ is also a root of f. Therefore, those complex roots which are not real must occur in conjugate pairs, and the entire set of roots has the form $\{t_1, \ldots, t_k, c_1, \bar{c}_1, \ldots, c_r, \bar{c}_r\}$ where $t_1, \ldots, t_k$ are real and $c_1, \ldots, c_r$ are non-real complex numbers. Thus f factors

$$f = c(x - t_1) \cdots (x - t_k)p_1 \cdots p_r$$

where p_i is the quadratic polynomial

$$p_i = (x - c_i)(x - \bar{c}_i).$$

These polynomials p_i have real coefficients. We conclude that every irreducible polynomial over the real number field has degree 1 or 2. Each polynomial over R is the product of certain linear factors, obtained from the real roots of f, and certain irreducible quadratic polynomials.

Exercises

1. Let p be a monic polynomial over the field F, and let f and g be relatively prime polynomials over F. Prove that the g.c.d. of pf and pg is p.

2. Assuming the Fundamental Theorem of Algebra, prove the following. If f and g are polynomials over the field of complex numbers, then g.c.d. $(f, g) = 1$ if and only if f and g have no common root.

3. Let D be the differentiation operator on the space of polynomials over the field of complex numbers. Let f be a monic polynomial over the field of complex numbers. Prove that

$$f = (x - c_1) \cdots (x - c_k)$$

where $c_1, \ldots, c_k$ are *distinct* complex numbers if and only if f and Df are relatively prime. In other words, f has no repeated root if and only if f and Df have no common root. (Assume the Fundamental Theorem of Algebra.)

4. Prove the following generalization of Taylor's formula. Let f, g, and h be polynomials over a subfield of the complex numbers, with $\deg f \leq n$. Then

$$f(g) = \sum_{k=0}^{n} \frac{1}{k!} f^{(k)}(h)(g - h)^k.$$

(Here $f(g)$ denotes 'f of g.')

For the remaining exercises, we shall need the following definition. If f, g, and p are polynomials over the field F with $p \neq 0$, we say that f is **congruent to** g **modulo** p if $(f - g)$ is divisible by p. If f is congruent to g modulo p, we write

$$f \equiv g \bmod p.$$

5. Prove, for any non-zero polynomial p, that congruence modulo p is an equivalence relation.

(a) It is reflexive: $f \equiv f \bmod p$.
(b) It is symmetric: if $f \equiv g \bmod p$, then $g \equiv f \bmod p$.
(c) It is transitive: if $f \equiv g \bmod p$ and $g \equiv h \bmod p$, then $f \equiv h \bmod p$.

6. Suppose $f \equiv g \bmod p$ and $f_1 \equiv g_1 \bmod p$.

(a) Prove that $f + f_1 \equiv g + g_1 \bmod p$.
(b) Prove that $ff_1 \equiv gg_1 \bmod p$.

7. Use Exercise 7 to prove the following. If f, g, h, and p are polynomials over the field F and $p \neq 0$, and if $f \equiv g \bmod p$, then $h(f) \equiv h(g) \bmod p$.

8. If p is an irreducible polynomial and $fg \equiv 0 \bmod p$, prove that either $f \equiv 0 \bmod p$ or $g \equiv 0 \bmod p$. Give an example which shows that this is false if p is not irreducible.

5. *Determinants*

5.1. *Commutative Rings*

In this chapter we shall prove the essential facts about determinants of square matrices. We shall do this not only for matrices over a field, but also for matrices with entries which are 'scalars' of a more general type. There are two reasons for this generality. First, at certain points in the next chapter, we shall find it necessary to deal with determinants of matrices with polynomial entries. Second, in the treatment of determinants which we present, one of the axioms for a field plays no role, namely, the axiom which guarantees a multiplicative inverse for each non-zero element. For these reasons, it is appropriate to develop the theory of determinants for matrices, the entries of which are elements from a commutative ring with identity.

Definition. *A* **ring** *is a set* K, *together with two operations* $(x, y) \rightarrow x + y$ *and* $(x, y) \rightarrow xy$ *satisfying*

(a) K *is a commutative group under the operation* $(x, y) \rightarrow x + y$ (K *is a commutative group under addition*);

(b) $(xy)z = x(yz)$ (*multiplication is associative*);

(c) $x(y + z) = xy + xz$; $(y + z)x = yx + zx$ (*the two distributive laws hold*).

If $xy = yx$ *for all* x *and* y *in* K, *we say that the ring* K *is* **commutative.** *If there is an element* 1 *in* K *such that* $1x = x1 = x$ *for each* x, K *is said to be a* **ring with identity,** *and* 1 *is called the* **identity** *for* K.

We are interested here in commutative rings with identity. Such a ring can be described briefly as a set K, together with two operations which satisfy all the axioms for a field given in Chapter 1, except possibly for axiom (8) and the condition $1 \neq 0$. Thus, a field is a commutative ring with non-zero identity such that to each non-zero x there corresponds an element x^{-1} with $xx^{-1} = 1$. The set of integers, with the usual operations, is a commutative ring with identity which is not a field. Another commutative ring with identity is the set of all polynomials over a field, together with the addition and multiplication which we have defined for polynomials.

If K is a commutative ring with identity, we define an $m \times n$ matrix over K to be a function A from the set of pairs (i, j) of integers, $1 \leq i \leq m$, $1 \leq j \leq n$, into K. As usual we represent such a matrix by a rectangular array having m rows and n columns. The sum and product of matrices over K are defined as for matrices over a field

$$\begin{aligned}(A + B)_{ij} &= A_{ij} + B_{ij}\\ (AB)_{ij} &= \sum_k A_{ik}B_{kj}\end{aligned}$$

the sum being defined when A and B have the same number of rows and the same number of columns, the product being defined when the number of columns of A is equal to the number of rows of B. The basic algebraic properties of these operations are again valid. For example,

$$A(B + C) = AB + AC, \qquad (AB)C = A(BC), \qquad \text{etc.}$$

As in the case of fields, we shall refer to the elements of K as scalars. We may then define linear combinations of the rows or columns of a matrix as we did earlier. Roughly speaking, all that we previously did for matrices over a field is valid for matrices over K, excluding those results which depended upon the ability to 'divide' in K.

5.2. Determinant Functions

Let K be a commutative ring with identity. We wish to assign to each $n \times n$ (square) matrix over K a scalar (element of K) to be known as the determinant of the matrix. It is possible to define the determinant of a square matrix A by simply writing down a formula for this determinant in terms of the entries of A. One can then deduce the various properties of determinants from this formula. However, such a formula is rather complicated, and to gain some technical advantage we shall proceed as follows. We shall define a 'determinant function' on $K^{n \times n}$ as a function which assigns to each $n \times n$ matrix over K a scalar, the function having these special properties. It is linear as a function of each of the rows of the matrix: its value is 0 on any matrix having two equal rows; and its value on the $n \times n$ identity matrix is 1. We shall prove that such a function exists, and then that it is unique, i.e., that there is precisely one such function. As we prove the uniqueness, an explicit formula for the determinant will be obtained, along with many of its useful properties.

This section will be devoted to the definition of 'determinant function' and to the proof that at least one such function exists.

Definition. *Let* K *be a commutative ring with identity,* n *a positive integer, and let* D *be a function which assigns to each* $n \times n$ *matrix* A *over* K *a scalar* D(A) *in* K. *We say that* D *is* **n-linear** *if for each* i, $1 \leq i \leq n$, D *is a linear function of the* ith *row when the other* $(n-1)$ *rows are held fixed.*

This definition requires some clarification. If D is a function from $K^{n\times n}$ into K, and if $\alpha_1, \ldots, \alpha_n$ are the rows of the matrix A, let us also write

$$D(A) = D(\alpha_1, \ldots, \alpha_n)$$

that is, let us also think of D as the function of the rows of A. The statement that D is n-linear then means

$$D(\alpha_1, \ldots, c\alpha_i + \alpha_i', \ldots, \alpha_n) = cD(\alpha_1, \ldots, \alpha_i, \ldots, \alpha_n) + D(\alpha_1, \ldots, \alpha_i', \ldots, \alpha_n). \tag{5-1}$$

If we fix all rows except row i and regard D as a function of the ith row, it is often convenient to write $D(\alpha_i)$ for $D(A)$. Thus, we may abbreviate (5-1) to

$$D(c\alpha_i + \alpha_i') = cD(\alpha_i) + D(\alpha_i')$$

so long as it is clear what the meaning is.

Example 1. Let $k_1, \ldots, k_n$ be positive integers, $1 \leq k_i \leq n$, and let a be an element of K. For each $n \times n$ matrix A over K, define

$$D(A) = aA(1, k_1) \cdots A(n, k_n). \tag{5-2}$$

Then the function D defined by (5-2) is n-linear. For, if we regard D as a function of the ith row of A, the others being fixed, we may write

$$D(\alpha_i) = A(i, k_i)b$$

where b is some fixed element of K. Let $\alpha_i' = (A_{i1}', \ldots, A_{in}')$. Then we have

$$\begin{aligned} D(c\alpha_i + \alpha_i') &= [cA(i, k_i) + A'(i, k_i)]b \\ &= cD(\alpha_i) + D(\alpha_i'). \end{aligned}$$

Thus D is a linear function of each of the rows of A.

A particular n-linear function of this type is

$$D(A) = A_{11}A_{22} \cdots A_{nn}.$$

In other words, the 'product of the diagonal entries' is an n-linear function on $K^{n\times n}$.

EXAMPLE 2. Let us find all 2-linear functions on 2×2 matrices over K. Let D be such a function. If we denote the rows of the 2×2 identity matrix by ϵ_1, ϵ_2, we have

$$D(A) = D(A_{11}\epsilon_1 + A_{12}\epsilon_2, A_{21}\epsilon_1 + A_{22}\epsilon_2).$$

Using the fact that D is 2-linear, (5-1), we have

$$\begin{aligned} D(A) &= A_{11}D(\epsilon_1, A_{21}\epsilon_1 + A_{22}\epsilon_2) + A_{12}D(\epsilon_2, A_{21}\epsilon_1 + A_{22}\epsilon_2) \\ &= A_{11}A_{21}D(\epsilon_1, \epsilon_1) + A_{11}A_{22}D(\epsilon_1, \epsilon_2) \\ &\qquad + A_{12}A_{21}D(\epsilon_2, \epsilon_1) + A_{12}A_{22}D(\epsilon_2, \epsilon_2). \end{aligned}$$

Thus D is completely determined by the four scalars

$$D(\epsilon_1, \epsilon_1), \qquad D(\epsilon_1, \epsilon_2), \qquad D(\epsilon_2, \epsilon_1), \qquad \text{and} \qquad D(\epsilon_2, \epsilon_2).$$

The reader should find it easy to verify the following. If a, b, c, d are any four scalars in K and if we define

$$D(A) = A_{11}A_{21}a + A_{11}A_{22}b + A_{12}A_{21}c + A_{12}A_{22}d$$

then D is a 2-linear function on 2×2 matrices over K and

$$\begin{aligned} D(\epsilon_1, \epsilon_1) &= a, \qquad & D(\epsilon_1, \epsilon_2) &= b \\ D(\epsilon_2, \epsilon_1) &= c, & D(\epsilon_2, \epsilon_2) &= d. \end{aligned}$$

Lemma. *A linear combination of* n*-linear functions is* n*-linear.*

Proof. It suffices to prove that a linear combination of two n-linear functions is n-linear. Let D and E be n-linear functions. If a and b belong to K, the linear combination $aD + bE$ is of course defined by

$$(aD + bE)(A) = aD(A) + bE(A).$$

Hence, if we fix all rows except row i

$$\begin{aligned} (aD + bE)(c\alpha_i + \alpha_i') &= aD(c\alpha_i + \alpha_i') + bE(c\alpha_i + \alpha_i') \\ &= acD(\alpha_i) + aD(\alpha_i') + bcE(\alpha_i) + bE(\alpha_i') \\ &= c(aD + bE)(\alpha_i) + (aD + bE)(\alpha_i'). \quad \blacksquare \end{aligned}$$

If K is a field and V is the set of $n \times n$ matrices over K, the above lemma says the following. The set of n-linear functions on V is a subspace of the space of all functions from V into K.

EXAMPLE 3. Let D be the function defined on 2×2 matrices over K by

$$D(A) = A_{11}A_{22} - A_{12}A_{21}. \tag{5-3}$$

Now D is the sum of two functions of the type described in Example 1:

$$\begin{aligned} D &= D_1 + D_2 \\ D_1(A) &= A_{11}A_{22} \\ D_2(A) &= -A_{12}A_{21}. \end{aligned}$$

By the above lemma, D is a 2-linear function. The reader who has had any experience with determinants will not find this surprising, since he will recognize (5-3) as the usual definition of the determinant of a 2×2 matrix. Of course the function D we have just defined is not a typical 2-linear function. It has many special properties. Let us note some of these properties. First, if I is the 2×2 identity matrix, then $D(I) = 1$, i.e., $D(\epsilon_1, \epsilon_2) = 1$. Second, if the two rows of A are equal, then

$$D(A) = A_{11}A_{12} - A_{12}A_{11} = 0.$$

Third, if A' is the matrix obtained from a 2×2 matrix A by interchanging its rows, then $D(A') = -D(A)$; for

$$\begin{aligned} D(A') &= A'_{11}A'_{22} - A'_{12}A'_{21} \\ &= A_{21}A_{12} - A_{22}A_{11} \\ &= -D(A). \end{aligned}$$

Definition. *Let* D *be an* n-*linear function. We say* D *is* **alternating** (*or* **alternate**) *if the following two conditions are satisfied:*

(a) D(A) = 0 *whenever two rows of* A *are equal.*

(b) *If* A′ *is a matrix obtained from* A *by interchanging two rows of* A, *then* D(A′) = −D(A).

We shall prove below that any n-linear function D which satisfies (a) automatically satisfies (b). We have put both properties in the definition of alternating n-linear function as a matter of convenience. The reader will probably also note that if D satisfies (b) and A is a matrix with two equal rows, then $D(A) = -D(A)$. It is tempting to conclude that D satisfies condition (a) as well. This is true, for example, if K is a field in which $1 + 1 \neq 0$, but in general (a) is not a consequence of (b).

Definition. *Let* K *be a commutative ring with identity, and let* n *be a positive integer. Suppose* D *is a function from* n × n *matrices over* K *into* K. *We say that* D *is a* **determinant function** *if* D *is* n-*linear, alternating, and* D(I) = 1.

As we stated earlier, we shall ultimately show that there is exactly one determinant function on $n \times n$ matrices over K. This is easily seen for 1×1 matrices $A = [a]$ over K. The function D given by $D(A) = a$ is a determinant function, and clearly this is the only determinant function on 1×1 matrices. We are also in a position to dispose of the case $n = 2$. The function

$$D(A) = A_{11}A_{22} - A_{12}A_{21}$$

was shown in Example 3 to be a determinant function. Furthermore, the formula exhibited in Example 2 shows that D is the only determinant function on 2×2 matrices. For we showed that for any 2-linear function D

$$D(A) = A_{11}A_{21}D(\epsilon_1, \epsilon_1) + A_{11}A_{22}D(\epsilon_1, \epsilon_2) + A_{12}A_{21}D(\epsilon_2, \epsilon_1) + A_{12}A_{22}D(\epsilon_2, \epsilon_2).$$

If D is alternating, then

$$D(\epsilon_1, \epsilon_1) = D(\epsilon_2, \epsilon_2) = 0$$

and

$$D(\epsilon_2, \epsilon_1) = -D(\epsilon_1, \epsilon_2) = -D(I).$$

If D also satisfies $D(I) = 1$, then

$$D(A) = A_{11}A_{22} - A_{12}A_{21}.$$

EXAMPLE 4. Let F be a field and let D be any alternating 3-linear function on 3×3 matrices over the polynomial ring $F[x]$.

Let

$$A = \begin{bmatrix} x & 0 & -x^2 \\ 0 & 1 & 0 \\ 1 & 0 & x^3 \end{bmatrix}.$$

If we denote the rows of the 3×3 identity matrix by $\epsilon_1, \epsilon_2, \epsilon_3$, then

$$D(A) = D(x\epsilon_1 - x^2\epsilon_3, \epsilon_2, \epsilon_1 + x^3\epsilon_3).$$

Since D is linear as a function of each row,

$$\begin{aligned} D(A) &= xD(\epsilon_1, \epsilon_2, \epsilon_1 + x^3\epsilon_3) - x^2D(\epsilon_3, \epsilon_2, \epsilon_1 + x^3\epsilon_3) \\ &= xD(\epsilon_1, \epsilon_2, \epsilon_1) + x^4D(\epsilon_1, \epsilon_2, \epsilon_3) - x^2D(\epsilon_3, \epsilon_2, \epsilon_1) - x^5D(\epsilon_3, \epsilon_2, \epsilon_3). \end{aligned}$$

Because D is alternating it follows that

$$D(A) = (x^4 + x^2)D(\epsilon_1, \epsilon_2, \epsilon_3).$$

Lemma. *Let* D *be a 2-linear function with the property that* D(A) = 0 *for all* 2×2 *matrices* A *over* K *having equal rows. Then* D *is alternating.*

Proof. What we must show is that if A is a 2×2 matrix and A' is obtained by interchanging the rows of A, then $D(A') = -D(A)$. If the rows of A are α and β, this means we must show that $D(\beta, \alpha) = -D(\alpha, \beta)$. Since D is 2-linear,

$$D(\alpha + \beta, \alpha + \beta) = D(\alpha, \alpha) + D(\alpha, \beta) + D(\beta, \alpha) + D(\beta, \beta).$$

By our hypothesis $D(\alpha + \beta, \alpha + \beta) = D(\alpha, \alpha) = D(\beta, \beta) = 0$. So

$$0 = D(\alpha, \beta) + D(\beta, \alpha). \quad \blacksquare$$

Lemma. *Let* D *be an* n*-linear function on* $n \times n$ *matrices over* K. *Suppose* D *has the property that* D(A) = 0 *whenever two adjacent rows of* A *are equal. Then* D *is alternating.*

Proof. We must show that $D(A) = 0$ when any two rows of A are equal, and that $D(A') = -D(A)$ if A' is obtained by interchanging some two rows of A. First, let us suppose that A' is obtained by interchanging two adjacent rows of A. The reader should see that the argument used in the proof of the preceding lemma extends to the present case and gives us $D(A') = -D(A)$.

Now let B be obtained by interchanging rows i and j of A, where $i < j$. We can obtain B from A by a succession of interchanges of pairs of adjacent rows. We begin by interchanging row i with row $(i + 1)$ and continue until the rows are in the order

$$\alpha_1, \ldots, \alpha_{i-1}, \alpha_{i+1}, \ldots, \alpha_j, \alpha_i, \alpha_{j+1}, \ldots, \alpha_n.$$

This requires $k = j - i$ interchanges of adjacent rows. We now move α_j to the ith position using $(k - 1)$ interchanges of adjacent rows. We have thus obtained B from A by $k + (k - 1) = 2k - 1$ interchanges of adjacent rows. Thus

$$D(B) = (-1)^{2k-1}D(A) = -D(A).$$

Suppose A is any $n \times n$ matrix with two equal rows, say $\alpha_i = \alpha_j$ with $i < j$. If $j = i + 1$, then A has two equal and adjacent rows and $D(A) = 0$. If $j > i + 1$, we interchange α_{i+1} and α_j and the resulting matrix B has two equal and adjacent rows, so $D(B) = 0$. On the other hand, $D(B) = -D(A)$, hence $D(A) = 0$. ∎

Definition. *If* $n > 1$ *and* A *is an* $n \times n$ *matrix over* K, *we let* $A(i|j)$ *denote the* $(n - 1) \times (n - 1)$ *matrix obtained by deleting the* i*th row and* j*th column of* A. *If* D *is an* $(n - 1)$*-linear function and* A *is an* $n \times n$ *matrix, we put* $D_{ij}(A) = D[A(i|j)]$.

Theorem 1. *Let* $n > 1$ *and let* D *be an alternating* $(n - 1)$*-linear function on* $(n - 1) \times (n - 1)$ *matrices over* K. *For each* j, $1 \leq j \leq n$, *the function* E_j *defined by*

$$E_j(A) = \sum_{i=1}^{n} (-1)^{i+j}A_{ij}D_{ij}(A) \tag{5-4}$$

is an alternating n*-linear function on* $n \times n$ *matrices* A. *If* D *is a determinant function, so is each* E_j.

Proof. If A is an $n \times n$ matrix, $D_{ij}(A)$ is independent of the ith row of A. Since D is $(n - 1)$-linear, it is clear that D_{ij} is linear as a function of any row except row i. Therefore $A_{ij}D_{ij}(A)$ is an n-linear function of A. A linear combination of n-linear functions is n-linear; hence, E_j is n-linear. To prove that E_j is alternating, it will suffice to show that $E_j(A) = 0$ whenever A has two equal and adjacent rows. Suppose $\alpha_k = \alpha_{k+1}$. If $i \neq k$ and $i \neq k + 1$, the matrix $A(i|j)$ has two equal rows, and thus $D_{ij}(A) = 0$. Therefore

$$E_j(A) = (-1)^{k+j}A_{kj}D_{kj}(A) + (-1)^{k+1+j}A_{(k+1)j}D_{(k+1)j}(A).$$

Since $\alpha_k = \alpha_{k+1}$,

$$A_{kj} = A_{(k+1)j} \quad \text{and} \quad A(k|j) = A(k+1|j).$$

Clearly then $E_j(A) = 0$.

Now suppose D is a determinant function. If $I^{(n)}$ is the $n \times n$ identity matrix, then $I^{(n)}(j|j)$ is the $(n-1) \times (n-1)$ identity matrix $I^{(n-1)}$. Since $I_{ij}^{(n)} = \delta_{ij}$, it follows from (5-4) that

$$E_j(I^{(n)}) = D(I^{(n-1)}). \tag{5-5}$$

Now $D(I^{(n-1)}) = 1$, so that $E_j(I^{(n)}) = 1$ and E_j is a determinant function. ∎

Corollary. *Let* K *be a commutative ring with identity and let* n *be a positive integer. There exists at least one determinant function on* $K^{n \times n}$.

Proof. We have shown the existence of a determinant function on 1×1 matrices over K, and even on 2×2 matrices over K. Theorem 1 tells us explicitly how to construct a determinant function on $n \times n$ matrices, given such a function on $(n-1) \times (n-1)$ matrices. The corollary follows by induction. ∎

Example 5. If B is a 2×2 matrix over K, we let

$$|B| = B_{11}B_{22} - B_{12}B_{21}.$$

Then $|B| = D(B)$, where D is the determinant function on 2×2 matrices. We showed that this function on $K^{2 \times 2}$ is unique. Let

$$A = \begin{bmatrix} A_{11} & A_{12} & A_{13} \\ A_{21} & A_{22} & A_{23} \\ A_{31} & A_{32} & A_{33} \end{bmatrix}$$

be a 3×3 matrix over K. If we define E_1, E_2, E_3 as in (5-4), then

$$E_1(A) = A_{11}\begin{vmatrix} A_{22} & A_{23} \\ A_{32} & A_{33} \end{vmatrix} - A_{21}\begin{vmatrix} A_{12} & A_{13} \\ A_{32} & A_{33} \end{vmatrix} + A_{31}\begin{vmatrix} A_{12} & A_{13} \\ A_{22} & A_{23} \end{vmatrix} \tag{5-6}$$

$$E_2(A) = -A_{12}\begin{vmatrix} A_{21} & A_{23} \\ A_{31} & A_{33} \end{vmatrix} + A_{22}\begin{vmatrix} A_{11} & A_{13} \\ A_{31} & A_{33} \end{vmatrix} - A_{32}\begin{vmatrix} A_{11} & A_{13} \\ A_{21} & A_{23} \end{vmatrix} \tag{5-7}$$

$$E_3(A) = A_{13}\begin{vmatrix} A_{21} & A_{22} \\ A_{31} & A_{32} \end{vmatrix} - A_{23}\begin{vmatrix} A_{11} & A_{12} \\ A_{31} & A_{32} \end{vmatrix} + A_{33}\begin{vmatrix} A_{11} & A_{12} \\ A_{21} & A_{22} \end{vmatrix}. \tag{5-8}$$

It follows from Theorem 1 that E_1, E_2, and E_3 are determinant functions. Actually, as we shall show later, $E_1 = E_2 = E_3$, but this is not yet apparent even in this simple case. It could, however, be verified directly, by expanding each of the above expressions. Instead of doing this we give some specific examples.

(a) Let $K = R[x]$ and

$$A = \begin{bmatrix} x-1 & x^2 & x^3 \\ 0 & x-2 & 1 \\ 0 & 0 & x-3 \end{bmatrix}.$$

Then

$$E_1(A) = (x-1)\begin{vmatrix} x-2 & 1 \\ 0 & x-3 \end{vmatrix} = (x-1)(x-2)(x-3)$$

$$\begin{aligned} E_2(A) &= -x^2\begin{vmatrix} 0 & 1 \\ 0 & x-3 \end{vmatrix} + (x-2)\begin{vmatrix} x-1 & x^3 \\ 0 & x-3 \end{vmatrix} \\ &= (x-1)(x-2)(x-3) \end{aligned}$$

and

$$\begin{aligned} E_3(A) &= x^3\begin{vmatrix} 0 & x-2 \\ 0 & 0 \end{vmatrix} - \begin{vmatrix} x-1 & x^2 \\ 0 & 0 \end{vmatrix} + (x-3)\begin{vmatrix} x-1 & x^2 \\ 0 & x-2 \end{vmatrix} \\ &= (x-1)(x-2)(x-3). \end{aligned}$$

(b) Let $K = R$ and

$$A = \begin{bmatrix} 0 & 1 & 0 \\ 0 & 0 & 1 \\ 1 & 0 & 0 \end{bmatrix}.$$

Then

$$E_1(A) = \begin{vmatrix} 1 & 0 \\ 0 & 1 \end{vmatrix} = 1$$

$$E_2(A) = -\begin{vmatrix} 0 & 1 \\ 1 & 0 \end{vmatrix} = 1$$

$$E_3(A) = -\begin{vmatrix} 0 & 1 \\ 1 & 0 \end{vmatrix} = 1.$$

Exercises

1. Each of the following expressions defines a function D on the set of 3×3 matrices over the field of real numbers. In which of these cases is D a 3-linear function?

(a) $D(A) = A_{11} + A_{22} + A_{33}$;
(b) $D(A) = (A_{11})^2 + 3A_{11}A_{22}$;
(c) $D(A) = A_{11}A_{12}A_{33}$;
(d) $D(A) = A_{13}A_{22}A_{32} + 5A_{12}A_{22}A_{32}$;
(e) $D(A) = 0$;
(f) $D(A) = 1$.

2. Verify directly that the three functions E_1, E_2, E_3 defined by (5-6), (5-7), and (5-8) are identical.

3. Let K be a commutative ring with identity. If A is a 2×2 matrix over K, the **classical adjoint** of A is the 2×2 matrix adj A defined by

$$\text{adj } A = \begin{bmatrix} A_{22} & -A_{12} \\ -A_{21} & A_{11} \end{bmatrix}.$$

If det denotes the unique determinant function on 2×2 matrices over K, show that

(a) $(\text{adj } A)A = A(\text{adj } A) = (\det A)I$;
(b) $\det(\text{adj } A) = \det(A)$;
(c) $\text{adj}(A^t) = (\text{adj } A)^t$.
(A^t denotes the transpose of A.)

4. Let A be a 2×2 matrix over a field F. Show that A is invertible if and only if $\det A \neq 0$. When A is invertible, give a formula for A^{-1}.

5. Let A be a 2×2 matrix over a field F, and suppose that $A^2 = 0$. Show for each scalar c that $\det(cI - A) = c^2$.

6. Let K be a subfield of the complex numbers and n a positive integer. Let $j_1, \ldots, j_n$ and $k_1, \ldots, k_n$ be positive integers not exceeding n. For an $n \times n$ matrix A over K define

$$D(A) = A(j_1, k_1)A(j_2, k_2) \cdots A(j_n, k_n).$$

Prove that D is n-linear if and only if the integers $j_1, \ldots, j_n$ are distinct.

7. Let K be a commutative ring with identity. Show that the determinant function on 2×2 matrices A over K is alternating and 2-linear as a function of the columns of A.

8. Let K be a commutative ring with identity. Define a function D on 3×3 matrices over K by the rule

$$D(A) = A_{11} \det \begin{bmatrix} A_{22} & A_{23} \\ A_{32} & A_{33} \end{bmatrix} - A_{12} \det \begin{bmatrix} A_{21} & A_{23} \\ A_{31} & A_{33} \end{bmatrix} + A_{13} \det \begin{bmatrix} A_{21} & A_{22} \\ A_{31} & A_{32} \end{bmatrix}.$$

Show that D is alternating and 3-linear as a function of the columns of A.

9. Let K be a commutative ring with identity and D an alternating n-linear function on $n \times n$ matrices over K. Show that
(a) $D(A) = 0$, if one of the rows of A is 0.
(b) $D(B) = D(A)$, if B is obtained from A by adding a scalar multiple of one row of A to another.

10. Let F be a field, A a 2×3 matrix over F, and (c_1, c_2, c_3) the vector in F^3 defined by

$$c_1 = \begin{vmatrix} A_{12} & A_{13} \\ A_{22} & A_{23} \end{vmatrix}, \quad c_2 = \begin{vmatrix} A_{13} & A_{11} \\ A_{23} & A_{21} \end{vmatrix}, \quad c_3 = \begin{vmatrix} A_{11} & A_{12} \\ A_{21} & A_{22} \end{vmatrix}.$$

Show that
(a) $\text{rank}(A) = 2$ if and only if $(c_1, c_2, c_3) \neq 0$;
(b) if A has rank 2, then (c_1, c_2, c_3) is a basis for the solution space of the system of equations $AX = 0$.

11. Let K be a commutative ring with identity, and let D be an alternating 2-linear function on 2×2 matrices over K. Show that $D(A) = (\det A)D(I)$ for all A. Now use this result (no computations with the entries allowed) to show that $\det(AB) = (\det A)(\det B)$ for any 2×2 matrices A and B over K.

12. Let F be a field and D a function on $n \times n$ matrices over F (with values in F). Suppose $D(AB) = D(A)D(B)$ for all A, B. Show that either $D(A) = 0$ for all A, or $D(I) = 1$. In the latter case show that $D(A) \neq 0$ whenever A is invertible.

13. Let R be the field of real numbers, and let D be a function on 2×2 matrices over R, with values in R, such that $D(AB) = D(A)D(B)$ for all A, B. Suppose also that

$$D\left(\begin{bmatrix} 0 & 1 \\ 1 & 0 \end{bmatrix}\right) \neq D\left(\begin{bmatrix} 1 & 0 \\ 0 & 1 \end{bmatrix}\right).$$

Prove the following.

(a) $D(0) = 0$;

(b) $D(A) = 0$ if $A^2 = 0$;

(c) $D(B) = -D(A)$ if B is obtained by interchanging the rows (or columns) of A;

(d) $D(A) = 0$ if one row (or one column) of A is 0;

(e) $D(A) = 0$ whenever A is singular.

14. Let A be a 2×2 matrix over a field F. Then the set of all matrices of the form $f(A)$, where f is a polynomial over F, is a commutative ring K with identity. If B is a 2×2 matrix over K, the determinant of B is then a 2×2 matrix over F, of the form $f(A)$. Suppose I is the 2×2 identity matrix over F and that B is the 2×2 matrix over K

$$B = \begin{bmatrix} A - A_{11}I & -A_{12}I \\ -A_{21}I & A - A_{22}I \end{bmatrix}.$$

Show that $\det B = f(A)$, where $f = x^2 - (A_{11} + A_{22})x + \det A$, and also that $f(A) = 0$.

5.3. Permutations and the Uniqueness of Determinants

In this section we prove the uniqueness of the determinant function on $n \times n$ matrices over K. The proof will lead us quite naturally to consider permutations and some of their basic properties.

Suppose D is an alternating n-linear function on $n \times n$ matrices over K. Let A be an $n \times n$ matrix over K with rows $\alpha_1, \alpha_2, \cdots, \alpha_n$. If we denote the rows of the $n \times n$ identity matrix over K by $\epsilon_1, \epsilon_2, \cdots, \epsilon_n$, then

$$\text{(5-9)} \qquad \alpha_i = \sum_{j=1}^{n} A(i, j)\epsilon_j, \qquad 1 \leq i \leq n.$$

Hence

$$\begin{aligned} D(A) &= D\left(\sum_j A(1, j)\epsilon_j, \alpha_2, \ldots, \alpha_n\right) \\ &= \sum_j A(1, j)D(\epsilon_j, \alpha_2, \ldots, \alpha_n). \end{aligned}$$

If we now replace α_2 by $\sum_k A(2, k)\epsilon_k$, we see that

$$D(\epsilon_j, \alpha_2, \ldots, \alpha_n) = \sum_k A(2, k)D(\epsilon_j, \epsilon_k, \ldots, \alpha_n).$$

Thus

$$D(A) = \sum_{j,k} A(1, j)A(2, k)D(\epsilon_j, \epsilon_k, \ldots, \alpha_n).$$

In $D(\epsilon_j, \epsilon_k, \ldots, \alpha_n)$ we next replace α_3 by $\Sigma A(3, l)\epsilon_l$ and so on. We finally obtain a complicated but theoretically important expression for $D(A)$, namely

(5-10) $$D(A) = \sum_{k_1, k_2, \ldots, k_n} A(1, k_1)A(2, k_2) \cdots A(n, k_n)D(\epsilon_{k_1}, \epsilon_{k_2}, \ldots, \epsilon_{k_n}).$$

In (5-10) the sum is extended over all sequences $(k_1, k_2, \ldots, k_n)$ of positive integers not exceeding n. This shows that D is a finite sum of functions of the type described by (5-2). It should be noted that (5-10) is a consequence just of assumption that D is n-linear, and that a special case of (5-10) was obtained in Example 2. Since D is alternating,

$$D(\epsilon_{k_1}, \epsilon_{k_2}, \ldots, \epsilon_{k_n}) = 0$$

whenever two of the indices k_i are equal. A sequence $(k_1, k_2, \ldots, k_n)$ of positive integers not exceeding n, with the property that no two of the k_i are equal, is called a **permutation of degree n.** In (5-10) we need therefore sum only over those sequences which are permutations of degree n.

Since a finite sequence, or n-tuple, is a function defined on the first n positive integers, a permutation of degree n may be defined as a one-one function from the set $\{1, 2, \ldots, n\}$ onto itself. Such a function σ corresponds to the n-tuple $(\sigma 1, \sigma 2, \ldots, \sigma n)$ and is thus simply a rule for ordering $1, 2, \ldots, n$ in some well-defined way.

If D is an alternating n-linear function and A is an $n \times n$ matrix over K, we then have

(5-11) $$D(A) = \sum_{\sigma} A(1, \sigma 1) \cdots A(n, \sigma n)D(\epsilon_{\sigma 1}, \ldots, \epsilon_{\sigma n})$$

where the sum is extended over the distinct permutations σ of degree n.

Next we shall show that

(5-12) $$D(\epsilon_{\sigma 1}, \ldots, \epsilon_{\sigma n}) = \pm D(\epsilon_1, \ldots, \epsilon_n)$$

where the sign $\pm$ depends only on the permutation σ. The reason for this is as follows. The sequence $(\sigma 1, \sigma 2, \ldots, \sigma n)$ can be obtained from the sequence $(1, 2, \ldots, n)$ by a finite number of interchanges of pairs of elements. For example, if $\sigma 1 \neq 1$, we can transpose 1 and $\sigma 1$, obtaining $(\sigma 1, \ldots, 1, \ldots)$. Proceeding in this way we shall arrive at the sequence $(\sigma 1, \ldots, \sigma n)$ after n or less such interchanges of pairs. Since D is alternating, the sign of its value changes each time that we interchange two of the rows ϵ_i and ϵ_j. Thus, if we pass from $(1, 2, \ldots, n)$ to $(\sigma 1, \sigma 2, \ldots, \sigma n)$ by means of m interchanges of pairs (i, j), we shall have

$$D(\epsilon_{\sigma 1}, \ldots, \epsilon_{\sigma n}) = (-1)^m D(\epsilon_1, \ldots, \epsilon_n).$$

In particular, if D is a determinant function

(5-13) $$D(\epsilon_{\sigma 1}, \ldots, \epsilon_{\sigma n}) = (-1)^m$$

where m depends only upon σ, not upon D. Thus all determinant functions assign the same value to the matrix with rows $\epsilon_{\sigma 1}, \ldots, \epsilon_{\sigma n}$, and this value is either 1 or -1.

Now a basic fact about permutations is the following. If σ is a permutation of degree n, one can pass from the sequence $(1, 2, \ldots, n)$ to the sequence $(\sigma 1, \sigma 2, \ldots, \sigma n)$ by a succession of interchanges of pairs, and this can be done in a variety of ways; however, no matter how it is done, the number of interchanges used is either always even or always odd. The permutation is then called **even** or **odd**, respectively. One defines the **sign** of a permutation by

$$\operatorname{sgn} \sigma = \begin{cases} 1, & \text{if } \sigma \text{ is even} \\ -1, & \text{if } \sigma \text{ is odd} \end{cases}$$

the symbol '1' denoting here the integer 1.

We shall show below that this basic property of permutations can be deduced from what we already know about determinant functions. Let us assume this for the time being. Then the integer m occurring in (5-13) is always even if σ is an even permutation, and is always odd if σ is an odd permutation. For any alternating n-linear function D we then have

$$D(\epsilon_{\sigma 1}, \ldots, \epsilon_{\sigma n}) = (\operatorname{sgn} \sigma) D(\epsilon_1, \ldots, \epsilon_n)$$

and using (5-11)

$$D(A) = \left[\sum_{\sigma} (\operatorname{sgn} \sigma) A(1, \sigma 1) \cdots A(n, \sigma n) \right] D(I). \tag{5-14}$$

Of course I denotes the $n \times n$ identity matrix.

From (5-14) we see that there is precisely one determinant function on $n \times n$ matrices over K. If we denote this function by det, it is given by

$$\det(A) = \sum_{\sigma} (\operatorname{sgn} \sigma) A(1, \sigma 1) \cdots A(n, \sigma n) \tag{5-15}$$

the sum being extended over the distinct permutations σ of degree n. We can formally summarize as follows.

Theorem 2. *Let* K *be a commutative ring with identity and let* n *be a positive integer. There is precisely one determinant function on the set of* n $\times$ n *matrices over* K, *and it is the function* det *defined by (5-15). If* D *is any alternating* n*-linear function on* $K^{n \times n}$, *then for each* n $\times$ n *matrix* A

$$D(A) = (\det A) D(I).$$

This is the theorem we have been seeking, but we have left a gap in the proof. That gap is the proof that for a given permutation σ, when we pass from $(1, 2, \ldots, n)$ to $(\sigma 1, \sigma 2, \ldots, \sigma n)$ by interchanging pairs, the number of interchanges is always even or always odd. This basic combinatorial fact can be proved without any reference to determinants;

however, we should like to point out how it follows from the *existence* of a determinant function on $n \times n$ matrices.

Let us take K to be the ring of integers. Let D be a determinant function on $n \times n$ matrices over K. Let σ be a permutation of degree n, and suppose we pass from $(1, 2, \ldots, n)$ to $(\sigma 1, \sigma 2, \ldots, \sigma n)$ by m interchanges of pairs (i, j), $i \neq j$. As we showed in (5-13)

$$(-1)^m = D(\epsilon_{\sigma 1}, \ldots, \epsilon_{\sigma n})$$

that is, the number $(-1)^m$ must be the value of D on the matrix with rows $\epsilon_{\sigma 1}, \ldots, \epsilon_{\sigma n}$. If

$$D(\epsilon_{\sigma 1}, \ldots, \epsilon_{\sigma n}) = 1,$$

then m must be even. If

$$D(\epsilon_{\sigma 1}, \ldots, \epsilon_{\sigma n}) = -1,$$

then m must be odd.

Since we have an explicit formula for the determinant of an $n \times n$ matrix and this formula involves the permutations of degree n, let us conclude this section by making a few more observations about permutations. First, let us note that there are precisely $n! = 1 \cdot 2 \cdots n$ permutations of degree n. For, if σ is such a permutation, there are n possible choices for $\sigma 1$; when this choice has been made, there are $(n - 1)$ choices for $\sigma 2$, then $(n - 2)$ choices for $\sigma 3$, and so on. So there are

$$n(n-1)(n-2) \cdots 2 \cdot 1 = n!$$

permutations σ. The formula (5-15) for det (A) thus gives det (A) as a sum of $n!$ terms, one for each permutation of degree n. A given term is a product

$$A(1, \sigma 1) \cdots A(n, \sigma n)$$

of n entries of A, one entry from each row and one from each column, and is prefixed by a '+' or '−' sign according as σ is an even or odd permutation.

When permutations are regarded as one-one functions from the set $\{1, 2, \ldots, n\}$ onto itself, one can define a product of permutations. The product of σ and τ will simply be the composed function $\sigma\tau$ defined by

$$(\sigma\tau)(i) = \sigma(\tau(i)).$$

If ϵ denotes the identity permutation, $\epsilon(i) = i$, then each σ has an inverse σ^{-1} such that

$$\sigma\sigma^{-1} = \sigma^{-1}\sigma = \epsilon.$$

One can summarize these observations by saying that, under the operation of composition, the set of permutations of degree n is a group. This group is usually called the **symmetric group of degree n.**

From the point of view of products of permutations, the basic property of the sign of a permutation is that

$$\text{sgn}(\sigma\tau) = (\text{sgn}\,\sigma)(\text{sgn}\,\tau). \tag{5-16}$$

In other words, $\sigma\tau$ is an even permutation if σ and τ are either both even or both odd, while $\sigma\tau$ is odd if one of the two permutations is odd and the other is even. One can see this from the definition of the sign in terms of successive interchanges of pairs (i, j). It may also be instructive if we point out how $\operatorname{sgn}(\sigma\tau) = (\operatorname{sgn}\sigma)(\operatorname{sgn}\tau)$ follows from a fundamental property of determinants.

Let K be the ring of integers and let σ and τ be permutations of degree n. Let $\epsilon_1, \ldots, \epsilon_n$ be the rows of the $n \times n$ identity matrix over K, let A be the matrix with rows $\epsilon_{\tau 1}, \ldots, \epsilon_{\tau n}$, and let B be the matrix with rows $\epsilon_{\sigma 1}, \ldots, \epsilon_{\sigma n}$. The ith row of A contains exactly one non-zero entry, namely the 1 in column τi. From this it is easy to see that $\epsilon_{\sigma\tau i}$ is the ith row of the product matrix AB. Now

$$\det(A) = \operatorname{sgn}\tau, \qquad \det(B) = \operatorname{sgn}\sigma, \qquad \text{and} \qquad \det(AB) = \operatorname{sgn}(\sigma\tau).$$

So we shall have $\operatorname{sgn}(\sigma\tau) = (\operatorname{sgn}\sigma)(\operatorname{sgn}\tau)$ as soon as we prove the following.

Theorem 3. *Let* K *be a commutative ring with identity, and let* A *and* B *be* n $\times$ n *matrices over* K. *Then*

$$\det(AB) = (\det A)(\det B).$$

Proof. Let B be a fixed $n \times n$ matrix over K, and for each $n \times n$ matrix A define $D(A) = \det(AB)$. If we denote the rows of A by $\alpha_1, \ldots, \alpha_n$, then

$$D(\alpha_1, \ldots, \alpha_n) = \det(\alpha_1 B, \ldots, \alpha_n B).$$

Here $\alpha_j B$ denotes the $1 \times n$ matrix which is the product of the $1 \times n$ matrix α_j and the $n \times n$ matrix B. Since

$$(c\alpha_i + \alpha_i')B = c\alpha_i B + \alpha_i' B$$

and det is n-linear, it is easy to see that D is n-linear. If $\alpha_i = \alpha_j$, then $\alpha_i B = \alpha_j B$, and since det is alternating,

$$D(\alpha_1, \ldots, \alpha_n) = 0.$$

Hence, D is alternating. Now D is an alternating n-linear function, and by Theorem 2

$$D(A) = (\det A)D(I).$$

But $D(I) = \det(IB) = \det B$, so

$$\det(AB) = D(A) = (\det A)(\det B). \quad \blacksquare$$

The fact that $\operatorname{sgn}(\sigma\tau) = (\operatorname{sgn}\sigma)(\operatorname{sgn}\tau)$ is only one of many corollaries to Theorem 3. We shall consider some of these corollaries in the next section.

Exercises

1. If K is a commutative ring with identity and A is the matrix over K given by

$$A = \begin{bmatrix} 0 & a & b \\ -a & 0 & c \\ -b & -c & 0 \end{bmatrix}$$

show that $\det A = 0$.

2. Prove that the determinant of the Vandermonde matrix

$$\begin{bmatrix} 1 & a & a^2 \\ 1 & b & b^2 \\ 1 & c & c^2 \end{bmatrix}$$

is $(b - a)(c - a)(c - b)$.

3. List explicitly the six permutations of degree 3, state which are odd and which are even, and use this to give the complete formula (5-15) for the determinant of a 3×3 matrix.

4. Let σ and τ be the permutations of degree 4 defined by $\sigma 1 = 2$, $\sigma 2 = 3$, $\sigma 3 = 4$, $\sigma 4 = 1$, $\tau 1 = 3$, $\tau 2 = 1$, $\tau 3 = 2$, $\tau 4 = 4$.

(a) Is σ odd or even? Is τ odd or even?

(b) Find $\sigma\tau$ and $\tau\sigma$.

5. If A is an invertible $n \times n$ matrix over a field, show that $\det A \neq 0$.

6. Let A be a 2×2 matrix over a field. Prove that $\det (I + A) = 1 + \det A$ if and only if trace $(A) = 0$.

7. An $n \times n$ matrix A is called **triangular** if $A_{ij} = 0$ whenever $i > j$ or if $A_{ij} = 0$ whenever $i < j$. Prove that the determinant of a triangular matrix is the product $A_{11}A_{22} \cdots A_{nn}$ of its diagonal entries.

8. Let A be a 3×3 matrix over the field of complex numbers. We form the matrix $xI - A$ with polynomial entries, the i, j entry of this matrix being the polynomial $\delta_{ij}x - A_{ij}$. If $f = \det (xI - A)$, show that f is a monic polynomial of degree 3. If we write

$$f = (x - c_1)(x - c_2)(x - c_3)$$

with complex numbers c_1, c_2, and c_3, prove that

$$c_1 + c_2 + c_3 = \text{trace } (A) \quad \text{and} \quad c_1c_2c_3 = \det A.$$

9. Let n be a positive integer and F a field. If σ is a permutation of degree n, prove that the function

$$T(x_1, \ldots, x_n) = (x_{\sigma 1}, \ldots, x_{\sigma n})$$

is an invertible linear operator on F^n.

10. Let F be a field, n a positive integer, and S the set of $n \times n$ matrices over F. Let V be the vector space of all functions from S into F. Let W be the set of alternating n-linear functions on S. Prove that W is a subspace of V. What is the dimension of W?

11. Let T be a linear operator on F^n. Define

$$D_T(\alpha_1, \ldots, \alpha_n) = \det (T\alpha_1, \ldots, T\alpha_n).$$

(a) Show that D_T is an alternating n-linear function.

(b) If

$$c = \det (T\epsilon_1, \ldots, T\epsilon_n)$$

show that for any n vectors $\alpha_1, \ldots, \alpha_n$ we have

$$\det (T\alpha_1, \ldots, T\alpha_n) = c \det (\alpha_1, \ldots, \alpha_n).$$

(c) If $\mathcal{B}$ is any ordered basis for F^n and A is the matrix of T in the ordered basis $\mathcal{B}$, show that $\det A = c$.

(d) What do you think is a reasonable name for the scalar c?

12. If σ is a permutation of degree n and A is an $n \times n$ matrix over the field F with row vectors $\alpha_1, \ldots, \alpha_n$, let $\sigma(A)$ denote the $n \times n$ matrix with row vectors $\alpha_{\sigma 1}, \ldots, \alpha_{\sigma n}$.

(a) Prove that $\sigma(AB) = \sigma(A)B$, and in particular that $\sigma(A) = \sigma(I)A$.

(b) If T is the linear operator of Exercise 9, prove that the matrix of T in the standard ordered basis is $\sigma(I)$.

(c) Is $\sigma^{-1}(I)$ the inverse matrix of $\sigma(I)$?

(d) Is it true that $\sigma(A)$ is similar to A?

13. Prove that the sign function on permutations is unique in the following sense. If f is any function which assigns to each permutation of degree n an integer, and if $f(\sigma\tau) = f(\sigma)f(\tau)$, then f is identically 0, or f is identically 1, or f is the sign function.

5.4. Additional Properties of Determinants

In this section we shall relate some of the useful properties of the determinant function on $n \times n$ matrices. Perhaps the first thing we should point out is the following. In our discussion of $\det A$, the rows of A have played a privileged role. Since there is no fundamental difference between rows and columns, one might very well expect that $\det A$ is an alternating n-linear function of the columns of A. This is the case, and to prove it, it suffices to show that

$$\det (A^t) = \det (A) \tag{5-17}$$

where A^t denotes the transpose of A.

If σ is a permutation of degree n,

$$A^t(i, \sigma i) = A(\sigma i, i).$$

From the expression (5-15) one then has

$$\det (A^t) = \sum_\sigma (\operatorname{sgn} \sigma) A(\sigma 1, 1) \cdots A(\sigma n, n).$$

When $i = \sigma^{-1}j$, $A(\sigma i, i) = A(j, \sigma^{-1}j)$. Thus

$$A(\sigma 1, 1) \cdots A(\sigma n, n) = A(1, \sigma^{-1}1) \cdots A(n, \sigma^{-1}n).$$

Since $\sigma\sigma^{-1}$ is the identity permutation,

$$(\text{sgn } \sigma)(\text{sgn } \sigma^{-1}) = 1 \quad \text{or} \quad \text{sgn } (\sigma^{-1}) = \text{sgn } (\sigma).$$

Furthermore, as σ varies over all permutations of degree n, so does σ^{-1}. Therefore

$$\begin{aligned} \det (A^t) &= \sum_\sigma (\text{sgn } \sigma^{-1}) A(1, \sigma^{-1}1) \cdots A(n, \sigma^{-1}n) \\ &= \det A \end{aligned}$$

proving (5-17).

On certain occasions one needs to compute specific determinants. When this is necessary, it is frequently useful to take advantage of the following fact. *If B is obtained from A by adding a multiple of one row of A to another (or a multiple of one column to another), then*

(5-18) $$\det B = \det A.$$

We shall prove the statement about rows. Let B be obtained from A by adding $c\alpha_j$ to α_i, where $i < j$. Since det is linear as a function of the ith row

$$\begin{aligned} \det B &= \det A + c \det (\alpha_1, \ldots, \alpha_j, \ldots, \alpha_j, \ldots, \alpha_n) \\ &= \det A. \end{aligned}$$

Another useful fact is the following. Suppose we have an $n \times n$ matrix of the block form

$$\begin{bmatrix} A & B \\ 0 & C \end{bmatrix}$$

where A is an $r \times r$ matrix, C is an $s \times s$ matrix, B is $r \times s$, and 0 denotes the $s \times r$ zero matrix. Then

(5-19) $$\det \begin{bmatrix} A & B \\ 0 & C \end{bmatrix} = (\det A)(\det C).$$

To prove this, define

$$D(A, B, C) = \det \begin{bmatrix} A & B \\ 0 & C \end{bmatrix}.$$

If we fix A and B, then D is alternating and s-linear as a function of the rows of C. Thus, by Theorem 2

$$D(A, B, C) = (\det C) D(A, B, I)$$

where I is the $s \times s$ identity matrix. By subtracting multiples of the rows of I from the rows of B and using the statement above (5-18), we obtain

$$D(A, B, I) = D(A, 0, I).$$

Now $D(A, 0, I)$ is clearly alternating and r-linear as a function of the rows of A. Thus

$$D(A, 0, I) = (\det A) D(I, 0, I).$$

But $D(I, 0, I) = 1$, so

$$\begin{aligned} D(A, B, C) &= (\det C)D(A, B, I) \\ &= (\det C)D(A, 0, I) \\ &= (\det C)(\det A). \end{aligned}$$

By the same sort of argument, or by taking transposes

$$\det \begin{bmatrix} A & 0 \\ B & C \end{bmatrix} = (\det A)(\det C). \tag{5-20}$$

EXAMPLE 6. Suppose K is the field of rational numbers and we wish to compute the determinant of the 4×4 matrix

$$A = \begin{bmatrix} 1 & -1 & 2 & 3 \\ 2 & 2 & 0 & 2 \\ 4 & 1 & -1 & -1 \\ 1 & 2 & 3 & 0 \end{bmatrix}.$$

By subtracting suitable multiples of row 1 from rows 2, 3, and 4, we obtain the matrix

$$\begin{bmatrix} 1 & -1 & 2 & 3 \\ 0 & 4 & -4 & -4 \\ 0 & 5 & -9 & -13 \\ 0 & 3 & 1 & -3 \end{bmatrix}$$

which we know by (5-18) will have the same determinant as A. If we subtract $\frac{5}{4}$ of row 2 from row 3 and then subtract $\frac{3}{4}$ of row 2 from row 4, we obtain

$$B = \begin{bmatrix} 1 & -1 & 2 & 3 \\ 0 & 4 & -4 & -4 \\ 0 & 0 & -4 & -8 \\ 0 & 0 & 4 & 0 \end{bmatrix}$$

and again $\det B = \det A$. The block form of B tells us that

$$\det A = \det B = \begin{vmatrix} 1 & -1 \\ 0 & 4 \end{vmatrix} \begin{vmatrix} -4 & -8 \\ 4 & 0 \end{vmatrix} = 4(32) = 128.$$

Now let $n > 1$ and let A be an $n \times n$ matrix over K. In Theorem 1, we showed how to construct a determinant function on $n \times n$ matrices, given one on $(n - 1) \times (n - 1)$ matrices. Now that we have proved the uniqueness of the determinant function, the formula (5-4) tells us the following. If we fix any column index j,

$$\det A = \sum_{i=1}^{n} (-1)^{i+j} A_{ij} \det A(i|j).$$

The scalar $(-1)^{i+j} \det A(i|j)$ is usually called the i, j **cofactor** of A or the cofactor of the i, j entry of A. The above formula for det A is then called the expansion of det A by cofactors of the jth column (or sometimes the expansion by minors of the jth column). If we set

$$C_{ij} = (-1)^{i+j} \det A(i|j)$$

then the above formula says that for each j

$$\det A = \sum_{i=1}^{n} A_{ij}C_{ij}$$

where the cofactor C_{ij} is $(-1)^{i+j}$ times the determinant of the $(n-1) \times (n-1)$ matrix obtained by deleting the ith row and jth column of A.

If $j \neq k$, then

$$\sum_{i=1}^{n} A_{ik}C_{ij} = 0.$$

For, replace the jth column of A by its kth column, and call the resulting matrix B. Then B has two equal columns and so $\det B = 0$. Since $B(i|j) = A(i|j)$, we have

$$\begin{aligned} 0 &= \det B \\ &= \sum_{i=1}^{n} (-1)^{i+j}B_{ij} \det B(i|j) \\ &= \sum_{i=1}^{n} (-1)^{i+j}A_{ik} \det A(i|j) \\ &= \sum_{i=1}^{n} A_{ik}C_{ij}. \end{aligned}$$

These properties of the cofactors can be summarized by

(5-21) $$\sum_{i=1}^{n} A_{ik}C_{ij} = \delta_{jk} \det A.$$

The $n \times n$ matrix adj A, which is the transpose of the matrix of cofactors of A, is called the **classical adjoint** of A. Thus

(5-22) $$(\text{adj } A)_{ij} = C_{ji} = (-1)^{i+j} \det A(j|i).$$

The formulas (5-21) can be summarized in the matrix equation

(5-23) $$(\text{adj } A)A = (\det A)I.$$

We wish to see that $A(\text{adj } A) = (\det A)I$ also. Since $A^t(i|j) = A(j|i)^t$, we have

$$(-1)^{i+j} \det A^t(i|j) = (-1)^{j+i} \det A(j|i)$$

which simply says that the i, j cofactor of A^t is the j, i cofactor of A. Thus

(5-24) $$\text{adj } (A^t) = (\text{adj } A)^t$$

By applying (5-23) to A^t, we obtain

$$(\text{adj } A^t)A^t = (\det A^t)I = (\det A)I$$

and transposing

$$A(\text{adj } A^t)^t = (\det A)I.$$

Using (5-24), we have what we want:

$$A(\text{adj } A) = (\det A)I. \tag{5-25}$$

As for matrices over a field, an $n \times n$ matrix A over K is called **invertible over** K if there is an $n \times n$ matrix A^{-1} with entries in K such that $AA^{-1} = A^{-1}A = I$. If such an inverse matrix exists it is unique; for the same argument used in Chapter 1 shows that when $BA = AC = I$ we have $B = C$. The formulas (5-23) and (5-25) tell us the following about invertibility of matrices over K. If the element $\det A$ has a multiplicative inverse in K, then A is invertible and $A^{-1} = (\det A)^{-1} \text{ adj } A$ is the unique inverse of A. Conversely, it is easy to see that if A is invertible over K, the element $\det A$ is invertible in K. For, if $BA = I$ we have

$$1 = \det I = \det (AB) = (\det A)(\det B).$$

What we have proved is the following.

Theorem 4. *Let* A *be an* n $\times$ n *matrix over* K. *Then* A *is invertible over* K *if and only if det* A *is invertible in* K. *When* A *is invertible, the unique inverse for* A *is*

$$A^{-1} = (\det A)^{-1} \text{ adj } A.$$

In particular, an n $\times$ n *matrix over a field is invertible if and only if its determinant is different from zero.*

We should point out that this determinant criterion for invertibility proves that an $n \times n$ matrix with either a left or right inverse is invertible. This proof is completely independent of the proof which we gave in Chapter 1 for matrices over a field. We should also like to point out what invertibility means for matrices with polynomial entries. If K is the polynomial ring $F[x]$, the only elements of K which are invertible are the non-zero scalar polynomials. For if f and g are polynomials and $fg = 1$, we have $\deg f + \deg g = 0$ so that $\deg f = \deg g = 0$, i.e., f and g are scalar polynomials. So an $n \times n$ matrix over the polynomial ring $F[x]$ is invertible over $F[x]$ if and only if its determinant is a non-zero scalar polynomial.

EXAMPLE 7. Let $K = R[x]$, the ring of polynomials over the field of real numbers. Let

$$A = \begin{bmatrix} x^2 + x & x + 1 \\ x - 1 & 1 \end{bmatrix}, \qquad B = \begin{bmatrix} x^2 - 1 & x + 2 \\ x^2 - 2x + 3 & x \end{bmatrix}.$$

Then, by a short computation, $\det A = x + 1$ and $\det B = -6$. Thus A is not invertible over K, whereas B is invertible over K. Note that

$$\text{adj } A = \begin{bmatrix} 1 & -x - 1 \\ -x + 1 & x^2 + x \end{bmatrix}, \qquad \text{adj } B = \begin{bmatrix} x & -x - 2 \\ -x^2 + 2x - 3 & x^2 - 1 \end{bmatrix}$$

and $(\text{adj } A)A = (x + 1)I$, $(\text{adj } B)B = -6I$. Of course,

$$B^{-1} = -\frac{1}{6}\begin{bmatrix} x & -x-2 \\ -x^2+2x-3 & 1-x^2 \end{bmatrix}.$$

EXAMPLE 8. Let K be the ring of integers and

$$A = \begin{bmatrix} 1 & 2 \\ 3 & 4 \end{bmatrix}.$$

Then $\det A = -2$ and

$$\text{adj } A = \begin{bmatrix} 4 & -2 \\ -3 & 1 \end{bmatrix}.$$

Thus A is not invertible as a matrix over the ring of integers; however, we can also regard A as a matrix over the field of rational numbers. If we do, then A is invertible and

$$A^{-1} = -\frac{1}{2}\begin{bmatrix} 4 & -2 \\ -3 & 1 \end{bmatrix} = \begin{bmatrix} -2 & 1 \\ \frac{3}{2} & \frac{1}{2} \end{bmatrix}.$$

In connection with invertible matrices, we should like to mention one further elementary fact. Similar matrices have the same determinant, that is, if P is invertible over K and $B = P^{-1}AP$, then $\det B = \det A$. This is clear since

$$\det (P^{-1}AP) = (\det P^{-1})(\det A)(\det P) = \det A.$$

This simple observation makes it possible to define the determinant of a linear operator on a finite dimensional vector space. If T is a linear operator on V, we define the determinant of T to be the determinant of any $n \times n$ matrix which represents T in an ordered basis for V. Since all such matrices are similar, they have the same determinant and our definition makes sense. In this connection, see Exercise 11 of section 5.3.

We should like now to discuss **Cramer's rule** for solving systems of linear equations. Suppose A is an $n \times n$ matrix over the field F and we wish to solve the system of linear equations $AX = Y$ for some given n-tuple $(y_1, \ldots, y_n)$. If $AX = Y$, then

$$(\text{adj } A)AX = (\text{adj } A)Y$$

and so

$$(\det A)X = (\text{adj } A)Y.$$

Thus

$$\begin{aligned}(\det A)x_j &= \sum_{i=1}^{n} (\text{adj } A)_{ji}y_i \\ &= \sum_{i=1}^{n} (-1)^{i+j}y_i \det A(i|j).\end{aligned}$$

This last expression is the determinant of the $n \times n$ matrix obtained by replacing the jth column of A by Y. If $\det A = 0$, all this tells us nothing; however, if $\det A \neq 0$, we have what is known as Cramer's rule. Let A

be an $n \times n$ matrix over the field F such that $\det A \neq 0$. If $y_1, \ldots, y_n$ are any scalars in F, the unique solution $X = A^{-1}Y$ of the system of equations $AX = Y$ is given by

$$x_j = \frac{\det B_j}{\det A}, \qquad j = 1, \ldots, n$$

where B_j is the $n \times n$ matrix obtained from A by replacing the jth column of A by Y.

In concluding this chapter, we should like to make some comments which serve to place determinants in what we believe to be the proper perspective. From time to time it is necessary to compute specific determinants, and this section has been partially devoted to techniques which will facilitate such work. However, the principal role of determinants in this book is theoretical. There is no disputing the beauty of facts such as Cramer's rule. But Cramer's rule is an inefficient tool for solving systems of linear equations, chiefly because it involves too many computations. So one should concentrate on what Cramer's rule says, rather than on how to compute with it. Indeed, while reflecting on this entire chapter, we hope that the reader will place more emphasis on understanding what the determinant function is and how it behaves than on how to compute determinants of specific matrices.

Exercises

1. Use the classical adjoint formula to compute the inverses of each of the following 3×3 real matrices.

$$\begin{bmatrix} -2 & 3 & 2 \\ 6 & 0 & 3 \\ 4 & 1 & -1 \end{bmatrix}, \qquad \begin{bmatrix} \cos\theta & 0 & -\sin\theta \\ 0 & 1 & 0 \\ \sin\theta & 0 & \cos\theta \end{bmatrix}$$

2. Use Cramer's rule to solve each of the following systems of linear equations over the field of rational numbers.

(a)
$$\begin{aligned} x + y + z &= 11 \\ 2x - 6y - z &= 0 \\ 3x + 4y + 2z &= 0. \end{aligned}$$

(b)
$$\begin{aligned} 3x - 2y &= 7 \\ 3y - 2z &= 6 \\ 3z - 2x &= -1. \end{aligned}$$

3. An $n \times n$ matrix A over a field F is **skew-symmetric** if $A^t = -A$. If A is a skew-symmetric $n \times n$ matrix with complex entries and n is odd, prove that $\det A = 0$.

4. An $n \times n$ matrix A over a field F is called **orthogonal** if $AA^t = I$. If A is orthogonal, show that $\det A = \pm 1$. Give an example of an orthogonal matrix for which $\det A = -1$.

5. An $n \times n$ matrix A over the field of complex numbers is said to be **unitary** if $AA^* = I$ (A^* denotes the conjugate transpose of A). If A is unitary, show that $|\det A| = 1$.

6. Let T and U be linear operators on the finite dimensional vector space V. Prove
(a) $\det (TU) = (\det T)(\det U)$;
(b) T is invertible if and only if $\det T \neq 0$.

7. Let A be an $n \times n$ matrix over K, a commutative ring with identity. Suppose A has the block form

$$A = \begin{bmatrix} A_1 & 0 & \cdots & 0 \\ 0 & A_2 & \cdots & 0 \\ \vdots & \vdots & & \vdots \\ 0 & 0 & \cdots & A_k \end{bmatrix}$$

where A_j is an $r_j \times r_j$ matrix. Prove

$$\det A = (\det A_1)(\det A_2) \cdots (\det A_k).$$

8. Let V be the vector space of $n \times n$ matrices over the field F. Let B be a fixed element of V and let T_B be the linear operator on V defined by $T_B(A) = AB - BA$. Show that $\det T_B = 0$.

9. Let A be an $n \times n$ matrix over a field, $A \neq 0$. If r is any positive integer between 1 and n, an $r \times r$ **submatrix** of A is any $r \times r$ matrix obtained by deleting $(n - r)$ rows and $(n - r)$ columns of A. The **determinant rank** of A is the largest positive integer r such that some $r \times r$ submatrix of A has a non-zero determinant. Prove that the determinant rank of A is equal to the row rank of A (= column rank A).

10. Let A be an $n \times n$ matrix over the field F. Prove that there are at most n distinct scalars c in F such that $\det (cI - A) = 0$.

11. Let A and B be $n \times n$ matrices over the field F. Show that if A is invertible there are at most n scalars c in F for which the matrix $cA + B$ is not invertible.

12. If V is the vector space of $n \times n$ matrices over F and B is a fixed $n \times n$ matrix over F, let L_B and R_B be the linear operators on V defined by $L_B(A) = BA$ and $R_B(A) = AB$. Show that
(a) $\det L_B = (\det B)^n$;
(b) $\det R_B = (\det B)^n$.

13. Let V be the vector space of all $n \times n$ matrices over the field of complex numbers, and let B be a fixed $n \times n$ matrix over C. Define a linear operator M_B on V by $M_B(A) = BAB^*$, where $B^* = \overline{B^t}$. Show that

$$\det M_B = |\det B|^{2n}.$$

Now let H be the set of all Hermitian matrices in V, A being Hermitian if $A = A^*$. Then H is a vector space over the field of *real* numbers. Show that the function T_B defined by $T_B(A) = BAB^*$ is a linear operator on the real vector space H, and then show that $\det T_B = |\det B|^{2n}$. (*Hint:* In computing $\det T_B$, show that V has a basis consisting of Hermitian matrices and then show that $\det T_B = \det M_B$.)

14. Let A, B, C, D be *commuting* $n \times n$ matrices over the field F. Show that the determinant of the $2n \times 2n$ matrix

$$\begin{bmatrix} A & B \\ C & D \end{bmatrix}$$

is $\det (AD - BC)$.

5.5. Modules

If K is a commutative ring with identity, a module over K is an algebraic system which behaves like a vector space, with K playing the role of the scalar field. To be precise, we say that V is a **module over** K (or a K-**module**) if

1. there is an addition $(\alpha, \beta) \rightarrow \alpha + \beta$ on V, under which V is a commutative group;
2. there is a multiplication $(c, \alpha) \rightarrow c\alpha$ of elements α in V and c in K such that

$$\begin{aligned} (c_1 + c_2)\alpha &= c_1\alpha + c_2\alpha \\ c(\alpha_1 + \alpha_2) &= c\alpha_1 + c\alpha_2 \\ (c_1c_2)\alpha &= c_1(c_2\alpha) \\ 1\alpha &= \alpha. \end{aligned}$$

For us, the most important K-modules will be the n-tuple modules K^n. The matrix modules $K^{m \times n}$ will also be important. If V is any module, we speak of linear combinations, linear dependence and linear independence, just as we do in a vector space. We must be careful not to apply to V any vector space results which depend upon division by non-zero scalars, the one field operation which may be lacking in the ring K. For example, if $\alpha_1, \ldots, \alpha_k$ are linearly dependent, we cannot conclude that some α_i is a linear combination of the others. This makes it more difficult to find bases in modules.

A **basis** for the module V is a linearly independent subset which spans (or generates) the module. This is the same definition which we gave for vector spaces; and, the important property of a basis $\mathcal{B}$ is that each element of V can be expressed uniquely as a linear combination of (some finite number of) elements of $\mathcal{B}$. If one admits into mathematics the Axiom of Choice (see Appendix), it can be shown that every vector space has a basis. The reader is well aware that a basis exists in any vector space which is spanned by a finite number of vectors. But this is not the case for modules. Therefore we need special names for modules which have bases and for modules which are spanned by finite numbers of elements.

Definition. *The* K-*module* V *is called a* **free module** *if it has a basis. If* V *has a finite basis containing* n *elements, then* V *is called a* **free K-module with n generators.**

Definition. *The module* V *is* **finitely generated** *if it contains a finite subset which spans* V. *The* **rank** *of a finitely generated module is the smallest integer* k *such that some* k *elements span* V.

We repeat that a module may be finitely generated without having a finite basis. If V is a free K-module with n generators, then V is isomorphic to the module K^n. If $\{\beta_1, \ldots, \beta_n\}$ is a basis for V, there is an isomorphism which sends the vector $c_1\beta_1 + \cdots + c_n\beta_n$ onto the n-tuple $(c_1, \ldots, c_n)$ in K^n. It is not immediately apparent that the same module V could not also be a free module on k generators, with $k \neq n$. In other words, it is not obvious that any two bases for V must contain the same number of elements. The proof of that fact is an interesting application of determinants.

Theorem 5. *Let* K *be a commutative ring with identity. If* V *is a free* K*-module with* n *generators, then the rank of* V *is* n.

Proof. We are to prove that V cannot be spanned by less than n of its elements. Since V is isomorphic to K^n, we must show that, if $m < n$, the module K^n is not spanned by n-tuples $\alpha_1, \ldots, \alpha_m$. Let A be the matrix with rows $\alpha_1, \ldots, \alpha_m$. Suppose that each of the standard basis vectors $\epsilon_1, \ldots, \epsilon_n$ is a linear combination of $\alpha_1, \ldots, \alpha_m$. Then there exists a matrix P in $K^{n \times m}$ such that

$$PA = I$$

where I is the $n \times n$ identity matrix. Let $\tilde{A}$ be the $n \times n$ matrix obtained by adjoining $n - m$ rows of 0's to the bottom of A, and let $\tilde{P}$ be any $n \times n$ matrix which has the columns of P as its first n columns. Then

$$\tilde{P}\tilde{A} = I.$$

Therefore $\det \tilde{A} \neq 0$. But, since $m < n$, at least one row of $\tilde{A}$ has all 0 entries. This contradiction shows that $\alpha_1, \ldots, \alpha_m$ do not span K^n. ∎

It is interesting to note that Theorem 5 establishes the uniqueness of the dimension of a (finite-dimensional) vector space. The proof, based upon the existence of the determinant function, is quite different from the proof we gave in Chapter 2. From Theorem 5 we know that 'free module of rank n' is the same as 'free module with n generators.'

If V is a module over K, the **dual module** V^* consists of all linear functions f from V into K. If V is a free module of rank n, then V^* is also a free module of rank n. The proof is just the same as for vector spaces. If $\{\beta_1, \ldots, \beta_n\}$ is an ordered basis for V, there is an associated **dual basis** $\{f_1, \ldots, f_n\}$ for the module V^*. The function f_i assigns to each α in V its ith coordinate relative to $\{\beta_1, \ldots, \beta_n\}$:

$$\alpha = f_1(\alpha)\beta_1 + \cdots + f_n(\alpha)\beta_n.$$

If f is a linear function on V, then

$$f = f(\beta_1)f_1 + \cdots + f(\beta_n)f_n.$$

5.6. Multilinear Functions

The purpose of this section is to place our discussion of determinants in what we believe to be the proper perspective. We shall treat alternating multilinear forms on modules. These forms are the natural generalization of determinants as we presented them. The reader who has not read (or does not wish to read) the brief account of modules in Section 5.5 can still study this section profitably by consistently reading 'vector space over F of dimension n' for 'free module over K of rank n.'

Let K be a commutative ring with identity and let V be a module over K. If r is a positive integer, a function L from $V^r = V \times V \times \cdots \times V$ into K is called **multilinear** if $L(\alpha_1, \ldots, \alpha_r)$ is linear as a function of each α_i when the other α_j's are held fixed, that is, if for each i

$$L(\alpha_1, \ldots, c\alpha_i + \beta_i, \ldots, \alpha_r) = cL(\alpha_1, \ldots, \alpha_i, \ldots, \alpha_r + L(\alpha_1, \ldots, \beta_i, \ldots, \alpha_r).$$

A multilinear function on V^r will also be called an **r-linear form** on V or a **multilinear form of degree** r on V. Such functions are sometimes called **r-tensors** on V. The collection of all multilinear functions on V^r will be denoted by $M^r(V)$. If L and M are in $M^r(V)$, then the sum $L + M$:

$$(L + M)(\alpha_1, \ldots, \alpha_r) = L(\alpha_1, \ldots, \alpha_r) + M(\alpha_1, \ldots, \alpha_r)$$

is also multilinear; and, if c is an element of K, the product cL:

$$(cL)(\alpha_1, \ldots, \alpha_r) = cL(\alpha_1, \ldots, \alpha_r)$$

is multilinear. Therefore $M^r(V)$ is a K-module—a submodule of the module of all functions from V^r into K.

If $r = 1$ we have $M^1(V) = V^*$, the dual module of linear functions on V. Linear functions can also be used to construct examples of multilinear forms of higher order. If $f_1, \ldots, f_r$ are linear functions on V, define

$$L(\alpha_1, \ldots, \alpha_r) = f_1(\alpha_1)f_2(\alpha_2) \cdots f_r(\alpha_r).$$

Clearly L is an r-linear form on V.

Example 9. If V is a module, a 2-linear form on V is usually called a **bilinear form** on V. Let A be an $n \times n$ matrix with entries in K. Then

$$L(X, Y) = Y^tAX$$

defines a bilinear form L on the module $K^{n\times 1}$. Similarly,

$$M(\alpha, \beta) = \alpha A\beta^t$$

defines a bilinear form M on K^n.

Example 10. The determinant function associates with each $n \times n$

matrix A an element $\det A$ in K. If $\det A$ is considered as a function of the rows of A:

$$\det A = D(\alpha_1, \ldots, \alpha_n)$$

then D is an n-linear form on K^n.

EXAMPLE 11. It is easy to obtain an algebraic expression for the general r-linear form on the module K^n. If $\alpha_1, \ldots, \alpha_r$ are vectors in V and A is the $r \times n$ matrix with rows $\alpha_1, \ldots, \alpha_r$, then for any function L in $M^r(K^n)$,

$$\begin{aligned} L(\alpha_1, \ldots, \alpha_r) &= L\left(\sum_{j=1}^{n} A_{1j}\epsilon_j, \alpha_2, \ldots, \alpha_r\right) \\ &= \sum_{j=1}^{n} A_{1j}L(\epsilon_j, \alpha_2, \ldots, \alpha_r) \\ &= \sum_{j=1}^{n} A_{1j}L\left(\epsilon_j, \sum_{j=1}^{n} A_{2k}\epsilon_k, \ldots, \alpha_r\right) \\ &= \sum_{j=1}^{n} \sum_{k=1}^{n} A_{1j}A_{2k}L(\epsilon_j, \epsilon_k, \alpha_3, \ldots, \alpha_r) \\ &= \sum_{j,k=1}^{n} A_{1j}A_{2k}L(\epsilon_j, \epsilon_k, \alpha_3, \ldots, \alpha_r). \end{aligned}$$

If we replace $\alpha_3, \ldots, \alpha_r$ in turn by their expressions as linear combinations of the standard basis vectors, and if we write $A(i, j)$ for A_{ij}, we obtain the following:

$$\text{(5-26)} \qquad L(\alpha_1, \ldots, \alpha_r) = \sum_{j_1, \ldots, j_r = 1}^{n} A(1, j_1) \cdots A(r, j_r)L(\epsilon_{j_1}, \ldots \epsilon_{j_r}).$$

In (5-26), there is one term for each r-tuple $J = (j_1, \ldots, j_r)$ of positive integers between 1 and n. There are n^r such r-tuples. Thus L is completely determined by (5-26) and the particular values:

$$c_J = L(\epsilon_{j_1}, \ldots, \epsilon_{j_r})$$

assigned to the n^r elements $(\epsilon_{j_1}, \ldots, \epsilon_{j_r})$. It is also easy to see that if for each r-tuple J we choose an element c_J of K then

$$\text{(5-27)} \qquad L(\alpha_1, \ldots, \alpha_r) = \sum_J A(1, j_1) \cdots A(r, j_r)c_J$$

defines an r-linear form on K^n.

Suppose that L is a multilinear function on V^r and M is a multilinear function on V^s. We define a function $L \otimes M$ on V^{r+s} by

$$\text{(5-28)} \qquad (L \otimes M)(\alpha_1, \ldots, \alpha_{r+s}) = L(\alpha_1, \ldots, \alpha_r)M(\alpha_{r+1}, \ldots, \alpha_{r+s}).$$

If we think of V^{r+s} as $V^r \times V^s$, then for α in V^r and β in V^s

$$(L \otimes M)(\alpha, \beta) = L(\alpha)M(\beta).$$

It is clear that $L \otimes M$ is multilinear on V^{r+s}. The function $L \otimes M$ is called the **tensor product** of L and M. The tensor product is not commutative. In fact, $M \otimes L \neq L \otimes M$ unless $L = 0$ or $M = 0$; however, the tensor product does relate nicely to the module operations in M^r and M^s.

Lemma. *Let* L, L_1 *be* r*-linear forms on* V, *let* M, M_1 *be* s*-linear forms on* V *and let* c *be an element of* K.

(a) $(cL + L_1) \otimes M = c(L \otimes M) + L_1 \otimes M$;
(b) $L \otimes (cM + M_1) = c(L \otimes M) + L \otimes M_1$.

Proof. Exercise.

Tensoring is associative, i.e., if L, M and N are (respectively) r-, s- and t-linear forms on V, then

$$(L \otimes M) \otimes N = L \otimes (M \otimes N).$$

This is immediate from the fact that the multiplication in K is associative. Therefore, if $L_1, L_2, \ldots, L_k$ are multilinear functions on $V^{r_1}, \ldots, V^{r_k}$, then the tensor product

$$L = L_1 \otimes \cdots \otimes L_k$$

is unambiguously defined as a multilinear function on V^r, where $r = r_1 + \cdots + r_k$. We mentioned a particular case of this earlier. If $f_1, \ldots, f_r$ are linear functions on V, then the tensor product

$$L = f_1 \otimes \cdots \otimes f_r$$

is given by

$$L(\alpha_1, \ldots, \alpha_r) = f_1(\alpha_1) \cdots f_r(\alpha_r).$$

Theorem 6. *Let* K *be a commutative ring with identity. If* V *is a free* K*-module of rank* n *then* $M^r(V)$ *is a free* K*-module of rank* n^r; *in fact, if* $\{f_1, \ldots, f_n\}$ *is a basis for the dual module* V^*, *the* n^r *tensor products*

$$f_{j_1} \otimes \cdots \otimes f_{j_r}, \qquad 1 \leq j_1 \leq n, \ldots, 1 \leq j_r \leq n$$

form a basis for $M^r(V)$.

Proof. Let $\{f_1, \ldots, f_n\}$ be an ordered basis for V^* which is dual to the basis $\{\beta_1, \ldots, \beta_n\}$ for V. For each vector α in V we have

$$\alpha = f_1(\alpha)\beta_1 + \cdots + f_n(\alpha)\beta_n.$$

We now make the calculation carried out in Example 11. If L is an r-linear form on V and $\alpha_1, \ldots, \alpha_r$ are elements of V, then by (5-26)

$$L(\alpha_1, \ldots, \alpha_r) = \sum_{j_1, \ldots, j_r} f_{j_1}(\alpha_1) \cdots f_{j_r}(\alpha_r) L(\beta_{j_1}, \ldots, \beta_{j_r}).$$

In other words,

$$L = \sum_{j_1, \ldots, j_r} L(\beta_{j_1}, \ldots, \beta_{j_r}) f_{j_1} \otimes \cdots \otimes f_{j_r}. \tag{5-29}$$

This shows that the n^r tensor products

$$E_J = f_{j_1} \otimes \cdots \otimes f_{j_r} \tag{5-30}$$

given by the r-tuples $J = (j_1, \ldots, j_r)$ span the module $M^r(V)$. We see that the various r-forms E_J are independent, as follows. Suppose that for each J we have an element c_J in K and we form the multilinear function

$$L = \sum_J c_J E_J. \tag{5-31}$$

Notice that if $I = (i_1, \ldots, i_r)$, then

$$E_J(\beta_{i_1}, \ldots, \beta_{i_r}) = \begin{cases} 0, & I \neq J \\ 1, & I = J. \end{cases}$$

Therefore we see from (5-31) that

$$c_I = L(\beta_{i_1}, \ldots, \beta_{i_r}). \tag{5-32}$$

In particular, if $L = 0$ then $c_I = 0$ for each r-tuple I. ∎

Definition. *Let* L *be an* r*-linear form on a* K*-module* V. *We say that* L *is* **alternating** *if* $L(\alpha_1, \ldots, \alpha_r) = 0$ *whenever* $\alpha_i = \alpha_j$ *with* $i \neq j$.

If L is an alternating multilinear function on V^r, then

$$L(\alpha_1, \ldots, \alpha_i, \ldots, \alpha_j, \ldots, \alpha_r) = -L(\alpha_1, \ldots, \alpha_j, \ldots, \alpha_i, \ldots, \alpha_r).$$

In other words, if we transpose two of the vectors (with different indices) in the r-tuple $(\alpha_1, \ldots, \alpha_r)$ the associated value of L changes sign. Since every permutation σ is a product of transpositions, we see that $L(\alpha_{\sigma 1}, \ldots, \alpha_{\sigma r}) = (\operatorname{sgn} \sigma)\, L(\alpha_1, \ldots, \alpha_r)$.

We denote by $\Lambda^r(V)$ the collection of all alternating r-linear forms on V. It should be clear that $\Lambda^r(V)$ is a submodule of $M^r(V)$.

EXAMPLE 12. Earlier in this chapter, we showed that on the module K^n there is precisely one alternating n-linear form D with the property that $D(\epsilon_1, \ldots, \epsilon_n) = 1$. We also showed in Theorem 2 that if L is any form in $\Lambda^n(K^n)$ then

$$L = L(\epsilon_1, \ldots, \epsilon_n)D.$$

In other words, $\Lambda^n(K^n)$ is a free K-module of rank 1. We also developed an explicit formula (5-15) for D. In terms of the notation we are now using, that formula may be written

$$D = \sum_\sigma (\operatorname{sgn} \sigma)\, f_{\sigma 1} \otimes \cdots \otimes f_{\sigma n}. \tag{5-33}$$

where $f_1, \ldots, f_n$ are the standard coordinate functions on K^n and the sum is extended over the $n!$ different permutations σ of the set $\{1, \ldots, n\}$. If we write the determinant of a matrix A as

$$\det A = \sum_\sigma (\operatorname{sgn} \sigma)\, A(\sigma 1, 1) \cdots A(\sigma n, n)$$

then we obtain a different expression for D:

$$(5\text{-}34)\qquad D(\alpha_1, \ldots, \alpha_n) = \sum_\sigma (\operatorname{sgn} \sigma) f_1(\alpha_{\sigma 1}) \cdots f_n(\alpha_{\sigma n}) = \sum_\sigma (\operatorname{sgn} \sigma) L(\alpha_{\sigma 1}, \ldots, \alpha_{\sigma n})$$

where $L = f_1 \otimes \cdots \otimes f_n$.

There is a general method for associating an alternating form with a multilinear form. If L is an r-linear form on a module V and if σ is a permutation of $\{1, \ldots, r\}$, we obtain another r-linear function L_σ by defining

$$L_\sigma(\alpha_1, \ldots, \alpha_r) = L(\alpha_{\sigma 1}, \ldots, \alpha_{\sigma r}).$$

If L happens to be alternating, then $L_\sigma = (\operatorname{sgn} \sigma)L$. Now, for each L in $M^r(V)$ we define a function $\pi_r L$ in $M^r(V)$ by

$$(5\text{-}35)\qquad \pi_r L = \sum_\sigma (\operatorname{sgn} \sigma) L_\sigma$$

that is,

$$(5\text{-}36)\qquad (\pi_r L)(\alpha_1, \ldots, \alpha_r) = \sum_\sigma (\operatorname{sgn} \sigma) L(\alpha_{\sigma 1}, \ldots, \alpha_{\sigma r}).$$

Lemma. *π_r is a linear transformation from $M^r(V)$ into $\Lambda^r(V)$. If L is in $\Lambda^r(V)$ then $\pi_r L = r!L$.*

Proof. Let τ be any permutation of $\{1, \ldots, r\}$. Then

$$(\pi_r L)(\alpha_{\tau 1}, \ldots, \alpha_{\tau r}) = \sum_\sigma (\operatorname{sgn} \sigma) L(\alpha_{\tau\sigma 1}, \ldots, \alpha_{\tau\sigma r}) = (\operatorname{sgn} \tau) \sum_\sigma (\operatorname{sgn} \tau\sigma) L(\alpha_{\tau\sigma 1}, \ldots, \alpha_{\tau\sigma r}).$$

As σ runs (once) over all permutations of $\{1, \ldots, r\}$, so does $\tau\sigma$. Therefore,

$$(\pi_r L)(\alpha_{\tau 1}, \ldots, \alpha_{\tau r}) = (\operatorname{sgn} \tau)(\pi_r L)(\alpha_1, \ldots, \alpha_r).$$

Thus $\pi_r L$ is an alternating form.

If L is in $\Lambda^r(V)$, then $L(\alpha_{\sigma 1}, \ldots, \alpha_{\sigma r}) = (\operatorname{sgn} \sigma) L(\alpha_1, \ldots, \alpha_r)$ for each σ; hence $\pi_r L = r!L$. ∎

In (5-33) we showed that the determinant function D in $\Lambda^n(K^n)$ is

$$D = \pi_n(f_1 \otimes \cdots \otimes f_n)$$

where $f_1, \ldots, f_n$ are the standard coordinate functions on K^n. There is an important remark we should make in connection with the last lemma. If K is a field of characteristic zero, such that $r!$ is invertible in K, then π maps $M^r(V)$ onto $\Lambda^r(V)$. In fact, in that case it is more natural from one point of view to use the map $\pi_1 = (1/r!)\pi$ rather than π, because π_1 is a projection of $M^r(V)$ onto $\Lambda^r(V)$, i.e., a linear map of $M^r(V)$ onto $\Lambda^r(V)$ such that $\pi_1(L) = L$ if and only if L is in $\Lambda^r(V)$.

Theorem 7. *Let K be a commutative ring with identity and let V be a free K-module of rank n. If r > n, then* $\Lambda^r(V) = \{0\}$. *If* $1 \le r \le n$, *then* $\Lambda^r(V)$ *is a free K-module of rank* $\binom{n}{r}$.

Proof. Let $\{\beta_1, \ldots, \beta_n\}$ be an ordered basis for V with dual basis $\{f_1, \ldots, f_n\}$. If L is in $M^r(V)$, we have

$$L = \sum_J L(\beta_{j_1}, \ldots, \beta_{j_r}) f_{j_1} \otimes \cdots \otimes f_{j_r} \tag{5-37}$$

where the sum extends over all r-tuples $J = (j_1, \ldots, j_r)$ of integers between 1 and n. If L is alternating, then

$$L(\beta_{j_1}, \ldots, \beta_{j_r}) = 0$$

whenever two of the subscripts j_i are the same. If $r > n$, then in each r-tuple J some integer must be repeated. Thus $\Lambda^r(V) = \{0\}$ if $r > n$.

Now suppose $1 \le r \le n$. If L is in $\Lambda^r(V)$, the sum in (5-37) need be extended only over the r-tuples J for which $j_1, \ldots, j_r$ are distinct, because all other terms are 0. Each r-tuple of distinct integers between 1 and n is a permutation of an r-tuple $J = (j_1, \ldots, j_r)$ such that $j_1 < \cdots < j_r$. This special type of r-tuple is called an r-**shuffle** of $\{1, \ldots, n\}$. There are

$$\binom{n}{r} = \frac{n!}{r!(n-r)!}$$

such shuffles.

Suppose we fix an r-shuffle J. Let L_J be the sum of all the terms in (5-37) corresponding to permutations of the shuffle J. If σ is a permutation of $\{1, \ldots, r\}$, then

$$L(\beta_{j_{\sigma 1}}, \ldots, \beta_{j_{\sigma r}}) = (\operatorname{sgn} \sigma)\, L(\beta_{j_1}, \ldots, \beta_{j_r}).$$

Thus

$$L_J = L(\beta_{j_1}, \ldots, \beta_{j_r}) D_J \tag{5-38}$$

where

$$\begin{aligned} D_J &= \sum_\sigma (\operatorname{sgn} \sigma)\, f_{j_{\sigma 1}} \otimes \cdots \otimes f_{j_{\sigma r}} \\ &= \pi_r(f_{j_1} \otimes \cdots \otimes f_{j_r}). \end{aligned} \tag{5-39}$$

We see from (5-39) that each D_J is alternating and that

$$L = \sum_{\text{shuffles } J} L(\beta_{j_1}, \ldots, \beta_{j_r}) D_J \tag{5-40}$$

for every L in $\Lambda^r(V)$. The assertion is that the $\binom{n}{r}$ forms D_J constitute a basis for $\Lambda^r(V)$. We have seen that they span $\Lambda^r(V)$. It is easy to see that they are independent, as follows. If $I = (i_1, \ldots, i_r)$ and $J = (j_1, \ldots, j_r)$ are shuffles, then

$$D_J(\beta_{i_1}, \ldots, \beta_{i_r}) = \begin{cases} 1, & I = J \\ 0, & I \ne J \end{cases}. \tag{5-41}$$

Suppose we have a scalar c_J for each shuffle and we define

$$L = \sum_J c_J D_J.$$

From (5-40) and (5-41) we obtain

$$c_I = L(\beta_{i_1}, \ldots, \beta_{i_r}).$$

In particular, if $L = 0$ then $c_I = 0$ for each shuffle I. ∎

Corollary. *If V is a free K-module of rank n, then $\Lambda^n(V)$ is a free K-module of rank 1. If T is a linear operator on V, there is a unique element c in K such that*

$$L(T\alpha_1, \ldots, T\alpha_n) = cL(\alpha_1, \ldots, \alpha_n)$$

for every alternating n-linear form L on V.

Proof. If L is in $\Lambda^n(V)$, then clearly

$$L_T(\alpha_1, \ldots, \alpha_n) = L(T\alpha_1, \ldots, T\alpha_n)$$

defines an alternating n-linear form L_T. Let M be a generator for the rank 1 module $\Lambda^n(V)$. Each L in $\Lambda^n(V)$ is uniquely expressible as $L = aM$ for some a in K. In particular, $M_T = cM$ for a certain c. For $L = aM$ we have

$$\begin{aligned} L_T &= (aM)_T \\ &= aM_T \\ &= a(cM) \\ &= c(aM) \\ &= cL. \end{aligned}$$ ∎

Of course, the element c in the last corollary is called the **determinant** of T. From (5-39) for the case $r = n$ (when there is only one shuffle $J = (1, \ldots, n)$) we see that the determinant of T is the determinant of the matrix which represents T in any ordered basis $\{\beta_1, \ldots, \beta_n\}$. Let us see why. The representing matrix has i, j entry

$$A_{ij} = f_j(T\beta_i)$$

so that

$$\begin{aligned} D_J(T\beta_1, \ldots, T\beta_n) &= \sum_\sigma (\operatorname{sgn} \sigma) A(1, \sigma 1) \cdots A(n, \sigma n) \\ &= \det A. \end{aligned}$$

On the other hand,

$$\begin{aligned} D_J(T\beta_1, \ldots, T\beta_n) &= (\det T)\, D_J(\beta_1, \ldots, \beta_n) \\ &= \det T. \end{aligned}$$

The point of these remarks is that via Theorem 7 and its corollary we obtain a definition of the determinant of a linear operator which does not presume knowledge of determinants of matrices. Determinants of matrices can be defined in terms of determinants of operators instead of the other way around.

We want to say a bit more about the special alternating r-linear forms D_J, which we associated with a basis $\{f_1, \ldots, f_n\}$ for V^* in (5-39). It is important to understand that $D_J(\alpha_1, \ldots, \alpha_r)$ is the determinant of a certain $r \times r$ matrix. If

$$A_{ij} = f_j(\alpha_i), \qquad 1 \leq i \leq r, 1 \leq j \leq n,$$

that is, if

$$\alpha_i = A_{i1}\beta_1 + \cdots + A_{in}\beta_n, \qquad 1 \leq i \leq r$$

and J is the r-shuffle $(j_1, \ldots, j_r)$, then

$$(5\text{-}42) \qquad D_J(\alpha_1, \ldots, \alpha_r) = \sum_\sigma (\operatorname{sgn} \sigma)\, A(1, j_{\sigma 1}) \cdots A(n, j_{\sigma n})$$
$$= \det \begin{bmatrix} A(1, j_1) & \cdots & A(1, j_r) \\ \vdots & & \vdots \\ A(r, j_1) & \cdots & A(r, j_r) \end{bmatrix}.$$

Thus $D_J(\alpha_1, \ldots, \alpha_r)$ is the determinant of the $r \times r$ matrix formed from columns $j_1, \ldots, j_r$ of the $r \times n$ matrix which has (the coordinate n-tuples of) $\alpha_1, \ldots, \alpha_r$ as its rows. Another notation which is sometimes used for this determinant is

$$(5\text{-}43) \qquad D_J(\alpha_1, \ldots, \alpha_r) = \frac{\partial(\alpha_1, \ldots, \alpha_r)}{\partial(\beta_{j_1}, \ldots, \beta_{j_r})}.$$

In this notation, the proof of Theorem 7 shows that every alternating r-linear form L can be expressed relative to a basis $\{\beta_1, \ldots, \beta_n\}$ by the equation

$$(5\text{-}44) \qquad L(\alpha_1, \ldots, \alpha_r) = \sum_{j_1 < \cdots < j_r} \frac{\partial(\alpha_1, \ldots, \alpha_r)}{\partial(\beta_{j_1}, \ldots, \beta_{j_r})} L(\beta_{j_1}, \ldots, \beta_{j_r}).$$

5.7. The Grassman Ring

Many of the important properties of determinants and alternating multilinear forms are best described in terms of a multiplication operation on forms, called the exterior product. If L and M are, respectively, alternating r and s-linear forms on the module V, we have an associated product of L and M, the tensor product $L \otimes M$. This is not an alternating form unless $L = 0$ or $M = 0$; however, we have a natural way of projecting it into $\Lambda^{r+s}(V)$. It appears that

$$(5\text{-}45) \qquad L \cdot M = \pi_{r+s}(L \otimes M)$$

should be the 'natural' multiplication of alternating forms. But, is it?

Let us take a specific example. Suppose that V is the module K^n and $f_1, \ldots, f_n$ are the standard coordinate functions on K^n. If $i \neq j$, then

$$f_i \cdot f_j = \pi_2(f_i \otimes f_j)$$

is the (determinant) function

$$D_{ij} = f_i \otimes f_j - f_j \otimes f_i$$

given by (5-39). Now suppose k is an index different from i and j. Then

$$\begin{aligned} D_{ij} \cdot f_k &= \pi_3[(f_i \otimes f_j - f_j \otimes f_i) \otimes f_k] \\ &= \pi_3(f_i \otimes f_j \otimes f_k) - \pi_3(f_j \otimes f_i \otimes f_k). \end{aligned}$$

The proof of the lemma following equation (5-36) shows that for any r-linear form L and any permutation σ of $\{1, \ldots, r\}$

$$\pi_r(L_\sigma) = \operatorname{sgn} \sigma \, \pi_r(L)$$

Hence, $D_{ij} \cdot f_k = 2\pi_3(f_i \otimes f_j \otimes f_k)$. By a similar computation, $f_i \cdot D_{jk} = 2\pi_3(f_i \otimes f_j \otimes f_k)$. Thus we have

$$(f_i \cdot f_j) \cdot f_k = f_i \cdot (f_j \cdot f_k)$$

and all of this looks very promising. But there is a catch. Despite the computation that we have just completed, the putative multiplication in (5-45) is not associative. In fact, if l is an index different from i, j, k, then one can calculate that

$$D_{ij} \cdot D_{kl} = 4\pi_4(f_i \otimes f_j \otimes f_k \otimes f_l)$$

and that

$$(D_{ij} \cdot f_k) \cdot f_l = 6\pi_4(f_i \otimes f_j \otimes f_k \otimes f_l).$$

Thus, in general

$$(f_i \cdot f_j) \cdot (f_k \cdot f_l) \neq [(f_i \cdot f_j) \cdot f_k] \cdot f_l$$

and we see that our first attempt to find a multiplication has produced a non-associative operation.

The reader should not be surprised if he finds it rather tedious to give a direct verification of the two equations showing non-associativity. This is typical of the subject, and it is also typical that there is a general fact which considerably simplifies the work.

Suppose L is an r-linear form and that M is an s-linear form on the module V. Then

$$\begin{aligned} \pi_{r+s}((\pi_r L) \otimes (\pi_s M)) &= \pi_{r+s}(\sum_{\sigma,\tau} (\operatorname{sgn} \sigma)(\operatorname{sgn} \tau) L_\sigma \otimes M_\tau) \\ &= \sum_{\sigma,\tau} (\operatorname{sgn} \sigma)(\operatorname{sgn} \tau)\pi_{r+s}(L_\sigma \otimes M_\tau) \end{aligned}$$

where σ varies over the symmetric group, S_r, of all permutations of $\{1, \ldots, r\}$, and τ varies over S_s. Each pair σ, τ defines an element (σ, τ) of S_{r+s} which permutes the first r elements of $\{1, \ldots, r + s\}$ according to σ and the last s elements according to τ. It is clear that

$$\operatorname{sgn}(\sigma, \tau) = (\operatorname{sgn} \sigma)(\operatorname{sgn} \tau)$$

and that

$$(L \otimes M)_{(\sigma,\tau)} = L_\sigma \otimes L_\tau.$$

Therefore

$$\pi_{r+s}[(\pi_r L) \otimes (\pi_s M)] = \sum_{\sigma,\tau} \operatorname{sgn}(\sigma, \tau)\, \pi_{r+s}\, [(L \otimes M)_{(\sigma,\tau)}].$$

Now we have already observed that

$$\operatorname{sgn}(\sigma, \tau)\pi_{r+s}[(L \otimes M)_{(\sigma,\tau)}] = \pi_{r+s}(L \otimes M).$$

Thus, it follows that

$$\pi_{r+s}[(\pi_r L) \otimes (\pi_s M)] = r!s!\, \pi_{r+s}(L \otimes M). \tag{5-46}$$

This formula simplifies a number of computations. For example, suppose we have an r-shuffle $I = (i_1, \ldots, i_r)$ and s-shuffle $J = (j_1, \ldots, j_s)$. To make things simple, assume, in addition, that

$$i_1 < \cdots < i_r < j_1 < \cdots < j_s.$$

Then we have the associated determinant functions

$$D_I = \pi_r(E_I)$$
$$D_J = \pi_s(E_J)$$

where E_I and E_J are given by (5-30). Using (5-46), we see immediately that

$$\begin{aligned} D_I \cdot D_J &= \pi_{r+s}[\pi_r(E_I) \otimes \pi_s(E_J)] \\ &= r!s!\pi_{r+s}(E_I \otimes E_J). \end{aligned}$$

Since $E_I \otimes E_J = E_{I \cup J}$, it follows that

$$D_I \cdot D_J = r!s!\, D_{I \cup J}.$$

This suggests that the lack of associativity for the multiplication (5-45) results from the fact that $D_I \cdot D_J \neq D_{I \cup J}$. After all, the product of D_I and D_J ought to be $D_{I \cup J}$. To repair the situation, we should define a new product, the **exterior product** (or **wedge product**) of an alternating r-linear form L and an alternating s-linear form M by

$$L \wedge M = \frac{1}{r!s!}\, \pi_{r+s}(L \otimes M). \tag{5-47}$$

We then have

$$D_I \wedge D_J = D_{I \cup J}$$

for the determinant functions on K^n, and, if there is any justice at all, we must have found the proper multiplication of alternating multilinear forms. Unfortunately, (5-47) fails to make sense for the most general case under consideration, since we may not be able to divide by $r!s!$ in the ring K. If K is a field of characteristic zero, then (5-47) is meaningful, and one can proceed quite rapidly to show that the wedge product is associative.

Theorem 8. *Let* K *be a field of characteristic zero and* V *a vector space over* K. *Then the exterior product is an associative operation on the alternating multilinear forms on* V. *In other words, if* L, M, *and* N *are alternating multilinear forms on* V *of degrees* r, s, *and* t, *respectively, then*

$$(L \wedge M) \wedge N = L \wedge (M \wedge N).$$

Proof. It follows from (5-47) that $cd(L \wedge M) = cL \wedge dM$ for any scalars c and d. Hence

$$r!s!t![(L \wedge M) \wedge N] = r!s!(L \wedge M) \wedge t!N$$

and since $\pi_t(N) = t!N$, it results that

$$\begin{aligned} r!s!t![(L \wedge M) \wedge N] &= \pi_{r+s}(L \otimes M) \wedge \pi_t(N) \\ &= \frac{1}{(r+s)!}\frac{1}{t!}\pi_{r+s+t}[\pi_{r+s}(L \otimes M) \otimes \pi_t(N)]. \end{aligned}$$

From (5-46) we now see that

$$r!s!t[(L \wedge M) \wedge N] = \pi_{r+s+t}(L \otimes M \otimes N).$$

By a similar computation

$$r!s!t![L \wedge (M \wedge N)] = \pi_{r+s+t}(L \otimes M \otimes N)$$

and therefore, $(L \wedge M) \wedge N = L \wedge (M \wedge N)$. ∎

Now we return to the general case, in which it is only assumed that K is a commutative ring with identity. Our first problem is to replace (5-47) by an equivalent definition which works in general. If L and M are alternating multilinear forms of degrees r and s respectively, we shall construct a canonical alternating multilinear form $L \wedge M$ of degree $r + s$ such that

$$r!s!(L \wedge M) = \pi_{r+s}(L \otimes M).$$

Let us recall how we define $\pi_{r+s}(L \otimes M)$. With each permutation σ of $\{1, \ldots, r+s\}$ we associate the multilinear function

$$(\text{sgn } \sigma)(L \otimes M)_\sigma \tag{5-48}$$

where

$$(L \otimes M)_\sigma(\alpha_1, \ldots, \alpha_{r+s}) = (L \otimes M)(\alpha_{\sigma 1}, \ldots, \alpha_{\sigma(r+s)})$$

and we sum the functions (5-48) over all permutations σ. There are $(r+s)!$ permutations; however, since L and M are alternating, many of the functions (5-48) are the same. In fact there are at most

$$\frac{(r+s)!}{r!s!}$$

distinct functions (5-48). Let us see why. Let S_{r+s} be the set of permutations of $\{1, \ldots, r+s\}$, i.e., let S_{r+s} be the symmetric group of degree $r + s$. As in the proof of (5-46), we distinguish the subset G that consists of the permutations σ which permute the sets $\{1, \ldots, r\}$ and $\{r+1, \ldots, r+s\}$ within themselves. In other words, σ is in G if $1 \leq \sigma i \leq r$ for each i between 1 and r. (It necessarily follows that $r + 1 \leq \sigma j \leq r + s$ for each j between $r + 1$ and $r + s$.) Now G is a subgroup of S_{r+s}, that is, if σ and τ are in G then $\sigma\tau^{-1}$ is in G. Evidently G has $r!s!$ members.

We have a map

$$S_{r+s} \xrightarrow{\psi} M^{r+s}(V)$$

defined by

$$\psi(\sigma) = (\text{sgn } \sigma)(L \otimes M)_\sigma.$$

Since L and M are alternating,

$$\psi(\gamma) = L \otimes M$$

for every γ in G. Therefore, since $(N\sigma)\tau = N\tau\sigma$ for any $(r + s)$-linear form N on V, we have

$$\psi(\tau\gamma) = \psi(\tau), \qquad \tau \text{ in } S_{r+s}, \gamma \text{ in } G.$$

This says that the map ψ is constant on each (left) **coset** τG of the subgroup G. If τ_1 and τ_2 are in S_{r+s}, the cosets $\tau_1 G$ and $\tau_2 G$ are either identical or disjoint, according as $\tau_2^{-1}\tau_1$ is in G or is not in G. Each coset contains $r!s!$ elements; hence, there are

$$\frac{(r + s)!}{r!s!}$$

distinct cosets. If S_{r+s}/G denotes the collection of cosets then ψ defines a function on S_{r+s}/G, i.e., by what we have shown, there is a function $\bar{\psi}$ on that set so that

$$\psi(\tau) = \bar{\psi}(\tau G)$$

for every τ in S_{r+s}. If H is a left coset of G, then $\bar{\psi}(H) = \psi(\tau)$ for every τ in H.

We now define the **exterior product** of the alternating multilinear forms L and M of degrees r and s by setting

$$L \wedge M = \sum_H \bar{\psi}(H) \tag{5-49}$$

where H varies over S_{r+s}/G. Another way to phrase the definition of $L \wedge M$ is the following. Let S be any set of permutations of $\{1, \ldots, r + s\}$ which contains exactly one element from each left coset of G. Then

$$L \wedge M = \sum_\sigma (\text{sgn } \sigma)(L \otimes M)_\sigma \tag{5-50}$$

where σ varies over S. Clearly

$$r!s!\, L \wedge M = \pi_{r+s}(L \otimes M)$$

so that the new definition is equivalent to (5-47) when K is a field of characteristic zero.

Theorem 9. *Let* K *be a commutative ring with identity and let* V *be a module over* K*. Then the exterior product is an associative operation on the alternating multilinear forms on* V*. In other words, if* L, M, *and* N *are alternating multilinear forms on* V *of degrees* r, s, *and* t, *respectively, then*

$$(L \wedge M) \wedge N = L \wedge (M \wedge N).$$

Proof. Although the proof of Theorem 8 does not apply here, it does suggest how to handle the general case. Let $G(r, s, t)$ be the subgroup of S_{r+s+t} that consists of the permutations which permute the sets

$$\{1, \ldots, r\}, \{r + 1, \ldots, r + s\}, \{r + s + 1, \ldots, r + s + t\}$$

within themselves. Then $(\operatorname{sgn} \mu)(L \otimes M \otimes N)_\mu$ is the same multilinear function for all μ in a given left coset of $G(r, s, t)$. Choose one element from each left coset of $G(r, s, t)$, and let E be the sum of the corresponding terms $(\operatorname{sgn} \mu)(L \otimes M \otimes N)_\mu$. Then E is independent of the way in which the representatives μ are chosen, and

$$r!s!t!\, E = \pi_{r+s+t}(L \otimes M \otimes N).$$

We shall show that $(L \wedge M) \wedge N$ and $L \wedge (M \wedge N)$ are both equal to E.

Let $G(r + s, t)$ be the subgroup of S_{r+s+t} that permutes the sets

$$\{1, \ldots, r + s\}, \{r + s + 1, \ldots, r + s + t\}$$

within themselves. Let T be any set of permutations of $\{1, \ldots, r + s + t\}$ which contains exactly one element from each left coset of $G(r + s, t)$. By (5-50)

$$(L \wedge M) \wedge N = \sum_\tau (\operatorname{sgn} \tau)[(L \wedge M) \otimes N]_\tau$$

where the sum is extended over the permutations τ in T. Now let $G(r, s)$ be the subgroup of S_{r+s} that permutes the sets

$$\{1, \ldots, r\}, \{r + 1, \ldots, r + s\}$$

within themselves. Let S be any set of permutations of $\{1, \ldots, r + s\}$ which contains exactly one element from each left coset of $G(r, s)$. From (5-50) and what we have shown above, it follows that

$$(L \wedge M) \wedge N = \sum_{\sigma,\tau} (\operatorname{sgn} \sigma)(\operatorname{sgn} \tau)[(L \otimes M)_\sigma \otimes N]_\tau$$

where the sum is extended over all pairs σ, τ in $S \times T$. If we agree to identify each σ in S_{r+s} with the element of S_{r+s+t} which agrees with σ on $\{1, \ldots, r + s\}$ and is the identity on $\{r + s + 1, \ldots, r + s + t\}$, then we may write

$$(L \wedge M) \wedge N = \sum_{\sigma,\tau} \operatorname{sgn} (\sigma \tau)[(L \otimes M \otimes N)_\sigma]_\tau.$$

But,

$$[(L \otimes M \otimes N)_\sigma]_\tau = (L \otimes M \otimes N)_{\tau\sigma}.$$

Therefore

$$(L \wedge M) \wedge N = \sum_{\sigma,\tau} \operatorname{sgn} (\tau \sigma)(L \otimes M \otimes N)_{\tau\sigma}.$$

Now suppose we have

$$\tau_1\sigma_1 = \tau_2\sigma_2 \gamma$$

with σ_i in S, τ_i in T, and γ in $G(r, s, t)$. Then $\tau_2^{-1} \tau_1 = \sigma_2\gamma\sigma_1^{-1}$, and since

$\sigma_2\gamma\sigma_1^{-1}$ lies in $G(r + s, t)$, it follows that τ_1 and τ_2 are in the same left coset of $G(r + s, t)$. Therefore, $\tau_1 = \tau_2$, and $\sigma_1 = \sigma_2\gamma$. But this implies that σ_1 and σ_2 (regarded as elements of S_{r+s}) lie in the same coset of $G(r, s)$; hence $\sigma_1 = \sigma_2$. Therefore, the products $\tau\sigma$ corresponding to the

$$\frac{(r + s + t)!}{(r + s)!t!}\frac{(r + s)!}{r!s!}$$

pairs (τ, σ) in $T \times S$ are all distinct and lie in distinct cosets of $G(r, s, t)$. Since there are exactly

$$\frac{(r + s + t)!}{r!s!t!}$$

left cosets of $G(r, s, t)$ in S_{r+s+t}, it follows that $(L \wedge M) \wedge N = E$. By an analogous argument, $L \wedge (M \wedge N) = E$ as well. ∎

Example 13. The exterior product is closely related to certain formulas for evaluating determinants known as the **Laplace expansions.** Let K be a commutative ring with identity and n a positive integer. Suppose that $1 \leq r < n$, and let L be the alternating r-linear form on K^n defined by

$$L(\alpha_1, \dots, \alpha_r) = \det \begin{bmatrix} A_{11} & \cdots & A_{1r} \\ \vdots & & \vdots \\ A_{r1} & \cdots & A_{rr} \end{bmatrix}.$$

If $s = n - r$ and M is the alternating s-linear form

$$M(\alpha_1, \dots, \alpha_s) = \det \begin{bmatrix} A_{1(r+1)} & \cdots & A_{1n} \\ \vdots & & \vdots \\ A_{s(r+1)} & \cdots & A_{sn} \end{bmatrix}$$

then $L \wedge M = D$, the determinant function on K^n. This is immediate from the fact that $L \wedge M$ is an alternating n-linear form and (as can be seen)

$$(L \wedge M)(\epsilon_1, \dots, \epsilon_n) = 1.$$

If we now describe $L \wedge M$ in the correct way, we obtain one Laplace expansion for the determinant of an $n \times n$ matrix over K.

In the permutation group S_n, let G be the subgroup which permutes the sets $\{1, \dots, r\}$ and $\{r + 1, \dots, n\}$ within themselves. Each left coset of G contains precisely one permutation σ such that $\sigma 1 < \sigma 2 < \dots < \sigma r$ and $\sigma(r + 1) < \dots < \sigma n$. The sign of this permutation is given by

$$\operatorname{sgn} \sigma = (-1)^{\sigma 1 + \cdots + \sigma r + (r(r-1)/2)}.$$

The wedge product $L \wedge M$ is given by

$$(L \wedge M)(\alpha_1, \dots, \alpha_n) = \Sigma\, (\operatorname{sgn} \sigma) L(\alpha_{\sigma 1}, \dots, \alpha_{\sigma r}) M(\alpha_{\sigma(r+1)}, \dots, \alpha_{\sigma n})$$

where the sum is taken over a collection of σ's, one from each coset of G. Therefore,

$$(L \wedge M)(\alpha_1, \ldots, \alpha_n) = \sum_{j_1 < \cdots < j_r} e_J \, L(\alpha_{j_1}, \ldots, \alpha_{j_r}) M(\alpha_{k_1}, \ldots, \alpha_{k_s})$$

where

$$e_J = (-1)^{j_1 + \cdots + j_r + (r(r-1)/2)}$$
$$k_i = \sigma(r + i).$$

In other words,

$$\det A = \sum_{j_1 < \cdots < j_r} e_J \begin{vmatrix} A_{j_1,1} & \cdots & A_{j_1,r} \\ \vdots & & \vdots \\ A_{j_r,1} & \cdots & A_{j_r,r} \end{vmatrix} \begin{vmatrix} A_{k_1,r+1} & \cdots & A_{k_1,n} \\ \vdots & & \vdots \\ A_{k_r,r+1} & \cdots & A_{k_r,n} \end{vmatrix}$$

This is one Laplace expansion. Others may be obtained by replacing the sets $\{1, \ldots, r\}$ and $\{r + 1, \ldots, n\}$ by two different complementary sets of indices.

If V is a K-module, we may put the various form modules $\Lambda^r(V)$ together and use the exterior product to define a ring. For simplicity, we shall do this only for the case of a free K-module of rank n. The modules $\Lambda^r(V)$ are then trivial for $r > n$. We define

$$\Lambda(V) = \Lambda^0(V) \oplus \Lambda^1(V) \oplus \cdots \oplus \Lambda^n(V).$$

This is an external direct sum—something which we have not discussed previously. The elements of $\Lambda(V)$ are the $(n + 1)$-tuples $(L_0, \ldots, L_n)$ with L_r in $\Lambda^r(V)$. Addition and multiplication by elements of K are defined as one would expect for $(n + 1)$-tuples. Incidentally, $\Lambda^0(V) = K$. If we identify $\Lambda^r(K)$ with the $(n + 1)$-tuples $(0, \ldots, 0, L, 0, \ldots, 0)$ where L is in $\Lambda^r(K)$, then $\Lambda^r(K)$ is a submodule of $\Lambda(V)$ and the direct sum decomposition

$$\Lambda(V) = \Lambda^0(V) \oplus \cdots \oplus \Lambda^n(V)$$

holds in the usual sense. Since $\Lambda^r(V)$ is a free K-module of rank $\binom{n}{r}$, we see that $\Lambda(V)$ is a free K-module and

$$\text{rank } \Lambda(V) = \sum_{r=0}^{n} \binom{n}{r}$$
$$= 2^n.$$

The exterior product defines a multiplication in $\Lambda(V)$: Use the exterior product on forms and extend it linearly to $\Lambda(V)$. It distributes over the addition of $\Lambda(V)$ and gives $\Lambda(V)$ the structure of a ring. This ring is the **Grassman ring** over V^*. It is not a commutative ring, e.g., if L, M are respectively in Λ^r and Λ^s, then

$$L \wedge M = (-1)^{rs} M \wedge L.$$

But, the Grassman ring is important in several parts of mathematics.

6. Elementary Canonical Forms

6.1. Introduction

We have mentioned earlier that our principal aim is to study linear transformations on finite-dimensional vector spaces. By this time, we have seen many specific examples of linear transformations, and we have proved a few theorems about the general linear transformation. In the finite-dimensional case we have utilized ordered bases to represent such transformations by matrices, and this representation adds to our insight into their behavior. We have explored the vector space $L(V, W)$, consisting of the linear transformations from one space into another, and we have explored the linear algebra $L(V, V)$, consisting of the linear transformations of a space into itself.

In the next two chapters, we shall be preoccupied with linear operators. Our program is to select a single linear operator T on a finite-dimensional vector space V and to 'take it apart to see what makes it tick.' At this early stage, it is easiest to express our goal in matrix language: Given the linear operator T, find an ordered basis for V in which the matrix of T assumes an especially simple form.

Here is an illustration of what we have in mind. Perhaps the simplest matrices to work with, beyond the scalar multiples of the identity, are the diagonal matrices:

$$D = \begin{bmatrix} c_1 & 0 & 0 & \cdots & 0 \\ 0 & c_2 & 0 & \cdots & 0 \\ 0 & 0 & c_3 & \cdots & 0 \\ \vdots & \vdots & \vdots & & \vdots \\ 0 & 0 & 0 & \cdots & c_n \end{bmatrix}. \tag{6-1}$$

Let T be a linear operator on an n-dimensional space V. If we could find an ordered basis $\mathcal{B} = \{\alpha_1, \ldots, \alpha_n\}$ for V in which T were represented by a diagonal matrix D (6-1), we would gain considerable information about T. For instance, simple numbers associated with T, such as the rank of T or the determinant of T, could be determined with little more than a glance at the matrix D. We could describe explicitly the range and the null space of T. Since $[T]_{\mathcal{B}} = D$ if and only if

$$T\alpha_k = c_k\alpha_k, \qquad k = 1, \ldots, n \tag{6-2}$$

the range would be the subspace spanned by those α_k's for which $c_k \neq 0$ and the null space would be spanned by the remaining α_k's. Indeed, it seems fair to say that, if we knew a basis $\mathcal{B}$ and a diagonal matrix D such that $[T]_{\mathcal{B}} = D$, we could answer readily any question about T which might arise.

Can each linear operator T be represented by a diagonal matrix in some ordered basis? If not, for which operators T does such a basis exist? How can we find such a basis if there is one? If no such basis exists, what is the simplest type of matrix by which we can represent T? These are some of the questions which we shall attack in this (and the next) chapter. The form of our questions will become more sophisticated as we learn what some of the difficulties are.

6.2. Characteristic Values

The introductory remarks of the previous section provide us with a starting point for our attempt to analyze the general linear operator T. We take our cue from (6-2), which suggests that we should study vectors which are sent by T into scalar multiples of themselves.

Definition. *Let* V *be a vector space over the field* F *and let* T *be a linear operator on* V. *A* **characteristic value** *of* T *is a scalar* c *in* F *such that there is a non-zero vector* α *in* V *with* $T\alpha = c\alpha$. *If* c *is a characteristic value of* T, *then*

(a) *any* α *such that* $T\alpha = c\alpha$ *is called a* **characteristic vector** *of* T *associated with the characteristic value* c;

(b) *the collection of all* α *such that* $T\alpha = c\alpha$ *is called the* **characteristic space** *associated with* c.

Characteristic values are often called characteristic roots, latent roots, eigenvalues, proper values, or spectral values. In this book we shall use only the name 'characteristic values.'

If T is any linear operator and c is any scalar, the set of vectors α such that $T\alpha = c\alpha$ is a subspace of V. It is the null space of the linear transformation $(T - cI)$. We call c a characteristic value of T if this subspace is different from the zero subspace, i.e., if $(T - cI)$ fails to be 1:1. If the underlying space V is finite-dimensional, $(T - cI)$ fails to be 1:1 precisely when its determinant is different from 0. Let us summarize.

Theorem 1. *Let* T *be a linear operator on a finite-dimensional space* V *and let* c *be a scalar. The following are equivalent.*

(i) c *is a characteristic value of* T.
(ii) *The operator* (T − cI) *is singular (not invertible).*
(iii) *det* (T − cI) = 0.

The determinant criterion (iii) is very important because it tells us where to look for the characteristic values of T. Since $\det(T - cI)$ is a polynomial of degree n in the variable c, we will find the characteristic values as the roots of that polynomial. Let us explain carefully.

If $\mathcal{B}$ is any ordered basis for V and $A = [T]_{\mathcal{B}}$, then $(T - cI)$ is invertible if and only if the matrix $(A - cI)$ is invertible. Accordingly, we make the following definition.

Definition. *If* A *is an* n × n *matrix over the field* F, *a* **characteristic value of A in** F *is a scalar* c *in* F *such that the matrix* (A − cI) *is singular (not invertible).*

Since c is a characteristic value of A if and only if $\det(A - cI) = 0$, or equivalently if and only if $\det(cI - A) = 0$, we form the matrix $(xI - A)$ with polynomial entries, and consider the polynomial $f = \det(xI - A)$. Clearly the characteristic values of A in F are just the scalars c in F such that $f(c) = 0$. For this reason f is called the **characteristic polynomial** of A. It is important to note that f is a monic polynomial which has degree exactly n. This is easily seen from the formula for the determinant of a matrix in terms of its entries.

Lemma. *Similar matrices have the same characteristic polynomial.*

Proof. If $B = P^{-1}AP$, then

$$\begin{aligned}\det(xI - B) &= \det(xI - P^{-1}AP)\\ &= \det(P^{-1}(xI - A)P)\\ &= \det P^{-1} \cdot \det(xI - A) \cdot \det P\\ &= \det(xI - A). \quad \blacksquare\end{aligned}$$

This lemma enables us to define sensibly the characteristic polynomial of the operator T as the characteristic polynomial of any $n \times n$ matrix which represents T in some ordered basis for V. Just as for matrices, the characteristic values of T will be the roots of the characteristic polynomial for T. In particular, this shows us that T cannot have more than n distinct characteristic values. It is important to point out that T may not have any characteristic values.

EXAMPLE 1. Let T be the linear operator on R^2 which is represented in the standard ordered basis by the matrix

$$A = \begin{bmatrix} 0 & -1 \\ 1 & 0 \end{bmatrix}.$$

The characteristic polynomial for T (or for A) is

$$\det (xI - A) = \begin{vmatrix} x & 1 \\ -1 & x \end{vmatrix} = x^2 + 1.$$

Since this polynomial has no real roots, T has no characteristic values. If U is the linear operator on C^2 which is represented by A in the standard ordered basis, then U has two characteristic values, i and $-i$. Here we see a subtle point. In discussing the characteristic values of a matrix A, we must be careful to stipulate the field involved. The matrix A above has no characteristic values in R, but has the two characteristic values i and $-i$ in C.

EXAMPLE 2. Let A be the (real) 3×3 matrix

$$\begin{bmatrix} 3 & 1 & -1 \\ 2 & 2 & -1 \\ 2 & 2 & 0 \end{bmatrix}.$$

Then the characteristic polynomial for A is

$$\begin{vmatrix} x-3 & -1 & 1 \\ -2 & x-2 & 1 \\ -2 & -2 & x \end{vmatrix} = x^3 - 5x^2 + 8x - 4 = (x-1)(x-2)^2.$$

Thus the characteristic values of A are 1 and 2.

Suppose that T is the linear operator on R^3 which is represented by A in the standard basis. Let us find the characteristic vectors of T associated with the characteristic values, 1 and 2. Now

$$A - I = \begin{bmatrix} 2 & 1 & -1 \\ 2 & 1 & -1 \\ 2 & 2 & -1 \end{bmatrix}.$$

It is obvious at a glance that $A - I$ has rank equal to 2 (and hence $T - I$ has nullity equal to 1). So the space of characteristic vectors associated with the characteristic value 1 is one-dimensional. The vector $\alpha_1 = (1, 0, 2)$ spans the null space of $T - I$. Thus $T\alpha = \alpha$ if and only if α is a scalar multiple of α_1. Now consider

$$A - 2I = \begin{bmatrix} 1 & 1 & -1 \\ 2 & 0 & -1 \\ 2 & 2 & -2 \end{bmatrix}.$$

Evidently $A - 2I$ also has rank 2, so that the space of characteristic vectors associated with the characteristic value 2 has dimension 1. Evidently $T\alpha = 2\alpha$ if and only if α is a scalar multiple of $\alpha_2 = (1, 1, 2)$.

Definition. *Let* T *be a linear operator on the finite-dimensional space* V. *We say that* T *is* **diagonalizable** *if there is a basis for* V *each vector of which is a characteristic vector of* T.

The reason for the name should be apparent; for, if there is an ordered basis $\mathcal{B} = \{\alpha_1, \ldots, \alpha_n\}$ for V in which each α_i is a characteristic vector of T, then the matrix of T in the ordered basis $\mathcal{B}$ is diagonal. If $T\alpha_i = c_i\alpha_i$, then

$$[T]_{\mathcal{B}} = \begin{bmatrix} c_1 & 0 & \cdots & 0 \\ 0 & c_2 & \cdots & 0 \\ \vdots & \vdots & & \vdots \\ 0 & 0 & \cdots & c_n \end{bmatrix}.$$

We certainly do not require that the scalars $c_1, \ldots, c_n$ be distinct; indeed, they may all be the same scalar (when T is a scalar multiple of the identity operator).

One could also define T to be diagonalizable when the characteristic vectors of T span V. This is only superficially different from our definition, since we can select a basis out of any spanning set of vectors.

For Examples 1 and 2 we purposely chose linear operators T on R^n which are not diagonalizable. In Example 1, we have a linear operator on R^2 which is not diagonalizable, because it has no characteristic values. In Example 2, the operator T has characteristic values; in fact, the characteristic polynomial for T factors completely over the real number field: $f = (x - 1)(x - 2)^2$. Nevertheless T fails to be diagonalizable. There is only a one-dimensional space of characteristic vectors associated with each of the two characteristic values of T. Hence, we cannot possibly form a basis for R^3 which consists of characteristic vectors of T.

Suppose that T is a diagonalizable linear operator. Let $c_1, \ldots, c_k$ be the *distinct* characteristic values of T. Then there is an ordered basis $\mathcal{B}$ in which T is represented by a diagonal matrix which has for its diagonal entries the scalars c_i, each repeated a certain number of times. If c_i is repeated d_i times, then (we may arrange that) the matrix has the block form

$$(6\text{-}3) \qquad [T]_{\mathcal{B}} = \begin{bmatrix} c_1I_1 & 0 & \cdots & 0 \\ 0 & c_2I_2 & \cdots & 0 \\ \vdots & \vdots & & \vdots \\ 0 & 0 & \cdots & c_kI_k \end{bmatrix}$$

where I_j is the $d_j \times d_j$ identity matrix. From that matrix we see two things. First, the characteristic polynomial for T is the product of (possibly repeated) linear factors:

$$f = (x - c_1)^{d_1} \cdots (x - c_k)^{d_k}.$$

If the scalar field F is algebraically closed, e.g., the field of complex numbers, every polynomial over F can be so factored (see Section 4.5); however, if F is not algebraically closed, we are citing a special property of T when we say that its characteristic polynomial has such a factorization. The second thing we see from (6-3) is that d_i, the number of times which c_i is repeated as root of f, is equal to the dimension of the space of characteristic vectors associated with the characteristic value c_i. That is because the nullity of a diagonal matrix is equal to the number of zeros which it has on its main diagonal, and the matrix $[T - c_iI]_{\mathcal{B}}$ has d_i zeros on its main diagonal. This relation between the dimension of the characteristic space and the multiplicity of the characteristic value as a root of f does not seem exciting at first; however, it will provide us with a simpler way of determining whether a given operator is diagonalizable.

Lemma. *Suppose that* $T\alpha = c\alpha$. *If* f *is any polynomial, then* $f(T)\alpha = f(c)\alpha$.

Proof. Exercise.

Lemma. *Let* T *be a linear operator on the finite-dimensional space* V. *Let* $c_1, \ldots, c_k$ *be the distinct characteristic values of* T *and let* W_i *be the space of characteristic vectors associated with the characteristic value* c_i. *If* $W = W_1 + \cdots + W_k$, *then*

$$\dim W = \dim W_1 + \cdots + \dim W_k.$$

In fact, if $\mathcal{B}_i$ *is an ordered basis for* W_i, *then* $\mathcal{B} = (\mathcal{B}_1, \ldots, \mathcal{B}_k)$ *is an ordered basis for* W.

Proof. The space $W = W_1 + \cdots + W_k$ is the subspace spanned by all of the characteristic vectors of T. Usually when one forms the sum W of subspaces W_i, one expects that $\dim W < \dim W_1 + \cdots + \dim W_k$ because of linear relations which may exist between vectors in the various spaces. This lemma states that the characteristic spaces associated with different characteristic values are independent of one another.

Suppose that (for each i) we have a vector β_i in W_i, and assume that $\beta_1 + \cdots + \beta_k = 0$. We shall show that $\beta_i = 0$ for each i. Let f be any polynomial. Since $T\beta_i = c_i\beta_i$, the preceding lemma tells us that

$$\begin{aligned} 0 = f(T)0 &= f(T)\beta_1 + \cdots + f(T)\beta_k \\ &= f(c_1)\beta_1 + \cdots + f(c_k)\beta_k. \end{aligned}$$

Choose polynomials $f_1, \ldots, f_k$ such that

$$f_i(c_j) = \delta_{ij} = \begin{cases} 1, & i = j \\ 0, & i \neq j. \end{cases}$$

Then

$$0 = f_i(T)0 = \sum_j \delta_{ij}\beta_j$$
$$= \beta_i.$$

Now, let $\mathcal{B}_i$ be an ordered basis for W_i, and let $\mathcal{B}$ be the sequence $\mathcal{B} = (\mathcal{B}_1, \ldots, \mathcal{B}_k)$. Then $\mathcal{B}$ spans the subspace $W = W_1 + \cdots + W_k$. Also, $\mathcal{B}$ is a linearly independent sequence of vectors, for the following reason. Any linear relation between the vectors in $\mathcal{B}$ will have the form $\beta_1 + \cdots + \beta_k = 0$, where β_i is some linear combination of the vectors in $\mathcal{B}_i$. From what we just did, we know that $\beta_i = 0$ for each i. Since each $\mathcal{B}_i$ is linearly independent, we see that we have only the trivial linear relation between the vectors in $\mathcal{B}$. ∎

Theorem 2. *Let* T *be a linear operator on a finite-dimensional space* V. *Let* $c_1, \ldots, c_k$ *be the distinct characteristic values of* T *and let* W_i *be the null space of* $(T - c_iI)$. *The following are equivalent.*

(i) T *is diagonalizable.*

(ii) *The characteristic polynomial for* T *is*

$$f = (x - c_1)^{d_1} \cdots (x - c_k)^{d_k}$$

and dim $W_i = d_i$, $i = 1, \ldots, k$.

(iii) *dim* $W_1 + \cdots +$ *dim* $W_k =$ *dim* V.

Proof. We have observed that (i) implies (ii). If the characteristic polynomial f is the product of linear factors, as in (ii), then $d_1 + \cdots + d_k = \dim V$. For, the sum of the d_i's is the degree of the characteristic polynomial, and that degree is dim V. Therefore (ii) implies (iii). Suppose (iii) holds. By the lemma, we must have $V = W_1 + \cdots + W_k$, i.e., the characteristic vectors of T span V. ∎

The matrix analogue of Theorem 2 may be formulated as follows. Let A be an $n \times n$ matrix with entries in a field F, and let $c_1, \ldots, c_k$ be the distinct characteristic values of A in F. For each i, let W_i be the space of column matrices X (with entries in F) such that

$$(A - c_iI)X = 0,$$

and let $\mathcal{B}_i$ be an ordered basis for W_i. The bases $\mathcal{B}_1, \ldots, \mathcal{B}_k$ collectively string together to form the sequence of columns of a matrix P:

$$P = [P_1, P_2, \ldots] = (\mathcal{B}_1, \ldots, \mathcal{B}_k).$$

The matrix A is similar over F to a diagonal matrix if and only if P is a square matrix. When P is square, P is invertible and $P^{-1}AP$ is diagonal.

Example 3. Let T be the linear operator on R^3 which is represented in the standard ordered basis by the matrix

$$A = \begin{bmatrix} 5 & -6 & -6 \\ -1 & 4 & 2 \\ 3 & -6 & -4 \end{bmatrix}.$$

Let us indicate how one might compute the characteristic polynomial, using various row and column operations:

$$\begin{aligned}
\begin{vmatrix} x-5 & 6 & 6 \\ 1 & x-4 & -2 \\ -3 & 6 & x+4 \end{vmatrix} &= \begin{vmatrix} x-5 & 0 & 6 \\ 1 & x-2 & -2 \\ -3 & 2-x & x+4 \end{vmatrix} \\
&= (x-2)\begin{vmatrix} x-5 & 0 & 6 \\ 1 & 1 & -2 \\ -3 & -1 & x+4 \end{vmatrix} \\
&= (x-2)\begin{vmatrix} x-5 & 0 & 6 \\ 1 & 1 & -2 \\ -2 & 0 & x+2 \end{vmatrix} \\
&= (x-2)\begin{vmatrix} x-5 & 6 \\ -2 & x+2 \end{vmatrix} \\
&= (x-2)(x^2-3x+2) \\
&= (x-2)^2(x-1).
\end{aligned}$$

What are the dimensions of the spaces of characteristic vectors associated with the two characteristic values? We have

$$A - I = \begin{bmatrix} 4 & -6 & -6 \\ -1 & 3 & 2 \\ 3 & -6 & -5 \end{bmatrix}$$

$$A - 2I = \begin{bmatrix} 3 & -6 & -6 \\ -1 & 2 & 2 \\ 3 & -6 & -6 \end{bmatrix}.$$

We know that $A - I$ is singular and obviously rank $(A - I) \geq 2$. Therefore, rank $(A - I) = 2$. It is evident that rank $(A - 2I) = 1$.

Let W_1, W_2 be the spaces of characteristic vectors associated with the characteristic values 1, 2. We know that dim $W_1 = 1$ and dim $W_2 = 2$. By Theorem 2, T is diagonalizable. It is easy to exhibit a basis for R^3 in which T is represented by a diagonal matrix. The null space of $(T - I)$ is spanned by the vector $\alpha_1 = (3, -1, 3)$ and so $\{\alpha_1\}$ is a basis for W_1. The null space of $T - 2I$ (i.e., the space W_2) consists of the vectors (x_1, x_2, x_3) with $x_1 = 2x_2 + 2x_3$. Thus, one example of a basis for W_2 is

$$\begin{aligned} \alpha_2 &= (2, 1, 0) \\ \alpha_3 &= (2, 0, 1). \end{aligned}$$

If $\mathcal{B} = \{\alpha_1, \alpha_2, \alpha_3\}$, then $[T]_{\mathcal{B}}$ is the diagonal matrix

$$D = \begin{bmatrix} 1 & 0 & 0 \\ 0 & 2 & 0 \\ 0 & 0 & 2 \end{bmatrix}.$$

The fact that T is diagonalizable means that the original matrix A is similar (over R) to the diagonal matrix D. The matrix P which enables us to change coordinates from the basis $\mathcal{B}$ to the standard basis is (of course) the matrix which has the transposes of $\alpha_1, \alpha_2, \alpha_3$ as its column vectors:

$$P = \begin{bmatrix} 3 & 2 & 2 \\ -1 & 1 & 0 \\ 3 & 0 & 1 \end{bmatrix}.$$

Furthermore, $AP = PD$, so that

$$P^{-1}AP = D.$$

Exercises

1. In each of the following cases, let T be the linear operator on R^2 which is represented by the matrix A in the standard ordered basis for R^2, and let U be the linear operator on C^2 represented by A in the standard ordered basis. Find the characteristic polynomial for T and that for U, find the characteristic values of each operator, and for each such characteristic value c find a basis for the corresponding space of characteristic vectors.

$$A = \begin{bmatrix} 1 & 0 \\ 0 & 0 \end{bmatrix}, \quad A = \begin{bmatrix} 2 & 3 \\ -1 & 1 \end{bmatrix}, \quad A = \begin{bmatrix} 1 & 1 \\ 1 & 1 \end{bmatrix}.$$

2. Let V be an n-dimensional vector space over F. What is the characteristic polynomial of the identity operator on V? What is the characteristic polynomial for the zero operator?

3. Let A be an $n \times n$ triangular matrix over the field F. Prove that the characteristic values of A are the diagonal entries of A, i.e., the scalars A_{ii}.

4. Let T be the linear operator on R^3 which is represented in the standard ordered basis by the matrix

$$\begin{bmatrix} -9 & 4 & 4 \\ -8 & 3 & 4 \\ -16 & 8 & 7 \end{bmatrix}.$$

Prove that T is diagonalizable by exhibiting a basis for R^3, each vector of which is a characteristic vector of T.

5. Let

$$A = \begin{bmatrix} 6 & -3 & -2 \\ 4 & -1 & -2 \\ 10 & -5 & -3 \end{bmatrix}.$$

Is A similar over the field R to a diagonal matrix? Is A similar over the field C to a diagonal matrix?

6. Let T be the linear operator on R^4 which is represented in the standard ordered basis by the matrix

$$\begin{bmatrix} 0 & 0 & 0 & 0 \\ a & 0 & 0 & 0 \\ 0 & b & 0 & 0 \\ 0 & 0 & c & 0 \end{bmatrix}.$$

Under what conditions on a, b, and c is T diagonalizable?

7. Let T be a linear operator on the n-dimensional vector space V, and suppose that T has n *distinct* characteristic values. Prove that T is diagonalizable.

8. Let A and B be $n \times n$ matrices over the field F. Prove that if $(I - AB)$ is invertible, then $I - BA$ is invertible and

$$(I - BA)^{-1} = I + B(I - AB)^{-1}A.$$

9. Use the result of Exercise 8 to prove that, if A and B are $n \times n$ matrices over the field F, then AB and BA have precisely the same characteristic values in F.

10. Suppose that A is a 2×2 matrix with real entries which is symmetric ($A^t = A$). Prove that A is similar over R to a diagonal matrix.

11. Let N be a 2×2 complex matrix such that $N^2 = 0$. Prove that either $N = 0$ or N is similar over C to

$$\begin{bmatrix} 0 & 0 \\ 1 & 0 \end{bmatrix}.$$

12. Use the result of Exercise 11 to prove the following: If A is a 2×2 matrix with complex entries, then A is similar over C to a matrix of one of the two types

$$\begin{bmatrix} a & 0 \\ 0 & b \end{bmatrix} \qquad \begin{bmatrix} a & 0 \\ 1 & a \end{bmatrix}.$$

13. Let V be the vector space of all functions from R into R which are continuous, i.e., the space of continuous real-valued functions on the real line. Let T be the linear operator on V defined by

$$(Tf)(x) = \int_0^x f(t)\, dt.$$

Prove that T has no characteristic values.

14. Let A be an $n \times n$ *diagonal* matrix with characteristic polynomial

$$(x - c_1)^{d_1} \cdots (x - c_k)^{d_k},$$

where $c_1, \ldots, c_k$ are distinct. Let V be the space of $n \times n$ matrices B such that $AB = BA$. Prove that the dimension of V is $d_1^2 + \cdots + d_k^2$.

15. Let V be the space of $n \times n$ matrices over F. Let A be a fixed $n \times n$ matrix over F. Let T be the linear operator 'left multiplication by A' on V. Is it true that A and T have the same characteristic values?

6.3. Annihilating Polynomials

In attempting to analyze a linear operator T, one of the most useful things to know is the class of polynomials which annihilate T. Specifically, suppose T is a linear operator on V, a vector space over the field F. If p is a polynomial over F, then $p(T)$ is again a linear operator on V. If q is another polynomial over F, then

$$\begin{aligned}(p + q)(T) &= p(T) + q(T)\\ (pq)(T) &= p(T)q(T).\end{aligned}$$

Therefore, the collection of polynomials p which annihilate T, in the sense that

$$p(T) = 0,$$

is an ideal in the polynomial algebra $F[x]$. It may be the zero ideal, i.e., it may be that T is not annihilated by any non-zero polynomial. But, that cannot happen if the space V is finite-dimensional.

Suppose T is a linear operator on the n-dimensional space V. Look at the first $(n^2 + 1)$ powers of T:

$$I, T, T^2, \ldots, T^{n^2}.$$

This is a sequence of $n^2 + 1$ operators in $L(V, V)$, the space of linear operators on V. The space $L(V, V)$ has dimension n^2. Therefore, that sequence of $n^2 + 1$ operators must be linearly dependent, i.e., we have

$$c_0I + c_1T + \cdots + c_{n^2}T^{n^2} = 0$$

for some scalars c_i, not all zero. So, the ideal of polynomials which annihilate T contains a non-zero polynomial of degree n^2 or less.

According to Theorem 5 of Chapter 4, every polynomial ideal consists of all multiples of some fixed monic polynomial, the generator of the ideal. Thus, there corresponds to the operator T a monic polynomial p with this property: If f is a polynomial over F, then $f(T) = 0$ if and only if $f = pg$, where g is some polynomial over F.

Definition. *Let* T *be a linear operator on a finite-dimensional vector space* V *over the field* F. *The* **minimal polynomial** *for* T *is the (unique) monic generator of the ideal of polynomials over* F *which annihilate* T.

The name 'minimal polynomial' stems from the fact that the generator of a polynomial ideal is characterized by being the monic polynomial of minimum degree in the ideal. That means that the minimal polynomial p for the linear operator T is uniquely determined by these three properties:

(1) p is a monic polynomial over the scalar field F.

(2) $p(T) = 0$.

(3) No polynomial over F which annihilates T has smaller degree than p has.

If A is an $n \times n$ matrix over F, we define the **minimal polynomial** for A in an analogous way, as the unique monic generator of the ideal of all polynomials over F which annihilate A. If the operator T is represented in some ordered basis by the matrix A, then T and A have the same minimal polynomial. That is because $f(T)$ is represented in the basis by the matrix $f(A)$, so that $f(T) = 0$ if and only if $f(A) = 0$.

From the last remark about operators and matrices it follows that similar matrices have the same minimal polynomial. That fact is also clear from the definitions because

$$f(P^{-1}AP) = P^{-1}f(A)P$$

for every polynomial f.

There is another basic remark which we should make about minimal polynomials of matrices. Suppose that A is an $n \times n$ matrix with entries in the field F. Suppose that F_1 is a field which contains F as a subfield. (For example, A might be a matrix with rational entries, while F_1 is the field of real numbers. Or, A might be a matrix with real entries, while F_1 is the field of complex numbers.) We may regard A either as an $n \times n$ matrix over F or as an $n \times n$ matrix over F_1. On the surface, it might appear that we obtain two different minimal polynomials for A. Fortunately that is not the case; and we must see why. What is the definition of the minimal polynomial for A, regarded as an $n \times n$ matrix over the field F? We consider all monic polynomials with coefficients in F which annihilate A, and we choose the one of least degree. If f is a monic polynomial over F:

$$f = x^k + \sum_{j=0}^{k-1} a_j x^j \tag{6-4}$$

then $f(A) = 0$ merely says that we have a linear relation between the powers of A:

$$A^k + a_{k-1}A^{k-1} + \cdots + a_1A + a_0I = 0. \tag{6-5}$$

The degree of the minimal polynomial is the least positive integer k such that there is a linear relation of the form (6-5) between the powers I, $A, \ldots, A^k$. Furthermore, by the uniqueness of the minimal polynomial, there is for that k one and only one relation of the form (6-5); i.e., once the minimal k is determined, there are unique scalars $a_0, \ldots, a_{k-1}$ in F such that (6-5) holds. They are the coefficients of the minimal polynomial.

Now (for each k) we have in (6-5) a system of n^2 linear equations for the 'unknowns' $a_0, \ldots, a_{k-1}$. Since the entries of A lie in F, the coefficients of the system of equations (6-5) are in F. Therefore, if the system has a solution with $a_0, \ldots, a_{k-1}$ in F_1 it has a solution with $a_0, \ldots, a_{k-1}$ in F. (See the end of Section 1.4.) It should now be clear that the two minimal polynomials are the same.

What do we know thus far about the minimal polynomial for a linear operator on an n-dimensional space? Only that its degree does not exceed n^2. That turns out to be a rather poor estimate, since the degree cannot exceed n. We shall prove shortly that the operator is annihilated by its characteristic polynomial. First, let us observe a more elementary fact.

Theorem 3. *Let* T *be a linear operator on an* n*-dimensional vector space* V *[or, let* A *be an* n × n *matrix]. The characteristic and minimal polynomials for* T *[for* A*] have the same roots, except for multiplicities.*

Proof. Let p be the minimal polynomial for T. Let c be a scalar. What we want to show is that $p(c) = 0$ if and only if c is a characteristic value of T.

First, suppose $p(c) = 0$. Then

$$p = (x - c)q$$

where q is a polynomial. Since $\deg q < \deg p$, the definition of the minimal polynomial p tells us that $q(T) \neq 0$. Choose a vector β such that $q(T)\beta \neq 0$. Let $\alpha = q(T)\beta$. Then

$$\begin{aligned} 0 &= p(T)\beta \\ &= (T - cI)q(T)\beta \\ &= (T - cI)\alpha \end{aligned}$$

and thus, c is a characteristic value of T.

Now, suppose that c is a characteristic value of T, say, $T\alpha = c\alpha$ with $\alpha \neq 0$. As we noted in a previous lemma,

$$p(T)\alpha = p(c)\alpha.$$

Since $p(T) = 0$ and $\alpha \neq 0$, we have $p(c) = 0$. ∎

Let T be a diagonalizable linear operator and let $c_1, \ldots, c_k$ be the distinct characteristic values of T. Then it is easy to see that the minimal polynomial for T is the polynomial

$$p = (x - c_1) \cdots (x - c_k).$$

If α is a characteristic vector, then one of the operators $T - c_1I, \ldots, T - c_kI$ sends α into 0. Therefore

$$(T - c_1I) \cdots (T - c_kI)\alpha = 0$$

for every characteristic vector α. There is a basis for the underlying space which consists of characteristic vectors of T; hence

$$p(T) = (T - c_1I) \cdots (T - c_kI) = 0.$$

What we have concluded is this. If T is a diagonalizable linear operator, then the minimal polynomial for T is a product of distinct linear factors. As we shall soon see, that property characterizes diagonalizable operators.

EXAMPLE 4. Let's try to find the minimal polynomials for the operators in Examples 1, 2, and 3. We shall discuss them in reverse order. The operator in Example 3 was found to be diagonalizable with characteristic polynomial

$$f = (x - 1)(x - 2)^2.$$

From the preceding paragraph, we know that the minimal polynomial for T is

$$p = (x - 1)(x - 2).$$

The reader might find it reassuring to verify directly that

$$(A - I)(A - 2I) = 0.$$

In Example 2, the operator T also had the characteristic polynomial $f = (x - 1)(x - 2)^2$. But, this T is not diagonalizable, so we don't know that the minimal polynomial is $(x - 1)(x - 2)$. What do we know about the minimal polynomial in this case? From Theorem 3 we know that its roots are 1 and 2, with some multiplicities allowed. Thus we search for p among polynomials of the form $(x - 1)^k(x - 2)^l$, $k \geq 1, l \geq 1$. Try $(x - 1)(x - 2)$:

$$(A - I)(A - 2I) = \begin{bmatrix} 2 & 1 & -1 \\ 2 & 1 & -1 \\ 2 & 2 & -1 \end{bmatrix}\begin{bmatrix} 1 & 1 & -1 \\ 2 & 0 & -1 \\ 2 & 2 & -2 \end{bmatrix} = \begin{bmatrix} 2 & 0 & -1 \\ 2 & 0 & -1 \\ 4 & 0 & -2 \end{bmatrix}.$$

Thus, the minimal polynomial has degree at least 3. So, next we should try either $(x - 1)^2(x - 2)$ or $(x - 1)(x - 2)^2$. The second, being the characteristic polynomial, would seem a less random choice. One can readily compute that $(A - I)(A - 2I)^2 = 0$. Thus the minimal polynomial for T is its characteristic polynomial.

In Example 1 we discussed the linear operator T on R^2 which is represented in the standard basis by the matrix

$$A = \begin{bmatrix} 0 & -1 \\ 1 & 0 \end{bmatrix}.$$

The characteristic polynomial is $x^2 + 1$, which has no real roots. To determine the minimal polynomial, forget about T and concentrate on A. As a complex 2×2 matrix, A has the characteristic values i and $-i$. Both roots must appear in the minimal polynomial. Thus the minimal polynomial is divisible by $x^2 + 1$. It is trivial to verify that $A^2 + I = 0$. So the minimal polynomial is $x^2 + 1$.

Theorem 4 (Cayley-Hamilton). *Let* T *be a linear operator on a finite dimensional vector space* V. *If* f *is the characteristic polynomial for* T, *then* $f(T) = 0$; *in other words, the minimal polynomial divides the characteristic polynomial for* T.

Proof. Later on we shall give two proofs of this result independent of the one to be given here. The present proof, although short, may be difficult to understand. Aside from brevity, it has the virtue of providing an illuminating and far from trivial application of the general theory of determinants developed in Chapter 5.

Let K be the commutative ring with identity consisting of all polynomials in T. Of course, K is actually a commutative algebra with identity over the scalar field. Choose an ordered basis $\{\alpha_1, \ldots, \alpha_n\}$ for V, and let A be the matrix which represents T in the given basis. Then

$$T\alpha_i = \sum_{j=1}^{n} A_{ji}\alpha_j, \qquad 1 \leq i \leq n.$$

These equations may be written in the equivalent form

$$\sum_{j=1}^{n} (\delta_{ij}T - A_{ji}I)\alpha_j = 0, \qquad 1 \leq i \leq n.$$

Let B denote the element of $K^{n\times n}$ with entries

$$B_{ij} = \delta_{ij}T - A_{ji}I.$$

When $n = 2$

$$B = \begin{bmatrix} T - A_{11}I & -A_{21}I \\ -A_{12}I & T - A_{22}I \end{bmatrix}$$

and

$$\begin{aligned} \det B &= (T - A_{11}I)(T - A_{22}I) - A_{12}A_{21}I \\ &= T^2 - (A_{11} + A_{22})T + (A_{11}A_{22} - A_{12}A_{21})I \\ &= f(T) \end{aligned}$$

where f is the characteristic polynomial:

$$f = x^2 - (\text{trace } A)x + \det A.$$

For the case $n > 2$, it is also clear that

$$\det B = f(T)$$

since f is the determinant of the matrix $xI - A$ whose entries are the polynomials

$$(xI - A)_{ij} = \delta_{ij}x - A_{ji}.$$

We wish to show that $f(T) = 0$. In order that $f(T)$ be the zero operator, it is necessary and sufficient that $(\det B)\alpha_k = 0$ for $k = 1, \ldots, n$. By the definition of B, the vectors $\alpha_1, \ldots, \alpha_n$ satisfy the equations

$$\sum_{j=1}^{n} B_{ij}\alpha_j = 0, \qquad 1 \leq i \leq n. \tag{6-6}$$

When $n = 2$, it is suggestive to write (6-6) in the form

$$\begin{bmatrix} T - A_{11}I & -A_{21}I \\ -A_{12}I & T - A_{22}I \end{bmatrix} \begin{bmatrix} \alpha_1 \\ \alpha_2 \end{bmatrix} = \begin{bmatrix} 0 \\ 0 \end{bmatrix}.$$

In this case, the classical adjoint, adj B is the matrix

$$\tilde{B} = \begin{bmatrix} T - A_{22}I & A_{21}I \\ A_{12}I & T - A_{11}I \end{bmatrix}$$

and

$$\tilde{B}B = \begin{bmatrix} \det B & 0 \\ 0 & \det B \end{bmatrix}.$$

Hence, we have

$$\begin{aligned}(\det B)\begin{bmatrix}\alpha_1\\ \alpha_2\end{bmatrix} &= (\tilde{B}B)\begin{bmatrix}\alpha_1\\ \alpha_2\end{bmatrix}\\ &= \tilde{B}\left(B\begin{bmatrix}\alpha_1\\ \alpha_2\end{bmatrix}\right)\\ &= \begin{bmatrix}0\\ 0\end{bmatrix}.\end{aligned}$$

In the general case, let $\tilde{B} = \text{adj } B$. Then by (6-6)

$$\sum_{j=1}^{n} \tilde{B}_{ki}B_{ij}\alpha_j = 0$$

for each pair k, i, and summing on i, we have

$$\begin{aligned}0 &= \sum_{i=1}^{n}\sum_{j=1}^{n} \tilde{B}_{ki}B_{ij}\alpha_j\\ &= \sum_{j=1}^{n}\left(\sum_{i=1}^{n} \tilde{B}_{ki}B_{ij}\right)\alpha_j.\end{aligned}$$

Now $\tilde{B}B = (\det B)I$, so that

$$\sum_{i=1}^{n} \tilde{B}_{ki}B_{ij} = \delta_{kj} \det B.$$

Therefore

$$\begin{aligned}0 &= \sum_{j=1}^{n} \delta_{kj}(\det B)\alpha_j\\ &= (\det B)\alpha_k, \qquad 1 \le k \le n. \quad \blacksquare\end{aligned}$$

The Cayley-Hamilton theorem is useful to us at this point primarily because it narrows down the search for the minimal polynomials of various operators. If we know the matrix A which represents T in some ordered basis, then we can compute the characteristic polynomial f. We know that the minimal polynomial p divides f and that the two polynomials have the same roots. There is no method for computing precisely the roots of a polynomial (unless its degree is small); however, if f factors

(6-7) $\quad f = (x - c_1)^{d_1} \cdots (x - c_k)^{d_k}, \qquad c_1, \ldots, c_k \text{ distinct}, d_i \ge 1$

then

(6-8) $\quad p = (x - c_1)^{r_1} \cdots (x - c_k)^{r_k}, \qquad 1 \le r_j \le d_j.$

That is all we can say in general. If f is the polynomial (6-7) and has degree n, then for every polynomial p as in (6-8) we can find an $n \times n$ matrix which has f as its characteristic polynomial and p as its minimal polynomial. We shall not prove this now. But, we want to emphasize the fact that the knowledge that the characteristic polynomial has the form (6-7) tells us that the minimal polynomial has the form (6-8), and it tells us nothing else about p.

EXAMPLE 5. Let A be the 4×4 (rational) matrix

$$A = \begin{bmatrix} 0 & 1 & 0 & 1 \\ 1 & 0 & 1 & 0 \\ 0 & 1 & 0 & 1 \\ 1 & 0 & 1 & 0 \end{bmatrix}.$$

The powers of A are easy to pute:

$$A^2 = \begin{bmatrix} 2 & 0 & 2 & 0 \\ 0 & 2 & 0 & 2 \\ 2 & 0 & 2 & 0 \\ 0 & 2 & 0 & 2 \end{bmatrix}$$

$$A^3 = \begin{bmatrix} 0 & 4 & 0 & 4 \\ 4 & 0 & 4 & 0 \\ 0 & 4 & 0 & 4 \\ 4 & 0 & 4 & 0 \end{bmatrix}.$$

Thus $A^3 = 4A$, i.e., if $p = x^3 - 4x = x(x + 2)(x - 2)$, then $p(A) = 0$. The minimal polynomial for A must divide p. That minimal polynomial is obviously not of degree 1, since that would mean that A was a scalar multiple of the identity. Hence, the candidates for the minimal polynomial are: p, $x(x + 2)$, $x(x - 2)$, $x^2 - 4$. The three quadratic polynomials can be eliminated because it is obvious at a glance that $A^2 \neq -2A$, $A^2 \neq 2A$, $A^2 \neq 4I$. Therefore p is the minimal polynomial for A. In particular 0, 2, and -2 are the characteristic values of A. One of the factors x, $x - 2$, $x + 2$ must be repeated twice in the characteristic polynomial. Evidently, rank $(A) = 2$. Consequently there is a two-dimensional space of characteristic vectors associated with the characteristic value 0. From Theorem 2, it should now be clear that the characteristic polynomial is $x^2(x^2 - 4)$ and that A is similar over the field of rational numbers to the matrix

$$\begin{bmatrix} 0 & 0 & 0 & 0 \\ 0 & 0 & 0 & 0 \\ 0 & 0 & 2 & 0 \\ 0 & 0 & 0 & -2 \end{bmatrix}.$$

Exercises

1. Let V be a finite-dimensional vector space. What is the minimal polynomial for the identity operator on V? What is the minimal polynomial for the zero operator?

2. Let a, b, and c be elements of a field F, and let A be the following 3×3 matrix over F:

$$A = \begin{bmatrix} 0 & 0 & c \\ 1 & 0 & b \\ 0 & 1 & a \end{bmatrix}.$$

Prove that the characteristic polynomial for A is $x^3 - ax^2 - bx - c$ and that this is also the minimal polynomial for A.

3. Let A be the 4×4 real matrix

$$A = \begin{bmatrix} 1 & 1 & 0 & 0 \\ -1 & -1 & 0 & 0 \\ -2 & -2 & 2 & 1 \\ 1 & 1 & -1 & 0 \end{bmatrix}.$$

Show that the characteristic polynomial for A is $x^2(x-1)^2$ and that it is also the minimal polynomial.

4. Is the matrix A of Exercise 3 similar over the field of complex numbers to a diagonal matrix?

5. Let V be an n-dimensional vector space and let T be a linear operator on V. Suppose that there exists some positive integer k so that $T^k = 0$. Prove that $T^n = 0$.

6. Find a 3×3 matrix for which the minimal polynomial is x^2.

7. Let n be a positive integer, and let V be the space of polynomials over R which have degree at most n (throw in the 0-polynomial). Let D be the differentiation operator on V. What is the minimal polynomial for D?

8. Let P be the operator on R^2 which projects each vector onto the x-axis, parallel to the y-axis: $P(x, y) = (x, 0)$. Show that P is linear. What is the minimal polynomial for P?

9. Let A be an $n \times n$ matrix with characteristic polynomial

$$f = (x - c_1)^{d_1} \cdots (x - c_k)^{d_k}.$$

Show that

$$c_1d_1 + \cdots + c_kd_k = \text{trace }(A).$$

10. Let V be the vector space of $n \times n$ matrices over the field F. Let A be a fixed $n \times n$ matrix. Let T be the linear operator on V defined by

$$T(B) = AB.$$

Show that the minimal polynomial for T is the minimal polynomial for A.

11. Let A and B be $n \times n$ matrices over the field F. According to Exercise 9 of Section 6.1, the matrices AB and BA have the same characteristic values. Do they have the same characteristic polynomial? Do they have the same minimal polynomial?

6.4. Invariant Subspaces

In this section, we shall introduce a few concepts which are useful in attempting to analyze a linear operator. We shall use these ideas to obtain characterizations of diagonalizable (and triangulable) operators in terms of their minimal polynomials.

Definition. *Let* V *be a vector space and* T *a linear operator on* V. *If* W *is a subspace of* V, *we say that* W *is* **invariant under** T *if for each vector* α *in* W *the vector* Tα *is in* W, *i.e., if* T(W) *is contained in* W.

EXAMPLE 6. If T is any linear operator on V, then V is invariant under T, as is the zero subspace. The range of T and the null space of T are also invariant under T.

EXAMPLE 7. Let F be a field and let D be the differentiation operator on the space $F[x]$ of polynomials over F. Let n be a positive integer and let W be the subspace of polynomials of degree not greater than n. Then W is invariant under D. This is just another way of saying that D is 'degree decreasing.'

EXAMPLE 8. Here is a very useful generalization of Example 6. Let T be a linear operator on V. Let U be any linear operator on V which commutes with T, i.e., $TU = UT$. Let W be the range of U and let N be the null space of U. Both W and N are invariant under T. If α is in the range of U, say $\alpha = U\beta$, then $T\alpha = T(U\beta) = U(T\beta)$ so that $T\alpha$ is in the range of U. If α is in N, then $U(T\alpha) = T(U\alpha) = T(0) = 0$; hence, $T\alpha$ is in N.

A particular type of operator which commutes with T is an operator $U = g(T)$, where g is a polynomial. For instance, we might have $U = T - cI$, where c is a characteristic value of T. The null space of U is familiar to us. We see that this example includes the (obvious) fact that the space of characteristic vectors of T associated with the characteristic value c is invariant under T.

EXAMPLE 9. Let T be the linear operator on R^2 which is represented in the standard ordered basis by the matrix

$$A = \begin{bmatrix} 0 & -1 \\ 1 & 0 \end{bmatrix}.$$

Then the only subspaces of R^2 which are invariant under T are R^2 and the zero subspace. Any other invariant subspace would necessarily have dimension 1. But, if W is the subspace spanned by some non-zero vector α, the fact that W is invariant under T means that α is a characteristic vector, but A has no real characteristic values.

When the subspace W is invariant under the operator T, then T induces a linear operator T_W on the space W. The linear operator T_W is defined by $T_W(\alpha) = T(\alpha)$, for α in W, but T_W is quite a different object from T since its domain is W not V.

When V is finite-dimensional, the invariance of W under T has a simple matrix interpretation, and perhaps we should mention it at this point. Suppose we choose an ordered basis $\mathcal{B} = \{\alpha_1, \ldots, \alpha_n\}$ for V such that $\mathcal{B}' = \{\alpha_1, \ldots, \alpha_r\}$ is an ordered basis for W ($r = \dim W$). Let $A = [T]_{\mathcal{B}}$ so that

$$T\alpha_j = \sum_{i=1}^{n} A_{ij}\alpha_i.$$

Since W is invariant under T, the vector $T\alpha_j$ belongs to W for $j \leq r$. This means that

$$T\alpha_j = \sum_{i=1}^{r} A_{ij}\alpha_i, \qquad j \leq r. \tag{6-9}$$

In other words, $A_{ij} = 0$ if $j \leq r$ and $i > r$.

Schematically, A has the block form

$$A = \begin{bmatrix} B & C \\ 0 & D \end{bmatrix} \tag{6-10}$$

where B is an $r \times r$ matrix, C is an $r \times (n - r)$ matrix, and D is an $(n - r) \times (n - r)$ matrix. The reader should note that according to (6-9) the matrix B is precisely the matrix of the induced operator T_W in the ordered basis $\mathcal{B}'$.

Most often, we shall carry out arguments about T and T_W without making use of the block form of the matrix A in (6-10). But we should note how certain relations between T_W and T are apparent from that block form.

Lemma. *Let* W *be an invariant subspace for* T. *The characteristic polynomial for the restriction operator* T_W *divides the characteristic polynomial for* T. *The minimal polynomial for* T_W *divides the minimal polynomial for* T.

Proof. We have

$$A = \begin{bmatrix} B & C \\ 0 & D \end{bmatrix}$$

where $A = [T]_{\mathcal{B}}$ and $B = [T_W]_{\mathcal{B}'}$. Because of the block form of the matrix

$$\det (xI - A) = \det (xI - B) \det (xI - D).$$

That proves the statement about characteristic polynomials. Notice that we used I to represent identity matrices of three different sizes.

The kth power of the matrix A has the block form

$$A^k = \begin{bmatrix} B^k & C_k \\ 0 & D^k \end{bmatrix}$$

where C_k is some $r \times (n - r)$ matrix. Therefore, any polynomial which annihilates A also annihilates B (and D too). So, the minimal polynomial for B divides the minimal polynomial for A. ∎

EXAMPLE 10. Let T be any linear operator on a finite-dimensional space V. Let W be the subspace spanned by *all* of the characteristic vectors of T. Let $c_1, \ldots, c_k$ be the distinct characteristic values of T. For each i, let W_i be the space of characteristic vectors associated with the characteristic value c_i, and let $\mathcal{B}_i$ be an ordered basis for W_i. The lemma before Theorem 2 tells us that $\mathcal{B}' = (\mathcal{B}_1, \ldots, \mathcal{B}_k)$ is an ordered basis for W. In particular,

$$\dim W = \dim W_1 + \cdots + \dim W_k.$$

Let $\mathcal{B}' = \{\alpha_1, \ldots, \alpha_r\}$ so that the first few α's form the basis $\mathcal{B}_1$, the next few $\mathcal{B}_2$, and so on. Then

$$T\alpha_i = t_i\alpha_i, \qquad i = 1, \ldots, r$$

where $(t_1, \ldots, t_r) = (c_1, c_1, \ldots, c_1, \ldots, c_k, c_k, \ldots, c_k)$ with c_i repeated $\dim W_i$ times.

Now W is invariant under T, since for each α in W we have

$$\begin{aligned} \alpha &= x_1\alpha_1 + \cdots + x_r\alpha_r \\ T\alpha &= t_1x_1\alpha_1 + \cdots + t_rx_r\alpha_r. \end{aligned}$$

Choose any other vectors $\alpha_{r+1}, \ldots, \alpha_n$ in V such that $\mathcal{B} = \{\alpha_1, \ldots, \alpha_n\}$ is a basis for V. The matrix of T relative to $\mathcal{B}$ has the block form (6-10), and the matrix of the restriction operator T_W relative to the basis $\mathcal{B}'$ is

$$B = \begin{bmatrix} t_1 & 0 & \cdots & 0 \\ 0 & t_2 & \cdots & 0 \\ \vdots & \vdots & & \vdots \\ 0 & 0 & \cdots & t_r \end{bmatrix}.$$

The characteristic polynomial of B (i.e., of T_W) is

$$g = (x - c_1)^{e_1} \cdots (x - c_k)^{e_k}$$

where $e_i = \dim W_i$. Furthermore, g divides f, the characteristic polynomial for T. Therefore, the multiplicity of c_i as a root of f is at least $\dim W_i$.

All of this should make Theorem 2 transparent. It merely says that T is diagonalizable if and only if $r = n$, if and only if $e_1 + \cdots + e_k = n$. It does not help us too much with the non-diagonalizable case, since we don't know the matrices C and D of (6-10).

Definition. *Let* W *be an invariant subspace for* T *and let* α *be a vector in* V. *The* **T-conductor of α into** W *is the set* $S_T(\alpha; W)$, *which consists of all polynomials* g *(over the scalar field) such that* $g(T)\alpha$ *is in* W.

Since the operator T will be fixed throughout most discussions, we shall usually drop the subscript T and write $S(\alpha; W)$. The authors usually call that collection of polynomials the 'stuffer' (*das einstopfende Ideal*). 'Conductor' is the more standard term, preferred by those who envision a less aggressive operator $g(T)$, gently leading the vector α into W. In the special case $W = \{0\}$ the conductor is called the **T-annihilator of** α.

Lemma. *If W is an invariant subspace for T, then W is invariant under every polynomial in T. Thus, for each α in V, the conductor S(α; W) is an ideal in the polynomial algebra F[x].*

Proof. If β is in W, then $T\beta$ is in W. Consequently, $T(T\beta) = T^2\beta$ is in W. By induction, $T^k\beta$ is in W for each k. Take linear combinations to see that $f(T)\beta$ is in W for every polynomial f.

The definition of $S(\alpha; W)$ makes sense if W is any subset of V. If W is a subspace, then $S(\alpha; W)$ is a subspace of $F[x]$, because

$$(cf + g)(T) = cf(T) + g(T).$$

If W is also invariant under T, let g be a polynomial in $S(\alpha; W)$, i.e., let $g(T)\alpha$ be in W. If f is any polynomial, then $f(T)[g(T)\alpha]$ will be in W. Since

$$(fg)(T) = f(T)g(T)$$

fg is in $S(\alpha; W)$. Thus the conductor absorbs multiplication by any polynomial. ∎

The unique monic generator of the ideal $S(\alpha; W)$ is also called the **T-conductor of α into W** (the **T-annihilator** in case $W = \{0\}$). The T-conductor of α into W is the monic polynomial g of least degree such that $g(T)\alpha$ is in W. A polynomial f is in $S(\alpha; W)$ if and only if g divides f. Note that the conductor $S(\alpha; W)$ always contains the minimal polynomial for T; hence, *every* T-*conductor divides the minimal polynomial for* T.

As the first illustration of how to use the conductor $S(\alpha; W)$, we shall characterize triangulable operators. The linear operator T is called **triangulable** if there is an ordered basis in which T is represented by a triangular matrix.

Lemma. *Let V be a finite-dimensional vector space over the field F. Let T be a linear operator on V such that the minimal polynomial for T is a product of linear factors*

$$p = (x - c_1)^{r_1} \cdots (x - c_k)^{r_k}, \qquad c_i \textit{ in } F.$$

Let W be a proper (W $\neq$ V) subspace of V which is invariant under T. There exists a vector α in V such that

(a) *α is not in W;*
(b) *(T − cI)α is in W, for some characteristic value c of the operator T.*

Proof. What (a) and (b) say is that the T-conductor of α into W is a linear polynomial. Let β be any vector in V which is not in W. Let g be the T-conductor of β into W. Then g divides p, the minimal polynomial for T. Since β is not in W, the polynomial g is not constant. Therefore,

$$g = (x - c_1)^{e_1} \cdots (x - c_k)^{e_k}$$

where at least one of the integers e_i is positive. Choose j so that $e_j > 0$. Then $(x - c_j)$ divides g:

$$g = (x - c_j)h.$$

By the definition of g, the vector $\alpha = h(T)\beta$ cannot be in W. But

$$\begin{aligned}(T - c_jI)\alpha &= (T - c_jI)h(T)\beta \\ &= g(T)\beta\end{aligned}$$

is in W. ▌

Theorem 5. *Let* V *be a finite-dimensional vector space over the field* F *and let* T *be a linear operator on* V. *Then* T *is triangulable if and only if the minimal polynomial for* T *is a product of linear polynomials over* F.

Proof. Suppose that the minimal polynomial factors

$$p = (x - c_1)^{r_1} \cdots (x - c_k)^{r_k}.$$

By repeated application of the lemma above, we shall arrive at an ordered basis $\mathcal{B} = \{\alpha_1, \ldots, \alpha_n\}$ in which the matrix representing T is upper-triangular:

$$[T]_{\mathcal{B}} = \begin{bmatrix} a_{11} & a_{12} & a_{13} & \cdots & a_{1n} \\ 0 & a_{22} & a_{23} & \cdots & a_{2n} \\ 0 & 0 & a_{33} & \cdots & a_{3n} \\ \vdots & \vdots & \vdots & & \vdots \\ 0 & 0 & 0 & \cdots & a_{nn} \end{bmatrix}. \tag{6-11}$$

Now (6-11) merely says that

$$T\alpha_j = a_{1j}\alpha_1 + \cdots + a_{jj}\alpha_j, \qquad 1 \leq j \leq n \tag{6-12}$$

that is, $T\alpha_j$ is in the subspace spanned by $\alpha_1, \ldots, \alpha_j$. To find $\alpha_1, \ldots, \alpha_n$, we start by applying the lemma to the subspace $W = \{0\}$, to obtain the vector α_1. Then apply the lemma to W_1, the space spanned by α_1, and we obtain α_2. Next apply the lemma to W_2, the space spanned by α_1 and α_2. Continue in that way. One point deserves comment. After $\alpha_1, \ldots, \alpha_i$ have been found, it is the triangular-type relations (6-12) for $j = 1, \ldots, i$ which ensure that the subspace spanned by $\alpha_1, \ldots, \alpha_i$ is invariant under T.

If T is triangulable, it is evident that the characteristic polynomial for T has the form

$$f = (x - c_1)^{d_1} \cdots (x - c_k)^{d_k}, \qquad c_i \text{ in } F.$$

Just look at the triangular matrix (6-11). The diagonal entries $a_{11}, \ldots, a_{1n}$ are the characteristic values, with c_i repeated d_i times. But, if f can be so factored, so can the minimal polynomial p, because it divides f. ▌

Corollary. *Let* F *be an algebraically closed field, e.g., the complex number field. Every* n × n *matrix over* F *is similar over* F *to a triangular matrix.*

Theorem 6. *Let V be a finite-dimensional vector space over the field F and let T be a linear operator on V. Then T is diagonalizable if and only if the minimal polynomial for T has the form*

$$p = (x - c_1) \cdots (x - c_k)$$

where $c_1, \ldots, c_k$ are distinct elements of F.

Proof. We have noted earlier that, if T is diagonalizable, its minimal polynomial is a product of distinct linear factors (see the discussion prior to Example 4). To prove the converse, let W be the subspace spanned by all of the characteristic vectors of T, and suppose $W \neq V$. By the lemma used in the proof of Theorem 5, there is a vector α not in W and a characteristic value c_j of T such that the vector

$$\beta = (T - c_jI)\alpha$$

lies in W. Since β is in W,

$$\beta = \beta_1 + \cdots + \beta_k$$

where $T\beta_i = c_i\beta_i$, $1 \leq i \leq k$, and therefore the vector

$$h(T)\beta = h(c_1)\beta_1 + \cdots + h(c_k)\beta_k$$

is in W, for every polynomial h.

Now $p = (x - c_j)q$, for some polynomial q. Also

$$q - q(c_j) = (x - c_j)h.$$

We have

$$q(T)\alpha - q(c_j)\alpha = h(T)(T - c_jI)\alpha = h(T)\beta.$$

But $h(T)\beta$ is in W and, since

$$0 = p(T)\alpha = (T - c_jI)q(T)\alpha$$

the vector $q(T)\alpha$ is in W. Therefore, $q(c_j)\alpha$ is in W. Since α is not in W, we have $q(c_j) = 0$. That contradicts the fact that p has distinct roots. ∎

At the end of Section 6.7, we shall give a different proof of Theorem 6. In addition to being an elegant result, Theorem 6 is useful in a computational way. Suppose we have a linear operator T, represented by the matrix A in some ordered basis, and we wish to know if T is diagonalizable. We compute the characteristic polynomial f. If we can factor f:

$$f = (x - c_1)^{d_1} \cdots (x - c_k)^{d_k}$$

we have two different methods for determining whether or not T is diagonalizable. One method is to see whether (for each i) we can find d_i independent characteristic vectors associated with the characteristic value c_i. The other method is to check whether or not $(T - c_1I) \cdots (T - c_kI)$ is the zero operator.

Theorem 5 provides a different proof of the Cayley-Hamilton theorem. That theorem is easy for a triangular matrix. Hence, via Theorem 5, we obtain the result for any matrix over an algebraically closed field. Any field is a subfield of an algebraically closed field. If one knows that result, one obtains a proof of the Cayley-Hamilton theorem for matrices over any field. If we at least admit into our discussion the Fundamental Theorem of Algebra (the complex number field is algebraically closed), then Theorem 5 provides a proof of the Cayley-Hamilton theorem for complex matrices, and that proof is independent of the one which we gave earlier.

Exercises

1. Let T be the linear operator on R^2, the matrix of which in the standard ordered basis is

$$A = \begin{bmatrix} 1 & -1 \\ 2 & 2 \end{bmatrix}.$$

(a) Prove that the only subspaces of R^2 invariant under T are R^2 and the zero subspace.

(b) If U is the linear operator on C^2, the matrix of which in the standard ordered basis is A, show that U has 1-dimensional invariant subspaces.

2. Let W be an invariant subspace for T. Prove that the minimal polynomial for the restriction operator T_W divides the minimal polynomial for T, without referring to matrices.

3. Let c be a characteristic value of T and let W be the space of characteristic vectors associated with the characteristic value c. What is the restriction operator T_W?

4. Let

$$A = \begin{bmatrix} 0 & 1 & 0 \\ 2 & -2 & 2 \\ 2 & -3 & 2 \end{bmatrix}.$$

Is A similar over the field of real numbers to a triangular matrix? If so, find such a triangular matrix.

5. Every matrix A such that $A^2 = A$ is similar to a diagonal matrix.

6. Let T be a diagonalizable linear operator on the n-dimensional vector space V, and let W be a subspace which is invariant under T. Prove that the restriction operator T_W is diagonalizable.

7. Let T be a linear operator on a finite-dimensional vector space over the field of complex numbers. Prove that T is diagonalizable if and only if T is annihilated by some polynomial over C which has distinct roots.

8. Let T be a linear operator on V. If every subspace of V is invariant under T, then T is a scalar multiple of the identity operator.

9. Let T be the indefinite integral operator

$$(Tf)(x) = \int_0^x f(t)\,dt$$

on the space of continuous functions on the interval [0, 1]. Is the space of polynomial functions invariant under T? The space of differentiable functions? The space of functions which vanish at $x = \frac{1}{2}$?

10. Let A be a 3×3 matrix with real entries. Prove that, if A is not similar over R to a triangular matrix, then A is similar over C to a diagonal matrix.

11. True or false? If the triangular matrix A is similar to a diagonal matrix, then A is already diagonal.

12. Let T be a linear operator on a finite-dimensional vector space over an algebraically closed field F. Let f be a polynomial over F. Prove that c is a characteristic value of $f(T)$ if and only if $c = f(t)$, where t is a characteristic value of T.

13. Let V be the space of $n \times n$ matrices over F. Let A be a fixed $n \times n$ matrix over F. Let T and U be the linear operators on V defined by

$$\begin{aligned} T(B) &= AB \\ U(B) &= AB - BA. \end{aligned}$$

(a) True or false? If A is diagonalizable (over F), then T is diagonalizable.
(b) True or false? If A is diagonalizable, then U is diagonalizable.

6.5. Simultaneous Triangulation; Simultaneous Diagonalization

Let V be a finite-dimensional space and let $\mathfrak{F}$ be a family of linear operators on V. We ask when we can simultaneously triangulate or diagonalize the operators in $\mathfrak{F}$, i.e., find one basis $\mathfrak{B}$ such that all of the matrices $[T]_{\mathfrak{B}}$, T in $\mathfrak{F}$, are triangular (or diagonal). In the case of diagonalization, it is necessary that $\mathfrak{F}$ be a commuting family of operators: $UT = TU$ for all T, U in $\mathfrak{F}$. That follows from the fact that all diagonal matrices commute. Of course, it is also necessary that each operator in $\mathfrak{F}$ be a diagonalizable operator. In order to simultaneously triangulate, each operator in $\mathfrak{F}$ must be triangulable. It is not necessary that $\mathfrak{F}$ be a commuting family; however, that condition is sufficient for simultaneous triangulation (if each T can be individually triangulated). These results follow from minor variations of the proofs of Theorems 5 and 6.

The subspace W is **invariant under** (the family of operators) $\mathfrak{F}$ if W is invariant under each operator in $\mathfrak{F}$.

Lemma. *Let $\mathfrak{F}$ be a commuting family of triangulable linear operators on V. Let W be a proper subspace of V which is invariant under $\mathfrak{F}$. There exists a vector α in V such that*

(a) *α is not in W;*
(b) *for each T in $\mathfrak{F}$, the vector $T\alpha$ is in the subspace spanned by α and W.*

Proof. It is no loss of generality to assume that $\mathfrak{F}$ contains only a finite number of operators, because of this observation. Let $\{T_1, \ldots, T_r\}$

be a maximal linearly independent subset of $\mathfrak{F}$, i.e., a basis for the subspace spanned by $\mathfrak{F}$. If α is a vector such that (b) holds for each T_i, then (b) will hold for every operator which is a linear combination of $T_1, \ldots, T_r$.

By the lemma before Theorem 5 (this lemma for a single operator), we can find a vector β_1 (not in W) and a scalar c_1 such that $(T_1 - c_1I)\beta_1$ is in W. Let V_1 be the collection of all vectors β in V such that $(T_1 - c_1I)\beta$ is in W. Then V_1 is a subspace of V which is properly larger than W. Furthermore, V_1 is invariant under $\mathfrak{F}$, for this reason. If T commutes with T_1, then

$$(T_1 - c_1I)(T\beta) = T(T_1 - c_1I)\beta.$$

If β is in V_1, then $(T_1 - c_1I)\beta$ is in W. Since W is invariant under each T in $\mathfrak{F}$, we have $T(T_1 - c_1I)\beta$ in W, i.e., $T\beta$ in V_1, for all β in V_1 and all T in $\mathfrak{F}$.

Now W is a proper subspace of V_1. Let U_2 be the linear operator on V_1 obtained by restricting T_2 to the subspace V_1. The minimal polynomial for U_2 divides the minimal polynomial for T_2. Therefore, we may apply the lemma before Theorem 5 to that operator and the invariant subspace W. We obtain a vector β_2 in V_1 (not in W) and a scalar c_2 such that $(T_2 - c_2I)\beta_2$ is in W. Note that

(a) β_2 is not in W;
(b) $(T_1 - c_1I)\beta_2$ is in W;
(c) $(T_2 - c_2I)\beta_2$ is in W.

Let V_2 be the set of all vectors β in V_1 such that $(T_2 - c_2I)\beta$ is in W. Then V_2 is invariant under $\mathfrak{F}$. Apply the lemma before Theorem 5 to U_3, the restriction of T_3 to V_2. If we continue in this way, we shall reach a vector $\alpha = \beta_r$ (not in W) such that $(T_j - c_jI)\alpha$ is in W, $j = 1, \ldots, r$. ∎

Theorem 7. *Let* V *be a finite-dimensional vector space over the field* F. *Let* $\mathfrak{F}$ *be a commuting family of triangulable linear operators on* V. *There exists an ordered basis for* V *such that every operator in* $\mathfrak{F}$ *is represented by a triangular matrix in that basis.*

Proof. Given the lemma which we just proved, this theorem has the same proof as does Theorem 5, if one replaces T by $\mathfrak{F}$. ∎

Corollary. *Let* $\mathfrak{F}$ *be a commuting family of* n × n *matrices over an algebraically closed field* F. *There exists a non-singular* n × n *matrix* P *with entries in* F *such that* $P^{-1}AP$ *is upper-triangular, for every matrix* A *in* $\mathfrak{F}$.

Theorem 8. *Let* $\mathfrak{F}$ *be a commuting family of diagonalizable linear operators on the finite-dimensional vector space* V. *There exists an ordered basis for* V *such that every operator in* $\mathfrak{F}$ *is represented in that basis by a diagonal matrix.*

Proof. We could prove this theorem by adapting the lemma before Theorem 7 to the diagonalizable case, just as we adapted the lemma before Theorem 5 to the diagonalizable case in order to prove Theorem 6. However, at this point it is easier to proceed by induction on the dimension of V.

If $\dim V = 1$, there is nothing to prove. Assume the theorem for vector spaces of dimension less than n, and let V be an n-dimensional space. Choose any T in $\mathcal{F}$ which is not a scalar multiple of the identity. Let $c_1, \ldots, c_k$ be the distinct characteristic values of T, and (for each i) let W_i be the null space of $T - c_iI$. Fix an index i. Then W_i is invariant under every operator which commutes with T. Let $\mathcal{F}_i$ be the family of linear operators on W_i obtained by restricting the operators in $\mathcal{F}$ to the (invariant) subspace W_i. Each operator in $\mathcal{F}_i$ is diagonalizable, because its minimal polynomial divides the minimal polynomial for the corresponding operator in $\mathcal{F}$. Since $\dim W_i < \dim V$, the operators in $\mathcal{F}_i$ can be simultaneously diagonalized. In other words, W_i has a basis $\mathcal{B}_i$ which consists of vectors which are simultaneously characteristic vectors for every operator in $\mathcal{F}_i$.

Since T is diagonalizable, the lemma before Theorem 2 tells us that $\mathcal{B} = (\mathcal{B}_1, \ldots, \mathcal{B}_k)$ is a basis for V. That is the basis we seek. ∎

Exercises

1. Find an invertible real matrix P such that $P^{-1}AP$ and $P^{-1}BP$ are both diagonal, where A and B are the real matrices

(a) $$A = \begin{bmatrix} 1 & 2 \\ 0 & 2 \end{bmatrix}, \quad B = \begin{bmatrix} 3 & -8 \\ 0 & -1 \end{bmatrix}$$

(b) $$A = \begin{bmatrix} 1 & 1 \\ 1 & 1 \end{bmatrix}, \quad B = \begin{bmatrix} 1 & a \\ a & 1 \end{bmatrix}.$$

2. Let $\mathcal{F}$ be a commuting family of 3×3 complex matrices. How many linearly independent matrices can $\mathcal{F}$ contain? What about the $n \times n$ case?

3. Let T be a linear operator on an n-dimensional space, and suppose that T has n distinct characteristic values. Prove that any linear operator which commutes with T is a polynomial in T.

4. Let A, B, C, and D be $n \times n$ complex matrices which commute. Let E be the $2n \times 2n$ matrix

$$E = \begin{bmatrix} A & B \\ C & D \end{bmatrix}.$$

Prove that $\det E = \det (AD - BC)$.

5. Let F be a field, n a positive integer, and let V be the space of $n \times n$ matrices over F. If A is a fixed $n \times n$ matrix over F, let T_A be the linear operator on V defined by $T_A(B) = AB - BA$. Consider the family of linear operators T_A obtained by letting A vary over all diagonal matrices. Prove that the operators in that family are simultaneously diagonalizable.

6.6. Direct-Sum Decompositions

As we continue with our analysis of a single linear operator, we shall formulate our ideas in a slightly more sophisticated way—less in terms of matrices and more in terms of subspaces. When we began this chapter, we described our goal this way: To find an ordered basis in which the matrix of T assumes an especially simple form. Now, we shall describe our goal as follows: To decompose the underlying space V into a sum of invariant subspaces for T such that the restriction operators on those subspaces are simple.

Definition. *Let* $W_1, \ldots, W_k$ *be subspaces of the vector space* V. *We say that* $W_1, \ldots, W_k$ *are* **independent** *if*

$$\alpha_1 + \cdots + \alpha_k = 0, \qquad \alpha_i \text{ in } W_i$$

implies that each α_i *is* 0.

For $k = 2$, the meaning of independence is $\{0\}$ intersection, i.e., W_1 and W_2 are independent if and only if $W_1 \cap W_2 = \{0\}$. If $k > 2$, the independence of $W_1, \ldots, W_k$ says much more than $W_1 \cap \cdots \cap W_k = \{0\}$. It says that each W_j intersects the sum of the other subspaces W_i only in the zero vector.

The significance of independence is this. Let $W = W_1 + \cdots + W_k$ be the subspace spanned by $W_1, \ldots, W_k$. Each vector α in W can be expressed as a sum

$$\alpha = \alpha_1 + \cdots + \alpha_k, \qquad \alpha_i \text{ in } W_i.$$

If $W_1, \ldots, W_k$ are independent, then that expression for α is unique; for if

$$\alpha = \beta_1 + \cdots + \beta_k, \qquad \beta_i \text{ in } W_i$$

then $0 = (\alpha_1 - \beta_1) + \cdots + (\alpha_k - \beta_k)$, hence $\alpha_i - \beta_i = 0$, $i = 1, \ldots, k$. Thus, when $W_1, \ldots, W_k$ are independent, we can operate with the vectors in W as k-tuples $(\alpha_1, \ldots, \alpha_k)$, α_i in W_i, in the same way as we operate with vectors in R^k as k-tuples of numbers.

Lemma. *Let* V *be a finite-dimensional vector space. Let* $W_1, \ldots, W_k$ *be subspaces of* V *and let* $W = W_1 + \cdots + W_k$. *The following are equivalent.*

(a) $W_1, \ldots, W_k$ *are independent.*

(b) *For each* j, $2 \leq j \leq k$, *we have*

$$W_j \cap (W_1 + \cdots + W_{j-1}) = \{0\}.$$

(c) *If* $\mathcal{B}_i$ *is an ordered basis for* W_i, $1 \leq i \leq k$, *then the sequence* $\mathcal{B} = (\mathcal{B}_1, \ldots, \mathcal{B}_k)$ *is an ordered basis for* W.

Proof. Assume (a). Let α be a vector in the intersection $W_j \cap (W_1 + \cdots + W_{j-1})$. Then there are vectors $\alpha_1, \ldots, \alpha_{j-1}$ with α_i in W_i such that $\alpha = \alpha_1 + \cdots + \alpha_{j-1}$. Since

$$\alpha_1 + \cdots + \alpha_{j-1} + (-\alpha) + 0 + \cdots + 0 = 0$$

and since $W_1, \ldots, W_k$ are independent, it must be that $\alpha_1 = \alpha_2 = \cdots = \alpha_{j-1} = \alpha = 0$.

Now, let us observe that (b) implies (a). Suppose

$$0 = \alpha_1 + \cdots + \alpha_k, \qquad \alpha_i \text{ in } W_i.$$

Let j be the largest integer i such that $\alpha_i \neq 0$. Then

$$0 = \alpha_1 + \cdots + \alpha_j, \qquad \alpha_j \neq 0.$$

Thus $\alpha_j = -\alpha_1 - \cdots - \alpha_{j-1}$ is a non-zero vector in $W_j \cap (W_1 + \cdots + W_{j-1})$.

Now that we know (a) and (b) are the same, let us see why (a) is equivalent to (c). Assume (a). Let $\mathfrak{B}_i$ be a basis for W_i, $1 \leq i \leq k$, and let $\mathfrak{B} = (\mathfrak{B}_1, \ldots, \mathfrak{B}_k)$. Any linear relation between the vectors in $\mathfrak{B}$ will have the form

$$\beta_1 + \cdots + \beta_k = 0$$

where β_i is some linear combination of the vectors in $\mathfrak{B}_i$. Since $W_1, \ldots, W_k$ are independent, each β_i is 0. Since each $\mathfrak{B}_i$ is independent, the relation we have between the vectors in $\mathfrak{B}$ is the trivial relation.

We relegate the proof that (c) implies (a) to the exercises (Exercise 2). ∎

If any (and hence all) of the conditions of the last lemma hold, we say that the sum $W = W_1 + \cdots + W_k$ is **direct** or that W is the **direct sum** of $W_1, \ldots, W_k$ and we write

$$W = W_1 \oplus \cdots \oplus W_k.$$

In the literature, the reader may find this direct sum referred to as an independent sum or the interior direct sum of $W_1, \ldots, W_k$.

EXAMPLE 11. Let V be a finite-dimensional vector space over the field F and let $\{\alpha_1, \ldots, \alpha_n\}$ be any basis for V. If W_i is the one-dimensional subspace spanned by α_i, then $V = W_1 \oplus \cdots \oplus W_n$.

EXAMPLE 12. Let n be a positive integer and F a subfield of the complex numbers, and let V be the space of all $n \times n$ matrices over F. Let W_1 be the subspace of all **symmetric** matrices, i.e., matrices A such that $A^t = A$. Let W_2 be the subspace of all **skew-symmetric** matrices, i.e., matrices A such that $A^t = -A$. Then $V = W_1 \oplus W_2$. If A is any matrix in V, the unique expression for A as a sum of matrices, one in W_1 and the other in W_2, is

$$A = A_1 + A_2$$
$$A_1 = \tfrac{1}{2}(A + A^t)$$
$$A_2 = \tfrac{1}{2}(A - A^t).$$

EXAMPLE 13. Let T be any linear operator on a finite-dimensional space V. Let $c_1, \ldots, c_k$ be the distinct characteristic values of T, and let W_i be the space of characteristic vectors associated with the characteristic value c_i. Then $W_1, \ldots, W_k$ are independent. See the lemma before Theorem 2. In particular, if T is diagonalizable, then $V = W_1 \oplus \cdots \oplus W_k$.

Definition. *If* V *is a vector space, a* **projection** *of* V *is a linear operator* E *on* V *such that* $E^2 = E$.

Suppose that E is a projection. Let R be the range of E and let N be the null space of E.

1. The vector β is in the range R if and only if $E\beta = \beta$. If $\beta = E\alpha$, then $E\beta = E^2\alpha = E\alpha = \beta$. Conversely, if $\beta = E\beta$, then (of course) β is in the range of E.
2. $V = R \oplus N$.
3. The unique expression for α as a sum of vectors in R and N is $\alpha = E\alpha + (\alpha - E\alpha)$.

From (1), (2), (3) it is easy to see the following. If R and N are subspaces of V such that $V = R \oplus N$, there is one and only one projection operator E which has range R and null space N. That operator is called the **projection on R along N**.

Any projection E is (trivially) diagonalizable. If $\{\alpha_1, \ldots, \alpha_r\}$ is a basis for R and $\{\alpha_{r+1}, \ldots, \alpha_n\}$ a basis for N, then the basis $\mathcal{B} = \{\alpha_1, \ldots, \alpha_n\}$ diagonalizes E:

$$[E]_{\mathcal{B}} = \begin{bmatrix} I & 0 \\ 0 & 0 \end{bmatrix}$$

where I is the $r \times r$ identity matrix. That should help explain some of the terminology connected with projections. The reader should look at various cases in the plane R^2 (or 3-space, R^3), to convince himself that the projection on R along N sends each vector into R by projecting it parallel to N.

Projections can be used to describe direct-sum decompositions of the space V. For, suppose $V = W_1 \oplus \cdots \oplus W_k$. For each j we shall define an operator E_j on V. Let α be in V, say $\alpha = \alpha_1 + \cdots + \alpha_k$ with α_i in W_i. Define $E_j\alpha = \alpha_j$. Then E_j is a well-defined rule. It is easy to see that E_j is linear, that the range of E_j is W_j, and that $E_j^2 = E_j$. The null space of E_j is the subspace

$$(W_1 + \cdots + W_{j-1} + W_{j+1} + \cdots + W_k)$$

for, the statement that $E_j\alpha = 0$ simply means $\alpha_j = 0$, i.e., that α is actually a sum of vectors from the spaces W_i with $i \neq j$. In terms of the projections E_j we have

$$\alpha = E_1\alpha + \cdots + E_k\alpha \tag{6-13}$$

for each α in V. What (6-13) says is that

$$I = E_1 + \cdots + E_k.$$

Note also that if $i \neq j$, then $E_iE_j = 0$, because the range of E_j is the subspace W_j which is contained in the null space of E_i. We shall now summarize our findings and state and prove a converse.

Theorem 9. *If* $V = W_1 \oplus \cdots \oplus W_k$, *then there exist* k *linear operators* $E_1, \ldots, E_k$ *on* V *such that*

(i) *each* E_i *is a projection* ($E_i^2 = E_i$);
(ii) $E_iE_j = 0$, *if* $i \neq j$;
(iii) $I = E_1 + \cdots + E_k$;
(iv) *the range of* E_i *is* W_i.

Conversely, if $E_1, \ldots, E_k$ *are* k *linear operators on* V *which satisfy conditions* (i), (ii), *and* (iii), *and if we let* W_i *be the range of* E_i, *then* $V = W_1 \oplus \cdots \oplus W_k$.

Proof. We have only to prove the converse statement. Suppose $E_1, \ldots, E_k$ are linear operators on V which satisfy the first three conditions, and let W_i be the range of E_i. Then certainly

$$V = W_1 + \cdots + W_k;$$

for, by condition (iii) we have

$$\alpha = E_1\alpha + \cdots + E_k\alpha$$

for each α in V, and $E_i\alpha$ is in W_i. This expression for α is unique, because if

$$\alpha = \alpha_1 + \cdots + \alpha_k$$

with α_i in W_i, say $\alpha_i = E_i\beta_i$, then using (i) and (ii) we have

$$\begin{aligned} E_j\alpha &= \sum_{i=1}^{k} E_j\alpha_i \\ &= \sum_{i=1}^{k} E_jE_i\beta_i \\ &= E_j^2\beta_j \\ &= E_j\beta_j \\ &= \alpha_j. \end{aligned}$$

This shows that V is the direct sum of the W_i. ∎

Exercises

1. Let V be a finite-dimensional vector space and let W_1 be any subspace of V. Prove that there is a subspace W_2 of V such that $V = W_1 \oplus W_2$.

2. Let V be a finite-dimensional vector space and let $W_1, \ldots, W_k$ be subspaces of V such that

$$V = W_1 + \cdots + W_k \quad \text{and} \quad \dim V = \dim W_1 + \cdots + \dim W_k.$$

Prove that $V = W_1 \oplus \cdots \oplus W_k$.

3. Find a projection E which projects R^2 onto the subspace spanned by $(1, -1)$ along the subspace spanned by $(1, 2)$.

4. If E_1 and E_2 are projections onto independent subspaces, then $E_1 + E_2$ is a projection. True or false?

5. If E is a projection and f is a polynomial, then $f(E) = aI + bE$. What are a and b in terms of the coefficients of f?

6. True or false? If a diagonalizable operator has only the characteristic values 0 and 1, it is a projection.

7. Prove that if E is the projection on R along N, then $(I - E)$ is the projection on N along R.

8. Let $E_1, \ldots, E_k$ be linear operators on the space V such that $E_1 + \cdots + E_k = I$.

(a) Prove that if $E_iE_j = 0$ for $i \neq j$, then $E_i^2 = E_i$ for each i.

(b) In the case $k = 2$, prove the converse of (a). That is, if $E_1 + E_2 = I$ and $E_1^2 = E_1$, $E_2^2 = E_2$, then $E_1E_2 = 0$.

9. Let V be a real vector space and E an idempotent linear operator on V, i.e., a projection. Prove that $(I + E)$ is invertible. Find $(I + E)^{-1}$.

10. Let F be a subfield of the complex numbers (or, a field of characteristic zero). Let V be a finite-dimensional vector space over F. Suppose that $E_1, \ldots, E_k$ are projections of V and that $E_1 + \cdots + E_k = I$. Prove that $E_iE_j = 0$ for $i \neq j$ (*Hint:* Use the trace function and ask yourself what the trace of a projection is.)

11. Let V be a vector space, let $W_1, \ldots, W_k$ be subspaces of V, and let

$$V_j = W_1 + \cdots + W_{j-1} + W_{j+1} + \cdots + W_k.$$

Suppose that $V = W_1 \oplus \cdots \oplus W_k$. Prove that the dual space V^* has the direct-sum decomposition $V^* = V_1^0 \oplus \cdots \oplus V_k^0$.

6.7. Invariant Direct Sums

We are primarily interested in direct-sum decompositions $V = W_1 \oplus \cdots \oplus W_k$, where each of the subspaces W_i is invariant under some given linear operator T. Given such a decomposition of V, T induces a linear operator T_i on each W_i by restriction. The action of T is then this.

If α is a vector in V, we have unique vectors $\alpha_1, \ldots, \alpha_k$ with α_i in W_i such that

$$\alpha = \alpha_1 + \cdots + \alpha_k$$

and then

$$T\alpha = T_1\alpha_1 + \cdots + T_k\alpha_k.$$

We shall describe this situation by saying that T is the **direct sum** of the operators $T_1, \ldots, T_k$. It must be remembered in using this terminology that the T_i are not linear operators on the space V but on the various subspaces W_i. The fact that $V = W_1 \oplus \cdots \oplus W_k$ enables us to associate with each α in V a unique k-tuple $(\alpha_1, \ldots, \alpha_k)$ of vectors α_i in W_i (by $\alpha = \alpha_1 + \cdots + \alpha_k$) in such a way that we can carry out the linear operations in V by working in the individual subspaces W_i. The fact that each W_i is invariant under T enables us to view the action of T as the independent action of the operators T_i on the subspaces W_i. Our purpose is to study T by finding invariant direct-sum decompositions in which the T_i are operators of an elementary nature.

Before looking at an example, let us note the matrix analogue of this situation. Suppose we select an ordered basis $\mathcal{B}_i$ for each W_i, and let $\mathcal{B}$ be the ordered basis for V consisting of the union of the $\mathcal{B}_i$ arranged in the order $\mathcal{B}_1, \ldots, \mathcal{B}_k$, so that $\mathcal{B}$ is a basis for V. From our discussion concerning the matrix analogue for a single invariant subspace, it is easy to see that if $A = [T]_{\mathcal{B}}$ and $A_i = [T_i]_{\mathcal{B}_i}$, then A has the block form

$$A = \begin{bmatrix} A_1 & 0 & \cdots & 0 \\ 0 & A_2 & \cdots & 0 \\ \vdots & \vdots & & \vdots \\ 0 & 0 & \cdots & A_k \end{bmatrix}. \tag{6-14}$$

In (6-14), A_i is a $d_i \times d_i$ matrix ($d_i = \dim W_i$), and the 0's are symbols for rectangular blocks of scalar 0's of various sizes. It also seems appropriate to describe (6-14) by saying that A is the **direct sum** of the matrices $A_1, \ldots, A_k$.

Most often, we shall describe the subspace W_i by means of the associated projections E_i (Theorem 9). Therefore, we need to be able to phrase the invariance of the subspaces W_i in terms of the E_i.

Theorem 10. *Let* T *be a linear operator on the space* V, *and let* $W_1, \ldots, W_k$ *and* $E_1, \ldots, E_k$ *be as in Theorem 9. Then a necessary and sufficient condition that each subspace* W_i *be invariant under* T *is that* T *commute with each of the projections* E_i, *i.e.,*

$$TE_i = E_iT, \qquad i = 1, \ldots, k.$$

Proof. Suppose T commutes with each E_i. Let α be in W_j. Then $E_j\alpha = \alpha$, and

$$\begin{aligned} T\alpha &= T(E_j\alpha) \\ &= E_j(T\alpha) \end{aligned}$$

which shows that $T\alpha$ is in the range of E_j, i.e., that W_j is invariant under T.

Assume now that each W_i is invariant under T. We shall show that $TE_j = E_jT$. Let α be any vector in V. Then

$$\begin{aligned} \alpha &= E_1\alpha + \cdots + E_k\alpha \\ T\alpha &= TE_1\alpha + \cdots + TE_k\alpha. \end{aligned}$$

Since $E_i\alpha$ is in W_i, which is invariant under T, we must have $T(E_i\alpha) = E_i\beta_i$ for some vector β_i. Then

$$\begin{aligned} E_jTE_i\alpha &= E_jE_i\beta_i \\ &= \begin{cases} 0, & \text{if } i \neq j \\ E_j\beta_j, & \text{if } i = j. \end{cases} \end{aligned}$$

Thus

$$\begin{aligned} E_jT\alpha &= E_jTE_1\alpha + \cdots + E_jTE_k\alpha \\ &= E_j\beta_j \\ &= TE_j\alpha. \end{aligned}$$

This holds for each α in V, so $E_jT = TE_j$. ∎

We shall now describe a diagonalizable operator T in the language of invariant direct sum decompositions (projections which commute with T). This will be a great help to us in understanding some deeper decomposition theorems later. The reader may feel that the description which we are about to give is rather complicated, in comparison to the matrix formulation or to the simple statement that the characteristic vectors of T span the underlying space. But, he should bear in mind that this is our first glimpse at a very effective method, by means of which various problems concerned with subspaces, bases, matrices, and the like can be reduced to algebraic calculations with linear operators. With a little experience, the efficiency and elegance of this method of reasoning should become apparent.

Theorem 11. *Let* T *be a linear operator on a finite-dimensional space* V. *If* T *is diagonalizable and if* $c_1, \ldots, c_k$ *are the distinct characteristic values of* T, *then there exist linear operators* $E_1, \ldots, E_k$ *on* V *such that*

(i) $T = c_1E_1 + \cdots + c_kE_k$;
(ii) $I = E_1 + \cdots + E_k$;
(iii) $E_iE_j = 0$, $i \neq j$;
(iv) $E_i^2 = E_i$ (E_i *is a projection*);
(v) *the range of* E_i *is the characteristic space for* T *associated with* c_i.

Conversely, if there exist k *distinct scalars* $c_1, \ldots, c_k$ *and* k *non-zero linear operators* $E_1, \ldots, E_k$ *which satisfy conditions* (i), (ii), *and* (iii), *then* T *is diagonalizable,* $c_1, \ldots, c_k$ *are the distinct characteristic values of* T, *and conditions* (iv) *and* (v) *are satisfied also.*

Proof. Suppose that T is diagonalizable, with distinct characteristic values $c_1, \ldots, c_k$. Let W_i be the space of characteristic vectors associated with the characteristic value c_i. As we have seen,

$$V = W_1 \oplus \cdots \oplus W_k.$$

Let $E_1, \ldots, E_k$ be the projections associated with this decomposition, as in Theorem 9. Then (ii), (iii), (iv) and (v) are satisfied. To verify (i), proceed as follows. For each α in V,

$$\alpha = E_1\alpha + \cdots + E_k\alpha$$

and so

$$\begin{aligned} T\alpha &= TE_1\alpha + \cdots + TE_k\alpha \\ &= c_1E_1\alpha + \cdots + c_kE_k\alpha. \end{aligned}$$

In other words, $T = c_1E_1 + \cdots + c_kE_k$.

Now suppose that we are given a linear operator T along with distinct scalars c_i and non-zero operators E_i which satisfy (i), (ii) and (iii). Since $E_iE_j = 0$ when $i \neq j$, we multiply both sides of $I = E_1 + \cdots + E_k$ by E_i and obtain immediately $E_i^2 = E_i$. Multiplying $T = c_1E_1 + \cdots + c_kE_k$ by E_i, we then have $TE_i = c_iE_i$, which shows that any vector in the range of E_i is in the null space of $(T - c_iI)$. Since we have assumed that $E_i \neq 0$, this proves that there is a non-zero vector in the null space of $(T - c_iI)$, i.e., that c_i is a characteristic value of T. Furthermore, the c_i are all of the characteristic values of T; for, if c is any scalar, then

$$T - cI = (c_1 - c)E_1 + \cdots + (c_k - c)E_k$$

so if $(T - cI)\alpha = 0$, we must have $(c_i - c)E_i\alpha = 0$. If α is not the zero vector, then $E_i\alpha \neq 0$ for some i, so that for this i we have $c_i - c = 0$.

Certainly T is diagonalizable, since we have shown that every non-zero vector in the range of E_i is a characteristic vector of T, and the fact that $I = E_1 + \cdots + E_k$ shows that these characteristic vectors span V. All that remains to be demonstrated is that the null space of $(T - c_iI)$ is exactly the range of E_i. But this is clear, because if $T\alpha = c_i\alpha$, then

$$\sum_{j=1}^{k} (c_j - c_i)E_j\alpha = 0$$

hence

$$(c_j - c_i)E_j\alpha = 0 \qquad \text{for each } j$$

and then

$$E_j\alpha = 0, \qquad j \neq i.$$

Since $\alpha = E_1\alpha + \cdots + E_k\alpha$, and $E_j\alpha = 0$ for $j \neq i$, we have $\alpha = E_i\alpha$, which proves that α is in the range of E_i. ∎

One part of Theorem 9 says that for a diagonalizable operator T, the scalars $c_1, \ldots, c_k$ and the operators $E_1, \ldots, E_k$ are uniquely determined by conditions (i), (ii), (iii), the fact that the c_i are distinct, and the fact that the E_i are non-zero. One of the pleasant features of the decomposition $T = c_1E_1 + \cdots + c_kE_k$ is that if g is any polynomial over the field F, then

$$g(T) = g(c_1)E_1 + \cdots + g(c_k)E_k.$$

We leave the details of the proof to the reader. To see how it is proved one need only compute T^r for each positive integer r. For example,

$$\begin{aligned} T^2 &= \sum_{i=1}^{k} c_iE_i \sum_{j=1}^{k} c_jE_j \\ &= \sum_{i=1}^{k} \sum_{j=1}^{k} c_ic_jE_iE_j \\ &= \sum_{i=1}^{k} c_i^2E_i^2 \\ &= \sum_{i=1}^{k} c_i^2E_i. \end{aligned}$$

The reader should compare this with $g(A)$ where A is a diagonal matrix; for then $g(A)$ is simply the diagonal matrix with diagonal entries $g(A_{11})$, $\ldots, g(A_{nn})$.

We should like in particular to note what happens when one applies the Lagrange polynomials corresponding to the scalars $c_1, \ldots, c_k$:

$$p_j = \prod_{i \neq j} \frac{(x - c_i)}{(c_j - c_i)}.$$

We have $p_j(c_i) = \delta_{ij}$, which means that

$$\begin{aligned} p_j(T) &= \sum_{i=1}^{k} \delta_{ij}E_i \\ &= E_j. \end{aligned}$$

Thus the projections E_j not only commute with T but are polynomials in T.

Such calculations with polynomials in T can be used to give an alternative proof of Theorem 6, which characterized diagonalizable operators in terms of their minimal polynomials. The proof is entirely independent of our earlier proof.

If T is diagonalizable, $T = c_1E_1 + \cdots + c_kE_k$, then

$$g(T) = g(c_1)E_1 + \cdots + g(c_k)E_k$$

for every polynomial g. Thus $g(T) = 0$ if and only if $g(c_i) = 0$ for each i. In particular, the minimal polynomial for T is

$$p = (x - c_1) \cdots (x - c_k).$$

Now suppose T is a linear operator with minimal polynomial $p = (x - c_1) \cdots (x - c_k)$, where $c_1, \ldots, c_k$ are distinct elements of the scalar field. We form the Lagrange polynomials

$$p_j = \prod_{i \neq j} \frac{(x - c_i)}{(c_j - c_i)}.$$

We recall from Chapter 4 that $p_j(c_i) = \delta_{ij}$ and for any polynomial g of degree less than or equal to $(k - 1)$ we have

$$g = g(c_1)p_1 + \cdots + g(c_k)p_k.$$

Taking g to be the scalar polynomial 1 and then the polynomial x, we have

$$\begin{aligned} 1 &= p_1 + \cdots + p_k \\ x &= c_1p_1 + \cdots + c_kp_k. \end{aligned} \tag{6-15}$$

(The astute reader will note that the application to x may not be valid because k may be 1. But if $k = 1$, T is a scalar multiple of the identity and hence diagonalizable.) Now let $E_j = p_j(T)$. From (6-15) we have

$$\begin{aligned} I &= E_1 + \cdots + E_k \\ T &= c_1E_1 + \cdots + c_kE_k. \end{aligned} \tag{6-16}$$

Observe that if $i \neq j$, then p_ip_j is divisible by the minimal polynomial p, because p_ip_j contains every $(x - c_r)$ as a factor. Thus

$$E_iE_j = 0, \qquad i \neq j. \tag{6-17}$$

We must note one further thing, namely, that $E_i \neq 0$ for each i. This is because p is the minimal polynomial for T and so we cannot have $p_i(T) = 0$ since p_i has degree less than the degree of p. This last comment, together with (6-16), (6-17), and the fact that the c_i are distinct enables us to apply Theorem 11 to conclude that T is diagonalizable. ∎

Exercises

1. Let E be a projection of V and let T be a linear operator on V. Prove that the range of E is invariant under T if and only if $ETE = TE$. Prove that both the range and null space of E are invariant under T if and only if $ET = TE$.

2. Let T be the linear operator on R^2, the matrix of which in the standard ordered basis is

$$\begin{bmatrix} 2 & 1 \\ 0 & 2 \end{bmatrix}.$$

Let W_1 be the subspace of R^2 spanned by the vector $\epsilon_1 = (1, 0)$.

(a) Prove that W_1 is invariant under T.

(b) Prove that there is no subspace W_2 which is invariant under T and which is complementary to W_1:

$$R^2 = W_1 \oplus W_2.$$

(Compare with Exercise 1 of Section 6.5.)

3. Let T be a linear operator on a finite-dimensional vector space V. Let R be the range of T and let N be the null space of T. Prove that R and N are independent if and only if $V = R \oplus N$.

4. Let T be a linear operator on V. Suppose $V = W_1 \oplus \cdots \oplus W_k$, where each W_i is invariant under T. Let T_i be the induced (restriction) operator on W_i.

(a) Prove that $\det(T) = \det(T_1) \cdots \det(T_k)$.

(b) Prove that the characteristic polynomial for f is the product of the characteristic polynomials for $f_1, \ldots, f_k$.

(c) Prove that the minimal polynomial for T is the least common multiple of the minimal polynomials for $T_1, \ldots, T_k$. (*Hint:* Prove and then use the corresponding facts about direct sums of matrices.)

5. Let T be the diagonalizable linear operator on R^3 which we discussed in Example 3 of Section 6.2. Use the Lagrange polynomials to write the representing matrix A in the form $A = E_1 + 2E_2$, $E_1 + E_2 = I$, $E_1E_2 = 0$.

6. Let A be the 4×4 matrix in Example 6 of Section 6.3. Find matrices E_1, E_2, E_3 such that $A = c_1E_1 + c_2E_2 + c_3E_3$, $E_1 + E_2 + E_3 = I$, and $E_iE_j = 0$, $i \neq j$.

7. In Exercises 5 and 6, notice that (for each i) the space of characteristic vectors associated with the characteristic value c_i is spanned by the column vectors of the various matrices E_j with $j \neq i$. Is that a coincidence?

8. Let T be a linear operator on V which commutes with every projection operator on V. What can you say about T?

9. Let V be the vector space of continuous real-valued functions on the interval $[-1, 1]$ of the real line. Let W_e be the subspace of even functions, $f(-x) = f(x)$, and let W_o be the subspace of odd functions, $f(-x) = -f(x)$.

(a) Show that $V = W_e \oplus W_o$.

(b) If T is the indefinite integral operator

$$(Tf)(x) = \int_0^x f(t)\,dt$$

are W_e and W_o invariant under T?

6.8. The Primary Decomposition Theorem

We are trying to study a linear operator T on the finite-dimensional space V, by decomposing T into a direct sum of operators which are in some sense elementary. We can do this through the characteristic values and vectors of T in certain special cases, i.e., when the minimal polynomial for T factors over the scalar field F into a product of distinct monic polynomials of degree 1. What can we do with the general T? If we try to study T using characteristic values, we are confronted with two problems. First, T may not have a single characteristic value; this is really a deficiency in the scalar field, namely, that it is not algebraically closed. Second, even if the characteristic polynomial factors completely over F into a product of polynomials of degree 1, there may not be enough characteristic vectors for T to span the space V; this is clearly a deficiency in T. The second situation is illustrated by the operator T on F^3 (F any field) represented in the standard basis by

$$A = \begin{bmatrix} 2 & 0 & 0 \\ 1 & 2 & 0 \\ 0 & 0 & -1 \end{bmatrix}.$$

The characteristic polynomial for A is $(x - 2)^2(x + 1)$ and this is plainly also the minimal polynomial for A (or for T). Thus T is not diagonalizable. One sees that this happens because the null space of $(T - 2I)$ has dimension 1 only. On the other hand, the null space of $(T + I)$ and the null space of $(T - 2I)^2$ together span V, the former being the subspace spanned by ϵ_3 and the latter the subspace spanned by ϵ_1 and ϵ_2.

This will be more or less our general method for the second problem. If (remember this is an assumption) the minimal polynomial for T decomposes

$$p = (x - c_1)^{r_1} \cdots (x - c_k)^{r_k}$$

where $c_1, \ldots, c_k$ are distinct elements of F, then we shall show that the space V is the direct sum of the null spaces of $(T - c_iI)^{r_i}$, $i = 1, \ldots, k$. The hypothesis about p is equivalent to the fact that T is triangulable (Theorem 5); however, that knowledge will not help us.

The theorem which we prove is more general than what we have described, since it works with the primary decomposition of the minimal polynomial, whether or not the primes which enter are all of first degree. The reader will find it helpful to think of the special case when the primes are of degree 1, and even more particularly, to think of the projection-type proof of Theorem 6, a special case of this theorem.

Theorem 12 (Primary Decomposition Theorem). *Let* T *be a linear operator on the finite-dimensional vector space* V *over the field* F. *Let* p *be the minimal polynomial for* T,

$$p = p_1^{r_1} \cdots p_k^{r_k}$$

where the p_i *are distinct irreducible monic polynomials over* F *and the* r_i *are positive integers. Let* W_i *be the null space of* $p_i(T)^{r_i}$, $i = 1, \ldots, k$. *Then*

(i) $V = W_1 \oplus \cdots \oplus W_k$;
(ii) *each* W_i *is invariant under* T;
(iii) *if* T_i *is the operator induced on* W_i *by* T, *then the minimal polynomial for* T_i *is* $p_i^{r_i}$.

Proof. The idea of the proof is this. If the direct-sum decomposition (i) is valid, how can we get hold of the projections $E_1, \ldots, E_k$ associated with the decomposition? The projection E_i will be the identity on W_i and zero on the other W_j. We shall find a polynomial h_i such that $h_i(T)$ is the identity on W_i and is zero on the other W_j, and so that $h_1(T) + \cdots + h_k(T) = I$, etc.

For each i, let

$$f_i = \frac{p}{p_i^{r_i}} = \prod_{j \neq i} p_j^{r_j}.$$

Since $p_1, \ldots, p_k$ are distinct prime polynomials, the polynomials $f_1, \ldots, f_k$ are relatively prime (Theorem 10, Chapter 4). Thus there are polynomials $g_1, \ldots, g_k$ such that

$$\sum_{i=1}^{n} f_i g_i = 1.$$

Note also that if $i \neq j$, then $f_i f_j$ is divisible by the polynomial p, because $f_i f_j$ contains each $p_m^{r_m}$ as a factor. We shall show that the polynomials $h_i = f_i g_i$ behave in the manner described in the first paragraph of the proof.

Let $E_i = h_i(T) = f_i(T)g_i(T)$. Since $h_1 + \cdots + h_k = 1$ and p divides $f_i f_j$ for $i \neq j$, we have

$$E_1 + \cdots + E_k = I$$
$$E_i E_j = 0, \quad \text{if } i \neq j.$$

Thus the E_i are projections which correspond to some direct-sum decomposition of the space V. We wish to show that the range of E_i is exactly the subspace W_i. It is clear that each vector in the range of E_i is in W_i, for if α is in the range of E_i, then $\alpha = E_i\alpha$ and so

$$\begin{aligned} p_i(T)^{r_i}\alpha &= p_i(T)^{r_i}E_i\alpha \\ &= p_i(T)^{r_i}f_i(T)g_i(T)\alpha \\ &= 0 \end{aligned}$$

because $p^{r_i}f_i g_i$ is divisible by the minimal polynomial p. Conversely, suppose that α is in the null space of $p_i(T)^{r_i}$. If $j \neq i$, then $f_j g_j$ is divisible by $p_i^{r_i}$ and so $f_j(T)g_j(T)\alpha = 0$, i.e., $E_j\alpha = 0$ for $j \neq i$. But then it is immediate that $E_i\alpha = \alpha$, i.e., that α is in the range of E_i. This completes the proof of statement (i).

It is certainly clear that the subspaces W_i are invariant under T. If T_i is the operator induced on W_i by T, then evidently $p_i(T_i)^{r_i} = 0$, because by definition $p_i(T)^{r_i}$ is 0 on the subspace W_i. This shows that the minimal polynomial for T_i divides $p_i^{r_i}$. Conversely, let g be any polynomial such that $g(T_i) = 0$. Then $g(T)f_i(T) = 0$. Thus gf_i is divisible by the minimal polynomial p of T, i.e., $p_i^{r_i}f_i$ divides gf_i. It is easily seen that $p_i^{r_i}$ divides g. Hence the minimal polynomial for T_i is $p_i^{r_i}$. ∎

Corollary. *If* $E_1, \ldots, E_k$ *are the projections associated with the primary decomposition of* T, *then each* E_i *is a polynomial in* T, *and accordingly if a linear operator* U *commutes with* T *then* U *commutes with each of the* E_i, *i.e., each subspace* W_i *is invariant under* U.

In the notation of the proof of Theorem 12, let us take a look at the special case in which the minimal polynomial for T is a product of first-degree polynomials, i.e., the case in which each p_i is of the form $p_i = x - c_i$. Now the range of E_i is the null space W_i of $(T - c_iI)^{r_i}$. Let us put $D = c_1E_1 + \cdots + c_kE_k$. By Theorem 11, D is a diagonalizable operator which we shall call the **diagonalizable part** of T. Let us look at the operator $N = T - D$. Now

$$T = TE_1 + \cdots + TE_k$$
$$D = c_1E_1 + \cdots + c_kE_k$$

so

$$N = (T - c_1I)E_1 + \cdots + (T - c_kI)E_k.$$

The reader should be familiar enough with projections by now so that he sees that

$$N^2 = (T - c_1I)^2E_1 + \cdots + (T - c_kI)^2E_k$$

and in general that

$$N^r = (T - c_1I)^rE_1 + \cdots + (T - c_kI)^rE_k.$$

When $r \geq r_i$ for each i, we shall have $N^r = 0$, because the operator $(T - c_iI)^r$ will then be 0 on the range of E_i.

Definition. *Let* N *be a linear operator on the vector space* V. *We say that* N *is* **nilpotent** *if there is some positive integer* r *such that* $N^r = 0$.

Theorem 13. *Let* T *be a linear operator on the finite-dimensional vector space* V *over the field* F. *Suppose that the minimal polynomial for* T *decomposes over* F *into a product of linear polynomials. Then there is a diagonalizable operator* D *on* V *and a nilpotent operator* N *on* V *such that*

(i) T = D + N,
(ii) DN = ND.

The diagonalizable operator D *and the nilpotent operator* N *are uniquely determined by* (i) *and* (ii) *and each of them is a polynomial in* T.

Proof. We have just observed that we can write $T = D + N$ where D is diagonalizable and N is nilpotent, and where D and N not only commute but are polynomials in T. Now suppose that we also have $T = D' + N'$ where D' is diagonalizable, N' is nilpotent, and $D'N' = N'D'$. We shall prove that $D = D'$ and $N = N'$.

Since D' and N' commute with one another and $T = D' + N'$, we see that D' and N' commute with T. Thus D' and N' commute with any polynomial in T; hence they commute with D and with N. Now we have

$$D + N = D' + N'$$

or

$$D - D' = N' - N$$

and all four of these operators commute with one another. Since D and D' are both diagonalizable and they commute, they are simultaneously diagonalizable, and $D - D'$ is diagonalizable. Since N and N' are both nilpotent and they commute, the operator $(N' - N)$ is nilpotent; for, using the fact that N and N' commute

$$(N' - N)^r = \sum_{j=0}^{r} \binom{r}{j} (N')^{r-j}(-N)^j$$

and so when r is sufficiently large every term in this expression for $(N' - N)^r$ will be 0. (Actually, a nilpotent operator on an n-dimensional space must have its nth power 0; if we take $r = 2n$ above, that will be large enough. It then follows that $r = n$ is large enough, but this is not obvious from the above expression.) Now $D - D'$ is a diagonalizable operator which is also nilpotent. Such an operator is obviously the zero operator; for since it is nilpotent, the minimal polynomial for this operator is of the form x^r for some $r \leq m$; but then since the operator is diagonalizable, the minimal polynomial cannot have a repeated root; hence $r = 1$ and the minimal polynomial is simply x, which says the operator is 0. Thus we see that $D = D'$ and $N = N'$. ∎

Corollary. *Let* V *be a finite-dimensional vector space over an algebraically closed field* F, *e.g., the field of complex numbers. Then every linear operator* T *on* V *can be written as the sum of a diagonalizable operator* D *and a nilpotent operator* N *which commute. These operators* D *and* N *are unique and each is a polynomial in* T.

From these results, one sees that the study of linear operators on vector spaces over an algebraically closed field is essentially reduced to the study of nilpotent operators. For vector spaces over non-algebraically closed fields, we still need to find some substitute for characteristic values and vectors. It is a very interesting fact that these two problems can be handled simultaneously and this is what we shall do in the next chapter.

In concluding this section, we should like to give an example which illustrates some of the ideas of the primary decomposition theorem. We have chosen to give it at the end of the section since it deals with differential equations and thus is not purely linear algebra.

Example 14. In the primary decomposition theorem, it is not necessary that the vector space V be finite dimensional, nor is it necessary for parts (i) and (ii) that p be the minimal polynomial for T. If T is a linear operator on an arbitrary vector space and *if* there is a monic polynomial p such that $p(T) = 0$, then parts (i) and (ii) of Theorem 12 are valid for T with the proof which we gave.

Let n be a positive integer and let V be the space of all n times continuously differentiable functions f on the real line which satisfy the differential equation

$$\text{(6-18)} \qquad \frac{d^n f}{dt^n} + a_{n-1}\frac{d^{n-1}f}{dt^{n-1}} + \cdots + a_1\frac{df}{dt} + a_0 f = 0$$

where $a_0, \ldots, a_{n-1}$ are some fixed constants. If C_n denotes the space of n times continuously differentiable functions, then the space V of solutions of this differential equation is a subspace of C_n. If D denotes the differentiation operator and p is the polynomial

$$p = x^n + a_{n-1}x^{n-1} + \cdots + a_1x + a_0$$

then V is the null space of the operator $p(D)$, because (6-18) simply says $p(D)f = 0$. Therefore, V is invariant under D. Let us now regard D as a linear operator on the subspace V. Then $p(D) = 0$.

If we are discussing differentiable complex-valued functions, then C_n and V are complex vector spaces, and $a_0, \ldots, a_{n-1}$ may be any complex numbers. We now write

$$p = (x - c_1)^{r_1} \cdots (x - c_k)^{r_k}$$

where $c_1, \ldots, c_k$ are distinct complex numbers. If W_j is the null space of $(D - c_jI)^{r_j}$, then Theorem 12 says that

$$V = W_1 \oplus \cdots \oplus W_k.$$

In other words, if f satisfies the differential equation (6-18), then f is uniquely expressible in the form

$$f = f_1 + \cdots + f_k$$

where f_j satisfies the differential equation $(D - c_jI)^{r_j}f_j = 0$. Thus, the study of the solutions to the equation (6-18) is reduced to the study of the space of solutions of a differential equation of the form

$$\text{(6-19)} \qquad (D - cI)^r f = 0.$$

This reduction has been accomplished by the general methods of linear algebra, i.e., by the primary decomposition theorem.

To describe the space of solutions to (6-19), one must know something about differential equations, that is, one must know something about D other than the fact that it is a linear operator. However, one does not need to know very much. It is very easy to establish by induction on r that if f is in C_r then

$$(D - cI)^r f = e^{ct}D^r(e^{-ct}f)$$

that is,

$$\frac{df}{dt} - cf(t) = e^{ct}\frac{d}{dt}(e^{-ct}f), \quad \text{etc.}$$

Thus $(D - cI)^r f = 0$ if and only if $D^r(e^{-ct}f) = 0$. A function g such that $D^r g = 0$, i.e., $d^r g/dt^r = 0$, must be a polynomial function of degree $(r - 1)$ or less:

$$g(t) = b_0 + b_1t + \cdots + b_{r-1}t^{r-1}.$$

Thus f satisfies (6-19) if and only if f has the form

$$f(t) = e^{ct}(b_0 + b_1t + \cdots + b_{r-1}t^{r-1}).$$

Accordingly, the 'functions' $e^{ct}, te^{ct}, \ldots, t^{r-1}e^{ct}$ span the space of solutions of (6-19). Since $1, t, \ldots, t^{r-1}$ are linearly independent functions and the exponential function has no zeros, these r functions t^je^{ct}, $0 \leq j \leq r - 1$, form a basis for the space of solutions.

Returning to the differential equation (6-18), which is

$$p(D)f = 0$$
$$p = (x - c_1)^{r_1} \cdots (x - c_k)^{r_k}$$

we see that the n functions $t^me^{c_jt}$, $0 \leq m \leq r_j - 1$, $1 \leq j \leq k$, form a basis for the space of solutions to (6-18). In particular, the space of solutions is finite-dimensional and has dimension equal to the degree of the polynomial p.

Exercises

1. Let T be a linear operator on R^3 which is represented in the standard ordered basis by the matrix

$$\begin{bmatrix} 6 & -3 & -2 \\ 4 & -1 & -2 \\ 10 & -5 & -3 \end{bmatrix}.$$

Express the minimal polynomial p for T in the form $p = p_1p_2$, where p_1 and p_2 are monic and irreducible over the field of real numbers. Let W_i be the null space of $p_i(T)$. Find bases $\mathcal{B}_i$ for the spaces W_1 and W_2. If T_i is the operator induced on W_i by T, find the matrix of T_i in the basis $\mathcal{B}_i$ (above).

2. Let T be the linear operator on R^3 which is represented by the matrix

$$\begin{bmatrix} 3 & 1 & -1 \\ 2 & 2 & -1 \\ 2 & 2 & 0 \end{bmatrix}$$

in the standard ordered basis. Show that there is a diagonalizable operator D on R^3 and a nilpotent operator N on R^3 such that $T = D + N$ and $DN = ND$. Find the matrices of D and N in the standard basis. (Just repeat the proof of Theorem 12 for this special case.)

3. If V is the space of all polynomials of degree less than or equal to n over a field F, prove that the differentiation operator on V is nilpotent.

4. Let T be a linear operator on the finite-dimensional space V with characteristic polynomial

$$f = (x - c_1)^{d_1} \cdots (x - c_k)^{d_k}$$

and minimal polynomial

$$p = (x - c_1)^{r_1} \cdots (x - c_k)^{r_k}.$$

Let W_i be the null space of $(T - c_iI)^{r_i}$.

(a) Prove that W_i is the set of all vectors α in V such that $(T - c_iI)^m\alpha = 0$ for some positive integer m (which may depend upon α).

(b) Prove that the dimension of W_i is d_i. (*Hint:* If T_i is the operator induced on W_i by T, then $T_i - c_iI$ is nilpotent; thus the characteristic polynomial for $T_i - c_iI$ must be x^{e_i} where e_i is the dimension of W_i (proof?); thus the characteristic polynomial of T_i is $(x - c_i)^{e_i}$; now use the fact that the characteristic polynomial for T is the product of the characteristic polynomials of the T_i to show that $e_i = d_i$.)

5. Let V be a finite-dimensional vector space over the field of complex numbers. Let T be a linear operator on V and let D be the diagonalizable part of T. Prove that if g is any polynomial with complex coefficients, then the diagonalizable part of $g(T)$ is $g(D)$.

6. Let V be a finite-dimensional vector space over the field F, and let T be a linear operator on V such that rank $(T) = 1$. Prove that either T is diagonalizable or T is nilpotent, not both.

7. Let V be a finite-dimensional vector space over F, and let T be a linear operator on V. Suppose that T commutes with every diagonalizable linear operator on V. Prove that T is a scalar multiple of the identity operator.

8. Let V be the space of $n \times n$ matrices over a field F, and let A be a fixed $n \times n$ matrix over F. Define a linear operator T on V by $T(B) = AB - BA$. Prove that if A is a nilpotent matrix, then T is a nilpotent operator.

9. Give an example of two 4×4 nilpotent matrices which have the same minimal polynomial (they necessarily have the same characteristic polynomial) but which are not similar.

10. Let T be a linear operator on the finite-dimensional space V, let $p = p_1^{r_1} \cdots p_k^{r_k}$ be the minimal polynomial for T, and let $V = W_1 \oplus \cdots \oplus W_k$ be the primary decomposition for T, i.e., W_j is the null space of $p_j(T)^{r_j}$. Let W be any subspace of V which is invariant under T. Prove that

$$W = (W \cap W_1) \oplus (W \cap W_2) \oplus \cdots \oplus (W \cap W_k).$$

11. What's wrong with the following proof of Theorem 13? Suppose that the minimal polynomial for T is a product of linear factors. Then, by Theorem 5, T is triangulable. Let $\mathcal{B}$ be an ordered basis such that $A = [T]_{\mathcal{B}}$ is upper-triangular. Let D be the diagonal matrix with diagonal entries $a_{11}, \ldots, a_{nn}$. Then $A = D + N$, where N is strictly upper-triangular. Evidently N is nilpotent.

12. If you thought about Exercise 11, think about it again, after you observe what Theorem 7 tells you about the diagonalizable and nilpotent parts of T.

13. Let T be a linear operator on V with minimal polynomial of the form p^n, where p is irreducible over the scalar field. Show that there is a vector α in V such that the T-annihilator of α is p^n.

14. Use the primary decomposition theorem and the result of Exercise 13 to prove the following. If T is any linear operator on a finite-dimensional vector space V, then there is a vector α in V with T-annihilator equal to the minimal polynomial for T.

15. If N is a nilpotent linear operator on an n-dimensional vector space V, then the characteristic polynomial for N is x^n.

7. The Rational and Jordan Forms

7.1. Cyclic Subspaces and Annihilators

Once again V is a finite-dimensional vector space over the field F and T is a fixed (but arbitrary) linear operator on V. If α is any vector in V, there is a smallest subspace of V which is invariant under T and contains α. This subspace can be defined as the intersection of all T-invariant subspaces which contain α; however, it is more profitable at the moment for us to look at things this way. If W is any subspace of V which is invariant under T and contains α, then W must also contain the vector $T\alpha$; hence W must contain $T(T\alpha) = T^2\alpha$, $T(T^2\alpha) = T^3\alpha$, etc. In other words W must contain $g(T)\alpha$ for every polynomial g over F. The set of all vectors of the form $g(T)\alpha$, with g in $F[x]$, is clearly invariant under T, and is thus the smallest T-invariant subspace which contains α.

Definition. *If α is any vector in* V, *the* **T-cyclic subspace generated by** α *is the subspace* $Z(\alpha; T)$ *of all vectors of the form* $g(T)\alpha$, g *in* F[x]. *If* $Z(\alpha; T) = V$, *then α is called a* **cyclic vector** *for* T.

Another way of describing the subspace $Z(\alpha; T)$ is that $Z(\alpha; T)$ is the subspace spanned by the vectors $T^k\alpha$, $k \geq 0$, and thus α is a cyclic vector for T if and only if these vectors span V. We caution the reader that the general operator T has no cyclic vectors.

EXAMPLE 1. For any T, the T-cyclic subspace generated by the zero vector is the zero subspace. The space $Z(\alpha; T)$ is one-dimensional if and only if α is a characteristic vector for T. For the identity operator, every non-zero vector generates a one-dimensional cyclic subspace; thus, if $\dim V > 1$, the identity operator has no cyclic vector. An example of an operator which has a cyclic vector is the linear operator T on F^2 which is represented in the standard ordered basis by the matrix

$$\begin{bmatrix} 0 & 0 \\ 1 & 0 \end{bmatrix}.$$

Here the cyclic vector (a cyclic vector) is ϵ_1; for, if $\beta = (a, b)$, then with $g = a + bx$ we have $\beta = g(T)\epsilon_1$. For this same operator T, the cyclic subspace generated by ϵ_2 is the one-dimensional space spanned by ϵ_2, because ϵ_2 is a characteristic vector of T.

For any T and α, we shall be interested in linear relations

$$c_0\alpha + c_1T\alpha + \cdots + c_kT^k\alpha = 0$$

between the vectors $T^j\alpha$, that is, we shall be interested in the polynomials $g = c_0 + c_1x + \cdots + c_kx^k$ which have the property that $g(T)\alpha = 0$. The set of all g in $F[x]$ such that $g(T)\alpha = 0$ is clearly an ideal in $F[x]$. It is also a non-zero ideal, because it contains the minimal polynomial p of the operator T ($p(T)\alpha = 0$ for every α in V).

Definition. *If α is any vector in V, the* **T-annihilator** *of α is the ideal $M(\alpha; T)$ in $F[x]$ consisting of all polynomials g over F such that $g(T)\alpha = 0$. The unique monic polynomial p_α which generates this ideal will also be called the* **T-annihilator** *of α.*

As we pointed out above, the T-annihilator p_α divides the minimal polynomial of the operator T. The reader should also note that $\deg (p_\alpha) > 0$ unless α is the zero vector.

Theorem 1. *Let α be any non-zero vector in V and let p_α be the T-annihilator of α.*

(i) The degree of p_α is equal to the dimension of the cyclic subspace $Z(\alpha; T)$.

(ii) If the degree of p_α is k, then the vectors $\alpha, T\alpha, T^2\alpha, \ldots, T^{k-1}\alpha$ form a basis for $Z(\alpha; T)$.

(iii) If U is the linear operator on $Z(\alpha; T)$ induced by T, then the minimal polynomial for U is p_α.

Proof. Let g be any polynomial over the field F. Write

$$g = p_\alpha q + r$$

where either $r = 0$ or $\deg (r) < \deg (p_\alpha) = k$. The polynomial $p_\alpha q$ is in the T-annihilator of α, and so

$$g(T)\alpha = r(T)\alpha.$$

Since $r = 0$ or $\deg (r) < k$, the vector $r(T)\alpha$ is a linear combination of the vectors $\alpha, T\alpha, \ldots, T^{k-1}\alpha$, and since $g(T)\alpha$ is a typical vector in

$Z(\alpha; T)$, this shows that these k vectors span $Z(\alpha; T)$. These vectors are certainly linearly independent, because any non-trivial linear relation between them would give us a non-zero polynomial g such that $g(T)\alpha = 0$ and $\deg(g) < \deg(p_\alpha)$, which is absurd. This proves (i) and (ii).

Let U be the linear operator on $Z(\alpha; T)$ obtained by restricting T to that subspace. If g is any polynomial over F, then

$$\begin{aligned} p_\alpha(U)g(T)\alpha &= p_\alpha(T)g(T)\alpha \\ &= g(T)p_\alpha(T)\alpha \\ &= g(T)0 \\ &= 0. \end{aligned}$$

Thus the operator $p_\alpha(U)$ sends every vector in $Z(\alpha; T)$ into 0 and is the zero operator on $Z(\alpha; T)$. Furthermore, if h is a polynomial of degree less than k, we cannot have $h(U) = 0$, for then $h(U)\alpha = h(T)\alpha = 0$, contradicting the definition of p_α. This shows that p_α is the minimal polynomial for U. ∎

A particular consequence of this theorem is the following: If α happens to be a cyclic vector for T, then the minimal polynomial for T must have degree equal to the dimension of the space V; hence, the Cayley-Hamilton theorem tells us that the minimal polynomial for T is the characteristic polynomial for T. We shall prove later that for any T there is a vector α in V which has the minimal polynomial of T for its annihilator. It will then follow that T has a cyclic vector if and only if the minimal and characteristic polynomials for T are identical. But it will take a little work for us to see this.

Our plan is to study the general T by using operators which have a cyclic vector. So, let us take a look at a linear operator U on a space W of dimension k which has a cyclic vector α. By Theorem 1, the vectors $\alpha, \ldots, U^{k-1}\alpha$ form a basis for the space W, and the annihilator p_α of α is the minimal polynomial for U (and hence also the characteristic polynomial for U). If we let $\alpha_i = U^{i-1}\alpha$, $i = 1, \ldots, k$, then the action of U on the ordered basis $\mathcal{B} = \{\alpha_1, \ldots, \alpha_k\}$ is

$$\begin{aligned} U\alpha_i &= \alpha_{i+1}, \qquad i = 1, \ldots, k-1 \\ U\alpha_k &= -c_0\alpha_1 - c_1\alpha_2 - \cdots - c_{k-1}\alpha_k \end{aligned} \tag{7-1}$$

where $p_\alpha = c_0 + c_1x + \cdots + c_{k-1}x^{k-1} + x^k$. The expression for $U\alpha_k$ follows from the fact that $p_\alpha(U)\alpha = 0$, i.e.,

$$U^k\alpha + c_{k-1}U^{k-1}\alpha + \cdots + c_1U\alpha + c_0\alpha = 0.$$

This says that the matrix of U in the ordered basis $\mathcal{B}$

$$\begin{bmatrix} 0 & 0 & 0 & \cdots & 0 & -c_0 \\ 1 & 0 & 0 & \cdots & 0 & -c_1 \\ 0 & 1 & 0 & \cdots & 0 & -c_2 \\ \vdots & \vdots & \vdots & & \vdots & \vdots \\ 0 & 0 & 0 & \cdots & 1 & -c_{k-1} \end{bmatrix}. \tag{7-2}$$

The matrix (7-2) is called the **companion matrix** of the monic polynomial p_α.

Theorem 2. *If* U *is a linear operator on the finite-dimensional space* W, *then* U *has a cyclic vector if and only if there is some ordered basis for* W *in which* U *is represented by the companion matrix of the minimal polynomial for* U.

Proof. We have just observed that if U has a cyclic vector, then there is such an ordered basis for W. Conversely, if we have some ordered basis $\{\alpha_1, \ldots, \alpha_k\}$ for W in which U is represented by the companion matrix of its minimal polynomial, it is obvious that α_1 is a cyclic vector for U. ∎

Corollary. *If* A *is the companion matrix of a monic polynomial* p, *then* p *is both the minimal and the characteristic polynomial of* A.

Proof. One way to see this is to let U be the linear operator on F^k which is represented by A in the standard ordered basis, and to apply Theorem 1 together with the Cayley-Hamilton theorem. Another method is to use Theorem 1 to see that p is the minimal polynomial for A and to verify by a direct calculation that p is the characteristic polynomial for A. ∎

One last comment—if T is any linear operator on the space V and α is any vector in V, then the operator U which T induces on the cyclic subspace $Z(\alpha; T)$ has a cyclic vector, namely, α. Thus $Z(\alpha; T)$ has an ordered basis in which U is represented by the companion matrix of p_α, the T-annihilator of α.

Exercises

1. Let T be a linear operator on F^2. Prove that any non-zero vector which is not a characteristic vector for T is a cyclic vector for T. Hence, prove that either T has a cyclic vector or T is a scalar multiple of the identity operator.

2. Let T be the linear operator on R^3 which is represented in the standard ordered basis by the matrix

$$\begin{bmatrix} 2 & 0 & 0 \\ 0 & 2 & 0 \\ 0 & 0 & -1 \end{bmatrix}.$$

Prove that T has no cyclic vector. What is the T-cyclic subspace generated by the vector $(1, -1, 3)$?

3. Let T be the linear operator on C^3 which is represented in the standard ordered basis by the matrix

$$\begin{bmatrix} 1 & i & 0 \\ -1 & 2 & -i \\ 0 & 1 & 1 \end{bmatrix}.$$

Find the T-annihilator of the vector $(1, 0, 0)$. Find the T-annihilator of $(1, 0, i)$.

4. Prove that if T^2 has a cyclic vector, then T has a cyclic vector. Is the converse true?

5. Let V be an n-dimensional vector space over the field F, and let N be a nilpotent linear operator on V. Suppose $N^{n-1} \neq 0$, and let α be any vector in V such that $N^{n-1}\alpha \neq 0$. Prove that α is a cyclic vector for N. What exactly is the matrix of N in the ordered basis $\{\alpha, N\alpha, \ldots, N^{n-1}\alpha\}$?

6. Give a direct proof that if A is the companion matrix of the monic polynomial p, then p is the characteristic polynomial for A.

7. Let V be an n-dimensional vector space, and let T be a linear operator on V. Suppose that T is *diagonalizable*.

(a) If T has a cyclic vector, show that T has n distinct characteristic values.

(b) If T has n distinct characteristic values, and if $\{\alpha_1, \ldots, \alpha_n\}$ is a basis of characteristic vectors for T, show that $\alpha = \alpha_1 + \cdots + \alpha_n$ is a cyclic vector for T.

8. Let T be a linear operator on the finite-dimensional vector space V. Suppose T has a cyclic vector. Prove that if U is any linear operator which commutes with T, then U is a polynomial in T.

7.2. Cyclic Decompositions and the Rational Form

The primary purpose of this section is to prove that if T is any linear operator on a finite-dimensional space V, then there exist vectors $\alpha_1, \ldots, \alpha_r$ in V such that

$$V = Z(\alpha_1; T) \oplus \cdots \oplus Z(\alpha_r; T).$$

In other words, we wish to prove that V is a direct sum of T-cyclic subspaces. This will show that T is the direct sum of a finite number of linear operators, each of which has a cyclic vector. The effect of this will be to reduce many questions about the general linear operator to similar questions about an operator which has a cyclic vector. The theorem which we prove (Theorem 3) is one of the deepest results in linear algebra and has many interesting corollaries.

The cyclic decomposition theorem is closely related to the following question. Which T-invariant subspaces W have the property that there exists a T-invariant subspace W' such that $V = W \oplus W'$? If W is any subspace of a finite-dimensional space V, then there exists a subspace W' such that $V = W \oplus W'$. Usually there are many such subspaces W' and each of these is called **complementary** to W. We are asking when a T-invariant subspace has a complementary subspace which is also invariant under T.

Let us suppose that $V = W \oplus W'$ where both W and W' are invariant under T and then see what we can discover about the subspace W. Each vector β in V is of the form $\beta = \gamma + \gamma'$ where γ is in W and γ' is in W'. If f is any polynomial over the scalar field, then

$$f(T)\beta = f(T)\gamma + f(T)\gamma'.$$

Since W and W' are invariant under T, the vector $f(T)\gamma$ is in W and $f(T)\gamma'$ is in W'. Therefore $f(T)\beta$ is in W if and only if $f(T)\gamma' = 0$. What interests us is the seemingly innocent fact that, if $f(T)\beta$ is in W, then $f(T)\beta = f(T)\gamma$.

Definition. *Let* T *be a linear operator on a vector space* V *and let* W *be a subspace of* V. *We say that* W *is* **T-admissible** *if*

(i) W *is invariant under* T;
(ii) *if* f(T)β *is in* W, *there exists a vector* γ *in* W *such that* f(T)β = f(T)γ.

As we just showed, if W is invariant and has a complementary invariant subspace, then W is admissible. One of the consequences of Theorem 3 will be the converse, so that admissibility characterizes those invariant subspaces which have complementary invariant subspaces.

Let us indicate how the admissibility property is involved in the attempt to obtain a decomposition

$$V = Z(\alpha_1; T) \oplus \cdots \oplus Z(\alpha_r; T).$$

Our basic method for arriving at such a decomposition will be to inductively select the vectors $\alpha_1, \ldots, \alpha_r$. Suppose that by some process or another we have selected $\alpha_1, \ldots, \alpha_j$ and the subspace

$$W_j = Z(\alpha_1; T) + \cdots + Z(\alpha_j; T)$$

is proper. We would like to find a non-zero vector α_{j+1} such that

$$W_j \cap Z(\alpha_{j+1}; T) = \{0\}$$

because the subspace $W_{j+1} = W_j \oplus Z(\alpha_{j+1}; T)$ would then come at least one dimension nearer to exhausting V. But, why should any such α_{j+1} exist? If $\alpha_1, \ldots, \alpha_j$ have been chosen so that W_j is a T-admissible subspace, then it is rather easy to see that we can find a suitable α_{j+1}. This is what will make our proof of Theorem 3 work, even if that is not how we phrase the argument.

Let W be a proper T-invariant subspace. Let us try to find a non-zero vector α such that

$$W \cap Z(\alpha; T) = \{0\}. \tag{7-3}$$

We can choose some vector β which is not in W. Consider the *T-conductor* $S(\beta; W)$, which consists of all polynomials g such that $g(T)\beta$ is in W. Recall that the monic polynomial $f = s(\beta; W)$ which generates the ideal $S(\beta; W)$ is also called the *T-conductor of* β *into* W. The vector $f(T)\beta$ is in W. Now, if W is T-admissible, there is a γ in W with $f(T)\beta = f(T)\gamma$. Let $\alpha = \beta - \gamma$ and let g be any polynomial. Since $\beta - \alpha$ is in W, $g(T)\beta$ will be in W if and only if $g(T)\alpha$ is in W; in other words, $S(\alpha; W) = S(\beta; W)$. Thus the polynomial f is also the T-conductor of α into W. But $f(T)\alpha = 0$. That tells us that $g(T)\alpha$ is in W if and only if $g(T)\alpha = 0$, i.e., the subspaces $Z(\alpha; T)$ and W are independent (7-3) and f is the T-annihilator of α.

Theorem 3 (Cyclic Decomposition Theorem). *Let* T *be a linear operator on a finite-dimensional vector space* V *and let* W_0 *be a proper* T-*admissible subspace of* V. *There exist non-zero vectors* $\alpha_1, \ldots, \alpha_r$ *in* V *with respective* T-*annihilators* $p_1, \ldots, p_r$ *such that*

(i) $V = W_0 \oplus Z(\alpha_1; T) \oplus \cdots \oplus Z(\alpha_r; T)$;

(ii) p_k *divides* p_{k-1}, $k = 2, \ldots, r$.

Furthermore, the integer r *and the annihilators* $p_1, \ldots, p_r$ *are uniquely determined by* (i), (ii), *and the fact that no* α_k *is* 0.

Proof. The proof is rather long; hence, we shall divide it into four steps. For the first reading it may seem easier to take $W_0 = \{0\}$, although it does not produce any substantial simplification. Throughout the proof, we shall abbreviate $f(T)\beta$ to $f\beta$.

Step 1. There exist non-zero vectors $\beta_1, \ldots, \beta_r$ *in* V *such that*

(a) $V = W_0 + Z(\beta_1; T) + \cdots + Z(\beta_r; T)$;

(b) *if* $1 \leq k \leq r$ *and*

$$W_k = W_0 + Z(\beta_1; T) + \cdots + Z(\beta_k; T)$$

then the conductor $p_k = s(\beta_k; W_{k-1})$ *has maximum degree among all* T-*conductors into the subspace* W_{k-1}, i.e., *for every* k

$$\deg p_k = \max_{\alpha \text{ in } V} \deg s(\alpha; W_{k-1}).$$

This step depends only upon the fact that W_0 is an invariant subspace. If W is a proper T-invariant subspace, then

$$0 < \max_{\alpha} \deg s(\alpha; W) \leq \dim V$$

and we can choose a vector β so that $\deg s(\beta; W)$ attains that maximum. The subspace $W + Z(\beta; T)$ is then T-invariant and has dimension larger than $\dim W$. Apply this process to $W = W_0$ to obtain β_1. If $W_1 = W_0 + Z(\beta_1; T)$ is still proper, then apply the process to W_1 to obtain β_2. Continue in that manner. Since $\dim W_k > \dim W_{k-1}$, we must reach $W_r = V$ in not more than $\dim V$ steps.

Step 2. Let $\beta_1, \ldots, \beta_r$ *be non-zero vectors which satisfy conditions* (a) *and* (b) *of Step 1. Fix* k, $1 \leq k \leq r$. *Let* β *be any vector in* V *and let* $f = s(\beta; W_{k-1})$. *If*

$$f\beta = \beta_0 + \sum_{1 \leq i < k} g_i\beta_i, \qquad \beta_i \text{ in } W_i$$

then f *divides each polynomial* g_i *and* $\beta_0 = f\gamma_0$, *where* γ_0 *is in* W_0.

If $k = 1$, this is just the statement that W_0 is T-admissible. In order to prove the assertion for $k > 1$, apply the division algorithm:

$$g_i = fh_i + r_i, \qquad r_i = 0 \quad \text{or} \quad \deg r_i < \deg f. \tag{7-4}$$

We wish to show that $r_i = 0$ for each i. Let

$$\gamma = \beta - \sum_{1}^{k-1} h_i\beta_i. \tag{7-5}$$

Since $\gamma - \beta$ is in W_{k-1},

$$s(\gamma; W_{k-1}) = s(\beta; W_{k-1}) = f.$$

Furthermore

$$f\gamma = \beta_0 + \sum_{1}^{k-1} r_i\beta_i. \tag{7-6}$$

Suppose that some r_i is different from 0. We shall deduce a contradiction. Let j be the largest index i for which $r_i \neq 0$. Then

$$f\gamma = \beta_0 + \sum_{1}^{j} r_i\beta_i, \qquad r_j \neq 0 \quad \text{and} \quad \deg r_j < \deg f. \tag{7-7}$$

Let $p = s(\gamma; W_{j-1})$. Since W_{k-1} contains W_{j-1}, the conductor $f = s(\gamma; W_{k-1})$ must divide p:

$$p = fg.$$

Apply $g(T)$ to both sides of (7-7):

$$p\gamma = gf\gamma = gr_j\beta_j + g\beta_0 + \sum_{1 \leq i < j} gr_i\beta_i. \tag{7-8}$$

By definition, $p\gamma$ is in W_{j-1}, and the last two terms on the right side of (7-8) are in W_{j-1}. Therefore, $gr_j\beta_j$ is in W_{j-1}. Now we use condition (b) of Step 1:

$$\begin{aligned} \deg (gr_j) &\geq \deg s(\beta_j; W_{j-1}) \\ &= \deg p_j \\ &\geq \deg s(\gamma; W_{j-1}) \\ &= \deg p \\ &= \deg (fg). \end{aligned}$$

Thus $\deg r_j \geq \deg f$, and that contradicts the choice of j. We now know that f divides each g_i and hence that $\beta_0 = f\gamma$. Since W_0 is T-admissible, $\beta_0 = f\gamma_0$ where γ_0 is in W_0. We remark in passing that Step 2 is a strengthened form of the assertion that each of the subspaces $W_1, W_2, \ldots, W_r$ is T-admissible.

Step 3. There exist non-zero vectors $\alpha_1, \ldots, \alpha_r$ in V which satisfy conditions (i) *and* (ii) *of Theorem 3.*

Start with vectors $\beta_1, \ldots, \beta_r$ as in Step 1. Fix k, $1 \leq k \leq r$. We apply Step 2 to the vector $\beta = \beta_k$ and the T-conductor $f = p_k$. We obtain

$$p_k\beta_k = p_k\gamma_0 + \sum_{1 \leq i < k} p_kh_i\beta_i \tag{7-9}$$

where γ_0 is in W_0 and $h_1, \ldots, h_{k-1}$ are polynomials. Let

$$\alpha_k = \beta_k - \gamma_0 - \sum_{1 \leq i < k} h_i\beta_i. \tag{7-10}$$

Since $\beta_k - \alpha_k$ is in W_{k-1},

$$s(\alpha_k; W_{k-1}) = s(\beta_k; W_{k-1}) = p_k \tag{7-11}$$

and since $p_k\alpha_k = 0$, we have

$$W_{k-1} \cap Z(\alpha_k; T) = \{0\}. \tag{7-12}$$

Because each α_k satisfies (7-11) and (7-12), it follows that

$$W_k = W_0 \oplus Z(\alpha_1; T) \oplus \cdots \oplus Z(\alpha_k; T)$$

and that p_k is the T-annihilator of α_k. In other words, the vectors $\alpha_1, \ldots, \alpha_r$ define the same sequence of subspaces $W_1, W_2, \ldots$ as do the vectors $\beta_1, \ldots, \beta_r$ and the T-conductors $p_k = s(\alpha_k, W_{k-1})$ have the same maximality properties (condition (b) of Step 1). The vectors $\alpha_1, \ldots, \alpha_r$ have the additional property that the subspaces W_0, $Z(\alpha_1; T)$, $Z(\alpha_2; T), \ldots$ are independent. It is therefore easy to verify condition (ii) in Theorem 3. Since $p_i\alpha_i = 0$ for each i, we have the trivial relation

$$p_k\alpha_k = 0 + p_1\alpha_1 + \cdots + p_{k-1}\alpha_{k-1}.$$

Apply Step 2 with $\beta_1, \ldots, \beta_k$ replaced by $\alpha_1, \ldots, \alpha_k$ and with $\beta = \alpha_k$. Conclusion: p_k divides each p_i with $i < k$.

Step 4. The number r *and the polynomials* $p_1, \ldots, p_r$ *are uniquely determined by the conditions of Theorem 3.*

Suppose that in addition to the vectors $\alpha_1, \ldots, \alpha_r$ in Theorem 3 we have non-zero vectors $\gamma_1, \ldots, \gamma_s$ with respective T-annihilators $g_1, \ldots, g_s$ such that

$$\begin{aligned} V &= W_0 \oplus Z(\gamma_1; T) \oplus \cdots \oplus Z(\gamma_s; T) \\ g_k &\text{ divides } g_{k-1}, \qquad k = 2, \ldots, s. \end{aligned} \tag{7-13}$$

We shall show that $r = s$ and $p_i = g_i$ for each i.

It is very easy to see that $p_1 = g_1$. The polynomial g_1 is determined from (7-13) as the T-conductor of V into W_0. Let $S(V; W_0)$ be the collection of polynomials f such that $f\beta$ is in W_0 for every β in V, i.e., polynomials f such that the range of $f(T)$ is contained in W_0. Then $S(V; W_0)$ is a non-zero ideal in the polynomial algebra. The polynomial g_1 is the monic generator of that ideal, for this reason. Each β in V has the form

$$\beta = \beta_0 + f_1\gamma_1 + \cdots + f_s\gamma_s$$

and so

$$g_1\beta = g_1\beta_0 + \sum_1^s g_1f_i\gamma_i.$$

Since each g_i divides g_1, we have $g_1\gamma_i = 0$ for all i and $g_1\beta = g_1\beta_0$ is in W_0. Thus g_1 is in $S(V; W_0)$. Since g_1 is the monic polynomial of least degree which sends γ_1 into W_0, we see that g_1 is the monic polynomial of least degree in the ideal $S(V; W_0)$. By the same argument, p_1 is the generator of that ideal, so $p_1 = g_1$.

If f is a polynomial and W is a subspace of V, we shall employ the shorthand fW for the set of all vectors $f\alpha$ with α in W. We have left to the exercises the proofs of the following three facts.

1. $fZ(\alpha; T) = Z(f\alpha; T)$.
2. If $V = V_1 \oplus \cdots \oplus V_k$, where each V_i is invariant under T, then $fV = fV_1 \oplus \cdots \oplus fV_k$.
3. If α and γ have the same T-annihilator, then $f\alpha$ and $f\gamma$ have the same T-annihilator and (therefore)

$$\dim Z(f\alpha; T) = \dim Z(f\gamma; T).$$

Now, we proceed by induction to show that $r = s$ and $p_i = g_i$ for $i = 2, \ldots, r$. The argument consists of counting dimensions in the right way. We shall give the proof that if $r \geq 2$ then $p_2 = g_2$, and from that the induction should be clear. Suppose that $r \geq 2$. Then

$$\dim W_0 + \dim Z(\alpha_1; T) < \dim V.$$

Since we know that $p_1 = g_1$, we know that $Z(\alpha_1; T)$ and $Z(\gamma_1; T)$ have the same dimension. Therefore,

$$\dim W_0 + \dim Z(\gamma_1; T) < \dim V$$

which shows that $s \geq 2$. Now it makes sense to ask whether or not $p_2 = g_2$. From the two decompositions of V, we obtain two decompositions of the subspace p_2V:

$$\begin{aligned} p_2V &= p_2W_0 \oplus Z(p_2\alpha_1; T) \\ p_2V &= p_2W_0 \oplus Z(p_2\gamma_1; T) \oplus \cdots \oplus Z(p_2\gamma_s; T). \end{aligned} \tag{7-14}$$

We have made use of facts (1) and (2) above and we have used the fact that $p_2\alpha_i = 0$, $i \geq 2$. Since we know that $p_1 = g_1$, fact (3) above tells us that $Z(p_2\alpha_1; T)$ and $Z(p_2\gamma_1; T)$ have the same dimension. Hence, it is apparent from (7-14) that

$$\dim Z(p_2\gamma_i; T) = 0, \qquad i \geq 2.$$

We conclude that $p_2\gamma_2 = 0$ and g_2 divides p_2. The argument can be reversed to show that p_2 divides g_2. Therefore $p_2 = g_2$. ∎

Corollary. *If* T *is a linear operator on a finite-dimensional vector space, then every* T*-admissible subspace has a complementary subspace which is also invariant under* T.

Proof. Let W_0 be an admissible subspace of V. If $W_0 = V$, the complement we seek is $\{0\}$. If W_0 is proper, apply Theorem 3 and let

$$W_0' = Z(\alpha_1; T) \oplus \cdots \oplus Z(\alpha_r; T).$$

Then W_0' is invariant under T and $V = W_0 \oplus W_0'$. ∎

Corollary. *Let* T *be a linear operator on a finite-dimensional vector space* V.

(a) *There exists a vector* α *in* V *such that the* T*-annihilator of* α *is the minimal polynomial for* T.

(b) T *has a cyclic vector if and only if the characteristic and minimal polynomials for* T *are identical.*

Proof. If $V = \{0\}$, the results are trivially true. If $V \neq \{0\}$, let

$$V = Z(\alpha_1; T) \oplus \cdots \oplus Z(\alpha_r; T) \tag{7-15}$$

where the T-annihilators $p_1, \ldots, p_r$ are such that p_{k+1} divides p_k, $1 \leq k \leq r - 1$. As we noted in the proof of Theorem 3, it follows easily that p_1 is the minimal polynomial for T, i.e., the T-conductor of V into $\{0\}$. We have proved (a).

We saw in Section 7.1 that, if T has a cyclic vector, the minimal polynomial for T coincides with the characteristic polynomial. The content of (b) is in the converse. Choose any α as in (a). If the degree of the minimal polynomial is dim V, then $V = Z(\alpha; T)$. ∎

Theorem 4 (Generalized Cayley-Hamilton Theorem). *Let* T *be a linear operator on a finite-dimensional vector space* V. *Let* p *and* f *be the minimal and characteristic polynomials for* T, *respectively.*

(i) p *divides* f.
(ii) p *and* f *have the same prime factors, except for multiplicities.*
(iii) *If*

$$\mathrm{p} = \mathrm{f}_1^{\mathrm{r}_1} \cdots \mathrm{f}_k^{\mathrm{r}_k} \tag{7-16}$$

is the prime factorization of p, *then*

$$\mathrm{f} = \mathrm{f}_1^{\mathrm{d}_1} \cdots \mathrm{f}_k^{\mathrm{d}_k} \tag{7-17}$$

where d_i *is the nullity of* $\mathrm{f}_i(\mathrm{T})^{\mathrm{r}_i}$ *divided by the degree of* f_i.

Proof. We disregard the trivial case $V = \{0\}$. To prove (i) and (ii), consider a cyclic decomposition (7-15) of V obtained from Theorem 3. As we noted in the proof of the second corollary, $p_1 = p$. Let U_i be the restriction of T to $Z(\alpha_i; T)$. Then U_i has a cyclic vector and so p_i is both the minimal polynomial and the characteristic polynomial for U_i. Therefore, the characteristic polynomial f is the product $f = p_1 \cdots p_r$. That is evident from the block form (6-14) which the matrix of T assumes in a suitable basis. Clearly $p_1 = p$ divides f, and this proves (i). Obviously any prime divisor of p is a prime divisor of f. Conversely, a prime divisor of $f = p_1 \cdots p_r$ must divide one of the factors p_i, which in turn divides p_1.

Let (7-16) be the prime factorization of p. We employ the primary decomposition theorem (Theorem 12 of Chapter 6). It tells us that, if V_i is the null space of $f_i(T)^{r_i}$, then

(7-18) $$V = V_1 \oplus \cdots \oplus V_k$$

and $f_i^{r_i}$ is the minimal polynomial of the operator T_i, obtained by restricting T to the (invariant) subspace V_i. Apply part (ii) of the present theorem to the operator T_i. Since its minimal polynomial is a power of the prime f_i, the characteristic polynomial for T_i has the form $f_i^{d_i}$, where $d_i \geq r_i$. Obviously

$$d_i = \frac{\dim V_i}{\deg f_i}$$

and (almost by definition) $\dim V_i = \text{nullity } f_i(T)^{r_i}$. Since T is the direct sum of the operators $T_1, \ldots, T_k$, the characteristic polynomial f is the product

$$f = f_1^{d_1} \cdots f_k^{d_k}. \quad \blacksquare$$

Corollary. *If* T *is a nilpotent linear operator on a vector space of dimension* n, *then the characteristic polynomial for* T *is* x^n.

Now let us look at the matrix analogue of the cyclic decomposition theorem. If we have the operator T and the direct-sum decomposition of Theorem 3, let $\mathcal{B}_i$ be the 'cyclic ordered basis'

$$\{\alpha_i, T\alpha_i, \ldots, T^{k_i-1}\alpha_i\}$$

for $Z(\alpha_i; T)$. Here k_i denotes the dimension of $Z(\alpha_i; T)$, that is, the degree of the annihilator p_i. The matrix of the induced operator T_i in the ordered basis $\mathcal{B}_i$ is the companion matrix of the polynomial p_i. Thus, if we let $\mathcal{B}$ be the ordered basis for V which is the union of the $\mathcal{B}_i$ arranged in the order $\mathcal{B}_1, \ldots, \mathcal{B}_r$, then the matrix of T in the ordered basis $\mathcal{B}$ will be

(7-19) $$A = \begin{bmatrix} A_1 & 0 & \cdots & 0 \\ 0 & A_2 & \cdots & 0 \\ \vdots & \vdots & & \vdots \\ 0 & 0 & \cdots & A_r \end{bmatrix}$$

where A_i is the $k_i \times k_i$ companion matrix of p_i. An $n \times n$ matrix A, which is the direct sum (7-19) of companion matrices of non-scalar monic polynomials $p_1, \ldots, p_r$ such that p_{i+1} divides p_i for $i = 1, \ldots, r-1$, will be said to be in **rational form.** The cyclic decomposition theorem tells us the following concerning matrices.

Theorem 5. *Let* F *be a field and let* B *be an* n $\times$ n *matrix over* F. *Then* B *is similar over the field* F *to one and only one matrix which is in rational form.*

Proof. Let T be the linear operator on F^n which is represented by B in the standard ordered basis. As we have just observed, there is some ordered basis for F^n in which T is represented by a matrix A in rational form. Then B is similar to this matrix A. Suppose B is similar over F to another matrix C which is in rational form. This means simply that there is some ordered basis for F^n in which the operator T is represented by the matrix C. If C is the direct sum of companion matrices C_i of monic polynomials $g_1, \ldots, g_s$ such that g_{i+1} divides g_i for $i = 1, \ldots, s-1$, then it is apparent that we shall have non-zero vectors $\beta_1, \ldots, \beta_s$ in V with T-annihilators $g_1, \ldots, g_s$ such that

$$V = Z(\beta_1; T) \oplus \cdots \oplus Z(\beta_s; T).$$

But then by the uniqueness statement in the cyclic decomposition theorem, the polynomials g_i are identical with the polynomials p_i which define the matrix A. Thus $C = A$. ∎

The polynomials $p_1, \ldots, p_r$ are called the **invariant factors** for the matrix B. In Section 7.4, we shall describe an algorithm for calculating the invariant factors of a given matrix B. The fact that it is possible to compute these polynomials by means of a finite number of rational operations on the entries of B is what gives the rational form its name.

Example 2. Suppose that V is a two-dimensional vector space over the field F and T is a linear operator on V. The possibilities for the cyclic subspace decomposition for T are very limited. For, if the minimal polynomial for T has degree 2, it is equal to the characteristic polynomial for T and T has a cyclic vector. Thus there is some ordered basis for V in which T is represented by the companion matrix of its characteristic polynomial. If, on the other hand, the minimal polynomial for T has degree 1, then T is a scalar multiple of the identity operator. If $T = cI$, then for any two linear independent vectors α_1 and α_2 in V we have

$$V = Z(\alpha_1; T) \oplus Z(\alpha_2; T)$$
$$p_1 = p_2 = x - c.$$

For matrices, this analysis says that every 2×2 matrix over the field F is similar over F to exactly one matrix of the types

$$\begin{bmatrix} c & 0 \\ 0 & c \end{bmatrix}, \quad \begin{bmatrix} 0 & -c_0 \\ 1 & -c_1 \end{bmatrix}.$$

Example 3. Let T be the linear operator on R^3 which is represented by the matrix

$$A = \begin{bmatrix} 5 & -6 & -6 \\ -1 & 4 & 2 \\ 3 & -6 & -4 \end{bmatrix}$$

in the standard ordered basis. We have computed earlier that the characteristic polynomial for T is $f = (x-1)(x-2)^2$ and the minimal polynomial for T is $p = (x-1)(x-2)$. Thus we know that in the cyclic decomposition for T the first vector α_1 will have p as its T-annihilator.

Since we are operating in a three-dimensional space, there can be only one further vector, α_2. It must generate a cyclic subspace of dimension 1, i.e., it must be a characteristic vector for T. Its T-annihilator p_2 must be $(x - 2)$, because we must have $pp_2 = f$. Notice that this tells us immediately that the matrix A is similar to the matrix

$$B = \begin{bmatrix} 0 & -2 & 0 \\ 1 & 3 & 0 \\ 0 & 0 & 2 \end{bmatrix}$$

that is, that T is represented by B in some ordered basis. How can we find suitable vectors α_1 and α_2? Well, we know that any vector which generates a T-cyclic subspace of dimension 2 is a suitable α_1. So let's just try ϵ_1. We have

$$T\epsilon_1 = (5, -1, 3)$$

which is not a scalar multiple of ϵ_1; hence $Z(\epsilon_1; T)$ has dimension 2. This space consists of all vectors $a\epsilon_1 + b(T\epsilon_1)$:

$$a(1, 0, 0) + b(5, -1, 3) = (a + 5b, -b, 3b)$$

or, all vectors (x_1, x_2, x_3) satisfying $x_3 = -3x_2$. Now what we want is a vector α_2 such that $T\alpha_2 = 2\alpha_2$ and $Z(\alpha_2; T)$ is disjoint from $Z(\epsilon_1; T)$. Since α_2 is to be a characteristic vector for T, the space $Z(\alpha_2; T)$ will simply be the one-dimensional space spanned by α_2, and so what we require is that α_2 not be in $Z(\epsilon_1; T)$. If $\alpha = (x_1, x_2, x_3)$, one can easily compute that $T\alpha = 2\alpha$ if and only if $x_1 = 2x_2 + 2x_3$. Thus $\alpha_2 = (2, 1, 0)$ satisfies $T\alpha_2 = 2\alpha_2$ and generates a T-cyclic subspace disjoint from $Z(\epsilon_1; T)$. The reader should verify directly that the matrix of T in the ordered basis

$$\{(1, 0, 0), (5, -1, 3), (2, 1, 0)\}$$

is the matrix B above.

Example 4. Suppose that T is a diagonalizable linear operator on V. It is interesting to relate a cyclic decomposition for T to a basis which diagonalizes the matrix of T. Let $c_1, \ldots, c_k$ be the distinct characteristic values of T and let V_i be the space of characteristic vectors associated with the characteristic value c_i. Then

$$V = V_1 \oplus \cdots \oplus V_k$$

and if $d_i = \dim V_i$ then

$$f = (x - c_1)^{d_1} \cdots (x - c_k)^{d_k}$$

is the characteristic polynomial for T. If α is a vector in V, it is easy to relate the cyclic subspace $Z(\alpha; T)$ to the subspaces $V_1, \ldots, V_k$. There are unique vectors $\beta_1, \ldots, \beta_k$ such that β_i is in V_i and

$$\alpha = \beta_1 + \cdots + \beta_k.$$

Since $T\beta_i = c_i\beta_i$, we have

$$f(T)\alpha = f(c_1)\beta_1 + \cdots + f(c_k)\beta_k \tag{7-20}$$

for every polynomial f. Given any scalars $t_1, \ldots, t_k$ there exists a polynomial f such that $f(c_i) = t_i$, $1 \leq i \leq k$. Therefore, $Z(\alpha; T)$ is just the subspace spanned by the vectors $\beta_1, \ldots, \beta_k$. What is the annihilator of α? According to (7-20), we have $f(T)\alpha = 0$ if and only if $f(c_i)\beta_i = 0$ for each i. In other words, $f(T)\alpha = 0$ provided $f(c_i) = 0$ for each i such that $\beta_i \neq 0$. Accordingly, the annihilator of α is the product

$$\prod_{\beta_i \neq 0} (x - c_i). \tag{7-21}$$

Now, let $\mathcal{B}_i = \{\beta_1^i, \ldots, \beta_{d_i}^i\}$ be an ordered basis for V_i. Let

$$r = \max_i d_i.$$

We define vectors $\alpha_1, \ldots, \alpha_r$ by

$$\alpha_j = \sum_{d_i \geq j} \beta_j^i, \qquad 1 \leq j \leq r. \tag{7-22}$$

The cyclic subspace $Z(\alpha_j; T)$ is the subspace spanned by the vectors β_j^i, as i runs over those indices for which $d_i \geq j$. The T-annihilator of α_j is

$$p_j = \prod_{d_i \geq j} (x - c_i). \tag{7-23}$$

We have

$$V = Z(\alpha_1; T) \oplus \cdots \oplus Z(\alpha_r; T)$$

because each β_j^i belongs to one and only one of the subspaces $Z(\alpha_1; T), \ldots, Z(\alpha_r; T)$ and $\mathcal{B} = (\mathcal{B}_1, \ldots, \mathcal{B}_k)$ is a basis for V. By (7-23), p_{j+1} divides p_j.

Exercises

1. Let T be the linear operator on F^2 which is represented in the standard ordered basis by the matrix

$$\begin{bmatrix} 0 & 0 \\ 1 & 0 \end{bmatrix}.$$

Let $\alpha_1 = (0, 1)$. Show that $F^2 \neq Z(\alpha_1; T)$ and that there is no non-zero vector α_2 in F^2 with $Z(\alpha_2; T)$ disjoint from $Z(\alpha_1; T)$.

2. Let T be a linear operator on the finite-dimensional space V, and let R be the range of T.

(a) Prove that R has a complementary T-invariant subspace if and only if R is independent of the null space N of T.

(b) If R and N are independent, prove that N is the unique T-invariant subspace complementary to R.

3. Let T be the linear operator on R^3 which is represented in the standard ordered basis by the matrix

$$\begin{bmatrix} 2 & 0 & 0 \\ 1 & 2 & 0 \\ 0 & 0 & 3 \end{bmatrix}.$$

Let W be the null space of $T - 2I$. Prove that W has no complementary T-invariant subspace. (*Hint:* Let $\beta = \epsilon_1$ and observe that $(T - 2I)\beta$ is in W. Prove there is no α in W with $(T - 2I)\beta = (T - 2I)\alpha$.)

4. Let T be the linear operator on F^4 which is represented in the standard ordered basis by the matrix

$$\begin{bmatrix} c & 0 & 0 & 0 \\ 1 & c & 0 & 0 \\ 0 & 1 & c & 0 \\ 0 & 0 & 1 & c \end{bmatrix}.$$

Let W be the null space of $T - cI$.

(a) Prove that W is the subspace spanned by ϵ_4.

(b) Find the monic generators of the ideals $S(\epsilon_4; W)$, $S(\epsilon_3; W)$, $S(\epsilon_2; W)$, $S(\epsilon_1; W)$.

5. Let T be a linear operator on the vector space V over the field F. If f is a polynomial over F and α is in V, let $f\alpha = f(T)\alpha$. If $V_1, \ldots, V_k$ are T-invariant subspaces and $V = V_1 \oplus \cdots \oplus V_k$, show that

$$fV = fV_1 \oplus \cdots \oplus fV_k.$$

6. Let T, V, and F be as in Exercise 5. Suppose α and β are vectors in V which have the same T-annihilator. Prove that, for any polynomial f, the vectors $f\alpha$ and $f\beta$ have the same T-annihilator.

7. Find the minimal polynomials and the rational forms of each of the following real matrices.

$$\begin{bmatrix} 0 & -1 & -1 \\ 1 & 0 & 0 \\ -1 & 0 & 0 \end{bmatrix}, \quad \begin{bmatrix} c & 0 & -1 \\ 0 & c & 1 \\ -1 & 1 & c \end{bmatrix}, \quad \begin{bmatrix} \cos\theta & \sin\theta \\ -\sin\theta & \cos\theta \end{bmatrix}.$$

8. Let T be the linear operator on R^3 which is represented in the standard ordered basis by

$$\begin{bmatrix} 3 & -4 & -4 \\ -1 & 3 & 2 \\ 2 & -4 & -3 \end{bmatrix}.$$

Find non-zero vectors $\alpha_1, \ldots, \alpha_r$ satisfying the conditions of Theorem 3.

9. Let A be the real matrix

$$A = \begin{bmatrix} 1 & 3 & 3 \\ 3 & 1 & 3 \\ -3 & -3 & -5 \end{bmatrix}.$$

Find an invertible 3×3 real matrix P such that $P^{-1}AP$ is in rational form.

10. Let F be a subfield of the complex numbers and let T be the linear operator on F^4 which is represented in the standard ordered basis by the matrix

$$\begin{bmatrix} 2 & 0 & 0 & 0 \\ 1 & 2 & 0 & 0 \\ 0 & a & 2 & 0 \\ 0 & 0 & b & 2 \end{bmatrix}.$$

Find the characteristic polynomial for T. Consider the cases $a = b = 1$; $a = b = 0$; $a = 0$, $b = 1$. In each of these cases, find the minimal polynomial for T and non-zero vectors $\alpha_1, \ldots, \alpha_r$ which satisfy the conditions of Theorem 3.

11. Prove that if A and B are 3×3 matrices over the field F, a necessary and sufficient condition that A and B be similar over F is that they have the same characteristic polynomial and the same minimal polynomial. Give an example which shows that this is false for 4×4 matrices.

12. Let F be a subfield of the field of complex numbers, and let A and B be $n \times n$ matrices over F. Prove that if A and B are similar over the field of complex numbers, then they are similar over F. (*Hint:* Prove that the rational form of A is the same whether A is viewed as a matrix over F or a matrix over C; likewise for B.)

13. Let A be an $n \times n$ matrix with complex entries. Prove that if every characteristic value of A is real, then A is similar to a matrix with real entries.

14. Let T be a linear operator on the finite-dimensional space V. Prove that there exists a vector α in V with this property. If f is a polynomial and $f(T)\alpha = 0$, then $f(T) = 0$. (Such a vector α is called a **separating vector** for the algebra of polynomials in T.) When T has a cyclic vector, give a direct proof that any cyclic vector is a separating vector for the algebra of polynomials in T.

15. Let F be a subfield of the field of complex numbers, and let A be an $n \times n$ matrix over F. Let p be the minimal polynomial for A. If we regard A as a matrix over C, then A has a minimal polynomial f as an $n \times n$ matrix over C. Use a theorem on linear equations to prove $p = f$. Can you also see how this follows from the cyclic decomposition theorem?

16. Let A be an $n \times n$ matrix with *real* entries such that $A^2 + I = 0$. Prove that n is even, and if $n = 2k$, then A is similar over the field of real numbers to a matrix of the block form

$$\begin{bmatrix} 0 & -I \\ I & 0 \end{bmatrix}$$

where I is the $k \times k$ identity matrix.

17. Let T be a linear operator on a finite-dimensional vector space V. Suppose that

(a) the minimal polynomial for T is a power of an irreducible polynomial;

(b) the minimal polynomial is equal to the characteristic polynomial.

Show that no non-trivial T-invariant subspace has a complementary T-invariant subspace.

18. If T is a diagonalizable linear operator, then every T-invariant subspace has a complementary T-invariant subspace.

19. Let T be a linear operator on the finite-dimensional space V. Prove that T has a cyclic vector if and only if the following is true: Every linear operator U which commutes with T is a polynomial in T.

20. Let V be a finite-dimensional vector space over the field F, and let T be a linear operator on V. We ask when it is true that every non-zero vector in V is a cyclic vector for T. Prove that this is the case if and only if the characteristic polynomial for T is irreducible over F.

21. Let A be an $n \times n$ matrix with *real* entries. Let T be the linear operator on R^n which is represented by A in the standard ordered basis, and let U be the linear operator on C^n which is represented by A in the standard ordered basis. Use the result of Exercise 20 to prove the following: If the only subspaces invariant under T are R^n and the zero subspace, then U is diagonalizable.

7.3. The Jordan Form

Suppose that N is a nilpotent linear operator on the finite-dimensional space V. Let us look at the cyclic decomposition for N which we obtain from Theorem 3. We have a positive integer r and r non-zero vectors $\alpha_1, \ldots, \alpha_r$ in V with N-annihilators $p_1, \ldots, p_r$, such that

$$V = Z(\alpha_1; N) \oplus \cdots \oplus Z(\alpha_r; N)$$

and p_{i+1} divides p_i for $i = 1, \ldots, r - 1$. Since N is nilpotent, the minimal polynomial is x^k for some $k \leq n$. Thus each p_i is of the form $p_i = x^{k_i}$, and the divisibility condition simply says that

$$k_1 \geq k_2 \geq \cdots \geq k_r.$$

Of course, $k_1 = k$ and $k_r \geq 1$. The companion matrix of x^{k_i} is the $k_i \times k_i$ matrix

$$A_i = \begin{bmatrix} 0 & 0 & \cdots & 0 & 0 \\ 1 & 0 & \cdots & 0 & 0 \\ 0 & 1 & \cdots & 0 & 0 \\ \vdots & \vdots & & \vdots & \vdots \\ 0 & 0 & \cdots & 1 & 0 \end{bmatrix}. \tag{7-24}$$

Thus Theorem 3 gives us an ordered basis for V in which the matrix of N is the direct sum of the elementary nilpotent matrices (7-24), the sizes of which decrease as i increases. One sees from this that associated with a nilpotent $n \times n$ matrix is a positive integer r and r positive integers $k_1, \ldots, k_r$ such that $k_1 + \cdots + k_r = n$ and $k_i \geq k_{i+1}$, and these positive integers determine the rational form of the matrix, i.e., determine the matrix up to similarity.

Here is one thing we should like to point out about the nilpotent operator N above. The positive integer r is precisely the nullity of N; in fact, the null space has as a basis the r vectors

$$N^{k_i - 1}\alpha_i. \tag{7-25}$$

For, let α be in the null space of N. We write α in the form

$$\alpha = f_1\alpha_1 + \cdots + f_r\alpha_r$$

where f_i is a polynomial, the degree of which we may assume is less than k_i. Since $N\alpha = 0$, for each i we have

$$\begin{aligned} 0 &= N(f_i\alpha_i) \\ &= Nf_i(N)\alpha_i \\ &= (xf_i)\alpha_i. \end{aligned}$$

Thus xf_i is divisible by x^{k_i}, and since deg $(f_i) > k_i$ this means that

$$f_i = c_i x^{k_i - 1}$$

where c_i is some scalar. But then

$$\alpha = c_1(x^{k_1-1}\alpha_1) + \cdots + c_r(x^{k_r-1}\alpha_r)$$

which shows us that the vectors (7-25) form a basis for the null space of N. The reader should note that this fact is also quite clear from the matrix point of view.

Now what we wish to do is to combine our findings about nilpotent operators or matrices with the primary decomposition theorem of Chapter 6. The situation is this: Suppose that T is a linear operator on V and that the characteristic polynomial for T factors over F as follows:

$$f = (x - c_1)^{d_1} \cdots (x - c_k)^{d_k}$$

where $c_1, \ldots, c_k$ are distinct elements of F and $d_i \geq 1$. Then the minimal polynomial for T will be

$$p = (x - c_1)^{r_1} \cdots (x - c_k)^{r_k}$$

where $1 \leq r_i \leq d_i$. If W_i is the null space of $(T - c_iI)^{r_i}$, then the primary decomposition theorem tells us that

$$V = W_1 \oplus \cdots \oplus W_k$$

and that the operator T_i induced on W_i by T has minimal polynomial $(x - c_i)^{r_i}$. Let N_i be the linear operator on W_i defined by $N_i = T_i - c_iI$. Then N_i is nilpotent and has minimal polynomial x^{r_i}. On W_i, T acts like N_i plus the scalar c_i times the identity operator. Suppose we choose a basis for the subspace W_i corresponding to the cyclic decomposition for the nilpotent operator N_i. Then the matrix of T_i in this ordered basis will be the direct sum of matrices

$$\begin{bmatrix} c & 0 & \cdots & 0 & 0 \\ 1 & c & \cdots & 0 & 0 \\ \vdots & \vdots & & \vdots & \vdots \\ & & & c & \\ 0 & 0 & \cdots & 1 & c \end{bmatrix} \tag{7-26}$$

each with $c = c_i$. Furthermore, the sizes of these matrices will decrease as one reads from left to right. A matrix of the form (7-26) is called an **elementary Jordan matrix with characteristic value** c. Now if we put all the bases for the W_i together, we obtain an ordered basis for V. Let us describe the matrix A of T in this ordered basis.

The matrix A is the direct sum

$$A = \begin{bmatrix} A_1 & 0 & \cdots & 0 \\ 0 & A_2 & \cdots & 0 \\ \vdots & \vdots & & \vdots \\ 0 & 0 & \cdots & A_k \end{bmatrix} \tag{7-27}$$

of matrices $A_1, \ldots, A_k$. Each A_i is of the form

$$A_i = \begin{bmatrix} J_1^{(i)} & 0 & \cdots & 0 \\ 0 & J_2^{(i)} & \cdots & 0 \\ \vdots & \vdots & & \vdots \\ 0 & 0 & \cdots & J_{n_i}^{(i)} \end{bmatrix}$$

where each $J_j^{(i)}$ is an elementary Jordan matrix with characteristic value c_i. Also, within each A_i, the sizes of the matrices $J_j^{(i)}$ decrease as j increases. An $n \times n$ matrix A which satisfies all the conditions described so far in this paragraph (for some distinct scalars $c_1, \ldots, c_k$) will be said to be in **Jordan form.**

We have just pointed out that if T is a linear operator for which the characteristic polynomial factors completely over the scalar field, then there is an ordered basis for V in which T is represented by a matrix which is in Jordan form. We should like to show now that this matrix is something uniquely associated with T, up to the order in which the characteristic values of T are written down. In other words, if two matrices are in Jordan form and they are similar, then they can differ only in that the order of the scalars c_i is different.

The uniqueness we see as follows. Suppose there is some ordered basis for V in which T is represented by the Jordan matrix A described in the previous paragraph. If A_i is a $d_i \times d_i$ matrix, then d_i is clearly the multiplicity of c_i as a root of the characteristic polynomial for A, or for T. In other words, the characteristic polynomial for T is

$$f = (x - c_1)^{d_1} \cdots (x - c_k)^{d_k}.$$

This shows that $c_1, \ldots, c_k$ and $d_1, \ldots, d_k$ are unique, up to the order in which we write them. The fact that A is the direct sum of the matrices A_i gives us a direct sum decomposition $V = W_1 \oplus \cdots \oplus W_k$ invariant under T. Now note that W_i must be the null space of $(T - c_iI)^n$, where $n = \dim V$; for, $A_i - c_iI$ is clearly nilpotent and $A_j - c_iI$ is non-singular for $j \neq i$. So we see that the subspaces W_i are unique. If T_i is the operator induced on W_i by T, then the matrix A_i is uniquely determined as the rational form for $(T_i - c_iI)$.

Now we wish to make some further observations about the operator T and the Jordan matrix A which represents T in some ordered basis. We shall list a string of observations:

(1) Every entry of A not on or immediately below the main diagonal

is 0. On the diagonal of A occur the k distinct characteristic values $c_1, \ldots, c_k$ of T. Also, c_i is repeated d_i times, where d_i is the multiplicity of c_i as a root of the characteristic polynomial, i.e., $d_i = \dim W_i$.

(2) For each i, the matrix A_i is the direct sum of n_i elementary Jordan matrices $J_j^{(i)}$ with characteristic value c_i. The number n_i is precisely the dimension of the space of characteristic vectors associated with the characteristic value c_i. For, n_i is the number of elementary nilpotent blocks in the rational form for $(T_i - c_iI)$, and is thus equal to the dimension of the null space of $(T - c_iI)$. In particular notice that T is diagonalizable if and only if $n_i = d_i$ for each i.

(3) For each i, the first block $J_1^{(i)}$ in the matrix A_i is an $r_i \times r_i$ matrix, where r_i is the multiplicity of c_i as a root of the *minimal* polynomial for T. This follows from the fact that the minimal polynomial for the nilpotent operator $(T_i - c_iI)$ is x^{r_i}.

Of course we have as usual the straight matrix result. If B is an $n \times n$ matrix over the field F and if the characteristic polynomial for B factors completely over F, then B is similar over F to an $n \times n$ matrix A in Jordan form, and A is unique up to a rearrangement of the order of its characteristic values. We call A the **Jordan form** of B.

Also, note that if F is an algebraically closed field, then the above remarks apply to every linear operator on a finite-dimensional space over F, or to every $n \times n$ matrix over F. Thus, for example, every $n \times n$ matrix over the field of complex numbers is similar to an essentially unique matrix in Jordan form.

EXAMPLE 5. Suppose T is a linear operator on C^2. The characteristic polynomial for T is either $(x - c_1)(x - c_2)$ where c_1 and c_2 are distinct complex numbers, or is $(x - c)^2$. In the former case, T is diagonalizable and is represented in some ordered basis by

$$\begin{bmatrix} c_1 & 0 \\ 0 & c_2 \end{bmatrix}.$$

In the latter case, the minimal polynomial for T may be $(x - c)$, in which case $T = cI$, or may be $(x - c)^2$, in which case T is represented in some ordered basis by the matrix

$$\begin{bmatrix} c & 0 \\ 1 & c \end{bmatrix}.$$

Thus every 2×2 matrix over the field of complex numbers is similar to a matrix of one of the two types displayed above, possibly with $c_1 = c_2$.

EXAMPLE 6. Let A be the complex 3×3 matrix

$$A = \begin{bmatrix} 2 & 0 & 0 \\ a & 2 & 0 \\ b & c & -1 \end{bmatrix}.$$

The characteristic polynomial for A is obviously $(x-2)^2(x+1)$. Either this is the minimal polynomial, in which case A is similar to

$$\begin{bmatrix} 2 & 0 & 0 \\ 1 & 2 & 0 \\ 0 & 0 & -1 \end{bmatrix}$$

or the minimal polynomial is $(x-2)(x+1)$, in which case A is similar to

$$\begin{bmatrix} 2 & 0 & 0 \\ 0 & 2 & 0 \\ 0 & 0 & -1 \end{bmatrix}.$$

Now

$$(A-2I)(A+I) = \begin{bmatrix} 0 & 0 & 0 \\ 3a & 0 & 0 \\ ac & 0 & 0 \end{bmatrix}$$

and thus A is similar to a diagonal matrix if and only if $a = 0$.

EXAMPLE 7. Let

$$A = \begin{bmatrix} 2 & 0 & 0 & 0 \\ 1 & 2 & 0 & 0 \\ 0 & 0 & 2 & 0 \\ 0 & 0 & a & 2 \end{bmatrix}.$$

The characteristic polynomial for A is $(x-2)^4$. Since A is the direct sum of two 2×2 matrices, it is clear that the minimal polynomial for A is $(x-2)^2$. Now if $a = 0$ or if $a = 1$, then the matrix A is in Jordan form. Notice that the two matrices we obtain for $a = 0$ and $a = 1$ have the same characteristic polynomial and the same minimal polynomial, but are not similar. They are not similar because for the first matrix the solution space of $(A - 2I)$ has dimension 3, while for the second matrix it has dimension 2.

EXAMPLE 8. Linear differential equations with constant coefficients (Example 14, Chapter 6) provide a nice illustration of the Jordan form. Let $a_0, \ldots, a_{n-1}$ be complex numbers and let V be the space of all n times differentiable functions f on an interval of the real line which satisfy the differential equation

$$\frac{d^n f}{dx^n} + a_{n-1}\frac{d^{n-1}f}{dx^{n-1}} + \cdots + a_1\frac{df}{dx} + a_0 f = 0.$$

Let D be the differentiation operator. Then V is invariant under D, because V is the null space of $p(D)$, where

$$p = x^n + \cdots + a_1 x + a_0.$$

What is the Jordan form for the differentiation operator on V?

Let $c_1, \ldots, c_k$ be the distinct complex roots of p:

$$p = (x - c_1)^{r_1} \cdots (x - c_k)^{r_k}.$$

Let V_i be the null space of $(D - c_iI)^{r_i}$, that is, the set of solutions to the differential equation

$$(D - c_iI)^{r_i}f = 0.$$

Then as we noted in Example 15, Chapter 6 the primary decomposition theorem tells us that

$$V = V_1 \oplus \cdots \oplus V_k.$$

Let N_i be the restriction of $D - c_iI$ to V_i. The Jordan form for the operator D (on V) is then determined by the rational forms for the nilpotent operators $N_1, \ldots, N_k$ on the spaces $V_1, \ldots, V_k$.

So, what we must know (for various values of c) is the rational form for the operator $N = (D - cI)$ on the space V_c, which consists of the solutions of the equation

$$(D - cI)^r f = 0.$$

How many elementary nilpotent blocks will there be in the rational form for N? The number will be the nullity of N, i.e., the dimension of the characteristic space associated with the characteristic value c. That dimension is 1, because any function which satisfies the differential equation

$$Df = cf$$

is a scalar multiple of the exponential function $h(x) = e^{cx}$. Therefore, the operator N (on the space V_c) has a cyclic vector. A good choice for a cyclic vector is $g = x^{r-1}h$:

$$g(x) = x^{r-1}e^{cx}.$$

This gives

$$\begin{aligned} Ng &= (r-1)x^{r-2}h \\ &\vdots \\ N^{r-1}g &= (r-1)!h \end{aligned}$$

The preceding paragraph shows us that the Jordan form for D (on the space V) is the direct sum of k elementary Jordan matrices, one for each root c_i.

Exercises

1. Let N_1 and N_2 be 3×3 nilpotent matrices over the field F. Prove that N_1 and N_2 are similar if and only if they have the same minimal polynomial.

2. Use the result of Exercise 1 and the Jordan form to prove the following: Let

A and B be $n \times n$ matrices over the field F which have the *same* characteristic polynomial

$$f = (x - c_1)^{d_1} \cdots (x - c_k)^{d_k}$$

and the same minimal polynomial. If no d_i is greater than 3, then A and B are similar.

3. If A is a complex 5×5 matrix with characteristic polynomial

$$f = (x - 2)^3(x + 7)^2$$

and minimal polynomial $p = (x - 2)^2(x + 7)$, what is the Jordan form for A?

4. How many possible Jordan forms are there for a 6×6 complex matrix with characteristic polynomial $(x + 2)^4(x - 1)^2$?

5. The differentiation operator on the space of polynomials of degree less than or equal to 3 is represented in the 'natural' ordered basis by the matrix

$$\begin{bmatrix} 0 & 1 & 0 & 0 \\ 0 & 0 & 2 & 0 \\ 0 & 0 & 0 & 3 \\ 0 & 0 & 0 & 0 \end{bmatrix}.$$

What is the Jordan form of this matrix? (F a subfield of the complex numbers.)

6. Let A be the complex matrix

$$\begin{bmatrix} 2 & 0 & 0 & 0 & 0 & 0 \\ 1 & 2 & 0 & 0 & 0 & 0 \\ -1 & 0 & 2 & 0 & 0 & 0 \\ 0 & 1 & 0 & 2 & 0 & 0 \\ 1 & 1 & 1 & 1 & 2 & 0 \\ 0 & 0 & 0 & 0 & 1 & -1 \end{bmatrix}.$$

Find the Jordan form for A.

7. If A is an $n \times n$ matrix over the field F with characteristic polynomial

$$f = (x - c_1)^{d_1} \cdots (x - c_k)^{d_k}$$

what is the trace of A?

8. Classify up to similarity all 3×3 complex matrices A such that $A^3 = I$.

9. Classify up to similarity all $n \times n$ complex matrices A such that $A^n = I$.

10. Let n be a positive integer, $n \geq 2$, and let N be an $n \times n$ matrix over the field F such that $N^n = 0$ but $N^{n-1} \neq 0$. Prove that N has no square root, i.e., that there is no $n \times n$ matrix A such that $A^2 = N$.

11. Let N_1 and N_2 be 6×6 nilpotent matrices over the field F. Suppose that N_1 and N_2 have the same minimal polynomial and the same nullity. Prove that N_1 and N_2 are similar. Show that this is not true for 7×7 nilpotent matrices.

12. Use the result of Exercise 11 and the Jordan form to prove the following: Let A and B be $n \times n$ matrices over the field F which have the same characteristic polynomial

$$f = (x - c_1)^{d_1} \cdots (x - c_k)^{d_k}$$

and the same minimal polynomial. Suppose also that for each i the solution spaces of $(A - c_iI)$ and $(B - c_iI)$ have the same dimension. If no d_i is greater than 6, then A and B are similar.

13. If N is a $k \times k$ elementary nilpotent matrix, i.e., $N^k = 0$ but $N^{k-1} \neq 0$, show that N^t is similar to N. Now use the Jordan form to prove that every complex $n \times n$ matrix is similar to its transpose.

14. What's wrong with the following proof? If A is a complex $n \times n$ matrix such that $A^t = -A$, then A is 0. (*Proof:* Let J be the Jordan form of A. Since $A^t = -A$, $J^t = -J$. But J is triangular so that $J^t = -J$ implies that every entry of J is zero. Since $J = 0$ and A is similar to J, we see that $A = 0$.) (Give an example of a non-zero A such that $A^t = -A$.)

15. If N is a nilpotent 3×3 matrix over C, prove that $A = I + \frac{1}{2}N - \frac{1}{8}N^2$ satisfies $A^2 = I + N$, i.e., A is a square root of $I + N$. Use the binomial series for $(1 + t)^{1/2}$ to obtain a similar formula for a square root of $I + N$, where N is any nilpotent $n \times n$ matrix over C.

16. Use the result of Exercise 15 to prove that if c is a non-zero complex number and N is a nilpotent complex matrix, then $(cI + N)$ has a square root. Now use the Jordan form to prove that every non-singular complex $n \times n$ matrix has a square root.

7.4. Computation of Invariant Factors

Suppose that A is an $n \times n$ matrix with entries in the field F. We wish to find a method for computing the invariant factors $p_1, \ldots, p_r$ which define the rational form for A. Let us begin with the very simple case in which A is the companion matrix (7.2) of a monic polynomial

$$p = x^n + c_{n-1}x^{n-1} + \cdots + c_1x + c_0.$$

In Section 7.1 we saw that p is both the minimal and the characteristic polynomial for the companion matrix A. Now, we want to give a direct calculation which shows that p is the characteristic polynomial for A. In this case,

$$xI - A = \begin{bmatrix} x & 0 & 0 & \cdots & 0 & c_0 \\ -1 & x & 0 & \cdots & 0 & c_1 \\ 0 & -1 & x & \cdots & 0 & c_2 \\ \vdots & \vdots & \vdots & & \vdots & \vdots \\ 0 & 0 & 0 & \cdots & x & c_{n-2} \\ 0 & 0 & 0 & \cdots & -1 & x + c_{n-1} \end{bmatrix}.$$

Add x times row n to row $(n - 1)$. This will remove the x in the $(n - 1, n - 1)$ place and it will not change the determinant. Then, add x times the new row $(n - 1)$ to row $(n - 2)$. Continue successively until all of the x's on the main diagonal have been removed by that process. The result is the matrix

$$\begin{bmatrix} 0 & 0 & 0 & \cdots & 0 & x^n + \cdots + c_1x + c_0 \\ -1 & 0 & 0 & \cdots & 0 & x^{n-1} + \cdots + c_2x + c_1 \\ 0 & -1 & 0 & \cdots & 0 & x^{n-2} + \cdots + c_3x + c_2 \\ \vdots & \vdots & \vdots & & \vdots & \vdots \\ 0 & 0 & 0 & \cdots & 0 & x^2 + c_{n-1}x + c_{n-2} \\ 0 & 0 & 0 & \cdots & -1 & x + c_{n-1} \end{bmatrix}$$

which has the same determinant as $xI - A$. The upper right-hand entry of this matrix is the polynomial p. We clean up the last column by adding to it appropriate multiples of the other columns:

$$\begin{bmatrix} 0 & 0 & 0 & \cdots & 0 & p \\ -1 & 0 & 0 & \cdots & 0 & 0 \\ 0 & -1 & 0 & \cdots & 0 & 0 \\ \vdots & \vdots & \vdots & & \vdots & \vdots \\ 0 & 0 & 0 & \cdots & 0 & 0 \\ 0 & 0 & 0 & \cdots & -1 & 0 \end{bmatrix}$$

Multiply each of the first $(n - 1)$ columns by -1 and then perform $(n - 1)$ interchanges of adjacent columns to bring the present column n to the first position. The total effect of the $2n - 2$ sign changes is to leave the determinant unaltered. We obtain the matrix

$$\begin{bmatrix} p & 0 & 0 & \cdots & 0 \\ 0 & 1 & 0 & \cdots & 0 \\ 0 & 0 & 1 & \cdots & 0 \\ \vdots & \vdots & \vdots & & \vdots \\ 0 & 0 & 0 & \cdots & 1 \end{bmatrix}. \tag{7-28}$$

It is then clear that $p = \det (xI - A)$.

We are going to show that, for any $n \times n$ matrix A, there is a succession of row and column operations which will transform $xI - A$ into a matrix much like (7-28), in which the invariant factors of A appear down the main diagonal. Let us be completely clear about the operations we shall use.

We shall be concerned with $F[x]^{m \times n}$, the collection of $m \times n$ matrices with entries which are polynomials over the field F. If M is such a matrix, an **elementary row operation** on M is one of the following

1. multiplication of one row of M by a non-zero scalar in F;
2. replacement of the rth row of M by row r plus f times row s, where f is any polynomial over F and $r \neq s$;
3. interchange of two rows of M.

The inverse operation of an elementary row operation is an elementary row operation of the same type. Notice that we could not make such an assertion if we allowed non-scalar polynomials in (1). An $m \times m$ **ele-**

mentary matrix, that is, an elementary matrix in $F[x]^{m\times m}$, is one which can be obtained from the $m \times m$ identity matrix by means of a single elementary row operation. Clearly each elementary row operation on M can be effected by multiplying M on the left by a suitable $m \times m$ elementary matrix; in fact, if e is the operation, then

$$e(M) = e(I)M.$$

Let M, N be matrices in $F[x]^{m\times n}$. We say that N is **row-equivalent** to M if N can be obtained from M by a finite succession of elementary row operations:

$$M = M_0 \rightarrow M_1 \rightarrow \cdots \rightarrow M_k = N.$$

Evidently N is row-equivalent to M if and only if M is row-equivalent to N, so that we may use the terminology 'M and N are row-equivalent.' If N is row-equivalent to M, then

$$N = PM$$

where the $m \times m$ matrix P is a product of elementary matrices:

$$P = E_1 \cdots E_k.$$

In particular, P is an invertible matrix with inverse

$$P^{-1} = E_k^{-1} \cdots E_1^{-1}.$$

Of course, the inverse of E_j comes from the inverse elementary row operation.

All of this is just as it is in the case of matrices with entries in F. It parallels the elementary results in Chapter 1. Thus, the next problem which suggests itself is to introduce a row-reduced echelon form for polynomial matrices. Here, we meet a new obstacle. How do we row-reduce a matrix? The first step is to single out the leading non-zero entry of row 1 and to divide every entry of row 1 by that entry. We cannot (necessarily) do that when the matrix has polynomial entries. As we shall see in the next theorem, we can circumvent this difficulty in certain cases; however, there is not any entirely suitable row-reduced form for the general matrix in $F[x]^{m\times n}$. If we introduce column operations as well and study the type of equivalence which results from allowing the use of both types of operations, we can obtain a very useful standard form for each matrix. The basic tool is the following.

Lemma. *Let* M *be a matrix in* $F[x]^{m\times n}$ *which has some non-zero entry in its first column, and let* p *be the greatest common divisor of the entries in column* 1 *of* M. *Then* M *is row-equivalent to a matrix* N *which has*

$$\begin{bmatrix} p \\ 0 \\ \vdots \\ 0 \end{bmatrix}$$

as its first column.

Proof. We shall prove something more than we have stated. We shall show that there is an algorithm for finding N, i.e., a prescription which a machine could use to calculate N in a finite number of steps. First, we need some notation.

Let M be any $m \times n$ matrix with entries in $F[x]$ which has a non-zero first column

$$M_1 = \begin{bmatrix} f_1 \\ \vdots \\ f_m \end{bmatrix}.$$

Define

$$l(M_1) = \min_{f_i \neq 0} \deg f_i \tag{7-29}$$
$$p(M_1) = \text{g.c.d.}\,(f_1, \ldots, f_m).$$

Let j be some index such that $\deg f_j = l(M_1)$. To be specific, let j be the smallest index i for which $\deg f_i = l(M_1)$. Attempt to divide each f_i by f_j:

$$f_i = f_j g_i + r_i, \qquad r_i = 0 \quad \text{or} \quad \deg r_i < \deg f_j. \tag{7-30}$$

For each i different from j, replace row i of M by row i minus g_i times row j. Multiply row j by the reciprocal of the leading coefficient of f_j and then interchange rows j and 1. The result of all these operations is a matrix M' which has for its first column

$$M_1' = \begin{bmatrix} \tilde{f}_j \\ r_2 \\ \vdots \\ r_{j-1} \\ r_1 \\ r_{j+1} \\ \vdots \\ r_m \end{bmatrix}. \tag{7-31}$$

where $\tilde{f}_j$ is the monic polynomial obtained by normalizing f_j to have leading coefficient 1. We have given a well-defined procedure for associating with each M a matrix M' with these properties.

(a) M' is row-equivalent to M.
(b) $p(M_1') = p(M_1)$.
(c) Either $l(M_1') < l(M_1)$ or

$$M_1' = \begin{bmatrix} p(M_1) \\ 0 \\ \vdots \\ 0 \end{bmatrix}.$$

It is easy to verify (b) and (c) from (7-30) and (7-31). Property (c) is just another way of stating that either there is some i such that $r_i \neq 0$ and $\deg r_i < \deg f_j$ or else $r_i = 0$ for all i and f_j is (therefore) the greatest common divisor of $f_1, \ldots, f_m$.

The proof of the lemma is now quite simple. We start with the matrix M and apply the above procedure to obtain M'. Property (c) tells us that either M' will serve as the matrix N in the lemma or $l(M_1') < l(M_1)$. In the latter case, we apply the procedure to M' to obtain the matrix $M^{(2)} = (M')'$. If $M^{(2)}$ is not a suitable N, we form $M^{(3)} = (M^{(2)})'$, and so on. The point is that the strict inequalities

$$l(M_1) > l(M_1') > l(M_1^{(2)}) > \cdots$$

cannot continue for very long. After not more than $l(M_1)$ iterations of our procedure, we must arrive at a matrix $M^{(k)}$ which has the properties we seek. ∎

Theorem 6. *Let* P *be an* m × m *matrix with entries in the polynomial algebra* F[x]. *The following are equivalent.*

(i) P *is invertible.*
(ii) *The determinant of* P *is a non-zero scalar polynomial.*
(iii) P *is row-equivalent to the* m × m *identity matrix.*
(iv) P *is a product of elementary matrices.*

Proof. Certainly (i) implies (ii) because the determinant function is multiplicative and the only polynomials invertible in $F[x]$ are the non-zero scalar ones. As a matter of fact, in Chapter 5 we used the classical adjoint to show that (i) and (ii) are equivalent. Our argument here provides a different proof that (i) follows from (ii). We shall complete the merry-go-round

$$\begin{array}{ccc} \text{(i)} & \rightarrow & \text{(ii)} \\ \uparrow & & \downarrow \\ \text{(iv)} & \leftarrow & \text{(iii).} \end{array}$$

The only implication which is not obvious is that (iii) follows from (ii).

Assume (ii) and consider the first column of P. It contains certain polynomials $p_1, \ldots, p_m$, and

$$\text{g.c.d.}\ (p_1, \ldots, p_m) = 1$$

because any common divisor of $p_1, \ldots, p_m$ must divide (the scalar) $\det P$. Apply the previous lemma to P to obtain a matrix

$$Q = \begin{bmatrix} 1 & a_2 & \cdots & a_m \\ 0 & & & \\ \vdots & & B & \\ 0 & & & \end{bmatrix} \tag{7-32}$$

which is row-equivalent to P. An elementary row operation changes the determinant of a matrix by (at most) a non-zero scalar factor. Thus $\det Q$

is a non-zero scalar polynomial. Evidently the $(m-1) \times (m-1)$ matrix B in (7-32) has the same determinant as does Q. Therefore, we may apply the last lemma to B. If we continue this way for m steps, we obtain an upper-triangular matrix

$$R = \begin{bmatrix} 1 & a_2 & \cdots & a_m \\ 0 & 1 & \cdots & b_m \\ \vdots & \vdots & & \vdots \\ 0 & 0 & \cdots & 1 \end{bmatrix}$$

which is row-equivalent to R. Obviously R is row-equivalent to the $m \times m$ identity matrix. ∎

Corollary. *Let M and N be m × n matrices with entries in the polynomial algebra F[x]. Then N is row-equivalent to M if and only if*

$$N = PM$$

where P is an invertible m × m matrix with entries in F[x].

We now define **elementary column operations** and **column-equivalence** in a manner analogous to row operations and row-equivalence. We do not need a new concept of elementary matrix because the class of matrices which can be obtained by performing one elementary column operation on the identity matrix is the same as the class obtained by using a single elementary row operation.

Definition. *The matrix N is* **equivalent** *to the matrix M if we can pass from M to N by means of a sequence of operations*

$$M = M_0 \rightarrow M_1 \rightarrow \cdots \rightarrow M_k = N$$

each of which is an elementary row operation or an elementary column operation.

Theorem 7. *Let M and N be m × n matrices with entries in the polynomial algebra F[x]. Then N is equivalent to M if and only if*

$$N = PMQ$$

where P is an invertible matrix in $F[x]^{m \times m}$ and Q is an invertible matrix in $F[x]^{n \times n}$.

Theorem 8. *Let A be an n × n matrix with entries in the field F, and let $p_1, \ldots, p_r$ be the invariant factors for A. The matrix $xI - A$ is equivalent to the n × n diagonal matrix with diagonal entries $p_1, \ldots, p_r$, 1, 1, . . . , 1.*

Proof. There exists an invertible $n \times n$ matrix P, with entries in F, such that PAP^{-1} is in rational form, that is, has the block form

$$PAP^{-1} = \begin{bmatrix} A_1 & 0 & \cdots & 0 \\ 0 & A_2 & \cdots & 0 \\ \vdots & \vdots & \ddots & \vdots \\ 0 & 0 & \cdots & A_r \end{bmatrix}$$

where A_i is the companion matrix of the polynomial p_i. According to Theorem 7, the matrix

$$P(xI - A)P^{-1} = xI - PAP^{-1} \tag{7-33}$$

is equivalent to $xI - A$. Now

$$xI - PAP^{-1} = \begin{bmatrix} xI - A_1 & 0 & \cdots & 0 \\ 0 & xI - A_2 & \cdots & 0 \\ \vdots & \vdots & & \vdots \\ 0 & 0 & \cdots & xI - A_r \end{bmatrix} \tag{7-34}$$

where the various I's we have used are identity matrices of appropriate sizes. At the beginning of this section, we showed that $xI - A_i$ is equivalent to the matrix

$$\begin{bmatrix} p_i & 0 & \cdots & 0 \\ 0 & 1 & \cdots & 0 \\ \vdots & \vdots & & \vdots \\ 0 & 0 & \cdots & 1 \end{bmatrix}.$$

From (7-33) and (7-34) it is then clear that $xI - A$ is equivalent to a diagonal matrix which has the polynomials p_i and $(n - r)$ 1's on its main diagonal. By a succession of row and column interchanges, we can arrange those diagonal entries in any order we choose, for example: $p_1, \ldots, p_r, 1, \ldots, 1$. ∎

Theorem 8 does not give us an effective way of calculating the elementary divisors $p_1, \ldots, p_r$ because our proof depends upon the cyclic decomposition theorem. We shall now give an explicit algorithm for reducing a polynomial matrix to diagonal form. Theorem 8 suggests that we may also arrange that successive elements on the main diagonal divide one another.

Definition. *Let* N *be a matrix in* $F[x]^{m \times n}$. *We say that* N *is in* (Smith) **normal form** *if*

(a) *every entry off the main diagonal of* N *is* 0;

(b) *on the main diagonal of* N *there appear (in order) polynomials* $f_1, \ldots, f_l$ *such that* f_k *divides* f_{k+1}, $1 \leq k \leq l - 1$.

In the definition, the number l is $l = \min(m, n)$. The main diagonal entries are $f_k = N_{kk}$, $k = 1, \ldots, l$.

Theorem 9. *Let* M *be an* $m \times n$ *matrix with entries in the polynomial algebra* F[x]. *Then* M *is equivalent to a matrix* N *which is in normal form.*

Proof. If $M = 0$, there is nothing to prove. If $M \neq 0$, we shall give an algorithm for finding a matrix M' which is equivalent to M and which has the form

$$M' = \begin{bmatrix} f_1 & 0 & \cdots & 0 \\ 0 & & & \\ \vdots & & R & \\ 0 & & & \end{bmatrix} \tag{7-35}$$

where R is an $(m-1) \times (n-1)$ matrix and f_1 divides every entry of R. We shall then be finished, because we can apply the same procedure to R and obtain f_2, etc.

Let $l(M)$ be the minimum of the degrees of the non-zero entries of M. Find the first column which contains an entry with degree $l(M)$ and interchange that column with column 1. Call the resulting matrix $M^{(0)}$. We describe a procedure for finding a matrix of the form

$$\begin{bmatrix} g & 0 & \cdots & 0 \\ 0 & & & \\ \vdots & & S & \\ 0 & & & \end{bmatrix} \tag{7-36}$$

which is equivalent to $M^{(0)}$. We begin by applying to the matrix $M^{(0)}$ the procedure of the lemma before Theorem 6, a procedure which we shall call PL6. There results a matrix

$$M^{(1)} = \begin{bmatrix} p & a & \cdots & b \\ 0 & c & \cdots & d \\ \vdots & \vdots & & \vdots \\ 0 & e & \cdots & f \end{bmatrix}. \tag{7-37}$$

If the entries $a, \ldots, b$ are all 0, fine. If not, we use the analogue of PL6 for the first row, a procedure which we might call PL6′. The result is a matrix

$$M^{(2)} = \begin{bmatrix} q & 0 & \cdots & 0 \\ a' & c' & \cdots & e' \\ \vdots & \vdots & & \vdots \\ b' & d' & \cdots & f' \end{bmatrix} \tag{7-38}$$

where q is the greatest common divisor of $p, a, \ldots, b$. In producing $M^{(2)}$, we may or may not have disturbed the nice form of column 1. If we did, we can apply PL6 once again. Here is the point. In not more than $l(M)$ steps:

$$M^{(0)} \xrightarrow{\text{PL6}} M^{(1)} \xrightarrow{\text{PL6}'} M^{(2)} \xrightarrow{\text{PL6}} \cdots \longrightarrow M^{(t)}$$

we must arrive at a matrix $M^{(t)}$ which has the form (7-36), because at each successive step we have $l(M^{(k+1)}) < l(M^{(k)})$. We name the process which we have just defined P7-36:

$$M^{(0)} \xrightarrow{\text{P7-36}} M^{(t)}.$$

In (7-36), the polynomial g may or may not divide every entry of S. If it does not, find the first column which has an entry not divisible by g and add that column to column 1. The new first column contains both g and an entry $gh + r$ where $r \neq 0$ and $\deg r < \deg g$. Apply process P7-36 and the result will be another matrix of the form (7-36), where the degree of the corresponding g has decreased.

It should now be obvious that in a finite number of steps we will obtain (7-35), i.e., we will reach a matrix of the form (7-36) where the degree of g cannot be further reduced. ∎

We want to show that the normal form associated with a matrix M is unique. Two things we have seen provide clues as to how the polynomials $f_1, \ldots, f_l$ in Theorem 9 are uniquely determined by M. First, elementary row and column operations do not change the determinant of a square matrix by more than a non-zero scalar factor. Second, elementary row and column operations do not change the greatest common divisor of the entries of a matrix.

Definition. *Let* M *be an* m $\times$ n *matrix with entries in* F[x]. *If* $1 \leq k \leq \min(m, n)$, *we define* $\delta_k(M)$ *to be the greatest common divisor of the determinants of all* k $\times$ k *submatrices of* M.

Recall that a $k \times k$ submatrix of M is one obtained by deleting some $m - k$ rows and some $n - k$ columns of M. In other words, we select certain k-tuples

$$\begin{aligned} I &= (i_1, \ldots, i_k), \qquad 1 \leq i_1 < \cdots < i_k \leq m \\ J &= (j_1, \ldots, j_k), \qquad 1 \leq j_1 < \cdots < j_k \leq n \end{aligned}$$

and look at the matrix formed using those rows and columns of M. We are interested in the determinants

$$D_{I,J}(M) = \det \begin{bmatrix} M_{i_1j_1} & \cdots & M_{i_1j_k} \\ \vdots & & \vdots \\ M_{i_kj_1} & \cdots & M_{i_kj_k} \end{bmatrix}. \tag{7-39}$$

The polynomial $\delta_k(M)$ is the greatest common divisor of the polynomials $D_{I,J}(M)$, as I and J range over the possible k-tuples.

Theorem 10. *If* M *and* N *are equivalent* m $\times$ n *matrices with entries in* F[x], *then*

$$\delta_k(M) = \delta_k(N), \qquad 1 \leq k \leq \min(m, n). \tag{7-40}$$

Proof. It will suffice to show that a single elementary row operation e does not change δ_k. Since the inverse of e is also an elementary row operation, it will suffice to show this: If a polynomial f divides every $D_{I,J}(M)$, then f divides $D_{I,J}(e(M))$ for all k-tuples I and J.

Since we are considering a row operation, let $\alpha_1, \ldots, \alpha_m$ be the rows of M and let us employ the notation

$$D_J(\alpha_{i_1}, \ldots, \alpha_{i_k}) = D_{I,J}(M).$$

Given I and J, what is the relation between $D_{I,J}(M)$ and $D_{I,J}(e(M))$? Consider the three types of operations e:

(a) multiplication of row r by a non-zero scalar c;
(b) replacement of row r by row r plus g times row s, $r \neq s$;
(c) interchange of rows r and s, $r \neq s$.

Forget about type (c) operations for the moment, and concentrate on types (a) and (b), which change only row r. If r is not one of the indices $i_1, \ldots, i_k$, then

$$D_{I,J}(e(M)) = D_{I,J}(M).$$

If r is among the indices $i_1, \ldots, i_k$, then in the two cases we have

$$\begin{aligned}\text{(a)}\quad D_{I,J}(e(M)) &= D_J(\alpha_{i_1}, \ldots, c\alpha_r, \ldots, \alpha_{i_k})\\ &= cD_J(\alpha_{i_1}, \ldots, \alpha_r, \ldots, \alpha_{i_k})\\ &= cD_{I,J}(M);\end{aligned}$$

$$\begin{aligned}\text{(b)}\quad D_{I,J}(e(M)) &= D_J(\alpha_{i_1}, \ldots, \alpha_r + g\alpha_s, \ldots, \alpha_{i_k})\\ &= D_{I,J}(M) + gD_J(\alpha_{i_1}, \ldots, \alpha_s, \ldots, \alpha_{i_k}).\end{aligned}$$

For type (a) operations, it is clear that any f which divides $D_{I,J}(M)$ also divides $D_{I,J}(e(M))$. For the case of a type (c) operation, notice that

$$\begin{aligned}D_J(\alpha_{i_1}, \ldots, \alpha_s, \ldots, \alpha_{i_k}) &= 0, &&\text{if } s = i_j \text{ for some } j\\ D_J(\alpha_{i_1}, \ldots, \alpha_s, \ldots, \alpha_{i_k}) &= \pm D_{I',J}(M), &&\text{if } s \neq i_j \text{ for all } j.\end{aligned}$$

The I' in the last equation is the k-tuple $(i_1, \ldots, s, \ldots, i_k)$ arranged in increasing order. It should now be apparent that, if f divides every $D_{I,J}(M)$, then f divides every $D_{I,J}(e(M))$.

Operations of type (c) can be taken care of by roughly the same argument or by using the fact that such an operation can be effected by a sequence of operations of types (a) and (b). ∎

Corollary. *Each matrix* M *in* $F[x]^{m\times n}$ *is equivalent to precisely one matrix* N *which is in normal form. The polynomials* $f_1, \ldots, f_l$ *which occur on the main diagonal of* N *are*

$$f_k = \frac{\delta_k(M)}{\delta_{k-1}(M)}, \qquad 1 \leq k \leq \min\,(m, n)$$

where, for convenience, we define $\delta_0(M) = 1$.

Proof. If N is in normal form with diagonal entries $f_1, \ldots, f_l$, it is quite easy to see that

$$\delta_k(N) = f_1 f_2 \cdots f_k. \quad ∎$$

Of course, we call the matrix N in the last corollary the **normal form** of M. The polynomials $f_1, \ldots, f_l$ are often called the **invariant factors** of M.

Suppose that A is an $n \times n$ matrix with entries in F, and let $p_1, \ldots, p_r$ be the invariant factors for A. We now see that the normal form of the matrix $xI - A$ has diagonal entries $1, 1, \ldots, 1, p_r, \ldots, p_1$. The last corollary tells us what $p_1, \ldots, p_r$ are, in terms of submatrices of $xI - A$. The number $n - r$ is the largest k such that $\delta_k(xI - A) = 1$. The minimal polynomial p_1 is the characteristic polynomial for A divided by the greatest common divisor of the determinants of all $(n-1) \times (n-1)$ submatrices of $xI - A$, etc.

Exercises

1. True or false? Every matrix in $F[x]^{n \times n}$ is row-equivalent to an upper-triangular matrix.

2. Let T be a linear operator on a finite-dimensional vector space and let A be the matrix of T in some ordered basis. Then T has a cyclic vector if and only if the determinants of the $(n-1) \times (n-1)$ submatrices of $xI - A$ are relatively prime.

3. Let A be an $n \times n$ matrix with entries in the field F and let $f_1, \ldots, f_n$ be the diagonal entries of the normal form of $xI - A$. For which matrices A is $f_1 \neq 1$?

4. Construct a linear operator T with minimal polynomial $x^2(x-1)^2$ and characteristic polynomial $x^3(x-1)^4$. Describe the primary decomposition of the vector space under T and find the projections on the primary components. Find a basis in which the matrix of T is in Jordan form. Also find an explicit direct sum decomposition of the space into T-cyclic subspaces as in Theorem 3 and give the invariant factors.

5. Let T be the linear operator on R^8 which is represented in the standard basis by the matrix

$$A = \begin{bmatrix} 1 & 1 & 1 & 1 & 1 & 1 & 1 & 1 \\ 0 & 0 & 0 & 0 & 0 & 0 & 0 & 1 \\ 0 & 0 & 0 & 0 & 0 & 0 & 0 & -1 \\ 0 & 1 & 1 & 0 & 0 & 0 & 0 & 1 \\ 0 & 0 & 0 & 1 & 1 & 0 & 0 & 0 \\ 0 & 1 & 1 & 1 & 1 & 1 & 0 & 1 \\ 0 & -1 & -1 & -1 & -1 & 0 & 1 & -1 \\ 0 & 0 & 0 & 0 & 0 & 0 & 0 & 0 \end{bmatrix}.$$

(a) Find the characteristic polynomial and the invariant factors.

(b) Find the primary decomposition of R^8 under T and the projections on the primary components. Find cyclic decompositions of each primary component as in Theorem 3.

(c) Find the Jordan form of A.

(d) Find a direct-sum decomposition of R^8 into T-cyclic subspaces as in Theorem 3. (*Hint:* One way to do this is to use the results in (b) and an appropriate generalization of the ideas discussed in Example 4.)

7.5. Summary; Semi-Simple Operators

In the last two chapters, we have been dealing with a single linear operator T on a finite-dimensional vector space V. The program has been to decompose T into a direct sum of linear operators of an elementary nature, for the purpose of gaining detailed information about how T 'operates' on the space V. Let us review briefly where we stand.

We began to study T by means of characteristic values and characteristic vectors. We introduced diagonalizable operators, the operators which can be completely described in terms of characteristic values and vectors. We then observed that T might not have a single characteristic vector. Even in the case of an algebraically closed scalar field, when every linear operator does have at least one characteristic vector, we noted that the characteristic vectors of T need not span the space.

We then proved the cyclic decomposition theorem, expressing any linear operator as the direct sum of operators with a cyclic vector, with no assumption about the scalar field. If U is a linear operator with a cyclic vector, there is a basis $\{\alpha_1, \ldots, \alpha_n\}$ with

$$\begin{aligned} U\alpha_j &= \alpha_{j+1}, \qquad j = 1, \ldots, n-1 \\ U\alpha_n &= -c_0\alpha_1 - c_1\alpha_2 - \cdots - c_{n-1}\alpha_n. \end{aligned}$$

The action of U on this basis is then to shift each α_j to the next vector α_{j+1}, except that $U\alpha_n$ is some prescribed linear combination of the vectors in the basis. Since the general linear operator T is the direct sum of a finite number of such operators U, we obtained an explicit and reasonably elementary description of the action of T.

We next applied the cyclic decomposition theorem to nilpotent operators. For the case of an algebraically closed scalar field, we combined this with the primary decomposition theorem to obtain the Jordan form. The Jordan form gives a basis $\{\alpha_1, \ldots, \alpha_n\}$ for the space V such that, for each j, either $T\alpha_j$ is a scalar multiple of α_j or $T\alpha_j = c\alpha_j + \alpha_{j+1}$. Such a basis certainly describes the action of T in an explicit and elementary manner.

The importance of the rational form (or the Jordan form) derives from the fact that it exists, rather than from the fact that it can be computed in specific cases. Of course, if one is given a specific linear operator T and can compute its cyclic or Jordan form, that is the thing to do; for, having such a form, one can reel off vast amounts of information about T. Two different types of difficulties arise in the computation of such standard forms. One difficulty is, of course, the length of the computations. The other difficulty is that there may not be any method for doing the computations, even if one has the necessary time and patience. The second difficulty arises in, say, trying to find the Jordan form of a complex matrix. There simply is no well-defined method for factoring the characteristic polynomial, and thus one is stopped at the outset. The rational form does not suffer from this difficulty. As we showed in Section 7.4, there is a well-defined method for finding the rational form of a given $n \times n$ matrix; however, such computations are usually extremely lengthy.

In our summary of the results of these last two chapters, we have not yet mentioned one of the theorems which we proved. This is the theorem which states that if T is a linear operator on a finite-dimensional vector space over an algebraically closed field, then T is uniquely expressible as the sum of a diagonalizable operator and a nilpotent operator which commute. This was proved from the primary decomposition theorem and certain information about diagonalizable operators. It is not as deep a theorem as the cyclic decomposition theorem or the existence of the Jordan form, but it does have important and useful applications in certain parts of mathematics. In concluding this chapter, we shall prove an analogous theorem, without assuming that the scalar field is algebraically closed. We begin by defining the operators which will play the role of the diagonalizable operators.

Definition. *Let* V *be a finite-dimensional vector space over the field* F, *and let* T *be a linear operator on* V. *We say that* T *is* **semi-simple** *if every* T-*invariant subspace has a complementary* T-*invariant subspace.*

What we are about to prove is that, with some restriction on the field F, every linear operator T is uniquely expressible in the form $T = S + N$, where S is semi-simple, N is nilpotent, and $SN = NS$. First, we are going to characterize semi-simple operators by means of their minimal polynomials, and this characterization will show us that, when F is algebraically closed, an operator is semi-simple if and only if it is diagonalizable.

Lemma. *Let* T *be a linear operator on the finite-dimensional vector space* V, *and let* $V = W_1 \oplus \cdots \oplus W_k$ *be the primary decomposition for* T. *In other words, if* p *is the minimal polynomial for* T *and* $p = p_1^{r_1} \cdots p_k^{r_k}$ *is the prime factorization of* p, *then* W_j *is the null space of* $p_j(T)^{r_j}$. *Let* W *be any subspace of* V *which is invariant under* T. *Then*

$$W = (W \cap W_1) \oplus \cdots \oplus (W \cap W_k)$$

Proof. For the proof we need to recall a corollary to our proof of the primary decomposition theorem in Section 6.8. If $E_1, \ldots, E_k$ are the projections associated with the decomposition $V = W_1 \oplus \cdots \oplus W_k$, then each E_j is a polynomial in T. That is, there are polynomials $h_1, \ldots, h_k$ such that $E_j = h_j(T)$.

Now let W be a subspace which is invariant under T. If α is any vector in W, then $\alpha = \alpha_1 + \cdots + \alpha_k$, where α_j is in W_j. Now $\alpha_j = E_j\alpha = h_j(T)\alpha$, and since W is invariant under T, each α_j is also in W. Thus each vector α in W is of the form $\alpha = \alpha_1 + \cdots + \alpha_k$, where α_j is in the intersection $W \cap W_j$. This expression is unique, since $V = W_1 \oplus \cdots \oplus W_k$. Therefore

$$W = (W \cap W_1) \oplus \cdots \oplus (W \cap W_k). \quad \blacksquare$$

Lemma. *Let* T *be a linear operator on* V, *and suppose that the minimal polynomial for* T *is irreducible over the scalar field* F. *Then* T *is semi-simple.*

Proof. Let W be a subspace of V which is invariant under T. We must prove that W has a complementary T-invariant subspace. According to a corollary of Theorem 3, it will suffice to prove that if f is a polynomial and β is a vector in V such that $f(T)\beta$ is in W, then there is a vector α in W with $f(T)\beta = f(T)\alpha$. So suppose β is in V and f is a polynomial such that $f(T)\beta$ is in W. If $f(T)\beta = 0$, we let $\alpha = 0$ and then α is a vector in W with $f(T)\beta = f(T)\alpha$. If $f(T)\beta \neq 0$, the polynomial f is not divisible by the minimal polynomial p of the operator T. Since p is prime, this means that f and p are relatively prime, and there exist polynomials g and h such that $fg + ph = 1$. Because $p(T) = 0$, we then have $f(T)g(T) = I$. From this it follows that the vector β must itself be in the subspace W; for

$$\begin{aligned} \beta &= g(T)f(T)\beta \\ &= g(T)(f(T)\beta) \end{aligned}$$

while $f(T)\beta$ is in W and W is invariant under T. Take $\alpha = \beta$. $\blacksquare$

Theorem 11. *Let* T *be a linear operator on the finite-dimensional vector space* V. *A necessary and sufficient condition that* T *be semi-simple is that the minimal polynomial* p *for* T *be of the form* $p = p_1 \cdots p_k$, *where* $p_1, \ldots, p_k$ *are distinct irreducible polynomials over the scalar field* F.

Proof. Suppose T is semi-simple. We shall show that no irreducible polynomial is repeated in the prime factorization of the minimal polynomial p. Suppose the contrary. Then there is some non-scalar monic polynomial g such that g^2 divides p. Let W be the null space of the operator $g(T)$. Then W is invariant under T. Now $p = g^2h$ for some polynomial h. Since g is not a scalar polynomial, the operator $g(T)h(T)$ is not the zero operator, and there is some vector β in V such that $g(T)h(T)\beta \neq 0$, i.e., $(gh)\beta \neq 0$. Now $(gh)\beta$ is in the subspace W, since $g(gh\beta) = g^2h\beta = p\beta = 0$. But there is no vector α in W such that $gh\beta = gh\alpha$; for, if α is in W

$$(gh)\alpha = (hg)\alpha = h(g\alpha) = h(0) = 0.$$

Thus, W cannot have a complementary T-invariant subspace, contradicting the hypothesis that T is semi-simple.

Now suppose the prime factorization of p is $p = p_1 \cdots p_k$, where $p_1, \ldots, p_k$ are distinct irreducible (non-scalar) monic polynomials. Let W be a subspace of V which is invariant under T. We shall prove that W has a complementary T-invariant subspace. Let $V = W_1 \oplus \cdots \oplus W_k$ be the primary decomposition for T, i.e., let W_j be the null space of $p_j(T)$. Let T_j be the linear operator induced on W_j by T, so that the minimal polynomial for T_j is the prime p_j. Now $W \cap W_j$ is a subspace of W_j which is invariant under T_j (or under T). By the last lemma, there is a subspace V_j of W_j such that $W_j = (W \cap W_j) \oplus V_j$ and V_j is invariant under T_j (and hence under T). Then we have

$$\begin{aligned} V &= W_1 \oplus \cdots \oplus W_k \\ &= (W \cap W_1) \oplus V_1 \oplus \cdots \oplus (W \cap W_k) \oplus V_k \\ &= (W \cap W_1) + \cdots + (W \cap W_k) \oplus V_1 \oplus \cdots \oplus V_k. \end{aligned}$$

By the first lemma above, $W = (W \cap W_1) \oplus \cdots \oplus (W \cap W_k)$, so that if $W' = V_1 \oplus \cdots \oplus V_k$, then $V = W \oplus W'$ and W' is invariant under T. ∎

Corollary. *If* T *is a linear operator on a finite-dimensional vector space over an algebraically closed field, then* T *is semi-simple if and only if* T *is diagonalizable.*

Proof. If the scalar field F is algebraically closed, the monic primes over F are the polynomials $x - c$. In this case, T is semi-simple if and only if the minimal polynomial for T is $p = (x - c_1) \cdots (x - c_k)$, where $c_1, \ldots, c_k$ are distinct elements of F. This is precisely the criterion for T to be diagonalizable, which we established in Chapter 6. ∎

We should point out that T is semi-simple if and only if there is some polynomial f, which is a product of distinct primes, such that $f(T) = 0$. This is only superficially different from the condition that the minimal polynomial be a product of distinct primes.

We turn now to expressing a linear operator as the sum of a semi-simple operator and a nilpotent operator which commute. In this, we shall restrict the scalar field to a subfield of the complex numbers. The informed reader will see that what is important is that the field F be a field of characteristic zero, that is, that for each positive integer n the sum $1 + \cdots + 1$ (n times) in F should not be 0. For a polynomial f over F, we denote by $f^{(k)}$ the kth formal derivative of f. In other words, $f^{(k)} = D^k f$, where D is the differentiation operator on the space of polynomials. If g is another polynomial, $f(g)$ denotes the result of substituting g in f, i.e., the polynomial obtained by applying f to the element g in the linear algebra $F[x]$.

Lemma (Taylor's Formula). *Let* F *be a field of characteristic zero and let* g *and* h *be polynomials over* F. *If* f *is any polynomial over* F *with* deg f $\leq$ n, *then*

$$\mathrm{f(g) = f(h) + f^{(1)}(h)(g - h) + \frac{f^{(2)}(h)}{2!}(g - h)^2 + \cdots + \frac{f^{(n)}(h)}{n!}(g - h)^n}.$$

Proof. What we are proving is a generalized Taylor formula. The reader is probably used to seeing the special case in which $h = c$, a scalar polynomial, and $g = x$. Then the formula says

$$f = f(x) = f(c) + f^{(1)}(c)(x - c) + \frac{f^{(2)}(c)}{2!}(x - c)^2 + \cdots + \frac{f^{(n)}(c)}{n!}(x - c)^n.$$

The proof of the general formula is just an application of the binomial theorem

$$(a + b)^k = a^k + ka^{k-1}b + \frac{k(k-1)}{2!}a^{k-2}b^2 + \cdots + b^k.$$

For the reader should see that, since substitution and differentiation are linear processes, one need only prove the formula when $f = x^k$. The formula for $f = \sum_{k=0}^{n} c_k x^k$ follows by a linear combination. In the case $f = x^k$ with $k \leq n$, the formula says

$$g^k = h^k + kh^{k-1}(g - h) + \frac{k(k-1)}{2!}h^{k-2}(g - h)^2 + \cdots + (g - h)^k$$

which is just the binomial expansion of

$$g^k = [h + (g - h)]^k. \quad \blacksquare$$

Lemma. *Let* F *be a subfield of the complex numbers, let* f *be a polynomial over* F, *and let* f′ *be the derivative of* f. *The following are equivalent:*

(a) f *is the product of distinct polynomials irreducible over* F.
(b) f *and* f′ *are relatively prime.*
(c) *As a polynomial with complex coefficients,* f *has no repeated root.*

Proof. Let us first prove that (a) and (b) are equivalent statements about f. Suppose in the prime factorization of f over the field F that some (non-scalar) prime polynomial p is repeated. Then $f = p^2h$ for some h in $F[x]$. Then

$$f' = p^2h' + 2pp'h$$

and p is also a divisor of f'. Hence f and f' are not relatively prime. We conclude that (b) implies (a).

Now suppose $f = p_1 \cdots p_k$, where $p_1, \ldots, p_k$ are distinct non-scalar irreducible polynomials over F. Let $f_j = f/p_j$. Then

$$f' = p_1'f_1 + p_2'f_2 + \cdots + p_k'f_k.$$

Let p be a prime polynomial which divides both f and f'. Then $p = p_i$ for some i. Now p_i divides f_j for $j \neq i$, and since p_i also divides

$$f' = \sum_{j=1}^{k} p_j' f_j$$

we see that p_i must divide $p_i' f_i$. Therefore p_i divides either f_i or p_i'. But p_i does not divide f_i since $p_1, \ldots, p_k$ are distinct. So p_i divides p_i'. This is not possible, since p_i' has degree one less than the degree of p_i. We conclude that no prime divides both f and f', or that $(f, f') = 1$.

To see that statement (c) is equivalent to (a) and (b), we need only observe the following: Suppose f and g are polynomials over F, a subfield of the complex numbers. We may also regard f and g as polynomials with complex coefficients. The statement that f and g are relatively prime as polynomials over F is equivalent to the statement that f and g are relatively prime as polynomials over the field of complex numbers. We leave the proof of this as an exercise. We use this fact with $g = f'$. Note that (c) is just (a) when f is regarded as a polynomial over the field of complex numbers. Thus (b) and (c) are equivalent, by the same argument that we used above. ∎

We can now prove a theorem which makes the relation between semi-simple operators and diagonalizable operators even more apparent.

Theorem 12. *Let* F *be a subfield of the field of complex numbers, let* V *be a finite-dimensional vector space over* F, *and let* T *be a linear operator on* V. *Let* $\mathfrak{B}$ *be an ordered basis for* V *and let* A *be the matrix of* T *in the ordered basis* $\mathfrak{B}$. *Then* T *is semi-simple if and only if the matrix* A *is similar over the field of complex numbers to a diagonal matrix.*

Proof. Let p be the minimal polynomial for T. According to Theorem 11, T is semi-simple if and only if $p = p_1 \cdots p_k$ where $p_1, \ldots, p_k$ are distinct irreducible polynomials over F. By the last lemma, we see that T is semi-simple if and only if p has no repeated complex root.

Now p is also the minimal polynomial for the matrix A. We know that A is similar over the field of complex numbers to a diagonal matrix if and only if its minimal polynomial has no repeated complex root. This proves the theorem. ∎

Theorem 13. *Let* F *be a subfield of the field of complex numbers, let* V *be a finite-dimensional vector space over* F, *and let* T *be a linear operator on* V. *There is a semi-simple operator* S *on* V *and a nilpotent operator* N *on* V *such that*

(i) T = S + N;
(ii) SN = NS.

Furthermore, the semi-simple S *and nilpotent* N *satisfying* (i) *and* (ii) *are unique, and each is a polynomial in* T.

Proof. Let $p_1^{r_1} \cdots p_k^{r_k}$ be the prime factorization of the minimal polynomial for T, and let $f = p_1 \cdots p_k$. Let r be the greatest of the positive integers $r_1, \ldots, r_k$. Then the polynomial f is a product of distinct primes, f^r is divisible by the minimal polynomial for T, and so

$$f(T)^r = 0.$$

We are going to construct a sequence of polynomials: $g_0, g_1, g_2, \ldots$ such that

$$f\left(x - \sum_{j=0}^{n} g_j f^j\right)$$

is divisible by f^{n+1}, $n = 0, 1, 2, \ldots$. We take $g_0 = 0$ and then $f(x - g_0 f^0) = f(x) = f$ is divisible by f. Suppose we have chosen $g_0, \ldots, g_{n-1}$. Let

$$h = x - \sum_{j=0}^{n-1} g_j f^j$$

so that, by assumption, $f(h)$ is divisible by f^n. We want to choose g_n so that

$$f(h - g_n f^n)$$

is divisible by f^{n+1}. We apply the general Taylor formula and obtain

$$f(h - g_n f^n) = f(h) - g_n f^n f'(h) + f^{n+1} b$$

where b is some polynomial. By assumption $f(h) = qf^n$. Thus, we see that to have $f(h - g_n f^n)$ divisible by f^{n+1} we need only choose g_n in such a way that $(q - g_n f')$ is divisible by f. This can be done, because f has no repeated prime factors and so f and f' are relatively prime. If a and e are polynomials such that $af + ef' = 1$, and if we let $g_n = eq$, then $q - g_n f'$ is divisible by f.

Now we have a sequence $g_0, g_1, \ldots$ such that f^{n+1} divides $f\left(x - \sum_{j=0}^{n} g_j f^j\right)$. Let us take $n = r - 1$ and then since $f(T)^r = 0$

$$f\left(T - \sum_{j=0}^{r-1} g_j(T) f(T)^j\right) = 0.$$

Let

$$N = \sum_{j=1}^{r-1} g_j(T) f(T)^j = \sum_{j=0}^{r-1} g_j(T) f(T)^j.$$

Since $\sum_{j=1}^{n} g_j f^j$ is divisible by f, we see that $N^r = 0$ and N is nilpotent. Let $S = T - N$. Then $f(S) = f(T - N) = 0$. Since f has distinct prime factors, S is semi-simple.

Now we have $T = S + N$ where S is semi-simple, N is nilpotent, and each is a polynomial in T. To prove the uniqueness statement, we shall pass from the scalar field F to the field of complex numbers. Let $\mathcal{B}$ be some ordered basis for the space V. Then we have

$$[T]_{\mathcal{B}} = [S]_{\mathcal{B}} + [N]_{\mathcal{B}}$$

while $[S]_{\mathcal{B}}$ is diagonalizable over the complex numbers and $[N]_{\mathcal{B}}$ is nilpotent. This diagonalizable matrix and nilpotent matrix which commute are uniquely determined, as we have shown in Chapter 6. ∎

Exercises

1. If N is a nilpotent linear operator on V, show that for any polynomial f the semi-simple part of $f(N)$ is a scalar multiple of the identity operator (F a subfield of C).

2. Let F be a subfield of the complex numbers, V a finite-dimensional vector space over F, and T a semi-simple linear operator on V. If f is any polynomial over F, prove that $f(T)$ is semi-simple.

3. Let T be a linear operator on a finite-dimensional space over a subfield of C. Prove that T is semi-simple if and only if the following is true: If f is a polynomial and $f(T)$ is nilpotent, then $f(T) = 0$.

8. Inner Product Spaces

8.1. Inner Products

Throughout this chapter we consider only real or complex vector spaces, that is, vector spaces over the field of real numbers or the field of complex numbers. Our main object is to study vector spaces in which it makes sense to speak of the 'length' of a vector and of the 'angle' between two vectors. We shall do this by studying a certain type of scalar-valued function on pairs of vectors, known as an inner product. One example of an inner product is the scalar or dot product of vectors in R^3. The scalar product of the vectors

$$\alpha = (x_1, x_2, x_3) \quad \text{and} \quad \beta = (y_1, y_2, y_3)$$

in R^3 is the real number

$$(\alpha|\beta) = x_1y_1 + x_2y_2 + x_3y_3.$$

Geometrically, this dot product is the product of the length of α, the length of β, and the cosine of the angle between α and β. It is therefore possible to define the geometric concepts of 'length' and 'angle' in R^3 by means of the algebraically defined scalar product.

An inner product on a vector space is a function with properties similar to the dot product in R^3, and in terms of such an inner product one can also define 'length' and 'angle.' Our comments about the general notion of angle will be restricted to the concept of perpendicularity (or orthogonality) of vectors. In this first section we shall say what an inner product is, consider some particular examples, and establish a few basic properties of inner products. Then we turn to the task of discussing length and orthogonality.

Definition. *Let F be the field of real numbers or the field of complex numbers, and V a vector space over F. An* **inner product** *on V is a function which assigns to each ordered pair of vectors α, β in V a scalar $(\alpha|\beta)$ in F in such a way that for all α, β, γ in V and all scalars c*

(a) $(\alpha + \beta|\gamma) = (\alpha|\gamma) + (\beta|\gamma)$;

(b) $(c\alpha|\beta) = c(\alpha|\beta)$;

(c) $(\beta|\alpha) = \overline{(\alpha|\beta)}$, *the bar denoting complex conjugation;*

(d) $(\alpha|\alpha) > 0$ *if* $\alpha \neq 0$.

It should be observed that conditions (a), (b), and (c) imply that

(e) $$(\alpha|c\beta + \gamma) = \bar{c}(\alpha|\beta) + (\alpha|\gamma).$$

One other point should be made. When F is the field R of real numbers, the complex conjugates appearing in (c) and (e) are superfluous; however, in the complex case they are necessary for the consistency of the conditions. Without these complex conjugates, we would have the contradiction:

$$(\alpha|\alpha) > 0 \quad \text{and} \quad (i\alpha|i\alpha) = -1(\alpha|\alpha) > 0.$$

In the examples that follow and throughout the chapter, F is either the field of real numbers or the field of complex numbers.

EXAMPLE 1. On F^n there is an inner product which we call the **standard inner product.** It is defined on $\alpha = (x_1, \ldots, x_n)$ and $\beta = (y_1, \ldots, y_n)$ by

(8-1) $$(\alpha|\beta) = \sum_j x_j\bar{y}_j.$$

When $F = R$, this may also be written

$$(\alpha|\beta) = \sum_j x_jy_j.$$

In the real case, the standard inner product is often called the dot or scalar product and denoted by $\alpha \cdot \beta$.

EXAMPLE 2. For $\alpha = (x_1, x_2)$ and $\beta = (y_1, y_2)$ in R^2, let

$$(\alpha|\beta) = x_1y_1 - x_2y_1 - x_1y_2 + 4x_2y_2.$$

Since $(\alpha|\alpha) = (x_1 - x_2)^2 + 3x_2^2$, it follows that $(\alpha|\alpha) > 0$ if $\alpha \neq 0$. Conditions (a), (b), and (c) of the definition are easily verified.

EXAMPLE 3. Let V be $F^{n\times n}$, the space of all $n \times n$ matrices over F. Then V is isomorphic to F^{n^2} in a natural way. It therefore follows from Example 1 that the equation

$$(A|B) = \sum_{j,k} A_{jk}\bar{B}_{jk}$$

defines an inner product on V. Furthermore, if we introduce the **conjugate transpose** matrix B^*, where $B^*_{kj} = \overline{B}_{jk}$, we may express this inner product on $F^{n\times n}$ in terms of the trace function:

$$(A|B) = \text{tr}(AB^*) = \text{tr}(B^*A).$$

For

$$\begin{aligned}\text{tr}(AB^*) &= \sum_j (AB^*)_{jj} \\ &= \sum_j \sum_k A_{jk}B^*_{kj} \\ &= \sum_j \sum_k A_{jk}\overline{B}_{jk}.\end{aligned}$$

Example 4. Let $F^{n\times 1}$ be the space of $n \times 1$ (column) matrices over F, and let Q be an $n \times n$ invertible matrix over F. For X, Y in $F^{n\times 1}$ set

$$(X|Y) = Y^*Q^*QX.$$

We are identifying the 1×1 matrix on the right with its single entry. When Q is the identity matrix, this inner product is essentially the same as that in Example 1; we call it the **standard inner product** on $F^{n\times 1}$. The reader should note that the terminology 'standard inner product' is used in two special contexts. For a general finite-dimensional vector space over F, there is no obvious inner product that one may call standard.

Example 5. Let V be the vector space of all continuous complex-valued functions on the unit interval, $0 \leq t \leq 1$. Let

$$(f|g) = \int_0^1 f(t)\overline{g(t)}\,dt.$$

The reader is probably more familiar with the space of real-valued continuous functions on the unit interval, and for this space the complex conjugate on g may be omitted.

Example 6. This is really a whole class of examples. One may construct new inner products from a given one by the following method. Let V and W be vector spaces over F and suppose $(\ |\ )$ is an inner product on W. If T is a non-singular linear transformation from V into W, then the equation

$$p_T(\alpha, \beta) = (T\alpha|T\beta)$$

defines an inner product p_T on V. The inner product in Example 4 is a special case of this situation. The following are also special cases.

(a) Let V be a finite-dimensional vector space, and let

$$\mathcal{B} = \{\alpha_1, \ldots, \alpha_n\}$$

be an ordered basis for V. Let $\epsilon_1, \ldots, \epsilon_n$ be the standard basis vectors in F^n, and let T be the linear transformation from V into F^n such that $T\alpha_j = \epsilon_j$, $j = 1, \ldots, n$. In other words, let T be the 'natural' isomorphism of V onto F^n that is determined by $\mathcal{B}$. If we take the standard inner product on F^n, then

$$p_T(\sum_j x_j\alpha_j, \sum_k y_k\alpha_k) = \sum_{j=1}^{n} x_j\bar{y}_j.$$

Thus, for any basis for V there is an inner product on V with the property $(\alpha_j|\alpha_k) = \delta_{jk}$; in fact, it is easy to show that there is exactly one such inner product. Later we shall show that every inner product on V is determined by some basis $\mathcal{B}$ in the above manner.

(b) We look again at Example 5 and take $V = W$, the space of continuous functions on the unit interval. Let T be the linear operator 'multiplication by t,' that is, $(Tf)(t) = tf(t)$, $0 \leq t \leq 1$. It is easy to see that T is linear. Also T is non-singular; for suppose $Tf = 0$. Then $tf(t) = 0$ for $0 \leq t \leq 1$; hence $f(t) = 0$ for $t > 0$. Since f is continuous, we have $f(0) = 0$ as well, or $f = 0$. Now using the inner product of Example 5, we construct a new inner product on V by setting

$$\begin{aligned} p_T(f, g) &= \int_0^1 (Tf)(t)\overline{(Tg)(t)}\,dt \\ &= \int_0^1 f(t)\overline{g(t)}t^2\,dt. \end{aligned}$$

We turn now to some general observations about inner products. Suppose V is a complex vector space with an inner product. Then for all α, β in V

$$(\alpha|\beta) = \text{Re}\,(\alpha|\beta) + i\,\text{Im}\,(\alpha|\beta)$$

where $\text{Re}\,(\alpha|\beta)$ and $\text{Im}\,(\alpha|\beta)$ are the real and imaginary parts of the complex number $(\alpha|\beta)$. If z is a complex number, then $\text{Im}\,(z) = \text{Re}\,(-iz)$. It follows that

$$\text{Im}\,(\alpha|\beta) = \text{Re}\,[-i(\alpha|\beta)] = \text{Re}\,(\alpha|i\beta).$$

Thus the inner product is completely determined by its 'real part' in accordance with

$$(\alpha|\beta) = \text{Re}\,(\alpha|\beta) + i\,\text{Re}\,(\alpha|i\beta). \tag{8-2}$$

Occasionally it is very useful to know that an inner product on a real or complex vector space is determined by another function, the so-called quadratic form determined by the inner product. To define it, we first denote the positive square root of $(\alpha|\alpha)$ by $||\alpha||$; $||\alpha||$ is called the **norm** of α with respect to the inner product. By looking at the standard inner products in R^1, C^1, R^2, and R^3, the reader should be able to convince himself that it is appropriate to think of the norm of α as the 'length' or 'magnitude' of α. The **quadratic form** determined by the inner product is the function that assigns to each vector α the scalar $||\alpha||^2$. It follows from the properties of the inner product that

$$||\alpha \pm \beta||^2 = ||\alpha||^2 \pm 2 \text{ Re } (\alpha|\beta) + ||\beta||^2$$

for all vectors α and β. Thus in the real case

$$(\alpha|\beta) = \frac{1}{4} ||\alpha + \beta||^2 - \frac{1}{4} ||\alpha - \beta||^2. \tag{8-3}$$

In the complex case we use (8-2) to obtain the more complicated expression

$$(\alpha|\beta) = \frac{1}{4} ||\alpha + \beta||^2 - \frac{1}{4} ||\alpha - \beta||^2 + \frac{i}{4} ||\alpha + i\beta||^2 - \frac{i}{4} ||\alpha - i\beta||^2. \tag{8-4}$$

Equations (8-3) and (8-4) are called the **polarization identities.** Note that (8-4) may also be written as follows:

$$(\alpha|\beta) = \frac{1}{4} \sum_{n=1}^{4} i^n ||\alpha + i^n\beta||^2.$$

The properties obtained above hold for any inner product on a real or complex vector space V, regardless of its dimension. We turn now to the case in which V is finite-dimensional. As one might guess, an inner product on a finite-dimensional space may always be described in terms of an ordered basis by means of a matrix.

Suppose that V is finite-dimensional, that

$$\mathcal{B} = \{\alpha_1, \ldots, \alpha_n\}$$

is an ordered basis for V, and that we are given a particular inner product on V; we shall show that the inner product is completely determined by the values

$$G_{jk} = (\alpha_k|\alpha_j) \tag{8-5}$$

it assumes on pairs of vectors in $\mathcal{B}$. If $\alpha = \sum_k x_k\alpha_k$ and $\beta = \sum_j y_j\alpha_j$, then

$$\begin{aligned}
(\alpha|\beta) &= (\sum_k x_n\alpha_k|\beta) \\
&= \sum_k x_k(\alpha_k|\beta) \\
&= \sum_k x_k \sum_j \bar{y}_j(\alpha_k|\alpha_j) \\
&= \sum_{j,k} \bar{y}_j G_{jk} x_k \\
&= Y^*GX
\end{aligned}$$

where X, Y are the coordinate matrices of α, β in the ordered basis $\mathcal{B}$, and G is the matrix with entries $G_{jk} = (\alpha_k|\alpha_j)$. We call G the **matrix of the inner product in the ordered basis** $\mathcal{B}$. It follows from (8-5) that G is hermitian, i.e., that $G = G^*$; however, G is a rather special kind of hermitian matrix. For G must satisfy the additional condition

$$X^*GX > 0, \qquad X \neq 0. \tag{8-6}$$

In particular, G must be invertible. For otherwise there exists an $X \neq 0$ such that $GX = 0$, and for any such X, (8-6) is impossible. More explicitly, (8-6) says that for any scalars $x_1, \ldots, x_n$ not all of which are 0

$$\sum_{j,k} \bar{x}_j G_{jk} x_k > 0. \tag{8-7}$$

From this we see immediately that each diagonal entry of G must be positive; however, this condition on the diagonal entries is by no means sufficient to insure the validity of (8-6). Sufficient conditions for the validity of (8-6) will be given later.

The above process is reversible; that is, if G is any $n \times n$ matrix over F which satisfies (8-6) and the condition $G = G^*$, then G is the matrix in the ordered basis $\mathfrak{B}$ of an inner product on V. This inner product is given by the equation

$$(\alpha|\beta) = Y^*GX$$

where X and Y are the coordinate matrices of α and β in the ordered basis $\mathfrak{B}$.

Exercises

1. Let V be a vector space and $(\ |\ )$ an inner product on V.

(a) Show that $(0|\beta) = 0$ for all β in V.

(b) Show that if $(\alpha|\beta) = 0$ for all β in V, then $\alpha = 0$.

2. Let V be a vector space over F. Show that the sum of two inner products on V is an inner product on V. Is the difference of two inner products an inner product? Show that a positive multiple of an inner product is an inner product.

3. Describe explicitly all inner products on R^1 and on C^1.

4. Verify that the standard inner product on F^n is an inner product.

5. Let $(\ |\ )$ be the standard inner product on R^2.

(a) Let $\alpha = (1, 2)$, $\beta = (-1, 1)$. If γ is a vector such that $(\alpha|\gamma) = -1$ and $(\beta|\gamma) = 3$, find γ.

(b) Show that for any α in R^2 we have $\alpha = (\alpha|\epsilon_1)\epsilon_1 + (\alpha|\epsilon_2)\epsilon_2$.

6. Let $(\ |\ )$ be the standard inner product on R^2, and let T be the linear operator $T(x_1, x_2) = (-x_2, x_1)$. Now T is 'rotation through 90°' and has the property that $(\alpha|T\alpha) = 0$ for all α in R^2. Find all inner products $[\ |\ ]$ on R^2 such that $[\alpha|T\alpha] = 0$ for each α.

7. Let $(\ |\ )$ be the standard inner product on C^2. Prove that there is no non-zero linear operator on C^2 such that $(\alpha|T\alpha) = 0$ for every α in C^2. Generalize.

8. Let A be a 2×2 matrix with real entries. For X, Y in $R^{2\times 1}$ let

$$f_A(X, Y) = Y^tAX.$$

Show that f_A is an inner product on $R^{2\times 1}$ if and only if $A = A^t$, $A_{11} > 0$, $A_{22} > 0$, and $\det A > 0$.

9. Let V be a real or complex vector space with an inner product. Show that the quadratic form determined by the inner product satisfies the **parallelogram law**

$$||\alpha + \beta||^2 + ||\alpha - \beta||^2 = 2||\alpha||^2 + 2||\beta||^2.$$

10. Let $(\ |\ )$ be the inner product on R^2 defined in Example 2, and let $\mathcal{B}$ be the standard ordered basis for R^2. Find the matrix of this inner product relative to $\mathcal{B}$.

11. Show that the formula

$$\left(\sum_j a_jx^j \Big| \sum_k b_kx^k\right) = \sum_{j,k} \frac{a_jb_k}{j + k + 1}$$

defines an inner product on the space $R[x]$ of polynomials over the field R. Let W be the subspace of polynomials of degree less than or equal to n. Restrict the above inner product to W, and find the matrix of this inner product on W, relative to the ordered basis $\{1, x, x^2, \ldots, x^n\}$. (*Hint:* To show that the formula defines an inner product, observe that

$$(f|g) = \int_0^1 f(t)g(t)\,dt$$

and work with the integral.)

12. Let V be a finite-dimensional vector space and let $\mathcal{B} = \{\alpha_1, \ldots, \alpha_n\}$ be a basis for V. Let $(\ |\ )$ be an inner product on V. If $c_1, \ldots, c_n$ are any n scalars, show that there is exactly one vector α in V such that $(\alpha|\alpha_j) = c_j$, $j = 1, \ldots, n$.

13. Let V be a complex vector space. A function J from V into V is called a **conjugation** if $J(\alpha + \beta) = J(\alpha) + J(\beta)$, $J(c\alpha) = \bar{c}J(\alpha)$, and $J(J(\alpha)) = \alpha$, for all scalars c and all α, β in V. If J is a conjugation show that:

(a) The set W of all α in V such that $J\alpha = \alpha$ is a vector space over R with respect to the operations defined in V.

(b) For each α in V there exist unique vectors β, γ in W such that $\alpha = \beta + i\gamma$.

14. Let V be a complex vector space and W a subset of V with the following properties:

(a) W is a real vector space with respect to the operations defined in V.

(b) For each α in V there exist unique vectors β, γ in W such that $\alpha = \beta + i\gamma$.

Show that the equation $J\alpha = \beta - i\gamma$ defines a conjugation on V such that $J\alpha = \alpha$ if and only if α belongs to W, and show also that J is the only conjugation on V with this property.

15. Find all conjugations on C^1 and C^2.

16. Let W be a finite-dimensional real subspace of a complex vector space V. Show that W satisfies condition (b) of Exercise 14 if and only if every basis of W is also a basis of V.

17. Let V be a complex vector space, J a conjugation on V, W the set of α in V such that $J\alpha = \alpha$, and f an inner product on W. Show that:

(a) There is a unique inner product g on V such that $g(\alpha, \beta) = f(\alpha, \beta)$ for all α, β in W,

(b) $g(J\alpha, J\beta) = g(\beta, \alpha)$ for all α, β in V.

What does part (a) say about the relation between the standard inner products on R^1 and C^1, or on R^n and C^n?

8.2. Inner Product Spaces

Now that we have some idea of what an inner product is, we shall turn our attention to what can be said about the combination of a vector space and some particular inner product on it. Specifically, we shall establish the basic properties of the concepts of 'length' and 'orthogonality' which are imposed on the space by the inner product.

Definition. *An* **inner product space** *is a real or complex vector space, together with a specified inner product on that space.*

A finite-dimensional real inner product space is often called a **Euclidean space.** A complex inner product space is often referred to as a **unitary space.**

Theorem 1. *If* V *is an inner product space, then for any vectors* α, β *in* V *and any scalar* c

(i) $||c\alpha|| = |c|\ ||\alpha||$;

(ii) $||\alpha|| > 0$ *for* $\alpha \neq 0$;

(iii) $|(\alpha|\beta)| \leq ||\alpha||\ ||\beta||$;

(iv) $||\alpha + \beta|| \leq ||\alpha|| + ||\beta||$.

Proof. Statements (i) and (ii) follow almost immediately from the various definitions involved. The inequality in (iii) is clearly valid when $\alpha = 0$. If $\alpha \neq 0$, put

$$\gamma = \beta - \frac{(\beta|\alpha)}{||\alpha||^2}\alpha.$$

Then $(\gamma|\alpha) = 0$ and

$$\begin{aligned} 0 \leq ||\gamma||^2 &= \left(\beta - \frac{(\beta|\alpha)}{||\alpha||^2}\alpha \,\middle|\, \beta - \frac{(\beta|\alpha)}{||\alpha||^2}\alpha\right) \\ &= (\beta|\beta) - \frac{(\beta|\alpha)(\alpha|\beta)}{||\alpha||^2} \\ &= ||\beta||^2 - \frac{|(\alpha|\beta)|^2}{||\alpha||^2}. \end{aligned}$$

Hence $|(\alpha|\beta)|^2 \leq ||\alpha||^2||\beta||^2$. Now using (c) we find that

$$\begin{aligned} ||\alpha + \beta||^2 &= ||\alpha||^2 + (\alpha|\beta) + (\beta|\alpha) + ||\beta||^2 \\ &= ||\alpha||^2 + 2 \text{ Re } (\alpha|\beta) + ||\beta||^2 \\ &\leq ||\alpha||^2 + 2 \, ||\alpha|| \, ||\beta|| + ||\beta||^2 \\ &= (||\alpha|| + ||\beta||)^2. \end{aligned}$$

Thus, $||\alpha + \beta|| \leq ||\alpha|| + ||\beta||$. ∎

The inequality in (iii) is called the **Cauchy-Schwarz inequality.** It has a wide variety of applications. The proof shows that if (for example) α is non-zero, then $|(\alpha|\beta)| < ||\alpha|| \, ||\beta||$ unless

$$\beta = \frac{(\beta|\alpha)}{||\alpha||^2}\alpha.$$

Thus, equality occurs in (iii) if and only if α and β are linearly dependent.

Example 7. If we apply the Cauchy-Schwarz inequality to the inner products given in Examples 1, 2, 3, and 5, we obtain the following:

(a) $$|\Sigma \, x_k\bar{y}_k| \leq (\Sigma \, |x_k|^2)^{1/2}(\Sigma \, |y_k|^2)^{1/2}$$

(b) $$|x_1y_1 - x_2y_1 - x_1y_2 + 4x_2y_2| \leq ((x_1 - x_2)^2 + 3x_2^2)^{1/2}((y_1 - y_2)^2 + 3y_2^2)^{1/2}$$

(c) $$|\text{tr } (AB^*)| \leq (\text{tr } (AA^*))^{1/2}(\text{tr } (BB^*))^{1/2}$$

(d) $$\left|\int_0^1 f(x)\overline{g(x)} \, dx\right| \leq \left(\int_0^1 |f(x)|^2 \, dx\right)^{1/2}\left(\int_0^1 |g(x)|^2 \, dx\right)^{1/2}.$$

Definitions. *Let α and β be vectors in an inner product space* V. *Then α is* **orthogonal** *to β if $(\alpha|\beta) = 0$; since this implies β is orthogonal to α, we often simply say that α and β are orthogonal. If* S *is a set of vectors in* V, S *is called an* **orthogonal set** *provided all pairs of distinct vectors in* S *are orthogonal. An* **orthonormal set** *is an orthogonal set* S *with the additional property that $||\alpha|| = 1$ for every α in* S.

The zero vector is orthogonal to every vector in V and is the only vector with this property. It is appropriate to think of an orthonormal set as a set of mutually perpendicular vectors, each having length 1.

Example 8. The standard basis of either R^n or C^n is an orthonormal set with respect to the standard inner product.

Example 9. The vector (x, y) in R^2 is orthogonal to $(-y, x)$ with respect to the standard inner product, for

$$((x, y)|(-y, x)) = -xy + yx = 0.$$

However, if R^2 is equipped with the inner product of Example 2, then (x, y) and $(-y, x)$ are orthogonal if and only if

$$y = \tfrac{1}{2}(-3 \pm \sqrt{13})x.$$

EXAMPLE 10. Let V be $C^{n \times n}$, the space of complex $n \times n$ matrices, and let E^{pq} be the matrix whose only non-zero entry is a 1 in row p and column q. Then the set of all such matrices E^{pq} is orthonormal with respect to the inner product given in Example 3. For

$$(E^{pq}|E^{rs}) = \operatorname{tr}(E^{pq}E^{sr}) = \delta_{qs}\operatorname{tr}(E^{pr}) = \delta_{qs}\delta_{pr}.$$

EXAMPLE 11. Let V be the space of continuous complex-valued (or real-valued) functions on the interval $0 \leq x \leq 1$ with the inner product

$$(f|g) = \int_0^1 f(x)\overline{g(x)}\,dx.$$

Suppose $f_n(x) = \sqrt{2}\cos 2\pi nx$ and that $g_n(x) = \sqrt{2}\sin 2\pi nx$. Then $\{1, f_1, g_1, f_2, g_2, \ldots\}$ is an infinite orthonormal set. In the complex case, we may also form the linear combinations

$$\frac{1}{\sqrt{2}}(f_n + ig_n), \qquad n = 1, 2, \ldots.$$

In this way we get a new orthonormal set S which consists of all functions of the form

$$h_n(x) = e^{2\pi inx}, \qquad n = \pm 1, \pm 2, \ldots.$$

The set S' obtained from S by adjoining the constant function 1 is also orthonormal. We assume here that the reader is familiar with the calculation of the integrals in question.

The orthonormal sets given in the examples above are all linearly independent. We show now that this is necessarily the case.

***Theorem* 2.** *An orthogonal set of non-zero vectors is linearly independent.*

Proof. Let S be a finite or infinite orthogonal set of non-zero vectors in a given inner product space. Suppose $\alpha_1, \alpha_2, \ldots, \alpha_m$ are distinct vectors in S and that

$$\beta = c_1\alpha_1 + c_2\alpha_2 + \cdots + c_m\alpha_m.$$

Then

$$\begin{aligned}(\beta|\alpha_k) &= (\sum_j c_j\alpha_j|\alpha_k)\\ &= \sum_j c_j(\alpha_j|\alpha_k)\\ &= c_k(\alpha_k|\alpha_k).\end{aligned}$$

Since $(\alpha_k|\alpha_k) \neq 0$, it follows that

$$c_k = \frac{(\beta|\alpha_k)}{\|\alpha_k\|^2}, \qquad 1 \leq k \leq m.$$

Thus when $\beta = 0$, each $c_k = 0$; so S is an independent set. ∎

Corollary. *If a vector β is a linear combination of an orthogonal sequence of non-zero vectors $\alpha_1, \ldots, \alpha_m$, then β is the particular linear combination*

$$\beta = \sum_{k=1}^{m} \frac{(\beta|\alpha_k)}{\|\alpha_k\|^2} \alpha_k. \tag{8-8}$$

This corollary follows from the proof of the theorem. There is another corollary which although obvious, should be mentioned. If $\{\alpha_1, \ldots, \alpha_m\}$ is an orthogonal set of non-zero vectors in a finite-dimensional inner product space V, then $m \leq \dim V$. This says that the number of mutually orthogonal directions in V cannot exceed the algebraically defined dimension of V. The maximum number of mutually orthogonal directions in V is what one would intuitively regard as the geometric dimension of V, and we have just seen that this is not greater than the algebraic dimension. The fact that these two dimensions are equal is a particular corollary of the next result.

Theorem 3. *Let V be an inner product space and let $\beta_1, \ldots, \beta_n$ be any independent vectors in V. Then one may construct orthogonal vectors $\alpha_1, \ldots, \alpha_n$ in V such that for each $k = 1, 2, \ldots, n$ the set*

$$\{\alpha_1, \ldots, \alpha_k\}$$

is a basis for the subspace spanned by $\beta_1, \ldots, \beta_k$.

Proof. The vectors $\alpha_1, \ldots, \alpha_n$ will be obtained by means of a construction known as the **Gram-Schmidt orthogonalization process.** First let $\alpha_1 = \beta_1$. The other vectors are then given inductively as follows: Suppose $\alpha_1, \ldots, \alpha_m$ $(1 \leq m < n)$ have been chosen so that for every k

$$\{\alpha_1, \ldots, \alpha_k\}, \qquad 1 \leq k \leq m$$

is an orthogonal basis for the subspace of V that is spanned by $\beta_1, \ldots, \beta_k$. To construct the next vector α_{m+1}, let

$$\alpha_{m+1} = \beta_{m+1} - \sum_{k=1}^{m} \frac{(\beta_{m+1}|\alpha_k)}{\|\alpha_k\|^2} \alpha_k. \tag{8-9}$$

Then $\alpha_{m+1} \neq 0$. For otherwise β_{m+1} is a linear combination of $\alpha_1, \ldots, \alpha_m$ and hence a linear combination of $\beta_1, \ldots, \beta_m$. Furthermore, if $1 \leq j \leq m$, then

$$\begin{aligned}(\alpha_{m+1}|\alpha_j) &= (\beta_{m+1}|\alpha_j) - \sum_{k=1}^{m} \frac{(\beta_{m+1}|\alpha_k)}{\|\alpha_k\|^2} (\alpha_k|\alpha_j) \\ &= (\beta_{m+1}|\alpha_j) - (\beta_{m+1}|\alpha_j) \\ &= 0.\end{aligned}$$

Therefore $\{\alpha_1, \ldots, \alpha_{m+1}\}$ is an orthogonal set consisting of $m + 1$ non-zero vectors in the subspace spanned by $\beta_1, \ldots, \beta_{m+1}$. By Theorem 2, it is a basis for this subspace. Thus the vectors $\alpha_1, \ldots, \alpha_n$ may be constructed one after the other in accordance with (8-9). In particular, when $n = 4$, we have

$$\begin{aligned}
\alpha_1 &= \beta_1 \\
\alpha_2 &= \beta_2 - \frac{(\beta_2|\alpha_1)}{||\alpha_1||^2}\alpha_1 \\
\alpha_3 &= \beta_3 - \frac{(\beta_3|\alpha_1)}{||\alpha_1||^2}\alpha_1 - \frac{(\beta_3|\alpha_2)}{||\alpha_2||^2}\alpha_2 \\
\alpha_4 &= \beta_4 - \frac{(\beta_4|\alpha_1)}{||\alpha_1||^2}\alpha_1 - \frac{(\beta_4|\alpha_2)}{||\alpha_2||^2}\alpha_2 - \frac{(\beta_4|\alpha_3)}{||\alpha_3||^2}\alpha_3.
\end{aligned} \tag{8-10}$$ ∎

Corollary. *Every finite-dimensional inner product space has an orthonormal basis.*

Proof. Let V be a finite-dimensional inner product space and $\{\beta_1, \ldots, \beta_n\}$ a basis for V. Apply the Gram-Schmidt process to construct an orthogonal basis $\{\alpha_1, \ldots, \alpha_n\}$. Then to obtain an orthonormal basis, simply replace each vector α_k by $\alpha_k/||\alpha_k||$. ∎

One of the main advantages which orthonormal bases have over arbitrary bases is that computations involving coordinates are simpler. To indicate in general terms why this is true, suppose that V is a finite-dimensional inner product space. Then, as in the last section, we may use Equation (8-5) to associate a matrix G with every ordered basis $\mathfrak{B} = \{\alpha_1, \ldots, \alpha_n\}$ of V. Using this matrix

$$G_{jk} = (\alpha_k|\alpha_j),$$

we may compute inner products in terms of coordinates. If $\mathfrak{B}$ is an orthonormal basis, then G is the identity matrix, and for any scalars x_j and y_k

$$(\sum_j x_j\alpha_j|\sum_k y_k\alpha_k) = \sum_j x_j\bar{y}_j.$$

Thus in terms of an orthonormal basis, the inner product in V looks like the standard inner product in F^n.

Although it is of limited practical use for computations, it is interesting to note that the Gram-Schmidt process may also be used to test for linear dependence. For suppose $\beta_1, \ldots, \beta_n$ are linearly dependent vectors in an inner product space V. To exclude a trivial case, assume that $\beta_1 \neq 0$. Let m be the largest integer for which $\beta_1, \ldots, \beta_m$ are independent. Then $1 \leq m < n$. Let $\alpha_1, \ldots, \alpha_m$ be the vectors obtained by applying the orthogonalization process to $\beta_1, \ldots, \beta_m$. Then the vector α_{m+1} given by (8-9) is necessarily 0. For α_{m+1} is in the subspace spanned by $\alpha_1, \ldots, \alpha_m$ and orthogonal to each of these vectors; hence it is 0 by (8-8). Conversely, if $\alpha_1, \ldots, \alpha_m$ are different from 0 and $\alpha_{m+1} = 0$, then $\beta_1, \ldots, \beta_{m+1}$ are linearly dependent.

EXAMPLE 12. Consider the vectors

$$\begin{aligned} \beta_1 &= (3, 0, 4) \\ \beta_2 &= (-1, 0, 7) \\ \beta_3 &= (2, 9, 11) \end{aligned}$$

in R^3 equipped with the standard inner product. Applying the Gram-Schmidt process to $\beta_1, \beta_2, \beta_3$, we obtain the following vectors.

$$\begin{aligned} \alpha_1 &= (3, 0, 4) \\ \alpha_2 &= (-1, 0, 7) - \frac{((-1, 0, 7)|(3, 0, 4))}{25}(3, 0, 4) \\ &= (-1, 0, 7) - (3, 0, 4) \\ &= (-4, 0, 3) \\ \alpha_3 &= (2, 9, 11) - \frac{((2, 9, 11)|(3, 0, 4))}{25}(3, 0, 4) \\ &\qquad - \frac{((2, 9, 11)|(-4, 0, 3))}{25}(-4, 0, 3) \\ &= (2, 9, 11) - 2(3, 0, 4) - -4, 0, 3) \\ &= (0, 9, 0). \end{aligned}$$

These vectors are evidently non-zero and mutually orthogonal. Hence $\{\alpha_1, \alpha_2, \alpha_3\}$ is an orthogonal basis for R^3. To express an arbitrary vector (x_1, x_2, x_3) in R^3 as a linear combination of $\alpha_1, \alpha_2, \alpha_3$ it is *not* necessary to solve any linear equations. For it suffices to use (8-8). Thus

$$(x_1, x_2, x_3) = \frac{3x_1 + 4x_3}{25}\alpha_1 + \frac{-4x_1 + 3x_3}{25}\alpha_2 + \frac{x_2}{9}\alpha_3$$

as is readily verified. In particular,

$$(1, 2, 3) = \tfrac{3}{5}(3, 0, 4) + \tfrac{1}{5}(-4, 0, 3) + \tfrac{2}{9}(0, 9, 0).$$

To put this point in another way, what we have shown is the following: The basis $\{f_1, f_2, f_3\}$ of $(R^3)^*$ which is dual to the basis $\{\alpha_1, \alpha_2, \alpha_3\}$ is defined explicitly by the equations

$$\begin{aligned} f_1(x_1, x_2, x_3) &= \frac{3x_1 + 4x_3}{25} \\ f_2(x_1, x_2, x_3) &= \frac{-4x_1 + 3x_3}{25} \\ f_3(x_1, x_2, x_3) &= \frac{x_2}{9} \end{aligned}$$

and these equations may be written more generally in the form

$$f_j(x_1, x_2, x_3) = \frac{((x_1, x_2, x_3)|\alpha_j)}{||\alpha_j||^2}.$$

Finally, note that from α_1, α_2, α_3 we get the orthonormal basis

$$\tfrac{1}{5}(3, 0, 4), \quad \tfrac{1}{5}(-4, 0, 3), \quad (0, 1, 0).$$

EXAMPLE 13. Let $A = \begin{bmatrix} a & b \\ c & d \end{bmatrix}$ where a, b, c, and d are complex numbers. Set $\beta_1 = (a, b)$, $\beta_2 = (c, d)$, and suppose that $\beta_1 \neq 0$. If we apply the orthogonalization process to β_1, β_2, using the standard inner product in C^2, we obtain the following vectors:

$$\begin{aligned}
\alpha_1 &= (a, b) \\
\alpha_2 &= (c, d) - \frac{((c, d)|(a, b))}{|a|^2 + |b|^2}(a, b) \\
&= (c, d) - \frac{(c\bar{a} + d\bar{b})}{|a|^2 + |b|^2}(a, b) \\
&= \left(\frac{cb\bar{b} - d\bar{b}a}{|a|^2 + |b|^2}, \frac{d\bar{a}a - c\bar{a}b}{|a|^2 + |b|^2}\right) \\
&= \frac{\det A}{|a|^2 + |b|^2}(-\bar{b}, \bar{a}).
\end{aligned}$$

Now the general theory tells us that $\alpha_2 \neq 0$ if and only if β_1, β_2 are linearly independent. On the other hand, the formula for α_2 shows that this is the case if and only if $\det A \neq 0$.

In essence, the Gram-Schmidt process consists of repeated applications of a basic geometric operation called orthogonal projection, and it is best understood from this point of view. The method of orthogonal projection also arises naturally in the solution of an important approximation problem.

Suppose W is a subspace of an inner product space V, and let β be an arbitrary vector in V. The problem is to find a best possible approximation to β by vectors in W. This means we want to find a vector α for which $||\beta - \alpha||$ is as small as possible subject to the restriction that α should belong to W. Let us make our language precise.

A **best approximation** to β by vectors in W is a vector α in W such that

$$||\beta - \alpha|| \leq ||\beta - \gamma||$$

for every vector γ in W.

By looking at this problem in R^2 or in R^3, one sees intuitively that a best approximation to β by vectors in W ought to be a vector α in W such that $\beta - \alpha$ is perpendicular (orthogonal) to W and that there ought to be exactly one such α. These intuitive ideas are correct for finite-dimensional subspaces and for some, but not all, infinite-dimensional subspaces. Since the precise situation is too complicated to treat here, we shall prove only the following result.

Theorem 4. *Let* W *be a subspace of an inner product space* V *and let* β *be a vector in* V.

(i) *The vector* α *in* W *is a best approximation to* β *by vectors in* W *if and only if* $\beta - \alpha$ *is orthogonal to every vector in* W.

(ii) *If a best approximation to* β *by vectors in* W *exists, it is unique.*

(iii) *If* W *is finite-dimensional and* $\{\alpha_1, \ldots, \alpha_n\}$ *is any orthonormal basis for* W, *then the vector*

$$\alpha = \sum_k \frac{(\beta|\alpha_k)}{||\alpha_k||^2} \alpha_k$$

is the (unique) best approximation to β *by vectors in* W.

Proof. First note that if γ is any vector in V, then $\beta - \gamma = (\beta - \alpha) + (\alpha - \gamma)$, and

$$||\beta - \gamma||^2 = ||\beta - \alpha||^2 + 2 \operatorname{Re} (\beta - \alpha|\alpha - \gamma) + ||\alpha - \gamma||^2.$$

Now suppose $\beta - \alpha$ is orthogonal to every vector in W, that γ is in W and that $\gamma \neq \alpha$. Then, since $\alpha - \gamma$ is in W, it follows that

$$\begin{aligned} ||\beta - \gamma||^2 &= ||\beta - \alpha||^2 + ||\alpha - \gamma||^2 \\ &> ||\beta - \alpha||^2. \end{aligned}$$

Conversely, suppose that $||\beta - \gamma|| \geq ||\beta - \alpha||$ for every γ in W. Then from the first equation above it follows that

$$2 \operatorname{Re} (\beta - \alpha|\alpha - \gamma) + ||\alpha - \gamma||^2 \geq 0$$

for all γ in W. Since every vector in W may be expressed in the form $\alpha - \gamma$ with γ in W, we see that

$$2 \operatorname{Re} (\beta - \alpha|\tau) + ||\tau||^2 \geq 0$$

for every τ in W. In particular, if γ is in W and $\gamma \neq \alpha$, we may take

$$\tau = -\frac{(\beta - \alpha|\alpha - \gamma)}{||\alpha - \gamma||^2} (\alpha - \gamma).$$

Then the inequality reduces to the statement

$$-2 \frac{|(\beta - \alpha|\alpha - \gamma)|^2}{||\alpha - \gamma||^2} + \frac{|(\beta - \alpha|\alpha - \gamma)|^2}{||\alpha - \gamma||^2} \geq 0.$$

This holds if and only if $(\beta - \alpha|\alpha - \gamma) = 0$. Therefore, $\beta - \alpha$ is orthogonal to every vector in W. This completes the proof of the equivalence of the two conditions on α given in (i). The orthogonality condition is evidently satisfied by at most one vector in W, which proves (ii).

Now suppose that W is a finite-dimensional subspace of V. Then we know, as a corollary of Theorem 3, that W has an orthogonal basis. Let $\{\alpha_1, \ldots, \alpha_n\}$ be any orthogonal basis for W and define α by (8-11). Then, by the computation in the proof of Theorem 3, $\beta - \alpha$ is orthogonal to each of the vectors α_k ($\beta - \alpha$ is the vector obtained at the last stage when the orthogonalization process is applied to $\alpha_1, \ldots, \alpha_n, \beta$). Thus $\beta - \alpha$ is orthogonal to every linear combination of $\alpha_1, \ldots, \alpha_n$, i.e., to every vector in W. If γ is in W and $\gamma \neq \alpha$, it follows that $||\beta - \gamma|| > ||\beta - \alpha||$. Therefore, α is the best approximation to β that lies in W. ∎

Definition. *Let V be an inner product space and S any set of vectors in V. The* **orthogonal complement** *of S is the set $S^\perp$ of all vectors in V which are orthogonal to every vector in S.*

The orthogonal complement of V is the zero subspace, and conversely $\{0\}^\perp = V$. If S is any subset of V, its orthogonal complement $S^\perp$ (S perp) is always a subspace of V. For S is non-empty, since it contains 0; and whenever α and β are in $S^\perp$ and c is any scalar,

$$\begin{aligned}(c\alpha + \beta|\gamma) &= c(\alpha|\gamma) + (\beta|\gamma)\\ &= c0 + 0\\ &= 0\end{aligned}$$

for every γ in S, thus $c\alpha + \beta$ also lies in S. In Theorem 4 the characteristic property of the vector α is that it is the only vector in W such that $\beta - \alpha$ belongs to $W^\perp$.

Definition. *Whenever the vector α in Theorem 4 exists it is called the* **orthogonal projection of β on W**. *If every vector in V has an orthogonal projection on W, the mapping that assigns to each vector in V its orthogonal projection on W is called the* **orthogonal projection of V on W**.

By Theorem 4, the orthogonal projection of an inner product space on a finite-dimensional subspace always exists. But Theorem 4 also implies the following result.

Corollary. *Let V be an inner product space, W a finite-dimensional subspace, and E the orthogonal projection of V on W. Then the mapping*

$$\beta \to \beta - E\beta$$

is the orthogonal projection of V on $W^\perp$.

Proof. Let β be an arbitrary vector in V. Then $\beta - E\beta$ is in $W^\perp$, and for any γ in $W^\perp$, $\beta - \gamma = E\beta + (\beta - E\beta - \gamma)$. Since $E\beta$ is in W and $\beta - E\beta - \gamma$ is in $W^\perp$, it follows that

$$\begin{aligned}||\beta - \gamma||^2 &= ||E\beta||^2 + ||\beta - E\beta - \gamma||^2 \\ &\geq ||\beta - (\beta - E\beta)||^2\end{aligned}$$

with strict inequality when $\gamma \neq \beta - E\beta$. Therefore, $\beta - E\beta$ is the best approximation to β by vectors in $W^\perp$. ∎

Example 14. Give R^3 the standard inner product. Then the orthogonal projection of $(-10, 2, 8)$ on the subspace W that is spanned by $(3, 12, -1)$ is the vector

$$\begin{aligned}\alpha &= \frac{((-10, 2, 8)|(3, 12, -1))}{9 + 144 + 1}(3, 12, -1) \\ &= \frac{-14}{154}(3, 12, -1).\end{aligned}$$

The orthogonal projection of R^3 on W is the linear transformation E defined by

$$(x_1, x_2, x_3) \to \left(\frac{3x_1 + 12x_2 - x_3}{154}\right)(3, 12, -1).$$

The rank of E is clearly 1; hence its nullity is 2. On the other hand,

$$E(x_1, x_2, x_3) = (0, 0, 0)$$

if and only if $3x_1 + 12x_2 - x_3 = 0$. This is the case if and only if (x_1, x_2, x_3) is in $W^\perp$. Therefore, $W^\perp$ is the null space of E, and $\dim (W^\perp) = 2$. Computing

$$(x_1, x_2, x_3) - \left(\frac{3x_1 + 12x_2 - x_3}{154}\right)(3, 12, -1)$$

we see that the orthogonal projection of R^3 on $W^\perp$ is the linear transformation $I - E$ that maps the vector (x_1, x_2, x_3) onto the vector

$$\frac{1}{154}(145x_1 - 36x_2 + 3x_3, -36x_1 + 10x_2 + 12x_3, 3x_1 + 12x_2 + 153x_3).$$

The observations made in Example 14 generalize in the following fashion.

Theorem 5. *Let* W *be a finite-dimensional subspace of an inner product space* V *and let* E *be the orthogonal projection of* V *on* W. *Then* E *is an idempotent linear transformation of* V *onto* W, $W^\perp$ *is the null space of* E, *and*

$$V = W \oplus W^\perp.$$

Proof. Let β be an arbitrary vector in V. Then $E\beta$ is the best approximation to β that lies in W. In particular, $E\beta = \beta$ when β is in W. Therefore, $E(E\beta) = E\beta$ for every β in V; that is, E is idempotent: $E^2 = E$. To prove that E is a linear transformation, let α and β be any vectors in

V and c an arbitrary scalar. Then, by Theorem 4, $\alpha - E\alpha$ and $\beta - E\beta$ are each orthogonal to every vector in W. Hence the vector

$$c(\alpha - E\alpha) + (\beta - E\beta) = (c\alpha + \beta) - (cE\alpha + E\beta)$$

also belongs to $W^\perp$. Since $cE\alpha + E\beta$ is a vector in W, it follows from Theorem 4 that

$$E(c\alpha + \beta) = cE\alpha + E\beta.$$

Of course, one may also prove the linearity of E by using (8-11). Again let β be any vector in V. Then $E\beta$ is the unique vector in W such that $\beta - E\beta$ is in $W^\perp$. Thus $E\beta = 0$ when β is in $W^\perp$. Conversely, β is in $W^\perp$ when $E\beta = 0$. Thus $W^\perp$ is the null space of E. The equation

$$\beta = E\beta + \beta - E\beta$$

shows that $V = W + W^\perp$; moreover, $W \cap W^\perp = \{0\}$. For if α is a vector in $W \cap W^\perp$, then $(\alpha|\alpha) = 0$. Therefore, $\alpha = 0$, and V is the direct sum of W and $W^\perp$. ∎

Corollary. *Under the conditions of the theorem,* I − E *is the orthogonal projection of* V *on* $W^\perp$. *It is an idempotent linear transformation of* V *onto* $W^\perp$ *with null space* W.

Proof. We have already seen that the mapping $\beta \to \beta - E\beta$ is the orthogonal projection of V on $W^\perp$. Since E is a linear transformation, this projection on $W^\perp$ is the linear transformation $I - E$. From its geometric properties one sees that $I - E$ is an idempotent transformation of V onto W. This also follows from the computation

$$\begin{aligned}(I - E)(I - E) &= I - E - E + E^2 \\ &= I - E.\end{aligned}$$

Moreover, $(I - E)\beta = 0$ if and only if $\beta = E\beta$, and this is the case if and only if β is in W. Therefore W is the null space of $I - E$. ∎

The Gram-Schmidt process may now be described geometrically in the following way. Given an inner product space V and vectors $\beta_1, \ldots, \beta_n$ in V, let P_k $(k > 1)$ be the orthogonal projection of V on the orthogonal complement of the subspace spanned by $\beta_1, \ldots, \beta_{k-1}$, and set $P_1 = I$. Then the vectors one obtains by applying the orthogonalization process to $\beta_1, \ldots, \beta_n$ are defined by the equations

(8-12) $$\alpha_k = P_k\beta_k, \qquad 1 \le k \le n.$$

Theorem 5 implies another result known as **Bessel's inequality**.

Corollary. *Let* $\{\alpha_1, \ldots, \alpha_n\}$ *be an orthogonal set of non-zero vectors in an inner product space* V. *If* β *is any vector in* V, *then*

$$\sum_k \frac{|(\beta|\alpha_k)|^2}{\|\alpha_k\|^2} \le \|\beta\|^2$$

and equality holds if and only if

$$\beta = \sum_k \frac{(\beta|\alpha_k)}{||\alpha_k||^2} \alpha_k.$$

Proof. Let $\gamma = \sum_k [(\beta|\alpha_k)/||\alpha_k||^2] \alpha_k$. Then $\beta = \gamma + \delta$ where $(\gamma|\delta) = 0$. Hence

$$||\beta||^2 = ||\gamma||^2 + ||\delta||^2.$$

It now suffices to prove that

$$||\gamma||^2 = \sum_k \frac{|(\beta|\alpha_k)|^2}{||\alpha_k||^2}.$$

This is straightforward computation in which one uses the fact that $(\alpha_j|\alpha_k) = 0$ for $j \neq k$. ∎

In the special case in which $\{\alpha_1, \ldots, \alpha_n\}$ is an orthonormal set, Bessel's inequality says that

$$\sum_k |(\beta|\alpha_k)|^2 \leq ||\beta||^2.$$

The corollary also tells us in this case that β is in the subspace spanned by $\alpha_1, \ldots, \alpha_n$ if and only if

$$\beta = \sum_k (\beta|\alpha_k) \alpha_k$$

or if and only if Bessel's inequality is actually an equality. Of course, in the event that V is finite dimensional and $\{\alpha_1, \ldots, \alpha_n\}$ is an orthogonal basis for V, the above formula holds for every vector β in V. In other words, if $\{\alpha_1, \ldots, \alpha_n\}$ is an orthonormal basis for V, the kth coordinate of β in the ordered basis $\{\alpha_1, \ldots, \alpha_n\}$ is $(\beta|\alpha_k)$.

EXAMPLE 15. We shall apply the last corollary to the orthogonal sets described in Example 11. We find that

(a) $$\sum_{k=-n}^{n} \left| \int_0^1 f(t)e^{-2\pi ikt}\, dt \right|^2 \leq \int_0^1 |f(t)|^2\, dt$$

(b) $$\int_0^1 \left| \sum_{k=-n}^{n} c_k e^{2\pi ikt} \right|^2 dt = \sum_{k=-n}^{n} |c_k|^2$$

(c) $$\int_0^1 (\sqrt{2} \cos 2\pi t + \sqrt{2} \sin 4\pi t)^2\, dt = 1 + 1 = 2.$$

Exercises

1. Consider R^4 with the standard inner product. Let W be the subspace of R^4 consisting of all vectors which are orthogonal to both $\alpha = (1, 0, -1, 1)$ and $\beta = (2, 3, -1, 2)$. Find a basis for W.

2. Apply the Gram-Schmidt process to the vectors $\beta_1 = (1, 0, 1)$, $\beta_2 = (1, 0, -1)$, $\beta_3 = (0, 3, 4)$, to obtain an orthonormal basis for R^3 with the standard inner product.

3. Consider C^3, with the standard inner product. Find an orthonormal basis for the subspace spanned by $\beta_1 = (1, 0, i)$ and $\beta_2 = (2, 1, 1 + i)$.

4. Let V be an inner product space. The **distance** between two vectors α and β in V is defined by

$$d(\alpha, \beta) = ||\alpha - \beta||.$$

Show that

(a) $d(\alpha, \beta) \geq 0$;
(b) $d(\alpha, \beta) = 0$ if and only if $\alpha = \beta$;
(c) $d(\alpha, \beta) = d(\beta, \alpha)$;
(d) $d(\alpha, \beta) \leq d(\alpha, \gamma) + d(\gamma, \beta)$.

5. Let V be an inner product space, and let α, β be vectors in V. Show that $\alpha = \beta$ if and only if $(\alpha|\gamma) = (\beta|\gamma)$ for every γ in V.

6. Let W be the subspace of R^2 spanned by the vector (3, 4). Using the standard inner product, let E be the orthogonal projection of R^2 onto W. Find

(a) a formula for $E(x_1, x_2)$;
(b) the matrix of E in the standard ordered basis;
(c) $W^{\perp}$;
(d) an orthonormal basis in which E is represented by the matrix

$$\begin{bmatrix} 1 & 0 \\ 0 & 0 \end{bmatrix}.$$

7. Let V be the inner product space consisting of R^2 and the inner product whose quadratic form is defined by

$$||(x_1, x_2||^2 = (x_1 - x_2)^2 + 3x_2^2.$$

Let E be the orthogonal projection of V onto the subspace W spanned by the vector (3, 4). Now answer the four questions of Exercise 6.

8. Find an inner product on R^2 such that $(\epsilon_1, \epsilon_2) = 2$.

9. Let V be the subspace of $R[x]$ of polynomials of degree at most 3. Equip V with the inner product

$$(f|g) = \int_0^1 f(t)g(t)\, dt.$$

(a) Find the orthogonal complement of the subspace of scalar polynomials.
(b) Apply the Gram-Schmidt process to the basis $\{1, x, x^2, x^3\}$.

10. Let V be the vector space of all $n \times n$ matrices over C, with the inner product $(A|B) = \operatorname{tr}(AB^*)$. Find the orthogonal complement of the subspace of diagonal matrices.

11. Let V be a finite-dimensional inner product space, and let $\{\alpha_1, \ldots, \alpha_n\}$ be an orthonormal basis for V. Show that for any vectors α, β in V

$$(\alpha|\beta) = \sum_{k=1}^{n} (\alpha|\alpha_k)\overline{(\beta|\alpha_k)}.$$

12. Let W be a finite-dimensional subspace of an inner product space V, and let E be the orthogonal projection of V on W. Prove that $(E\alpha|\beta) = (\alpha|E\beta)$ for all α, β in V.

13. Let S be a subset of an inner product space V. Show that $(S^{\perp})^{\perp}$ contains the subspace spanned by S. When V is finite-dimensional, show that $(S^{\perp})^{\perp}$ is the subspace spanned by S.

14. Let V be a finite-dimensional inner product space, and let $\mathcal{B} = \{\alpha_1, \ldots, \alpha_n\}$ be an orthonormal basis for V. Let T be a linear operator on V and A the matrix of T in the ordered basis $\mathcal{B}$. Prove that

$$A_{ij} = (T\alpha_j|\alpha_i).$$

15. Suppose $V = W_1 \oplus W_2$ and that f_1 and f_2 are inner products on W_1 and W_2, respectively. Show that there is a unique inner product f on V such that

(a) $W_2 = W_1^{\perp}$;

(b) $f(\alpha, \beta) = f_k(\alpha, \beta)$, when α, β are in W_k, $k = 1, 2$.

16. Let V be an inner product space and W a finite-dimensional subspace of V. There are (in general) many projections which have W as their range. One of these, the orthogonal projection on W, has the property that $||E\alpha|| \leq ||\alpha||$ for every α in V. Prove that if E is a projection with range W, such that $||E\alpha|| \leq ||\alpha||$ for all α in V, then E is the orthogonal projection on W.

17. Let V be the real inner product space consisting of the space of real-valued continuous functions on the interval, $-1 \leq t \leq 1$, with the inner product

$$(f|g) = \int_{-1}^{1} f(t)g(t)\,dt.$$

Let W be the subspace of odd functions, i.e., functions satisfying $f(-t) = -f(t)$. Find the orthogonal complement of W.

8.3. Linear Functionals and Adjoints

The first portion of this section treats linear functionals on an inner product space and their relation to the inner product. The basic result is that any linear functional f on a finite-dimensional inner product space is 'inner product with a fixed vector in the space,' i.e., that such an f has the form $f(\alpha) = (\alpha|\beta)$ for some fixed β in V. We use this result to prove the existence of the 'adjoint' of a linear operator T on V, this being a linear operator T^* such that $(T\alpha|\beta) = (\alpha|T^*\beta)$ for all α and β in V. Through the use of an orthonormal basis, this adjoint operation on linear operators (passing from T to T^*) is identified with the operation of forming the conjugate transpose of a matrix. We explore slightly the analogy between the adjoint operation and conjugation on complex numbers.

Let V be any inner product space, and let β be some fixed vector in V. We define a function f_β from V into the scalar field by

$$f_\beta(\alpha) = (\alpha|\beta).$$

This function f_β is a linear functional on V, because, by its very definition, $(\alpha|\beta)$ is linear as a function of α. If V is finite-dimensional, every linear functional on V arises in this way from some β.

Theorem 6. *Let* V *be a finite-dimensional inner product space, and* f *a linear functional on* V. *Then there exists a unique vector* β *in* V *such that* $f(\alpha) = (\alpha|\beta)$ *for all* α *in* V.

Proof. Let $\{\alpha_1, \alpha_2, \ldots, \alpha_n\}$ be an orthonormal basis for V. Put

$$\beta = \sum_{j=1}^{n} \overline{f(\alpha_j)}\alpha_j \tag{8-13}$$

and let f_β be the linear functional defined by

$$f_\beta(\alpha) = (\alpha|\beta).$$

Then

$$f_\beta(\alpha_k) = (\alpha_k|\sum_j \overline{f(\alpha_j)}\alpha_j) = f(\alpha_k).$$

Since this is true for each α_k, it follows that $f = f_\beta$. Now suppose γ is a vector in V such that $(\alpha|\beta) = (\alpha|\gamma)$ for all α. Then $(\beta - \gamma|\beta - \gamma) = 0$ and $\beta = \gamma$. Thus there is exactly one vector β determining the linear functional f in the stated manner. ∎

The proof of this theorem can be reworded slightly, in terms of the representation of linear functionals in a basis. If we choose an orthonormal basis $\{\alpha_1, \ldots, \alpha_n\}$ for V, the inner product of $\alpha = x_1\alpha_1 + \cdots + x_n\alpha_n$ and $\beta = y_1\alpha_1 + \cdots + y_n\alpha_n$ will be

$$(\alpha|\beta) = x_1\bar{y}_1 + \cdots + x_n\bar{y}_n.$$

If f is any linear functional on V, then f has the form

$$f(\alpha) = c_1x_1 + \cdots + c_nx_n$$

for some fixed scalars $c_1, \ldots, c_n$ determined by the basis. Of course $c_j = f(\alpha_j)$. If we wish to find a vector β in V such that $(\alpha|\beta) = f(\alpha)$ for all α, then clearly the coordinates y_j of β must satisfy $\bar{y}_j = c_j$ or $y_j = \overline{f(\alpha_j)}$. Accordingly,

$$\beta = \overline{f(\alpha_1)}\alpha_1 + \cdots + \overline{f(\alpha_n)}\alpha_n$$

is the desired vector.

Some further comments are in order. The proof of Theorem 6 that we have given is admirably brief, but it fails to emphasize the essential geometric fact that β lies in the orthogonal complement of the null space of f. Let W be the null space of f. Then $V = W + W^\perp$, and f is completely determined by its values on $W^\perp$. In fact, if P is the orthogonal projection of V on $W^\perp$, then

$$f(\alpha) = f(P\alpha)$$

for all α in V. Suppose $f \neq 0$. Then f is of rank 1 and $\dim (W^\perp) = 1$. If γ is any non-zero vector in $W^\perp$, it follows that

$$P\alpha = \frac{(\alpha|\gamma)}{||\gamma||^2}\gamma$$

for all α in V. Thus

$$f(\alpha) = (\alpha|\gamma) \cdot \frac{f(\gamma)}{||\gamma||^2}$$

for all α, and $\beta = [\overline{f(\gamma)}/||\gamma||^2]\,\gamma$.

EXAMPLE 16. We should give one example showing that Theorem 6 is not true without the assumption that V is finite dimensional. Let V be the vector space of polynomials over the field of complex numbers, with the inner product

$$(f|g) = \int_0^1 f(t)\overline{g(t)}\,dt.$$

This inner product can also be defined algebraically. If $f = \Sigma\, a_k x^k$ and $g = \Sigma\, b_k x^k$, then

$$(f|g) = \sum_{j,k} \frac{1}{j+k+1} a_j \overline{b_k}.$$

Let z be a fixed complex number, and let L be the linear functional 'evaluation at z':

$$L(f) = f(z).$$

Is there a polynomial g such that $(f|g) = L(f)$ for every f? The answer is no; for suppose we have

$$f(z) = \int_0^1 f(t)\overline{g(t)}\,dt$$

for every f. Let $h = x - z$, so that for any f we have $(hf)(z) = 0$. Then

$$0 = \int_0^1 h(t)f(t)\overline{g(t)}\,dt$$

for all f. In particular this holds when $f = \bar{h}g$ so that

$$\int_0^1 |h(t)|^2|g(t)|^2\,dt = 0$$

and so $hg = 0$. Since $h \neq 0$, it must be that $g = 0$. But L is not the zero functional; hence, no such g exists.

One can generalize the example somewhat, to the case where L is a linear combination of point evaluations. Suppose we select fixed complex numbers $z_1, \ldots, z_n$ and scalars $c_1, \ldots, c_n$ and let

$$L(f) = c_1 f(z_1) + \cdots + c_n f(z_n).$$

Then L is a linear functional on V, but there is no g with $L(f) = (f|g)$, unless $c_1 = c_2 = \cdots = c_n = 0$. Just repeat the above argument with $h = (x - z_1) \cdots (x - z_n)$.

We turn now to the concept of the adjoint of a linear operator.

Theorem 7. *For any linear operator* T *on a finite-dimensional inner product space* V, *there exists a unique linear operator* T* *on* V *such that*

$$(8\text{-}14) \qquad (\mathrm{T}\alpha|\beta) = (\alpha|\mathrm{T}^*\beta)$$

for all α, β *in* V.

Proof. Let β be any vector in V. Then $\alpha \rightarrow (T\alpha|\beta)$ is a linear functional on V. By Theorem 6 there is a unique vector β' in V such that $(T\alpha|\beta) = (\alpha|\beta')$ for every α in V. Let T^* denote the mapping $\beta \rightarrow \beta'$:

$$\beta' = T^*\beta.$$

We have (8-14), but we must verify that T^* is a linear operator. Let β, γ be in V and let c be a scalar. Then for any α,

$$\begin{aligned}(\alpha|T^*(c\beta + \gamma)) &= (T\alpha|c\beta + \gamma) \\ &= (T\alpha|c\beta) + (T\alpha|\gamma) \\ &= \bar{c}(T\alpha|\beta) + (T\alpha|\gamma) \\ &= \bar{c}(\alpha|T^*\beta) + (\alpha|T^*\gamma) \\ &= (\alpha|cT^*\beta) + (\alpha|T^*\gamma) \\ &= (\alpha|cT^*\beta + T^*\gamma).\end{aligned}$$

Thus $T^*(c\beta + \gamma) = cT^*\beta + T^*\gamma$ and T^* is linear.

The uniqueness of T^* is clear. For any β in V, the vector $T^*\beta$ is uniquely determined as the vector β' such that $(T\alpha|\beta) = (\alpha|\beta')$ for every α. ∎

Theorem 8. *Let* V *be a finite-dimensional inner product space and let* $\mathfrak{B} = \{\alpha_1, \ldots, \alpha_n\}$ *be an (ordered) orthonormal basis for* V. *Let* T *be a linear operator on* V *and let* A *be the matrix of* T *in the ordered basis* $\mathfrak{B}$. *Then* $A_{kj} = (T\alpha_j|\alpha_k)$.

Proof. Since $\mathfrak{B}$ is an orthonormal basis, we have

$$\alpha = \sum_{k=1}^{n} (\alpha|\alpha_k)\alpha_k.$$

The matrix A is defined by

$$T\alpha_j = \sum_{k=1}^{n} A_{kj}\alpha_k$$

and since

$$T\alpha_j = \sum_{k=1}^{n} (T\alpha_j|\alpha_k)\alpha_k$$

we have $A_{kj} = (T\alpha_j|\alpha_k)$. ∎

Corollary. *Let V be a finite-dimensional inner product space, and let T be a linear operator on V. In any orthonormal basis for V, the matrix of T* is the conjugate transpose of the matrix of T.*

Proof. Let $\mathcal{B} = \{\alpha_1, \ldots, \alpha_n\}$ be an orthonormal basis for V, let $A = [T]_{\mathcal{B}}$ and $B = [T^*]_{\mathcal{B}}$. According to Theorem 8,

$$A_{kj} = (T\alpha_j|\alpha_k)$$
$$B_{kj} = (T^*\alpha_j|\alpha_k).$$

By the definition of T^* we then have

$$\begin{aligned} B_{kj} &= (T^*\alpha_j|\alpha_k) \\ &= \overline{(\alpha_k|T^*\alpha_j)} \\ &= \overline{(T\alpha_k|\alpha_j)} \\ &= \overline{A_{jk}}. \quad \blacksquare \end{aligned}$$

EXAMPLE 17. Let V be a finite-dimensional inner product space and E the orthogonal projection of V on a subspace W. Then for any vectors α and β in V.

$$\begin{aligned} (E\alpha|\beta) &= (E\alpha|E\beta + (1 - E)\beta) \\ &= (E\alpha|E\beta) \\ &= (E\alpha + (1 - E)\alpha|E\beta) \\ &= (\alpha|E\beta). \end{aligned}$$

From the uniqueness of the operator E^* it follows that $E^* = E$. Now consider the projection E described in Example 14. Then

$$A = \frac{1}{154}\begin{bmatrix} 9 & 36 & -3 \\ 36 & 144 & -12 \\ -3 & -12 & 1 \end{bmatrix}$$

is the matrix of E in the standard orthonormal basis. Since $E = E^*$, A is also the matrix of E^*, and because $A = A^*$, this does not contradict the preceding corollary. On the other hand, suppose

$$\begin{aligned} \alpha_1 &= (154, 0, 0) \\ \alpha_2 &= (145, -36, 3) \\ \alpha_3 &= (-36, 10, 12). \end{aligned}$$

Then $\{\alpha_1, \alpha_2, \alpha_3\}$ is a basis, and

$$\begin{aligned} E\alpha_1 &= (9, 36, -3) \\ E\alpha_2 &= (0, 0, 0) \\ E\alpha_3 &= (0, 0, 0). \end{aligned}$$

Since $(9, 36, -3) = -(154, 0, 0) - (145, -36, 3)$, the matrix B of E in the basis $\{\alpha_1, \alpha_2, \alpha_3\}$ is defined by the equation

$$B = \begin{bmatrix} -1 & 0 & 0 \\ -1 & 0 & 0 \\ 0 & 0 & 0 \end{bmatrix}.$$

In this case $B \neq B^*$, and B^* is not the matrix of $E^* = E$ in the basis $\{\alpha_1, \alpha_2, \alpha_3\}$. Applying the corollary, we conclude that $\{\alpha_1, \alpha_2, \alpha_3\}$ is not an orthonormal basis. Of course this is quite obvious anyway.

Definition. *Let* T *be a linear operator on an inner product space* V. *Then we say that* T **has an adjoint on** V *if there exists a linear operator* T^* *on* V *such that* $(T\alpha|\beta) = (\alpha|T^*\beta)$ *for all* α *and* β *in* V.

By Theorem 7 every linear operator on a finite-dimensional inner product space V has an adjoint on V. In the infinite-dimensional case this is not always true. But in any case there is at most one such operator T^*; when it exists, we call it the **adjoint** of T.

Two comments should be made about the finite-dimensional case.

1. The adjoint of T depends not only on T but on the inner product as well.

2. As shown by Example 17, in an arbitrary ordered basis $\mathfrak{B}$, the relation between $[T]_{\mathfrak{B}}$ and $[T^*]_{\mathfrak{B}}$ is more complicated than that given in the corollary above.

EXAMPLE 18. Let V be $C^{n\times 1}$, the space of complex $n \times 1$ matrices, with inner product $(X|Y) = Y^*X$. If A is an $n \times n$ matrix with complex entries, the adjoint of the linear operator $X \to AX$ is the operator $X \to A^*X$. For

$$(AX|Y) = Y^*AX = (A^*Y)^*X = (X|A^*Y).$$

The reader should convince himself that this is really a special case of the last corollary.

EXAMPLE 19. This is similar to Example 18. Let V be $C^{n\times n}$ with the inner product $(A|B) = \operatorname{tr}(B^*A)$. Let M be a fixed $n \times n$ matrix over C. The adjoint of left multiplication by M is left multiplication by M^*. Of course, 'left multiplication by M' is the linear operator L_M defined by $L_M(A) = MA$.

$$\begin{aligned} (L_M(A)|B) &= \operatorname{tr}(B^*(MA)) \\ &= \operatorname{tr}(MAB^*) \\ &= \operatorname{tr}(AB^*M) \\ &= \operatorname{tr}(A(M^*B)^*) \\ &= (A|L_{M^*}(B)). \end{aligned}$$

Thus $(L_M)^* = L_{M^*}$. In the computation above, we twice used the characteristic property of the trace function: $\operatorname{tr}(AB) = \operatorname{tr}(BA)$.

Example 20. Let V be the space of polynomials over the field of complex numbers, with the inner product

$$(f|g) = \int_0^1 f(t)\overline{g(t)}\,dt.$$

If f is a polynomial, $f = \Sigma\, a_k x^k$, we let $\bar{f} = \Sigma\, \bar{a}_k x^k$. That is, $\bar{f}$ is the polynomial whose associated polynomial function is the complex conjugate of that for f:

$$\bar{f}(t) = \overline{f(t)}, \qquad t \text{ real}$$

Consider the operator 'multiplication by f,' that is, the linear operator M_f defined by $M_f(g) = fg$. Then this operator has an adjoint, namely, multiplication by $\bar{f}$. For

$$\begin{aligned}(M_f(g)|h) &= (fg|h)\\ &= \int_0^1 f(t)g(t)\overline{h(t)}\,dt\\ &= \int_0^1 g(t)\,[\overline{\bar{f}(t)h(t)}]\,dt\\ &= (g|\bar{f}h)\\ &= (g|M_{\bar{f}}(h))\end{aligned}$$

and so $(M_{\bar{f}})^* = M_f$.

Example 21. In Example 20, we saw that some linear operators on an infinite-dimensional inner product space do have an adjoint. As we commented earlier, some do not. Let V be the inner product space of Example 21, and let D be the differentiation operator on $C[x]$. Integration by parts shows that

$$(Df|g) = f(1)g(1) - f(0)g(0) - (f|Dg).$$

Let us fix g and inquire when there is a polynomial D^*g such that $(Df|g) = (f|D^*g)$ for all f. If such a D^*g exists, we shall have

$$(f|D^*g) = f(1)g(1) - f(0)g(0) - (f|Dg)$$

or

$$(f|D^*g + Dg) = f(1)g(1) - f(0)g(0).$$

With g fixed, $L(f) = f(1)g(1) - f(0)g(0)$ is a linear functional of the type considered in Example 16 and cannot be of the form $L(f) = (f|h)$ unless $L = 0$. If D^*g exists, then with $h = D^*g + Dg$ we do have $L(f) = (f|h)$, and so $g(0) = g(1) = 0$. The existence of a suitable polynomial D^*g implies $g(0) = g(1) = 0$. Conversely, if $g(0) = g(1) = 0$, the polynomial $D^*g = -Dg$ satisfies $(Df|g) = (f|D^*g)$ for all f. If we choose any g for which $g(0) \neq 0$ or $g(1) \neq 0$, we cannot suitably define D^*g, and so we conclude that D has no adjoint.

We hope that these examples enhance the reader's understanding of the adjoint of a linear operator. We see that the adjoint operation, passing from T to T^*, behaves somewhat like conjugation on complex numbers. The following theorem strengthens the analogy.

Theorem 9. *Let* V *be a finite-dimensional inner product space. If* T *and* U *are linear operators on* V *and* c *is a scalar,*

(i) $(T + U)^* = T^* + U^*$;
(ii) $(cT)^* = \bar{c}T^*$;
(iii) $(TU)^* = U^*T^*$;
(iv) $(T^*)^* = T$.

Proof. To prove (i), let α and β be any vectors in V. Then

$$\begin{aligned}((T + U)\alpha|\beta) &= (T\alpha + U\alpha|\beta)\\ &= (T\alpha|\beta) + (U\alpha|\beta)\\ &= (\alpha|T^*\beta) + (\alpha|U^*\beta)\\ &= (\alpha|T^*\beta + U^*\beta)\\ &= (\alpha|(T^* + U^*)\beta).\end{aligned}$$

From the uniqueness of the adjoint we have $(T + U)^* = T^* + U^*$. We leave the proof of (ii) to the reader. We obtain (iii) and (iv) from the relations

$$(TU\alpha|\beta) = (U\alpha|T^*\beta) = (\alpha|U^*T^*\beta)$$
$$(T^*\alpha|\beta) = \overline{(\beta|T^*\alpha)} = \overline{(T\beta|\alpha)} = (\alpha|T\beta). \quad \blacksquare$$

Theorem 9 is often phrased as follows: The mapping $T \rightarrow T^*$ is a conjugate-linear anti-isomorphism of period 2. The analogy with complex conjugation which we mentioned above is, of course, based upon the observation that complex conjugation has the properties $\overline{(z_1 + z_2)} = \bar{z}_1 + \bar{z}_2$, $\overline{(z_1z_2)} = \bar{z}_1\bar{z}_2$, $\bar{\bar{z}} = z$. One must be careful to observe the reversal of order in a product, which the adjoint operation imposes: $(UT)^* = T^*U^*$. We shall mention extensions of this analogy as we continue our study of linear operators on an inner product space. We might mention something along these lines now. A complex number z is real if and only if $z = \bar{z}$. One might expect that the linear operators T such that $T = T^*$ behave in some way like the real numbers. This is in fact the case. For example, if T is a linear operator on a finite-dimensional *complex* inner product space, then

$$T = U_1 + iU_2 \tag{8-15}$$

where $U_1 = U_1^*$ and $U_2 = U_2^*$. Thus, in some sense, T has a 'real part' and an 'imaginary part.' The operators U_1 and U_2 satisfying $U_1 = U_1^*$, and $U_2 = U_2^*$, and (8-15) are unique, and are given by

$$U_1 = \frac{1}{2}(T + T^*)$$

$$U_2 = \frac{1}{2i}(T - T^*).$$

A linear operator T such that $T = T^*$ is called **self-adjoint** (or **Hermitian**). If $\mathfrak{B}$ is an orthonormal basis for V, then

$$[T^*]_{\mathfrak{B}} = [T]^*_{\mathfrak{B}}$$

and so T is self-adjoint if and only if its matrix in every orthonormal basis is a self-adjoint matrix. Self-adjoint operators are important, not simply because they provide us with some sort of real and imaginary part for the general linear operator, but for the following reasons: (1) Self-adjoint operators have many special properties. For example, for such an operator there is an orthonormal basis of characteristic vectors. (2) Many operators which arise in practice are self-adjoint. We shall consider the special properties of self-adjoint operators later.

Exercises

1. Let V be the space C^2, with the standard inner product. Let T be the linear operator defined by $T\epsilon_1 = (1, -2)$, $T\epsilon_2 = (i, -1)$. If $\alpha = (x_1, x_2)$, find $T^*\alpha$.

2. Let T be the linear operator on C^2 defined by $T\epsilon_1 = (1 + i, 2)$, $T\epsilon_2 = (i, i)$. Using the standard inner product, find the matrix of T^* in the standard ordered basis. Does T commute with T^*?

3. Let V be C^3 with the standard inner product. Let T be the linear operator on V whose matrix in the standard ordered basis is defined by

$$A_{jk} = i^{j+k}, \qquad (i^2 = -1).$$

Find a basis for the null space of T^*.

4. Let V be a finite-dimensional inner product space and T a linear operator on V. Show that the range of T^* is the orthogonal complement of the null space of T.

5. Let V be a finite-dimensional inner product space and T a linear operator on V. If T is invertible, show that T^* is invertible and $(T^*)^{-1} = (T^{-1})^*$.

6. Let V be an inner product space and β, γ fixed vectors in V. Show that $T\alpha = (\alpha|\beta)\gamma$ defines a linear operator on V. Show that T has an adjoint, and describe T^* explicitly.

Now suppose V is C^n with the standard inner product, $\beta = (y_1, \ldots, y_n)$, and $\gamma = (x_1, \ldots, x_n)$. What is the j, k entry of the matrix of T in the standard ordered basis? What is the rank of this matrix?

7. Show that the product of two self-adjoint operators is self-adjoint if and only if the two operators commute.

8. Let V be the vector space of the polynomials over R of degree less than or equal to 3, with the inner product

$$(f|g) = \int_0^1 f(t)g(t)\,dt.$$

If t is a real number, find the polynomial g_t in V such that $(f|g_t) = f(t)$ for all f in V.

9. Let V be the inner product space of Exercise 8, and let D be the differentiation operator on V. Find D^*.

10. Let V be the space of $n \times n$ matrices over the complex numbers, with the inner product $(A, B) = \operatorname{tr}(AB^*)$. Let P be a fixed invertible matrix in V, and let T_P be the linear operator on V defined by $T_P(A) = P^{-1}AP$. Find the adjoint of T_P.

11. Let V be a finite-dimensional inner product space, and let E be an idempotent linear operator on V, i.e., $E^2 = E$. Prove that E is self-adjoint if and only if $EE^* = E^*E$.

12. Let V be a finite-dimensional *complex* inner product space, and let T be a linear operator on V. Prove that T is self-adjoint if and only if $(T\alpha|\alpha)$ is real for every α in V.

8.4. Unitary Operators

In this section, we consider the concept of an isomorphism between two inner product spaces. If V and W are vector spaces, an isomorphism of V onto W is a one-one linear transformation from V onto W, i.e., a one-one correspondence between the elements of V and those of W, which 'preserves' the vector space operations. Now an inner product space consists of a vector space and a specified inner product on that space. Thus, when V and W are inner product spaces, we shall require an isomorphism from V onto W not only to preserve the linear operations, but also to preserve inner products. An isomorphism of an inner product space onto itself is called a 'unitary operator' on that space. We shall consider various examples of unitary operators and establish their basic properties.

Definition. *Let* V *and* W *be inner product spaces over the same field, and let* T *be a linear transformation from* V *into* W. *We say that* T **preserves inner products** *if* $(T\alpha|T\beta) = (\alpha|\beta)$ *for all* α, β *in* V. *An* **isomorphism** *of* V *onto* W *is a vector space isomorphism* T *of* V *onto* W *which also preserves inner products.*

If T preserves inner products, then $||T\alpha|| = ||\alpha||$ and so T is necessarily non-singular. Thus an isomorphism from V onto W can also be defined as a linear transformation from V onto W which preserves inner products. If T is an isomorphism of V onto W, then T^{-1} is an isomorphism of W onto V; hence, when such a T exists, we shall simply say V and W are **isomorphic.** Of course, isomorphism of inner product spaces is an equivalence relation.

Theorem 10. *Let V and W be finite-dimensional inner product spaces over the same field, having the same dimension. If T is a linear transformation from V into W, the following are equivalent.*

(i) T *preserves inner products.*

(ii) T *is an (inner product space) isomorphism.*

(iii) T *carries every orthonormal basis for* V *onto an orthonormal basis for* W.

(iv) T *carries some orthonormal basis for* V *onto an orthonormal basis for* W.

Proof. (i) → (ii) If T preserves inner products, then $||T\alpha|| = ||\alpha||$ for all α in V. Thus T is non-singular, and since $\dim V = \dim W$, we know that T is a vector space isomorphism.

(ii) → (iii) Suppose T is an isomorphism. Let $\{\alpha_1, \ldots, \alpha_n\}$ be an orthonormal basis for V. Since T is a vector space isomorphism and $\dim W = \dim V$, it follows that $\{T\alpha_1, \ldots, T\alpha_n\}$ is a basis for W. Since T also preserves inner products, $(T\alpha_j|T\alpha_k) = (\alpha_j|\alpha_k) = \delta_{jk}$.

(iii) → (iv) This requires no comment.

(iv) → (i) Let $\{\alpha_1, \ldots, \alpha_n\}$ be an orthonormal basis for V such that $\{T\alpha_1, \ldots, T\alpha_n\}$ is an orthonormal basis for W. Then

$$(T\alpha_j|T\alpha_k) = (\alpha_j|\alpha_k) = \delta_{jk}.$$

For any $\alpha = x_1\alpha_1 + \cdots + x_n\alpha_n$ and $\beta = y_1\alpha_1 + \cdots + y_n\alpha_n$ in V, we have

$$\begin{aligned}(\alpha|\beta) &= \sum_{j=1}^{n} x_j\bar{y}_j \\ (T\alpha|T\beta) &= (\sum_j x_jT\alpha_j|\sum_k y_kT\alpha_k) \\ &= \sum_j\sum_k x_j\bar{y}_k(T\alpha_j|T\alpha_k) \\ &= \sum_{j=1}^{n} x_j\bar{y}_j\end{aligned}$$

and so T preserves inner products. ∎

Corollary. *Let V and W be finite-dimensional inner product spaces over the same field. Then V and W are isomorphic if and only if they have the same dimension.*

Proof. If $\{\alpha_1, \ldots, \alpha_n\}$ is an orthonormal basis for V and $\{\beta_1, \ldots, \beta_n\}$ is an orthonormal basis for W, let T be the linear transformation from V into W defined by $T\alpha_j = \beta_j$. Then T is an isomorphism of V onto W. ∎

EXAMPLE 22. If V is an n-dimensional inner product space, then each ordered orthonormal basis $\mathcal{B} = \{\alpha_1, \ldots, \alpha_n\}$ determines an isomorphism of V onto F^n with the standard inner product. The isomorphism is simply

$$T(x_1\alpha_1 + \cdots + x_n\alpha_n) = (x_1, \ldots, x_n).$$

There is the superficially different isomorphism which $\mathcal{B}$ determines of V onto the space $F^{n\times 1}$ with $(X|Y) = Y^*X$ as inner product. The isomorphism is

$$\alpha \to [\alpha]_{\mathcal{B}}$$

i.e., the transformation sending α into its coordinate matrix in the ordered basis $\mathcal{B}$. For any ordered basis $\mathcal{B}$, this is a vector space isomorphism; however, it is an isomorphism of the two inner product spaces if and only if $\mathcal{B}$ is orthonormal.

EXAMPLE 23. Here is a slightly less superficial isomorphism. Let W be the space of all 3×3 matrices A over R which are skew-symmetric, i.e., $A^t = -A$. We equip W with the inner product $(A|B) = \frac{1}{2}\operatorname{tr}(AB^t)$, the $\frac{1}{2}$ being put in as a matter of convenience. Let V be the space R^3 with the standard inner product. Let T be the linear transformation from V into W defined by

$$T(x_1, x_2, x_3) = \begin{bmatrix} 0 & -x_3 & x_2 \\ x_3 & 0 & -x_1 \\ -x_2 & x_1 & 0 \end{bmatrix}.$$

Then T maps V onto W, and putting

$$A = \begin{bmatrix} 0 & -x_3 & x_2 \\ x_3 & 0 & -x_1 \\ -x_2 & x_1 & 0 \end{bmatrix}, \qquad B = \begin{bmatrix} 0 & -y_3 & y_2 \\ y_3 & 0 & -y_1 \\ -y_2 & y_1 & 0 \end{bmatrix}$$

we have

$$\begin{aligned} \operatorname{tr}(AB^t) &= x_3y_3 + x_2y_2 + x_3y_3 + x_2y_2 + x_1y_1 \\ &= 2(x_1y_1 + x_2y_2 + x_3y_3). \end{aligned}$$

Thus $(\alpha|\beta) = (T\alpha|T\beta)$ and T is a vector space isomorphism. Note that T carries the standard basis $\{\epsilon_1, \epsilon_2, \epsilon_3\}$ onto the orthonormal basis consisting of the three matrices

$$\begin{bmatrix} 0 & 0 & 0 \\ 0 & 0 & -1 \\ 0 & 1 & 0 \end{bmatrix}, \quad \begin{bmatrix} 0 & 0 & 1 \\ 0 & 0 & 0 \\ -1 & 0 & 0 \end{bmatrix}, \quad \begin{bmatrix} 0 & -1 & 0 \\ 1 & 0 & 0 \\ 0 & 0 & 0 \end{bmatrix}.$$

EXAMPLE 24. It is not always particularly convenient to describe an isomorphism in terms of orthonormal bases. For example, suppose $G = P^*P$ where P is an invertible $n \times n$ matrix with complex entries. Let V be the space of complex $n \times 1$ matrices, with the inner product $[X|Y] = Y^*GX$.

Let W be the same vector space, with the standard inner product $(X|Y) = Y^*X$. We know that V and W are isomorphic inner product spaces. It would seem that the most convenient way to describe an isomorphism between V and W is the following: Let T be the linear transformation from V into W defined by $T(X) = PX$. Then

$$\begin{aligned}(TX|TY) &= (PX|PY)\\ &= (PY)^*(PX)\\ &= Y^*P^*PX\\ &= Y^*GX\\ &= [X|Y].\end{aligned}$$

Hence T is an isomorphism.

EXAMPLE 25. Let V be the space of all continuous real-valued functions on the unit interval, $0 \leq t \leq 1$, with the inner product

$$[f|g] = \int_0^1 f(t)g(t)t^2\,dt.$$

Let W be the same vector space with the inner product

$$(f|g) = \int_0^1 f(t)g(t)\,dt.$$

Let T be the linear transformation from V into W given by

$$(Tf)(t) = tf(t).$$

Then $(Tf|Tg) = [f|g]$, and so T preserves inner products; however, T is *not* an isomorphism of V onto W, because the range of T is not all of W. Of course, this happens because the underlying vector space is not finite-dimensional.

Theorem 11. *Let* V *and* W *be inner product spaces over the same field, and let* T *be a linear transformation from* V *into* W. *Then* T *preserves inner products if and only if* $||T\alpha|| = ||\alpha||$ *for every* α *in* V.

Proof. If T preserves inner products, T 'preserves norms.' Suppose $||T\alpha|| = ||\alpha||$ for every α in V. Then $||T\alpha||^2 = ||\alpha||^2$. Now using the appropriate polarization identity, (8-3) or (8-4), and the fact that T is linear, one easily obtains $(\alpha|\beta) = (T\alpha|T\beta)$ for all α, β in V. ∎

Definition. *A* **unitary operator** *on an inner product space is an isomorphism of the space onto itself.*

The product of two unitary operators is unitary. For, if U_1 and U_2 are unitary, then U_2U_1 is invertible and $||U_2U_1\alpha|| = ||U_1\alpha|| = ||\alpha||$ for each α. Also, the inverse of a unitary operator is unitary, since $||U\alpha|| = ||\alpha||$ says $||U^{-1}\beta|| = ||\beta||$, where $\beta = U\alpha$. Since the identity operator is clearly unitary, we see that the set of all unitary operators on an inner product space is a group, under the operation of composition.

If V is a finite-dimensional inner product space and U is a linear operator on V, Theorem 10 tells us that U is unitary if and only if $(U\alpha|U\beta) = (\alpha|\beta)$ for each α, β in V; or, if and only if for some (every) orthonormal basis $\{\alpha_1, \ldots, \alpha_n\}$ it is true that $\{U\alpha_1, \ldots, U\alpha_n\}$ is an orthonormal basis.

Theorem 12. *Let* U *be a linear operator on an inner product space* V. *Then* U *is unitary if and only if the adjoint* U* *of* U *exists and* UU* = U*U = I.

Proof. Suppose U is unitary. Then U is invertible and

$$(U\alpha|\beta) = (U\alpha|UU^{-1}\beta) = (\alpha|U^{-1}\beta)$$

for all α, β. Hence U^{-1} is the adjoint of U.

Conversely, suppose U^* exists and $UU^* = U^*U = I$. Then U is invertible, with $U^{-1} = U^*$. So, we need only show that U preserves inner products. We have

$$\begin{aligned}(U\alpha|U\beta) &= (\alpha|U^*U\beta) \\ &= (\alpha|I\beta) \\ &= (\alpha|\beta)\end{aligned}$$

for all α, β. ∎

EXAMPLE 26. Consider $C^{n\times 1}$ with the inner product $(X|Y) = Y^*X$. Let A be an $n \times n$ matrix over C, and let U be the linear operator defined by $U(X) = AX$. Then

$$(UX|UY) = (AX|AY) = Y^*A^*AX$$

for all X, Y. Hence, U is unitary if and only if $A^*A = I$.

Definition. *A complex* n × n *matrix* A *is called* **unitary,** *if* A*A = I.

Theorem 13. *Let* V *be a finite-dimensional inner product space and let* U *be a linear operator on* V. *Then* U *is unitary if and only if the matrix of* U *in some (or every) ordered orthonormal basis is a unitary matrix.*

Proof. At this point, this is not much of a theorem, and we state it largely for emphasis. If $\mathcal{B} = \{\alpha_1, \ldots, \alpha_n\}$ is an ordered orthonormal basis for V and A is the matrix of U relative to $\mathcal{B}$, then $A^*A = I$ if and only if $U^*U = I$. The result now follows from Theorem 12. ∎

Let A be an $n \times n$ matrix. The statement that A is unitary simply means

$$(A^*A)_{jk} = \delta_{jk}$$

or

$$\sum_{r=1}^{n} \overline{A_{rj}} A_{rk} = \delta_{jk}.$$

In other words, it means that the columns of A form an orthonormal set of column matrices, with respect to the standard inner product $(X|Y) = Y^*X$. Since $A^*A = I$ if and only if $AA^* = I$, we see that A is unitary exactly when the rows of A comprise an orthonormal set of n-tuples in C_n (with the standard inner product). So, using standard inner products, A is unitary if and only if the rows and columns of A are orthonormal sets. One sees here an example of the power of the theorem which states that a one-sided inverse for a matrix is a two-sided inverse. Applying this theorem as we did above, say to real matrices, we have the following: Suppose we have a square array of real numbers such that the sum of the squares of the entries in each row is 1 and distinct rows are orthogonal. Then the sum of the squares of the entries in each column is 1 and distinct columns are orthogonal. Write down the proof of this for a 3×3 array, without using any knowledge of matrices, and you should be reasonably impressed.

Definition. *A real or complex* $n \times n$ *matrix* A *is said to be* **orthogonal,** *if* $A^tA = I$.

A real orthogonal matrix is unitary; and, a unitary matrix is orthogonal if and only if each of its entries is real.

EXAMPLE 27. We give some examples of unitary and orthogonal matrices.

(a) A 1×1 matrix $[c]$ is orthogonal if and only if $c = \pm 1$, and unitary if and only if $\bar{c}c = 1$. The latter condition means (of course) that $|c| = 1$, or $c = e^{i\theta}$, where θ is real.

(b) Let

$$A = \begin{bmatrix} a & b \\ c & d \end{bmatrix}.$$

Then A is orthogonal if and only if

$$A^t = A^{-1} = \frac{1}{ad - bc}\begin{bmatrix} d & -b \\ -c & a \end{bmatrix}.$$

The determinant of any orthogonal matrix is easily seen to be ± 1. Thus A is orthogonal if and only if

$$A = \begin{bmatrix} a & b \\ -b & a \end{bmatrix}$$

or

$$A = \begin{bmatrix} a & b \\ b & -a \end{bmatrix}$$

where $a^2 + b^2 = 1$. The two cases are distinguished by the value of det A.

(c) The well-known relations between the trigonometric functions show that the matrix

$$A_\theta = \begin{bmatrix} \cos\theta & -\sin\theta \\ \sin\theta & \cos\theta \end{bmatrix}$$

is orthogonal. If θ is a real number, then A_θ is the matrix in the standard ordered basis for R^2 of the linear operator U_θ, rotation through the angle θ. The statement that A_θ is a real orthogonal matrix (hence unitary) simply means that U_θ is a unitary operator, i.e., preserves dot products.

(d) Let

$$A = \begin{bmatrix} a & b \\ c & d \end{bmatrix}$$

Then A is unitary if and only if

$$\begin{bmatrix} \bar{a} & \bar{c} \\ \bar{b} & \bar{d} \end{bmatrix} = \frac{1}{ad - bc}\begin{bmatrix} d & -b \\ -c & a \end{bmatrix}.$$

The determinant of a unitary matrix has absolute value 1, and is thus a complex number of the form $e^{i\theta}$, θ real. Thus A is unitary if and only if

$$A = \begin{bmatrix} a & b \\ -e^{i\theta}\bar{b} & e^{i\theta}\bar{a} \end{bmatrix} = \begin{bmatrix} 1 & 0 \\ 0 & e^{i\theta} \end{bmatrix}\begin{bmatrix} a & b \\ -\bar{b} & \bar{a} \end{bmatrix}$$

where θ is a real number, and a, b are complex numbers such that $|a|^2 + |b|^2 = 1$.

As noted earlier, the unitary operators on an inner product space form a group. From this and Theorem 13 it follows that the set $U(n)$ of all $n \times n$ unitary matrices is also a group. Thus the inverse of a unitary matrix and the product of two unitary matrices are again unitary. Of course this is easy to see directly. An $n \times n$ matrix A with complex entries is unitary if and only if $A^{-1} = A^*$. Thus, if A is unitary, we have $(A^{-1})^{-1} = A = (A^*)^{-1} = (A^{-1})^*$. If A and B are $n \times n$ unitary matrices, then $(AB)^{-1} = B^{-1}A^{-1} = B^*A^* = (AB)^*$.

The Gram-Schmidt process in C^n has an interesting corollary for matrices that involves the group $U(n)$.

Theorem 14. *For every invertible complex* $n \times n$ *matrix* B *there exists a unique lower-triangular matrix* M *with positive entries on the main diagonal such that* MB *is unitary.*

Proof. The rows $\beta_1, \ldots, \beta_n$ of B form a basis for C^n. Let $\alpha_1, \ldots, \alpha_n$ be the vectors obtained from $\beta_1, \ldots, \beta_n$ by the Gram-Schmidt process. Then, for $1 \le k \le n$, $\{\alpha_1, \ldots, \alpha_k\}$ is an orthogonal basis for the subspace spanned by $\{\beta_1, \ldots, \beta_k\}$, and

$$\alpha_k = \beta_k - \sum_{j<k} \frac{(\beta_k|\alpha_j)}{||\alpha_j||^2} \alpha_j.$$

Hence, for each k there exist unique scalars C_{kj} such that

$$\alpha_k = \beta_k - \sum_{j<k} C_{kj}\beta_j.$$

Let U be the unitary matrix with rows

$$\frac{\alpha_1}{||\alpha_1||}, \ldots, \frac{\alpha_n}{||\alpha_n||}$$

and M the matrix defined by

$$M_{kj} = \begin{cases} -\dfrac{1}{||\alpha_k||} \cdot C_{kj}, & \text{if } j < k \\ \dfrac{1}{||\alpha_k||}, & \text{if } j = k \\ 0, & \text{if } j > k. \end{cases}$$

Then M is lower-triangular, in the sense that its entries above the main diagonal are 0. The entries M_{kk} of M on the main diagonal are all > 0, and

$$\frac{\alpha_k}{||\alpha_k||} = \sum_{j=1}^{n} M_{kj}\beta_j, \qquad 1 \leq k \leq n.$$

Now these equations simply say that

$$U = MB.$$

To prove the uniqueness of M, let $T^+(n)$ denote the set of all complex $n \times n$ lower-triangular matrices with positive entries on the main diagonal. Suppose M_1 and M_2 are elements of $T^+(n)$ such that M_iB is in $U(n)$ for $i = 1, 2$. Then because $U(n)$ is a group

$$(M_1B)(M_2B)^{-1} = M_1M_2^{-1}$$

lies in $U(n)$. On the other hand, although it is not entirely obvious, $T^+(n)$ is also a group under matrix multiplication. One way to see this is to consider the geometric properties of the linear transformations

$$X \to MX, \qquad (M \text{ in } T^+(n))$$

on the space of column matrices. Thus M_2^{-1}, $M_1M_2^{-1}$, and $(M_1M_2^{-1})^{-1}$ are all in $T^+(n)$. But, since $M_1M_2^{-1}$ is in $U(n)$, $(M_1M_2^{-1})^{-1} = (M_1M_2^{-1})^*$. The transpose or conjugate transpose of any lower-triangular matrix is an upper-triangular matrix. Therefore, $M_1M_2^{-1}$ is simultaneously upper- and lower-triangular, i.e., diagonal. A diagonal matrix is unitary if and only if each of its entries on the main diagonal has absolute value 1; if the diagonal entries are all positive, they must equal 1. Hence $M_1M_2^{-1} = I$ and $M_1 = M_2$. ∎

Let $GL(n)$ denote the set of all invertible complex $n \times n$ matrices. Then $GL(n)$ is also a group under matrix multiplication. This group is called the **general linear group.** Theorem 14 is equivalent to the following result.

Corollary. *For each* B *in* GL(n) *there exist unique matrices* N *and* U *such that* N *is in* T⁺(n), U *is in* U(n), *and*

$$B = N \cdot U.$$

Proof. By the theorem there is a unique matrix M in $T^+(n)$ such that MB is in $U(n)$. Let $MB = U$ and $N = M^{-1}$. Then N is in $T^+(n)$ and $B = N \cdot U$. On the other hand, if we are given any elements N and U such that N is in $T^+(n)$, U is in $U(n)$, and $B = N \cdot U$, then $N^{-1}B$ is in $U(n)$ and N^{-1} is the unique matrix M which is characterized by the theorem; furthermore U is necessarily $N^{-1}B$. ∎

EXAMPLE 28. Let x_1 and x_2 be real numbers such that $x_1^2 + x_2^2 = 1$ and $x_1 \neq 0$. Let

$$B = \begin{bmatrix} x_1 & x_2 & 0 \\ 0 & 1 & 0 \\ 0 & 0 & 1 \end{bmatrix}.$$

Applying the Gram-Schmidt process to the rows of B, we obtain the vectors

$$\begin{aligned} \alpha_1 &= (x_1, x_2, 0) \\ \alpha_2 &= (0, 1, 0) - x_2(x_1, x_2, 0) \\ &= x_1(-x_2, x_1, 0) \\ \alpha_3 &= (0, 0, 1). \end{aligned}$$

Let U be the matrix with rows α_1, (α_2/x_1), α_3. Then U is unitary, and

$$U = \begin{bmatrix} x_1 & x_2 & 0 \\ -x_2 & x_1 & 0 \\ 0 & 0 & 1 \end{bmatrix} = \begin{bmatrix} 1 & 0 & 0 \\ -\dfrac{x_2}{x_1} & \dfrac{1}{x_1} & 0 \\ 0 & 0 & 1 \end{bmatrix} \begin{bmatrix} x_1 & x_2 & 0 \\ 0 & 1 & 0 \\ 0 & 0 & 1 \end{bmatrix}$$

Now multiplying by the inverse of

$$M = \begin{bmatrix} 1 & 0 & 0 \\ -\dfrac{x_2}{x_1} & \dfrac{1}{x_1} & 0 \\ 0 & 0 & 1 \end{bmatrix}$$

we find that

$$\begin{bmatrix} x_1 & x_2 & 0 \\ 0 & 1 & 0 \\ 0 & 0 & 1 \end{bmatrix} = \begin{bmatrix} 1 & 0 & 0 \\ x_2 & x_1 & 0 \\ 0 & 0 & 1 \end{bmatrix} \begin{bmatrix} x_1 & x_2 & 0 \\ -x_2 & x_1 & 0 \\ 0 & 0 & 1 \end{bmatrix}.$$

Let us now consider briefly change of coordinates in an inner product space. Suppose V is a finite-dimensional inner product space and that $\mathfrak{B} = \{\alpha_1, \ldots, \alpha_n\}$ and $\mathfrak{B}' = \{\alpha_1', \ldots, \alpha_n'\}$ are two ordered *orthonormal* bases for V. There is a unique (necessarily invertible) $n \times n$ matrix P such that

$$[\alpha]_{\mathfrak{B}'} = P^{-1}[\alpha]_{\mathfrak{B}}$$

for every α in V. If U is the unique linear operator on V defined by $U\alpha_j = \alpha_j'$, then P is the matrix of U in the ordered basis $\mathfrak{B}$:

$$\alpha_k' = \sum_{j=1}^{n} P_{jk}\alpha_j.$$

Since $\mathfrak{B}$ and $\mathfrak{B}'$ are orthonormal bases, U is a unitary operator and P is a unitary matrix. If T is any linear operator on V, then

$$[T]_{\mathfrak{B}'} = P^{-1}[T]_{\mathfrak{B}}P = P^*[T]_{\mathfrak{B}}P.$$

Definition. *Let* A *and* B *be complex* n × n *matrices. We say that* B *is* **unitarily equivalent to** A *if there is an* n × n *unitary matrix* P *such that* B = P⁻¹AP. *We say that* B *is* **orthogonally equivalent to** A *if there is an* n × n *orthogonal matrix* P *such that* B = P⁻¹AP.

With this definition, what we observed above may be stated as follows: If $\mathfrak{B}$ and $\mathfrak{B}'$ are two ordered orthonormal bases for V, then, for each linear operator T on V, the matrix $[T]_{\mathfrak{B}'}$ is unitarily equivalent to the matrix $[T]_{\mathfrak{B}}$. In case V is a real inner product space, these matrices are orthogonally equivalent, via a real orthogonal matrix.

Exercises

1. Find a unitary matrix which is not orthogonal, and find an orthogonal matrix which is not unitary.

2. Let V be the space of complex $n \times n$ matrices with inner product $(A|B) = \operatorname{tr}(AB^*)$. For each M in V, let T_M be the linear operator defined by $T_M(A) = MA$. Show that T_M is unitary if and only if M is a unitary matrix.

3. Let V be the set of complex numbers, regarded as a *real* vector space.

(a) Show that $(\alpha|\beta) = \operatorname{Re}(\alpha\bar{\beta})$ defines an inner product on V.

(b) Exhibit an (inner product space) isomorphism of V onto R^2 with the standard inner product.

(c) For each γ in V, let M_γ be the linear operator on V defined by $M_\gamma(\alpha) = \gamma\alpha$. Show that $(M_\gamma)^* = M_{\bar{\gamma}}$.

(d) For which complex numbers γ is M_γ self-adjoint?

(e) For which γ is M_γ unitary?

(f) For which γ is M_γ positive?

(g) What is det (M_γ)?

(h) Find the matrix of M_γ in the basis $\{1, i\}$.

(i) If T is a linear operator on V, find necessary and sufficient conditions for T to be an M_γ.

(j) Find a unitary operator on V which is not an M_γ.

4. Let V be R^2, with the standard inner product. If U is a unitary operator on V, show that the matrix of U in the standard ordered basis is either

$$\begin{bmatrix} \cos\theta & -\sin\theta \\ \sin\theta & \cos\theta \end{bmatrix} \quad \text{or} \quad \begin{bmatrix} \cos\theta & \sin\theta \\ \sin\theta & -\cos\theta \end{bmatrix}$$

for some real θ, $0 \leq \theta < 2\pi$. Let U_θ be the linear operator corresponding to the first matrix, i.e., U_θ is rotation through the angle θ. Now convince yourself that every unitary operator on V is either a rotation, or reflection about the ϵ_1-axis followed by a rotation.

(a) What is $U_\theta U_\phi$?

(b) Show that $U_\theta^* = U_{-\theta}$.

(c) Let ϕ be a fixed real number, and let $\mathcal{B} = \{\alpha_1, \alpha_2\}$ be the orthonormal basis obtained by rotating $\{\epsilon_1, \epsilon_2\}$ through the angle ϕ, i.e., $\alpha_j = U_\phi\epsilon_j$. If θ is another real number, what is the matrix of U_θ in the ordered basis $\mathcal{B}$?

5. Let V be R^3, with the standard inner product. Let W be the plane spanned by $\alpha = (1, 1, 1)$ and $\beta = (1, 1, -2)$. Let U be the linear operator defined, geometrically, as follows: U is rotation through the angle θ, about the straight line through the origin which is orthogonal to W. There are actually two such rotations —choose one. Find the matrix of U in the standard ordered basis. (Here is one way you might proceed. Find α_1 and α_2 which form an orthonormal basis for W. Let α_3 be a vector of norm 1 which is orthogonal to W. Find the matrix of U in the basis $\{\alpha_1, \alpha_2, \alpha_3\}$. Perform a change of basis.)

6. Let V be a finite-dimensional inner product space, and let W be a subspace of V. Then $V = W \oplus W^\perp$, that is, each α in V is uniquely expressible in the form $\alpha = \beta + \gamma$, with β in W and γ in $W^\perp$. Define a linear operator U by $U\alpha = \beta - \gamma$.

(a) Prove that U is both self-adjoint and unitary.

(b) If V is R^3 with the standard inner product and W is the subspace spanned by $(1, 0, 1)$, find the matrix of U in the standard ordered basis.

7. Let V be a *complex* inner product space and T a *self-adjoint* linear operator on V. Show that

(a) $||\alpha + iT\alpha|| = ||\alpha - iT\alpha||$ for every α in V.

(b) $\alpha + iT\alpha = \beta + iT\beta$ if and only if $\alpha = \beta$.

(c) $I + iT$ is non-singular.

(d) $I - iT$ is non-singular.

(e) Now suppose V is finite-dimensional, and prove that

$$U = (I - iT)(I + iT)^{-1}$$

is a unitary operator; U is called the **Cayley transform** of T. In a certain sense, $U = f(T)$, where $f(x) = (1 - ix)/(1 + ix)$.

8. If θ is a real number, prove that the following matrices are unitarily equivalent

$$\begin{bmatrix} \cos\theta & -\sin\theta \\ \sin\theta & \cos\theta \end{bmatrix}, \quad \begin{bmatrix} e^{i\theta} & 0 \\ 0 & e^{-i\theta} \end{bmatrix}.$$

9. Let V be a finite-dimensional inner product space and T a positive linear operator on V. Let p_T be the inner product on V defined by $p_T(\alpha, \beta) = (T\alpha|\beta)$. Let U be a linear operator on V and U^* its adjoint with respect to $(\ |\ )$. Prove that U is unitary with respect to the inner product p_T if and only if $T = U^*TU$.

10. Let V be a finite-dimensional inner product space. For each α, β in V, let $T_{\alpha,\beta}$ be the linear operator on V defined by $T_{\alpha,\beta}(\gamma) = (\gamma|\beta)\alpha$. Show that

(a) $T^*_{\alpha,\beta} = T_{\beta,\alpha}$.

(b) trace $(T_{\alpha,\beta}) = (\alpha|\beta)$.

(c) $T_{\alpha,\beta}T_{\gamma,\delta} = T_{\alpha,(\beta|\gamma)\delta}$.

(d) Under what conditions is $T_{\alpha,\beta}$ self-adjoint?

11. Let V be an n-dimensional inner product space over the field F, and let $L(V, V)$ be the space of linear operators on V. Show that there is a unique inner product on $L(V, V)$ with the property that $||T_{\alpha,\beta}||^2 = ||\alpha||^2||\beta||^2$ for all α, β in V. ($T_{\alpha,\beta}$ is the operator defined in Exercise 10.) Find an isomorphism between $L(V, V)$ with this inner product and the space of $n \times n$ matrices over F, with the inner product $(A|B) = \text{tr}\,(AB^*)$.

12. Let V be a finite-dimensional inner product space. In Exercise 6, we showed how to construct some linear operators on V which are both self-adjoint and unitary. Now prove that there are no others, i.e., that every self-adjoint unitary operator arises from some subspace W as we described in Exercise 6.

13. Let V and W be finite-dimensional inner product spaces having the same dimension. Let U be an isomorphism of V onto W. Show that:

(a) The mapping $T \to UTU^{-1}$ is an isomorphism of the vector space $L(V, V)$ onto the vector space $L(W, W)$.

(b) trace (UTU^{-1}) = trace (T) for each T in $L(V, V)$.

(c) $UT_{\alpha,\beta}U^{-1} = T_{U\alpha,U\beta}$ ($T_{\alpha,\beta}$ defined in Exercise 10).

(d) $(UTU^{-1})^* = UT^*U^{-1}$.

(e) If we equip $L(V, V)$ with inner product $(T_1|T_2)$ = trace $(T_1T_2^*)$, and similarly for $L(W, W)$, then $T \to UTU^{-1}$ is an inner product space isomorphism.

14. If V is an inner product space, a **rigid motion** is any function T from V into V (not necessarily linear) such that $||T\alpha - T\beta|| = ||\alpha - \beta||$ for all α, β in V. One example of a rigid motion is a linear unitary operator. Another example is translation by a fixed vector γ:

$$T_\gamma(\alpha) = \alpha + \gamma$$

(a) Let V be R^2 with the standard inner product. Suppose T is a rigid motion of V and that $T(0) = 0$. Prove that T is linear and a unitary operator.

(b) Use the result of part (a) to prove that every rigid motion of R^2 is composed of a translation, followed by a unitary operator.

(c) Now show that a rigid motion of R^2 is either a translation followed by a rotation, or a translation followed by a reflection followed by a rotation.

15. A unitary operator on R^4 (with the standard inner product) is simply a linear operator which preserves the quadratic form

$$||(x, y, z, t)||^2 = x^2 + y^2 + z^2 + t^2$$

that is, a linear operator U such that $||U\alpha||^2 = ||\alpha||^2$ for all α in R^4. In a certain part of the theory of relativity, it is of interest to find the linear operators T which preserve the form

$$||(x, y, z, t)||_L^2 = t^2 - x^2 - y^2 - z^2.$$

Now $|| \quad ||_L^2$ does not come from an inner product, but from something called the 'Lorentz metric' (which we shall not go into). For that reason, a linear operator T on R^4 such that $||T\alpha||_L^2 = ||\alpha||_L^2$, for every α in R^4, is called a **Lorentz transformation.**

(a) Show that the function U defined by

$$U(x, y, z, t) = \begin{bmatrix} t + x & y + iz \\ y - iz & t - x \end{bmatrix}$$

is an isomorphism of R^4 onto the real vector space H of all self-adjoint 2×2 complex matrices.

(b) Show that $||\alpha||_L^2 = \det (U\alpha)$.

(c) Suppose T is a (real) linear operator on the space H of 2×2 self-adjoint matrices. Show that $L = U^{-1}TU$ is a linear operator on R^4.

(d) Let M be any 2×2 complex matrix. Show that $T_M(A) = M^*AM$ defines a linear operator T_M on H. (Be sure you check that T_M maps H into H.)

(e) If M is a 2×2 matrix such that $|\det M| = 1$, show that $L_M = U^{-1}T_MU$ is a Lorentz transformation on R^4.

(f) Find a Lorentz transformation which is not an L_M.

8.5. Normal Operators

The principal objective in this section is the solution of the following problem. If T is a linear operator on a finite-dimensional inner product space V, under what conditions does V have an orthonormal basis consisting of characteristic vectors for T? In other words, when is there an *orthonormal* basis $\mathfrak{B}$ for V, such that the matrix of T in the basis $\mathfrak{B}$ is diagonal?

We shall begin by deriving some necessary conditions on T, which we shall subsequently show are sufficient. Suppose $\mathfrak{B} = \{\alpha_1, \ldots, \alpha_n\}$ is an orthonormal basis for V with the property

$$T\alpha_j = c_j\alpha_j, \qquad j = 1, \ldots, n. \tag{8-16}$$

This simply says that the matrix of T in the ordered basis $\mathfrak{B}$ is the diagonal matrix with diagonal entries $c_1, \ldots, c_n$. The adjoint operator T^* is represented in this same ordered basis by the conjugate transpose matrix, i.e., the diagonal matrix with diagonal entries $\bar{c}_1, \ldots, \bar{c}_n$. If V is a real inner product space, the scalars $c_1, \ldots, c_n$ are (of course) real, and so it must be that $T = T^*$ .In other words, if V is a finite-dimensional *real* inner probut space and T is a linear operator for which there is an orthonormal basis of characteristic vectos, then T must be self-adjoint. If V is a complex inner product space, the scalars $c_1, \ldots, c_n$ need not be real, i.e., T need not be self-adjoint. But notice that T must satisfy

$$TT^* = T^*T. \tag{8-17}$$

For, any two diagonal matrices commute, and since T and T^* are both represented by diagonal matrices in the ordered basis $\mathcal{B}$, we have (8-17). It is a rather remarkable fact that in the complex case this condition is also sufficient to imply the existence of an orthonormal basis of characteristic vectors.

Definition. *Let* V *be a finite-dimensional inner product space and* T *a linear operator on* V. *We say that* T *is* **normal** *if it commutes with its adjoint i.e.,* $TT^* = T^*T$.

Any self-adjoint operator is normal, as is any unitary operator. Any scalar multiple of a normal operator is normal; however, sums and products of normal operators are not generally normal. Although it is by no means necessary, we shall begin our study of normal operators by considering self-adjoint operators.

Theorem 15. *Let* V *be an inner product space and* T *a self-adjoint linear operator on* V. *Then each characteristic value of* T *is real, and characteristic vectors of* T *associated with distinct characteristic values are orthogonal.*

Proof. Suppose c is a characteristic value of T, i.e., that $T\alpha = c\alpha$ for some non-zero vector α. Then

$$\begin{aligned} c(\alpha|\alpha) &= (c\alpha|\alpha) \\ &= (T\alpha|\alpha) \\ &= (\alpha|T\alpha) \\ &= (\alpha|c\alpha) \\ &= \bar{c}(\alpha|\alpha). \end{aligned}$$

Since $(\alpha|\alpha) \neq 0$, we must have $c = \bar{c}$. Suppose we also have $T\beta = d\beta$ with $\beta \neq 0$. Then

$$\begin{aligned} c(\alpha|\beta) &= (T\alpha|\beta) \\ &= (\alpha|T\beta) \\ &= (\alpha|d\beta) \\ &= \bar{d}(\alpha|\beta) \\ &= d(\alpha|\beta). \end{aligned}$$

If $c \neq d$, then $(\alpha|\beta) = 0$. ∎

It should be pointed out that Theorem 15 says nothing about the existence of characteristic values or characteristic vectors.

Theorem 16. *On a finite-dimensional inner product space of positive dimension, every self-adjoint operator has a (non-zero) characteristic vector.*

Proof. Let V be an inner product space of dimension n, where $n > 0$, and let T be a self-adjoint operator on V. Choose an orthonormal basis $\mathfrak{B}$ for V and let $A = [T]_{\mathfrak{B}}$. Since $T = T^*$, we have $A = A^*$. Now let W be the space of $n \times 1$ matrices over C, with inner product $(X|Y) = Y^*X$. Then $U(X) = AX$ defines a self-adjoint linear operator U on W. The characteristic polynomial, det $(xI - A)$, is a polynomial of degree n over the complex numbers; every polynomial over C of positive degree has a root. Thus, there is a complex number c such that det $(cI - A) = 0$. This means that $A - cI$ is singular, or that there exists a non-zero X such that $AX = cX$. Since the operator U (multiplication by A) is self-adjoint, it follows from Theorem 15 that c is real. If V is a real vector space, we may choose X to have real entries. For then A and $A - cI$ have real entries, and since $A - cI$ is singular, the system $(A - cI)X = 0$ has a non-zero real solution X. It follows that there is a non-zero vector α in V such that $T\alpha = c\alpha$. ∎

There are several comments we should make about the proof.

(1) The proof of the existence of a non-zero X such that $AX = cX$ had nothing to do with the fact that A was Hermitian (self-adjoint). It shows that any linear operator on a finite-dimensional complex vector space has a characteristic vector. In the case of a real inner product space, the self-adjointness of A is used very heavily, to tell us that each characteristic value of A is real and hence that we can find a suitable X with real entries.

(2) The argument shows that the characteristic polynomial of a self-adjoint matrix has real coefficients, in spite of the fact that the matrix may not have real entries.

(3) The assumption that V is finite-dimensional is necessary for the theorem; a self-adjoint operator on an infinite-dimensional inner product space need not have a characteristic value.

EXAMPLE 29. Let V be the vector space of continuous complex-valued (or real-valued) continuous functions on the unit interval, $0 \leq t \leq 1$, with the inner product

$$(f|g) = \int_0^1 f(t)\overline{g(t)}\,dt.$$

The operator 'multiplication by t,' $(Tf)(t) = tf(t)$, is self-adjoint. Let us suppose that $Tf = cf$. Then

$$(t - c)f(t) = 0, \qquad 0 \le t \le 1$$

and so $f(t) = 0$ for $t \neq c$. Since f is continuous, $f = 0$. Hence T has no characteristic values (vectors).

Theorem 17. *Let V be a finite-dimensional inner product space, and let T be any linear operator on V. Suppose W is a subspace of V which is invariant under T. Then the orthogonal complement of W is invariant under T*.*

Proof. We recall that the fact that W is invariant under T does not mean that each vector in W is left fixed by T; it means that if α is in W then $T\alpha$ is in W. Let β be in $W^\perp$. We must show that $T^*\beta$ is in $W^\perp$, that is, that $(\alpha|T^*\beta) = 0$ for every α in W. If α is in W, then $T\alpha$ is in W, so $(T\alpha|\beta) = 0$. But $(T\alpha|\beta) = (\alpha|T^*\beta)$. ∎

Theorem 18. *Let V be a finite-dimensional inner product space, and let T be a self-adjoint linear operator on V. Then there is an orthonormal basis for V, each vector of which is a characteristic vector for T.*

Proof. We are assuming $\dim V > 0$. By Theorem 16, T has a characteristic vector α. Let $\alpha_1 = \alpha/||\alpha||$ so that α_1 is also a characteristic vector for T and $||\alpha_1|| = 1$. If $\dim V = 1$, we are done. Now we proceed by induction on the dimension of V. Suppose the theorem is true for inner product spaces of dimension less than $\dim V$. Let W be the one-dimensional subspace spanned by the vector α_1. The statement that α_1 is a characteristic vector for T simply means that W is invariant under T. By Theorem 17, the orthogonal complement $W^\perp$ is invariant under $T^* = T$. Now $W^\perp$, with the inner product from V, is an inner product space of dimension one less than the dimension of V. Let U be the linear operator induced on $W^\perp$ by T, that is, the restriction of T to $W^\perp$. Then U is self-adjoint, and by the induction hypothesis, $W^\perp$ has an orthonormal basis $\{\alpha_2, \ldots, \alpha_n\}$ consisting of characteristic vectors for U. Now each of these vectors is also a characteristic vector for T, and since $V = W \oplus W^\perp$, we conclude that $\{\alpha_1, \ldots, \alpha_n\}$ is the desired basis for V. ∎

Corollary. *Let A be an $n \times n$ Hermitian (self-adjoint) matrix. Then there is a unitary matrix P such that $P^{-1}AP$ is diagonal (A is unitarily equivalent to a diagonal matrix). If A is a real symmetric matrix, there is a real orthogonal matrix P such that $P^{-1}AP$ is diagonal.*

Proof. Let V be $C^{n\times 1}$, with the standard inner product, and let T be the linear operator on V which is represented by A in the standard ordered basis. Since $A = A^*$, we have $T = T^*$. Let $\mathcal{B} = \{\alpha_1, \ldots, \alpha_n\}$ be an ordered orthonormal basis for V, such that $T\alpha_j = c_j\alpha_j$, $j = 1, \ldots, n$. If $D = [T]_{\mathcal{B}}$, then D is the diagonal matrix with diagonal entries $c_1, \ldots, c_n$. Let P be the matrix with column vectors $\alpha_1, \ldots, \alpha_n$. Then $D = P^{-1}AP$.

In case each entry of A is real, we can take V to be R^n, with the standard inner product, and repeat the argument. In this case, P will be a unitary matrix with real entries, i.e., a real orthogonal matrix. ▌

Combining Theorem 18 with our comments at the beginning of this section, we have the following: If V is a finite-dimensional *real* inner product space and T is a linear operator on V, then V has an orthonormal basis of characteristic vectors for T if and only if T is self-adjoint. Equivalently, if A is an $n \times n$ matrix with *real* entries, there is a real orthogonal matrix P such that P^tAP is diagonal if and only if $A = A^t$. There is no such result for complex symmetric matrices. In other words, for complex matrices there is a significant difference between the conditions $A = A^t$ and $A = A^*$.

Having disposed of the self-adjoint case, we now return to the study of normal operators in general. We shall prove the analogue of Theorem 18 for normal operators, in the *complex* case. There is a reason for this restriction. A normal operator on a real inner product space may not have any non-zero characteristic vectors. This is true, for example, of all but two rotations in R^2.

Theorem 19. *Let* V *be a finite-dimensional inner product space and* T *a normal operator on* V. *Suppose* α *is a vector in* V. *Then* α *is a characteristic vector for* T *with characteristic value* c *if and only if* α *is a characteristic vector for* T* *with characteristic value* $\bar{c}$.

Proof. Suppose U is any normal operator on V. Then $||U\alpha|| = ||U^*\alpha||$. For using the condition $UU^* = U^*U$ one sees that

$$\begin{aligned}||U\alpha||^2 = (U\alpha|U\alpha) &= (\alpha|U^*U\alpha)\\ &= (\alpha|UU^*\alpha) = (U^*\alpha|U^*\alpha) = ||U^*\alpha||^2.\end{aligned}$$

If c is any scalar, the operator $U = T - cI$ is normal. For $(T - cI)^* = T^* - \bar{c}I$, and it is easy to check that $UU^* = U^*U$. Thus

$$||(T - cI)\alpha|| = ||(T^* - \bar{c}I)\alpha||$$

so that $(T - cI)\alpha = 0$ if and only if $(T^* - \bar{c}I)\alpha = 0$. ▌

Definition. *A complex* $n \times n$ *matrix* A *is called* **normal** *if* $AA^* = A^*A$.

It is not so easy to understand what normality of matrices or operators really means; however, in trying to develop some feeling for the concept, the reader might find it helpful to know that a triangular matrix is normal if and only if it is diagonal.

Theorem 20. *Let* V *be a finite-dimensional inner product space,* T *a linear operator on* V, *and* $\mathcal{B}$ *an orthonormal basis for* V. *Suppose that the* *matrix* A *of* T *in the basis* $\mathfrak{B}$ *is upper triangular. Then* T *is normal if and only if* A *is a diagonal matrix.*

Proof. Since $\mathfrak{B}$ is an orthonormal basis, A^* is the matrix of T^* in $\mathfrak{B}$. If A is diagonal, then $AA^* = A^*A$, and this implies $TT^* = T^*T$. Conversely, suppose T is normal, and let $\mathfrak{B} = \{\alpha_1, \ldots, \alpha_n\}$. Then, since A is upper-triangular, $T\alpha_1 = A_{11}\alpha_1$. By Theorem 19 this implies, $T^*\alpha_1 = \bar{A}_{11}\alpha_1$. On the other hand,

$$\begin{aligned} T^*\alpha_1 &= \sum_j (A^*)_{j1}\alpha_j \\ &= \sum_j \bar{A}_{1j}\alpha_j. \end{aligned}$$

Therefore, $A_{1j} = 0$ for every $j > 1$. In particular, $A_{12} = 0$, and since A is upper-triangular, it follows that

$$T\alpha_2 = A_{22}\alpha_2.$$

Thus $T^*\alpha_2 = \bar{A}_{22}\alpha_2$ and $A_{2j} = 0$ for all $j \neq 2$. Continuing in this fashion, we find that A is diagonal. ∎

Theorem 21. *Let* V *be a finite-dimensional complex inner product space and let* T *be any linear operator on* V. *Then there is an orthonormal basis for* V *in which the matrix of* T *is upper triangular.*

Proof. Let n be the dimension of V. The theorem is true when $n = 1$, and we proceed by induction on n, assuming the result is true for linear operators on complex inner product spaces of dimension $n - 1$. Since V is a finite-dimensional complex inner product space, there is a unit vector α in V and a scalar c such that

$$T^*\alpha = c\alpha.$$

Let W be the orthogonal complement of the subspace spanned by α and let S be the restriction of T to W. By Theorem 17, W is invariant under T. Thus S is a linear operator on W. Since W has dimension $n - 1$, our inductive assumption implies the existence of an orthonormal basis $\{\alpha_1, \ldots, \alpha_{n-1}\}$ for W in which the matrix of S is upper-triangular; let $\alpha_n = \alpha$. Then $\{\alpha_1, \ldots, \alpha_n\}$ is an orthonormal basis for V in which the matrix of T is upper-triangular. ∎

This theorem implies the following result for matrices.

Corollary. *For every complex* $n \times n$ *matrix* A *there is a unitary matrix* U *such that* $U^{-1}AU$ *is upper-triangular.*

Now combining Theorem 21 and Theorem 20, we immediately obtain the following analogue of Theorem 18 for normal operators.

Theorem 22. *Let V be a finite-dimensional complex inner product space and T a normal operator on V. Then V has an orthonormal basis consisting of characteristic vectors for T.*

Again there is a matrix interpretation.

Corollary. *For every normal matrix A there is a unitary matrix P such that $P^{-1}AP$ is a diagonal matrix.*

Exercises

1. For each of the following real symmetric matrices A, find a real orthogonal matrix P such that P^tAP is diagonal.

$$\begin{bmatrix} 1 & 1 \\ 1 & 1 \end{bmatrix}, \quad \begin{bmatrix} 1 & 2 \\ 2 & 1 \end{bmatrix}, \quad \begin{bmatrix} \cos\theta & \sin\theta \\ \sin\theta & -\cos\theta \end{bmatrix}$$

2. Is a complex symmetric matrix self-adjoint? Is it normal?

3. For

$$A = \begin{bmatrix} 1 & 2 & 3 \\ 2 & 3 & 4 \\ 3 & 4 & 5 \end{bmatrix}$$

there is a real orthogonal matrix P such that $P^tAP = D$ is diagonal. Find such a diagonal matrix D.

4. Let V be C^2, with the standard inner product. Let T be the linear operator on V which is represented in the standard ordered basis by the matrix

$$A = \begin{bmatrix} 1 & i \\ i & 1 \end{bmatrix}.$$

Show that T is normal, and find an orthonormal basis for V, consisting of characteristic vectors for T.

5. Give an example of a 2×2 matrix A such that A^2 is normal, but A is not normal.

6. Let T be a normal operator on a finite-dimensional complex inner product space. Prove that T is self-adjoint, positive, or unitary according as every characteristic value of T is real, positive, or of absolute value 1. (Use Theorem 22 to reduce to a similar question about diagonal matrices.)

7. Let T be a linear operator on the finite-dimensional inner product space V, and suppose T is both positive and unitary. Prove $T = I$.

8. Prove T is normal if and only if $T = T_1 + iT_2$, where T_1 and T_2 are self-adjoint operators which commute.

9. Prove that a real symmetric matrix has a real symmetric cube root; i.e., if A is real symmetric, there is a real symmetric B such that $B^3 = A$.

10. Prove that every positive matrix is the square of a positive matrix.

11. Prove that a normal and nilpotent operator is the zero operator.

12. If T is a normal operator, prove that characteristic vectors for T which are associated with distinct characteristic values are orthogonal.

13. Let T be a normal operator on a finite-dimensional complex inner product space. Prove that there is a polynomial f, with complex coefficients, such that $T^* = f(T)$. (Represent T by a diagonal matrix, and see what f must be.)

14. If two normal operators commute, prove that their product is normal.

9. Operators on Inner Product Spaces

9.1. Introduction

We regard most of the topics treated in Chapter 8 as fundamental, the material that everyone should know. The present chapter is for the more advanced student or for the reader who is eager to expand his knowledge concerning operators on inner product spaces. With the exception of the Principal Axis theorem, which is essentially just another formulation of Theorem 18 on the orthogonal diagonalization of self adjoint operators, and the other results on forms in Section 9.2, the material presented here is more sophisticated and generally more involved technically. We also make more demands of the reader, just as we did in the later parts of Chapters 5 and 7. The arguments and proofs are written in a more condensed style, and there are almost no examples to smooth the way; however, we have seen to it that the reader is well supplied with generous sets of exercises.

The first three sections are devoted to results concerning forms on inner product spaces and the relation between forms and linear operators. The next section deals with spectral theory, i.e., with the implications of Theorems 18 and 22 of Chapter 8 concerning the diagonalization of self-adjoint and normal operators. In the final section, we pursue the study of normal operators treating, in particular, the real case, and in so doing we examine what the primary decomposition theorem of Chapter 6 says about normal operators.

9.2. Forms on Inner Product Spaces

If T is a linear operator on a finite-dimensional inner product space V the function f defined on $V \times V$ by

$$f(\alpha, \beta) = (T\alpha|\beta)$$

may be regarded as a kind of substitute for T. Many questions about T are equivalent to questions concerning f. In fact, it is easy to see that f determines T. For if $\mathcal{B} = \{\alpha_1, \ldots, \alpha_n\}$ is an orthonormal basis for V, then the entries of the matrix of T in $\mathcal{B}$ are given by

$$A_{jk} = f(\alpha_k, \alpha_j).$$

It is important to understand why f determines T from a more abstract point of view. The crucial properties of f are described in the following definition.

Definition. *A* **(sesqui-linear) form** *on a real or complex vector space* V *is a function* f *on* $V \times V$ *with values in the field of scalars such that*

$$\text{(a)} \qquad f(c\alpha + \beta, \gamma) = cf(\alpha, \gamma) + f(\beta, \gamma)$$
$$\text{(b)} \qquad f(\alpha, c\beta + \gamma) = \bar{c}f(\alpha, \beta) + f(\alpha, \gamma)$$

for all α, β, γ *in* V *and all scalars* c.

Thus, a sesqui-linear form is a function on $V \times V$ such that $f(\alpha, \beta)$ is a linear function of α for fixed β and a conjugate-linear function of β for fixed α. In the real case, $f(\alpha, \beta)$ is linear as a function of each argument; in other words, f is a **bilinear form.** In the complex case, the sesqui-linear form f is not bilinear unless $f = 0$. In the remainder of this chapter, we shall omit the adjective 'sesqui-linear' unless it seems important to include it.

If f and g are forms on V and c is a scalar, it is easy to check that $cf + g$ is also a form. From this it follows that any linear combination of forms on V is again a form. Thus the set of all forms on V is a subspace of the vector space of all scalar-valued functions on $V \times V$.

Theorem 1. *Let* V *be a finite-dimensional inner product space and* f *a form on* V. *Then there is a unique linear operator* T *on* V *such that*

$$f(\alpha, \beta) = (T\alpha|\beta)$$

for all α, β *in* V, *and the map* $f \rightarrow T$ *is an isomorphism of the space of forms onto* $L(V, V)$.

Proof. Fix a vector β in V. Then $\alpha \rightarrow f(\alpha, \beta)$ is a linear function on V. By Theorem 6 there is a unique vector β' in V such that $f(\alpha, \beta) = (\alpha|\beta')$ for every α. We define a function U from V into V by setting $U\beta = \beta'$. Then

$$\begin{aligned}f(\alpha|c\beta+\gamma) &= (\alpha|U(c\beta+\gamma))\\ &= \bar{c}f(\alpha,\beta)+f(\alpha,\gamma)\\ &= \bar{c}(\alpha|U\beta)+(\alpha|U\gamma)\\ &= (\alpha|cU\beta+U\gamma)\end{aligned}$$

for all α, β, γ in V and all scalars c. Thus U is a linear operator on V and $T = U^*$ is an operator such that $f(\alpha, \beta) = (T\alpha|\beta)$ for all α and β. If we also have $f(\alpha, \beta) = (T'\alpha|\beta)$, then

$$(T\alpha - T'\alpha|\beta) = 0$$

for all α and β; so $T\alpha = T'\alpha$ for all α. Thus for each form f there is a unique linear operator T_f such that

$$f(\alpha, \beta) = (T_f\alpha|\beta)$$

for all α, β in V. If f and g are forms and c a scalar, then

$$\begin{aligned}(cf+g)(\alpha,\beta) &= (T_{cf+g}\alpha|\beta)\\ &= cf(\alpha,\beta)+g(\alpha,\beta)\\ &= c(T_f\alpha|\beta)+(T_g\alpha|\beta)\\ &= ((cT_f+T_g)\alpha|\beta)\end{aligned}$$

for all α and β in V. Therefore,

$$T_{cf+g} = cT_f + T_g$$

so $f \to T_f$ is a linear map. For each T in $L(V, V)$ the equation

$$f(\alpha, \beta) = (T\alpha|\beta)$$

defines a form such that $T_f = T$, and $T_f = 0$ if and only if $f = 0$. Thus $f \to T_f$ is an isomorphism. ∎

Corollary. *The equation*

$$(f|g) = \operatorname{tr}(T_f T_g^*)$$

defines an inner product on the space of forms with the property that

$$(f|g) = \sum_{j,k} f(\alpha_k, \alpha_j)\overline{g(\alpha_k, \alpha_j)}$$

for every orthonormal basis $\{\alpha_1, \ldots, \alpha_n\}$ *of* V.

Proof. It follows easily from Example 3 of Chapter 8 that $(T, U) \to \operatorname{tr}(TU^*)$ is an inner product on $L(V, V)$. Since $f \to T_f$ is an isomorphism, Example 6 of Chapter 8 shows that

$$(f|g) = \operatorname{tr}(T_f T_g^*)$$

is an inner product. Now suppose that A and B are the matrices of T_f and T_g in the orthonormal basis $\mathcal{B} = \{\alpha_1, \ldots, \alpha_n\}$. Then

$$A_{jk} = (T_f\alpha_k|\alpha_j) = f(\alpha_k, \alpha_j)$$

and $B_{jk} = (T_g\alpha_k|\alpha_j) = g(\alpha_k, \alpha_j)$. Since AB^* is the matrix of $T_fT_g^*$ in the basis $\mathcal{B}$, it follows that

$$(f|g) = \operatorname{tr}(AB^*) = \sum_{j,k} A_{jk}\bar{B}_{jk}. \quad \blacksquare$$

Definition. *If* f *is a form and* $\mathcal{B} = \{\alpha_1, \ldots, \alpha_n\}$ *an arbitrary ordered basis of* V, *the matrix* A *with entries*

$$A_{jk} = f(\alpha_k, \alpha_j)$$

is called the **matrix of f in the ordered basis** $\mathcal{B}$.

When $\mathcal{B}$ is an orthonormal basis, the matrix of f in $\mathcal{B}$ is also the matrix of the linear transformation T_f, but in general this is not the case.

If A is the matrix of f in the ordered basis $\mathcal{B} = \{\alpha_1, \ldots, \alpha_n\}$, it follows that

$$f(\sum_s x_s\alpha_s, \sum_r y_r\alpha_r) = \sum_{r,s} \bar{y}_r A_{rs} x_s \tag{9-1}$$

for all scalars x_s and y_r $(1 \le r, s \le n)$. In other words, the matrix A has the property that

$$f(\alpha, \beta) = Y^*AX$$

where X and Y are the respective coordinate matrices of α and β in the ordered basis $\mathcal{B}$.

The matrix of f in another basis

$$\alpha_j' = \sum_{i=1}^{n} P_{ij}\alpha_i, \qquad (1 \le j \le n)$$

is given by the equation

$$A' = P^*AP. \tag{9-2}$$

For

$$\begin{aligned} A'_{jk} &= f(\alpha_k', \alpha_j') \\ &= f(\sum_s P_{sk}\alpha_s, \sum_r P_{rj}\alpha_r) \\ &= \sum_{r,s} \overline{P_{rj}} A_{rs} P_{sk} \\ &= (P^*AP)_{jk}. \end{aligned}$$

Since $P^* = P^{-1}$ for unitary matrices, it follows from (9-2) that results concerning unitary equivalence may be applied to the study of forms.

Theorem 2. *Let* f *be a form on a finite-dimensional complex inner product space* V. *Then there is an orthonormal basis for* V *in which the matrix of* f *is upper-triangular.*

Proof. Let T be the linear operator on V such that $f(\alpha, \beta) = (T\alpha|\beta)$ for all α and β. By Theorem 21, there is an orthonormal basis

$\{\alpha_1, \ldots, \alpha_n\}$ in which the matrix of T is upper-triangular. Hence,

$$f(\alpha_k, \alpha_j) = (T\alpha_k|\alpha_j) = 0$$

when $j > k$. ∎

Definition. *A form* f *on a real or complex vector space* V *is called* **Hermitian** *if*

$$\mathrm{f}(\alpha, \beta) = \overline{\mathrm{f}(\beta, \alpha)}$$

for all α *and* β *in* V.

If T is a linear operator on a finite-dimensional inner product space V and f is the form

$$f(\alpha, \beta) = (T\alpha|\beta)$$

then $\overline{f(\beta, \alpha)} = (\alpha|T\beta) = (T^*\alpha|\beta)$; so f is Hermitian if and only if T is self-adjoint.

When f is Hermitian $f(\alpha, \alpha)$ is real for every α, and on complex spaces this property characterizes Hermitian forms.

Theorem 3. *Let* V *be a complex vector space and* f *a form on* V *such that* $\mathrm{f}(\alpha, \alpha)$ *is real for every* α. *Then* f *is Hermitian.*

Proof. Let α and β be vectors in V. We must show that $f(\alpha, \beta) = \overline{f(\beta, \alpha)}$. Now

$$f(\alpha + \beta, \alpha + \beta) = f(\alpha, \beta) + f(\alpha, \beta) + f(\beta, \alpha) + f(\beta, \beta).$$

Since $f(\alpha + \beta, \alpha + \beta)$, $f(\alpha, \alpha)$, and $f(\beta, \beta)$ are real, the number $f(\alpha, \beta) + f(\beta, \alpha)$ is real. Looking at the same argument with $\alpha + i\beta$ instead of $\alpha + \beta$, we see that $-if(\alpha, \beta) + if(\beta, \alpha)$ is real. Having concluded that two numbers are real, we set them equal to their complex conjugates and obtain

$$f(\alpha, \beta) + f(\beta, \alpha) = \overline{f(\alpha, \beta)} + \overline{f(\beta, \alpha)}$$
$$-if(\alpha, \beta) + if(\beta, \alpha) = i\overline{f(\alpha, \beta)} - i\overline{f(\beta, \alpha)}$$

If we multiply the second equation by i and add the result to the first equation, we obtain

$$2f(\alpha, \beta) = 2\overline{f(\beta, \alpha)}. \quad ∎$$

Corollary. *Let* T *be a linear operator on a complex finite-dimensional inner product space* V. *Then* T *is self-adjoint if and only if* $(\mathrm{T}\alpha|\alpha)$ *is real for every* α *in* V.

Theorem 4 (Principal Axis Theorem). *For every Hermitian form* f *on a finite-dimensional inner product space* V, *there is an orthonormal basis of* V *in which* f *is represented by a diagonal matrix with real entries.*

Proof. Let T be the linear operator such that $f(\alpha, \beta) = (T\alpha|\beta)$ for all α and β in V. Then, since $f(\alpha, \beta) = \overline{f(\beta, \alpha)}$ and $(\overline{T\beta|\alpha}) = (\alpha|T\beta)$, it follows that

$$(T\alpha|\beta) = \overline{f(\beta, \alpha)} = (\alpha|T\beta)$$

for all α and β; hence $T = T^*$. By Theorem 18 of Chapter 8, there is an orthonormal basis of V which consists of characteristic vectors for T. Suppose $\{\alpha_1, \ldots, \alpha_n\}$ is an orthonormal basis and that

$$T\alpha_j = c_j\alpha_j$$

for $1 \leq j \leq n$. Then

$$f(\alpha_k, \alpha_j) = (T\alpha_k|\alpha_j) = \delta_{kj}c_k$$

and by Theorem 15 of Chapter 8 each c_k is real. ∎

Corollary. *Under the above conditions*

$$f(\sum_j x_j\alpha_j, \sum_k y_k\alpha_k) = \sum_j c_j x_j \bar{y}_j.$$

Exercises

1. Which of the following functions f, defined on vectors $\alpha = (x_1, x_2)$ and $\beta = (y_1, y_2)$ in C^2, are (sesqui-linear) forms on C^2?

(a) $f(\alpha, \beta) = 1$.
(b) $f(\alpha, \beta) = (x_1 - \bar{y}_1)^2 + x_2\bar{y}_2$.
(c) $f(\alpha, \beta) = (x_1 + \bar{y}_1)^2 - (x_1 - \bar{y}_1)^2$.
(d) $f(\alpha, \beta) = x_1\bar{y}_2 - \bar{x}_2 y_1$.

2. Let f be the form on R^2 defined by

$$f((x_1, y_1), (x_2, y_2)) = x_1y_1 + x_2y_2.$$

Find the matrix of f in each of the following bases:

$$\{(1, 0), (0, 1)\}, \{(1, -1), (1, 1)\}, \{(1, 2), (3, 4)\}.$$

3. Let

$$A = \begin{bmatrix} 1 & i \\ -i & 2 \end{bmatrix}$$

and let g be the form (on the space of 2×1 complex matrices) defined by $g(X, Y) = Y^*AX$. Is g an inner product?

4. Let V be a complex vector space and let f be a (sesqui-linear) form on V which is symmetric: $f(\alpha, \beta) = f(\beta, \alpha)$. What is f?

5. Let f be the form on R^2 given by

$$f((x_1, x_2), (y_1, y_2)) = x_1y_1 + 4x_2y_2 + 2x_1y_2 + 2x_2y_1.$$

Find an ordered basis in which f is represented by a diagonal matrix.

6. Call the form f (left) **non-degenerate** if 0 is the only vector α such that $f(\alpha, \beta) = 0$ for all β. Let f be a form on an inner product space V. Prove that f is

non-degenerate if and only if the associated linear operator T_f (Theorem 1) is non-singular.

7. Let f be a form on a finite-dimensional vector space V. Look at the definition of left non-degeneracy given in Exercise 6. Define right non-degeneracy and prove that the form f is left non-degenerate if and only if f is right non-degenerate.

8. Let f be a non-degenerate form (Exercises 6 and 7) on a finite-dimensional space V. Let L be a linear functional on V. Show that there exists one and only one vector β in V such that $L(\alpha) = f(\alpha, \beta)$ for all α.

9. Let f be a non-degenerate form on a finite-dimensional space V. Show that each linear operator S has an 'adjoint relative to f,' i.e., an operator S' such that $f(S\alpha, \beta) = f(\alpha, S'\beta)$ for all α, β.

9.3. Positive Forms

In this section, we shall discuss non-negative (sesqui-linear) forms and their relation to a given inner product on the underlying vector space.

Definitions. *A form* f *on a real or complex vector space* V *is* **non-negative** *if it is Hermitian and* $f(\alpha, \alpha) \geq 0$ *for every* α *in* V. *The form* f *is* **positive** *if* f *is Hermitian and* $f(\alpha, \alpha) > 0$ *for all* $\alpha \neq 0$.

A positive form on V is simply an inner product on V. A non-negative form satisfies all of the properties of an inner product except that some non-zero vectors may be 'orthogonal' to themselves.

Let f be a form on the finite-dimensional space V. Let $\mathfrak{B} = \{\alpha_1, \ldots, \alpha_n\}$ be an ordered basis for V, and let A be the matrix of f in the basis $\mathfrak{B}$, that is, $A_{jk} = f(\alpha_k, \alpha_j)$. If $\alpha = x_1\alpha_1 + \cdots + x_n\alpha_n$, then

$$\begin{aligned} f(\alpha, \alpha) &= f(\sum_j x_j\alpha_j, \sum_k x_k\alpha_k) \\ &= \sum_j \sum_k x_j\bar{x}_k f(\alpha_j, \alpha_k) \\ &= \sum_j \sum_k A_{kj}x_j\bar{x}_k. \end{aligned}$$

So, we see that f is non-negative if and only if

$$A = A^*$$

and

$$\sum_j \sum_k A_{kj}x_j\bar{x}_k \geq 0 \quad \text{for all scalars} \quad x_1, \ldots, x_n. \tag{9-3}$$

In order that f should be positive, the inequality in (9-3) must be strict for all $(x_1, \ldots, x_n) \neq 0$. The conditions we have derived state that f is a positive form on V if and only if the function

$$g(X, Y) = Y^*AX$$

is a positive form on the space of $n \times 1$ column matrices over the scalar field.

Theorem 5. *Let F be the field of real numbers or the field of complex numbers. Let A be an n × n matrix over F. The function g defined by*

$$g(X, Y) = Y^*AX \tag{9-4}$$

is a positive form on the space $F^{n\times 1}$ *if and only if there exists an invertible* n × n *matrix* P *with entries in* F *such that* A = P*P.

Proof. For any $n \times n$ matrix A, the function g in (9-4) is a form on the space of column matrices. We are trying to prove that g is positive if and only if $A = P^*P$. First, suppose $A = P^*P$. Then g is Hermitian and

$$\begin{aligned} g(X, X) &= X^*P^*PX \\ &= (PX)^*PX \\ &\geq 0. \end{aligned}$$

If P is invertible and $X \neq 0$, then $(PX)^*PX > 0$.

Now, suppose that g is a positive form on the space of column matrices. Then it is an inner product and hence there exist column matrices $Q_1, \ldots, Q_n$ such that

$$\begin{aligned} \delta_{jk} &= g(Q_j, Q_k) \\ &= Q_k^*AQ_j. \end{aligned}$$

But this just says that, if Q is the matrix with columns $Q_1, \ldots, Q_n$, then $Q^*AQ = I$. Since $\{Q_1, \ldots, Q_n\}$ is a basis, Q is invertible. Let $P = Q^{-1}$ and we have $A = P^*P$. ∎

In practice, it is not easy to verify that a given matrix A satisfies the criteria for positivity which we have given thus far. One consequence of the last theorem is that if g is positive then $\det A > 0$, because $\det A = \det (P^*P) = \det P^* \det P = |\det P|^2$. The fact that $\det A > 0$ is by no means sufficient to guarantee that g is positive; however, there are n determinants associated with A which have this property: If $A = A^*$ and if each of those determinants is positive, then g is a positive form.

Definition. *Let* A *be an* n × n *matrix over the field* F. *The* **principal minors** *of* A *are the scalars* $\Delta_k(A)$ *defined by*

$$\Delta_k(A) = \det \begin{bmatrix} A_{11} & \cdots & A_{1k} \\ \vdots & & \vdots \\ A_{k1} & \cdots & A_{kk} \end{bmatrix}, \qquad 1 \leq k \leq n.$$

Lemma. *Let* A *be an invertible* n × n *matrix with entries in a field* F. *The following two statements are equivalent.*

(a) *There is an upper-triangular matrix* P *with* $P_{kk} = 1$ $(1 \le k \le n)$ *such that the matrix* $B = AP$ *is lower-triangular.*

(b) *The principal minors of* A *are all different from* 0.

Proof. Let P be any $n \times n$ matrix and set $B = AP$. Then

$$B_{jk} = \sum_r A_{jr}P_{rk}.$$

If P is upper-triangular and $P_{kk} = 1$ for every k, then

$$\sum_{r=1}^{k-1} A_{jr}P_{rk} = B_{jk} - A_{kk}, \qquad k > 1.$$

Now B is lower-triangular provided $B_{jk} = 0$ for $j < k$. Thus B will be lower-triangular if and only if

$$\text{(9-5)} \qquad \sum_{r=1}^{k-1} A_{jr}P_{rk} = -A_{kk}, \qquad \begin{array}{l} 1 \le j \le k-1 \\ 2 \le k \le n. \end{array}$$

So, we see that statement (a) in the lemma is equivalent to the statement that there exist scalars P_{rk}, $1 \le r \le k$, $1 \le k \le n$, which satisfy (9-5) and $P_{kk} = 1$, $1 \le k \le n$.

In (9-5), for each $k > 1$ we have a system of $k - 1$ linear equations for the unknowns $P_{1k}, P_{2k}, \ldots, P_{k-1,k}$. The coefficient matrix of that system is

$$\begin{bmatrix} A_{11} & \cdots & A_{1,k-1} \\ \vdots & & \vdots \\ A_{k-1} & \cdots & A_{k-1,k-1} \end{bmatrix}$$

and its determinant is the principal minor $\Delta_{k-1}(A)$. If each $\Delta_{k-1}(A) \neq 0$, the systems (9-5) have unique solutions. We have shown that statement (b) implies statement (a) and that the matrix P is unique.

Now suppose that (a) holds. Then, as we shall see,

$$\text{(9-6)} \qquad \begin{aligned} \Delta_k(A) &= \Delta_k(B) \\ &= B_{11}B_{22} \cdots B_{kk}, \qquad k = 1, \ldots, n. \end{aligned}$$

To verify (9-6), let $A_1, \ldots, A_n$ and $B_1, \ldots, B_n$ be the columns of A and B, respectively. Then

$$\text{(9-7)} \qquad \begin{aligned} B_1 &= A_1 \\ B_r &= \sum_{j=1}^{r-1} P_{jr}A_j + A_r, \qquad r > 1. \end{aligned}$$

Fix k, $1 \le k \le n$. From (9-7) we see that the rth column of the matrix

$$\begin{bmatrix} B_{11} & \cdots & B_{kk} \\ \vdots & & \vdots \\ B_{k1} & \cdots & B_{kk} \end{bmatrix}$$

is obtained by adding to the rth column of

$$\begin{bmatrix} A_{11} & \cdots & A_{1k} \\ \vdots & & \vdots \\ A_{k1} & \cdots & A_{kk} \end{bmatrix}$$

a linear combination of its other columns. Such operations do not change determinants. That proves (9-6), except for the trivial observation that because B is triangular $\Delta_k(B) = B_{11} \cdots B_{kk}$. Since A and P are invertible, B is invertible. Therefore,

$$\Delta(B) = B_{11} \cdots B_{nn} \neq 0$$

and so $\Delta_k(A) \neq 0$, $k = 1, \ldots, n$. ∎

Theorem 6. *Let f be a form on a finite-dimensional vector space V and let A be the matrix of f in an ordered basis $\mathcal{B}$. Then f is a positive form if and only if A = A* and the principal minors of A are all positive.*

Proof. Let's do the interesting half of the theorem first. Suppose that $A = A^*$ and $\Delta_k(A) > 0$, $1 \leq k \leq n$. By the lemma, there exists an (unique) upper-triangular matrix P with $P_{kk} = 1$ such that $B = AP$ is lower-triangular. The matrix P^* is lower-triangular, so that $P^*B = P^*AP$ is also lower-triangular. Since A is self-adjoint, the matrix $D = P^*AP$ is self-adjoint. A self-adjoint triangular matrix is necessarily a diagonal matrix. By the same reasoning which led to (9-6),

$$\begin{aligned} \Delta_k(D) &= \Delta_k(P^*B) \\ &= \Delta_k(B) \\ &= \Delta_k(A). \end{aligned}$$

Since D is diagonal, its principal minors are

$$\Delta_k(D) = D_{11} \cdots D_{kk}.$$

From $\Delta_k(D) > 0$, $1 \leq k \leq n$, we obtain $D_{kk} > 0$ for each k.

If A is the matrix of the form f in the ordered basis $\mathcal{B} = \{\alpha_1, \ldots, \alpha_n\}$, then $D = P^*AP$ is the matrix of f in the basis $\{\alpha_1', \ldots, \alpha_n'\}$ defined by

$$\alpha_j' = \sum_{i=1}^{n} P_{ij}\alpha_i.$$

See (9-2). Since D is diagonal with positive entries on its diagonal, it is obvious that

$$X^*DX > 0, \qquad X \neq 0$$

from which it follows that f is a positive form.

Now, suppose we start with a positive form f. We know that $A = A^*$. How do we show that $\Delta_k(A) > 0$, $1 \leq k \leq n$? Let V_k be the subspace spanned by $\alpha_1, \ldots, \alpha_k$ and let f_k be the restriction of f to $V_k \times V_k$. Evidently f_k is a positive form on V_k and, in the basis $\{\alpha_1, \ldots, \alpha_k\}$ it is represented by the matrix

$$\begin{bmatrix} A_{11} & \cdots & A_{1k} \\ \vdots & & \vdots \\ A_{k1} & \cdots & A_{kk} \end{bmatrix}.$$

As a consequence of Theorem 5, we noted that the positivity of a form implies that the determinant of any representing matrix is positive. ∎

There are some comments we should make, in order to complete our discussion of the relation between positive forms and matrices. What is it that characterizes the matrices which represent positive forms? If f is a form on a complex vector space and A is the matrix of f in some ordered basis, then f will be positive if and only if $A = A^*$ and

$$X^*AX > 0, \qquad \text{for all complex } X \neq 0. \tag{9-8}$$

It follows from Theorem 3 that the condition $A = A^*$ is redundant, i.e., that (9-8) implies $A = A^*$. On the other hand, if we are dealing with a real vector space the form f will be positive if and only if $A = A^t$ and

$$X^tAX > 0, \qquad \text{for all real } X \neq 0. \tag{9-9}$$

We want to emphasize that if a real matrix A satisfies (9-9), it does not follow that $A = A^t$. One thing which is true is that, *if* $A = A^t$ and (9-9) holds, then (9-8) holds as well. That is because

$$\begin{aligned}(X + iY)^*A(X + iY) &= (X^t - iY^t)A(X + iY) \\ &= X^tAX + Y^tAY + i[X^tAY - Y^tAX]\end{aligned}$$

and if $A = A^t$ then $Y^tAX = X^tAY$.

If A is an $n \times n$ matrix with complex entries and if A satisfies (9-9), we shall call A a **positive matrix.** The comments which we have just made may be summarized by saying this: In either the real or complex case, a form f is positive if and only if its matrix in some (in fact, every) ordered basis is a positive matrix.

Now suppose that V is a finite-dimensional inner product space. Let f be a non-negative form on V. There is a unique self-adjoint linear operator T on V such that

$$f(\alpha, \beta) = (T\alpha|\beta). \tag{9-10}$$

and T has the additional property that $(T\alpha|\alpha) \geq 0$.

Definition. *A linear operator* T *on a finite-dimensional inner product space* V *is* **non-negative** *if* $T = T^*$ *and* $(T\alpha|\alpha) \geq 0$ *for all* α *in* V. *A* **positive** *linear operator is one such that* $T = T^*$ *and* $(T\alpha|\alpha) > 0$ *for all* $\alpha \neq 0$.

If V is a finite-dimensional (real or complex) vector space and if $(\cdot|\cdot)$ is an inner product on V, there is an associated class of positive linear operators on V. Via (9-10) there is a one-one correspondence between that class of positive operators and the collection of all positive forms on V. We shall use the exercises for this section to emphasize the relationships between positive operators, positive forms, and positive matrices. The following summary may be helpful.

If A is an $n \times n$ matrix over the field of complex numbers, the following are equivalent.

(1) A is positive, i.e., $\sum_j \sum_k A_{kj} x_j \bar{x}_k > 0$ whenever $x_1, \ldots, x_n$ are complex numbers, not all 0.

(2) $(X|Y) = Y^*AX$ is an inner product on the space of $n \times 1$ complex matrices.

(3) Relative to the standard inner product $(X|Y) = Y^*X$ on $n \times 1$ matrices, the linear operator $X \rightarrow AX$ is positive.

(4) $A = P^*P$ for some invertible $n \times n$ matrix P over C.

(5) $A = A^*$, and the principal minors of A are positive.

If each entry of A is real, these are equivalent to:

(6) $A = A^t$, and $\sum_j \sum_k A_{kj} x_j x_k > 0$ whenever $x_1, \ldots, x_n$ are real numbers not all 0.

(7) $(X|Y) = Y^tAX$ is an inner product on the space of $n \times 1$ real matrices.

(8) Relative to the standard inner product $(X|Y) = Y^tX$ on $n \times 1$ real matrices, the linear operator $X \rightarrow AX$ is positive.

(9) There is an invertible $n \times n$ matrix P, with real entries, such that $A = P^tP$.

Exercises

1. Let V be C^2, with the standard inner product. For which vectors α in V is there a positive linear operator T such that $\alpha = T\epsilon_1$?

2. Let V be R^2, with the standard inner product. If θ is a real number, let T be the linear operator 'rotation through θ,'

$$T_\theta(x_1, x_2) = (x_1 \cos \theta - x_2 \sin \theta, x_1 \sin \theta + x_2 \cos \theta).$$

For which values of θ is T_θ a positive operator?

3. Let V be the space of $n \times 1$ matrices over C, with the inner product $(X|Y) = Y^*GX$ (where G is an $n \times n$ matrix such that this is an inner product). Let A be an $n \times n$ matrix and T the linear operator $T(X) = AX$. Find T^*. If Y is a fixed element of V, find the element Z of V which determines the linear functional $X \rightarrow Y^*X$. In other words, find Z such that $Y^*X = (X|Z)$ for all X in V.

4. Let V be a finite-dimensional inner product space. If T and U are positive linear operators on V, prove that $(T + U)$ is positive. Give an example which shows that TU need not be positive.

5. Let

$$A = \begin{bmatrix} 1 & \frac{1}{2} \\ \frac{1}{2} & \frac{1}{3} \end{bmatrix}.$$

(a) Show that A is positive.

(b) Let V be the space of 2×1 real matrices, with the inner product $(X|Y) = Y^tAX$. Find an orthonormal basis for V, by applying the Gram-Schmidt process to the basis $\{X_1, X_2\}$ defined by

$$X_1 = \begin{bmatrix} 1 \\ 0 \end{bmatrix}, \quad X_2 = \begin{bmatrix} 0 \\ 1 \end{bmatrix}.$$

(c) Find an invertible 2×2 real matrix P such that $A = P^tP$.

6. Which of the following matrices are positive?

$$\begin{bmatrix} 1 & 2 \\ 3 & 4 \end{bmatrix}, \quad \begin{bmatrix} 1 & 1+i \\ 1-i & 3 \end{bmatrix}, \quad \begin{bmatrix} 1 & -1 & 1 \\ 2 & -1 & 1 \\ 3 & -1 & 1 \end{bmatrix}, \quad \begin{bmatrix} 1 & \frac{1}{2} & \frac{1}{3} \\ \frac{1}{2} & \frac{1}{3} & \frac{1}{4} \\ \frac{1}{3} & \frac{1}{4} & \frac{1}{5} \end{bmatrix}.$$

7. Give an example of an $n \times n$ matrix which has all its principal minors positive, but which is not a positive matrix.

8. Does $((x_1, x_2)|(y_1, y_2)) = x_1\bar{y}_1 + 2x_2\bar{y}_1 + 2x_1\bar{y}_2 + x_2\bar{y}_2$ define an inner product on C^2?

9. Prove that every entry on the main diagonal of a positive matrix is positive.

10. Let V be a finite-dimensional inner product space. If T and U are linear operators on V, we write $T < U$ if $U - T$ is a positive operator. Prove the following:

(a) $T < U$ and $U < T$ is impossible.

(b) If $T < U$ and $U < S$, then $T < S$.

(c) If $T < U$ and $0 < S$, it need not be that $ST < SU$.

11. Let V be a finite-dimensional inner product space and E the orthogonal projection of V onto some subspace.

(a) Prove that, for any positive number c, the operator $cI + E$ is positive.

(b) Express in terms of E a self-adjoint linear operator T such that $T^2 = I + E$.

12. Let n be a positive integer and A the $n \times n$ matrix

$$A = \begin{bmatrix} 1 & \frac{1}{2} & \frac{1}{3} & \cdots & \frac{1}{n} \\ \frac{1}{2} & \frac{1}{3} & \frac{1}{4} & \cdots & \frac{1}{n+1} \\ \vdots & \vdots & \vdots & & \vdots \\ \frac{1}{n} & \frac{1}{n+1} & \frac{1}{n+2} & \cdots & \frac{1}{2n-1} \end{bmatrix}.$$

Prove that A is positive.

13. Let A be a self-adjoint $n \times n$ matrix. Prove that there is a real number c such that the matrix $cI + A$ is positive.

14. Prove that the product of two positive linear operators is positive if and only if they commute.

15. Let S and T be positive operators. Prove that every characteristic value of ST is positive.

9.4. More on Forms

This section contains two results which give more detailed information about (sesqui-linear) forms.

Theorem 7. *Let* f *be a form on a real or complex vector space* V *and* $\{\alpha_1, \ldots, \alpha_r\}$ *a basis for the finite-dimensional subspace* W *of* V. *Let* M *be the* r $\times$ r *matrix with entries*

$$M_{jk} = f(\alpha_k, \alpha_j)$$

and W′ *the set of all vectors* β *in* V *such that* $f(\alpha, \beta) = 0$ *for all* α *in* W. *Then* W′ *is a subspace of* V, *and* W $\cap$ W′ = {0} *if and only if* M *is invertible. When this is the case,* V = W + W′.

Proof. If β and γ are vectors in W' and c is a scalar, then for every α in W

$$\begin{aligned} f(\alpha, c\beta + \gamma) &= \bar{c}f(\alpha, \beta) + f(\alpha, \gamma) \\ &= 0. \end{aligned}$$

Hence, W' is a subspace of V.

Now suppose $\alpha = \sum_{k=1}^{r} x_k\alpha_k$ and that $\beta = \sum_{j=1}^{r} y_j\alpha_j$. Then

$$\begin{aligned} f(\alpha, \beta) &= \sum_{j,k} \bar{y}_j M_{jk} x_k \\ &= \sum_k \Big(\sum_j \bar{y}_j M_{jk}\Big) x_k. \end{aligned}$$

It follows from this that $W \cap W' \neq \{0\}$ if and only if the homogeneous system

$$\sum_{j=1}^{r} \bar{y}_j M_{jk} = 0, \qquad 1 \leq k \leq r$$

has a non-trivial solution $(y_1, \ldots, y_r)$. Hence $W \cap W' = \{0\}$ if and only if M^* is invertible. But the invertibility of M^* is equivalent to the invertibility of M.

Suppose that M is invertible and let

$$A = (M^*)^{-1} = (M^{-1})^*.$$

Define g_j on V by the equation

$$g_j(\beta) = \sum_{k=1}^{r} A_{jk}\overline{f(\alpha_k, \beta)}.$$

Then

$$\begin{aligned} g_j(c\beta + \gamma) &= \sum_k A_{jk}\overline{f(\alpha_k, c\beta + \gamma)} \\ &= c\sum_k A_{jk}f(\alpha_k, \beta) + \sum_k A_{jk}f(\alpha_k, \gamma) \\ &= cg_j(\beta) + g_j(\gamma). \end{aligned}$$

Hence, each g_j is a linear function on V. Thus we may define a linear operator E on V by setting

$$E\beta = \sum_{j=1}^{r} g_j(\beta)\alpha_j.$$

Since

$$\begin{aligned} g_j(\alpha_n) &= \sum_k A_{jk}\overline{f(\alpha_k, \alpha_n)} \\ &= \sum_k A_{jk}(M^*)_{kn} \\ &= \delta_{jn} \end{aligned}$$

it follows that $E(\alpha_n) = \alpha_n$ for $1 \leq n \leq r$. This implies $E\alpha = \alpha$ for every α in W. Therefore, E maps V onto W and $E^2 = E$. If β is an arbitrary vector in V, then

$$\begin{aligned} f(\alpha_n, E\beta) &= f\left(\alpha_n, \sum_j g_j(\beta)\alpha_j\right) \\ &= \sum_j \overline{g_j(\beta)}f(\alpha_n, \alpha_j) \\ &= \sum_j \left(\sum_k \overline{A}_{jk}f(\alpha_k, \beta)\right) f(\alpha_n, \alpha_j). \end{aligned}$$

Since $A^* = M^{-1}$, it follows that

$$\begin{aligned} f(\alpha_n, E\beta) &= \sum_k \left(\sum_j (M^{-1})_{kj}M_{jn}\right) f(\alpha_k, \beta) \\ &= \sum_k \delta_{kn}f(\alpha_k, \beta) \\ &= f(\alpha_n, \beta). \end{aligned}$$

This implies $f(\alpha, E\beta) = f(\alpha, \beta)$ for every α in W. Hence

$$f(\alpha, \beta - E\beta) = 0$$

for all α in W and β in V. Thus $I - E$ maps V into W'. The equation

$$\beta = E\beta + (I - E)\beta$$

shows that $V = W + W'$. One final point should be mentioned. Since $W \cap W' = \{0\}$, every vector in V is uniquely the sum of a vector in W

and a vector in W'. If β is in W', it follows that $E\beta = 0$. Hence $I - E$ maps V onto W'. ∎

The projection E constructed in the proof may be characterized as follows: $E\beta = \alpha$ if and only if α is in W and $\beta - \alpha$ belongs to W'. Thus E is independent of the basis of W that was used in its construction. Hence we may refer to E as the **projection of** V **on** W that is **determined by** the direct sum decomposition

$$V = W \oplus W'.$$

Note that E is an orthogonal projection if and only if $W' = W^{\perp}$.

Theorem 8. *Let* f *be a form on a real or complex vector space* V *and* A *the matrix of* f *in the ordered basis* $\{\alpha_1, \ldots, \alpha_n\}$ *of* V. *Suppose the principal minors of* A *are all different from* 0. *Then there is a unique upper-triangular matrix* P *with* $P_{kk} = 1$ $(1 \leq k \leq n)$ *such that*

$$P^*AP$$

is upper-triangular.

Proof. Since $\Delta_k(A^*) = \overline{\Delta_k(A)}$ $(1 \leq k \leq n)$, the principal minors of A^* are all different from 0. Hence, by the lemma used in the proof of Theorem 6, there exists an upper-triangular matrix P with $P_{kk} = 1$ such that A^*P is lower-triangular. Therefore, $P^*A = (A^*P)^*$ is upper-triangular. Since the product of two upper-triangular matrices is again upper-triangular, it follows that P^*AP is upper-triangular. This shows the existence but not the uniqueness of P. However, there is another more geometric argument which may be used to prove both the existence and uniqueness of P.

Let W_k be the subspace spanned by $\alpha_1, \ldots, \alpha_k$ and W_k' the set of all β in V such that $f(\alpha, \beta) = 0$ for every α in W_k. Since $\Delta_k(A) \neq 0$, the $k \times k$ matrix M with entries

$$M_{ij} = f(\alpha_j, \alpha_i) = A_{ij}$$

$(1 \leq i, j \leq k)$ is invertible. By Theorem 7

$$V = W_k \oplus W_k'.$$

Let E_k be the projection of V on W_k which is determined by this decomposition, and set $E_0 = 0$. Let

$$\beta_k = \alpha_k - E_{k-1}\alpha_k, \qquad (1 \leq k \leq n).$$

Then $\beta_1 = \alpha_1$, and $E_{k-1}\alpha_k$ belongs to W_{k-1} for $k > 1$. Thus when $k > 1$, there exist unique scalars P_{jk} such that

$$E_{k-1}\alpha_k = -\sum_{j=1}^{k-1} P_{jk}\alpha_j.$$

Setting $P_{kk} = 1$ and $P_{jk} = 0$ for $j > k$, we then have an $n \times n$ upper-triangular matrix P with $P_{kk} = 1$ and

$$\beta_k = \sum_{j=1}^{k} P_{jk}\alpha_j$$

for $k = 1, \ldots, n$. Suppose $1 \leq i < k$. Then β_i is in W_i and $W_i \subset W_{k-1}$. Since β_k belongs to W'_{k-1}, it follows that $f(\beta_i, \beta_k) = 0$. Let B denote the matrix of f in the ordered basis $\{\beta_1, \ldots, \beta_n\}$. Then

$$B_{ki} = f(\beta_i, \beta_k)$$

so $B_{ki} = 0$ when $k > i$. Thus B is upper-triangular. On the other hand,

$$B = P^*AP.$$

Conversely, suppose P is an upper-triangular matrix with $P_{kk} = 1$ such that P^*AP is upper-triangular. Set

$$\beta_k = \sum_j P_{jk}\alpha_j, \qquad (1 \leq k \leq n).$$

Then $\{\beta_1, \ldots, \beta_k\}$ is evidently a basis for W_k. Suppose $k > 1$. Then $\{\beta_1, \ldots, \beta_{k-1}\}$ is a basis for W_{k-1}, and since $f(\beta_i, \beta_k) = 0$ when $i < k$, we see that β_k is a vector in W'_{k-1}. The equation defining β_k implies

$$\alpha_k = -\left(\sum_{j=1}^{k-1} P_{jk}\alpha_j\right) + \beta_k.$$

Now $\sum_{j=1}^{k-1} P_{jk}\alpha_j$ belongs to W_{k-1} and β_k is in W'_{k-1}. Therefore, $P_{1k}, \ldots, P_{k-1k}$ are the unique scalars such that

$$E_{k-1}\alpha_k = -\sum_{j=1}^{k-1} P_{jk}\alpha_j$$

so that P is the matrix constructed earlier. ∎

9.5. Spectral Theory

In this section, we pursue the implications of Theorems 18 and 22 of Chapter 8 concerning the diagonalization of self-adjoint and normal operators.

Theorem 9 (Spectral Theorem). *Let T be a normal operator on a finite-dimensional complex inner product space V or a self-adjoint operator on a finite-dimensional real inner product space V. Let $c_1, \ldots, c_k$ be the distinct characteristic values of T. Let W_j be the characteristic space associated with c_j and E_j the orthogonal projection of V on W_j. Then W_j is orthogonal to W_i when $i \neq j$, V is the direct sum of $W_1, \ldots, W_k$, and*

(9-11) $$T = c_1E_1 + \cdots + c_kE_k.$$

Proof. Let α be a vector in W_j, β a vector in W_i, and suppose $i \neq j$. Then $c_j(\alpha|\beta) = (T\alpha|\beta) = (\alpha|T^*\beta) = (\alpha|\bar{c}_i\beta)$. Hence $(c_j - c_i)(\alpha|\beta) = 0$, and since $c_j - c_i \neq 0$, it follows that $(\alpha|\beta) = 0$. Thus W_j is orthogonal to W_i when $i \neq j$. From the fact that V has an orthonormal basis consisting of characteristic vectors (cf. Theorems 18 and 22 of Chapter 8), it follows that $V = W_1 + \cdots + W_k$. If α_j belongs to V_j $(1 \leq j \leq k)$ and $\alpha_1 + \cdots + \alpha_k = 0$, then

$$\begin{aligned} 0 &= (\alpha_i|\sum_j \alpha_j) = \sum_j (\alpha_i|\alpha_j) \\ &= ||\alpha_i||^2 \end{aligned}$$

for every i, so that V is the direct sum of $W_1, \ldots, W_k$. Therefore $E_1 + \cdots + E_k = I$ and

$$\begin{aligned} T &= TE_1 + \cdots + TE_k \\ &= c_1E_1 + \cdots + c_kE_k. \quad \blacksquare \end{aligned}$$

The decomposition (9-11) is called the **spectral resolution** of T. This terminology arose in part from physical applications which caused the **spectrum** of a linear operator on a finite-dimensional vector space to be defined as the set of characteristic values for the operator. It is important to note that the orthogonal projections $E_1, \ldots, E_k$ are canonically associated with T; in fact, they are polynomials in T.

Corollary. *If* $e_j = \prod_{i \neq j} \left(\frac{x - c_i}{c_j - c_i}\right)$, *then* $E_j = e_j(T)$ *for* $1 \leq j \leq k$.

Proof. Since $E_iE_j = 0$ when $i \neq j$, it follows that

$$T^2 = c_1^2E_1 + \cdots + c_k^2E_k$$

and by an easy induction argument that

$$T^n = c_1^nE_1 + \cdots + c_k^nE_k$$

for every integer $n \geq 0$. For an arbitrary polynomial

$$f = \sum_{n=0}^{r} a_nx^n$$

we have

$$\begin{aligned} f(T) &= \sum_{n=0}^{r} a_nT^n \\ &= \sum_{n=0}^{r} a_n \sum_{j=1}^{k} c_j^nE_j \\ &= \sum_{j=1}^{k} \left(\sum_{n=0}^{r} a_nc_j^n\right) E_j \\ &= \sum_{j=1}^{k} f(c_j)E_j. \end{aligned}$$

Since $e_j(c_m) = \delta_{jm}$, it follows that $e_j(T) = E_j$. $\blacksquare$

Because $E_1, \ldots, E_k$ are canonically associated with T and

$$I = E_1 + \cdots + E_k$$

the family of projections $\{E_1, \ldots, E_k\}$ is called the **resolution of the identity defined by** T.

There is a comment that should be made about the proof of the spectral theorem. We derived the theorem using Theorems 18 and 22 of Chapter 8 on the diagonalization of self-adjoint and normal operators. There is another, more algebraic, proof in which it must first be shown that the minimal polynomial of a normal operator is a product of distinct prime factors. Then one proceeds as in the proof of the primary decomposition theorem (Theorem 12, Chapter 6). We shall give such a proof in the next section.

In various applications it is necessary to know whether one may compute certain functions of operators or matrices, e.g., square roots. This may be done rather simply for diagonalizable normal operators.

Definition. *Let* T *be a diagonalizable normal operator on a finite-dimensional inner product space and*

$$\mathrm{T} = \sum_{j=1}^{k} \mathrm{c}_j \mathrm{E}_j$$

its spectral resolution. Suppose f *is a function whose domain includes the spectrum of* T *that has values in the field of scalars. Then the linear operator* f(T) *is defined by the equation*

$$\mathrm{f(T)} = \sum_{j=1}^{k} \mathrm{f(c_j)E_j}. \tag{9-12}$$

Theorem 10. *Let* T *be a diagonalizable normal operator with spectrum* S *on a finite-dimensional inner product space* V. *Suppose* f *is a function whose domain contains* S *that has values in the field of scalars. Then* f(T) *is a diagonalizable normal operator with spectrum* f(S). *If* U *is a unitary map of* V *onto* V′ *and* $\mathrm{T}' = \mathrm{UTU}^{-1}$, *then* S *is the spectrum of* T′ *and*

$$\mathrm{f(T')} = \mathrm{Uf(T)U^{-1}}.$$

Proof. The normality of $f(T)$ follows by a simple computation from (9-12) and the fact that

$$f(T)^* = \sum_j \overline{f(c_j)} E_j.$$

Moreover, it is clear that for every α in $E_j(V)$

$$f(T)\alpha = f(c_j)\alpha.$$

Thus, the set $f(S)$ of all $f(c)$ with c in S is contained in the spectrum of $f(T)$. Conversely, suppose $\alpha \neq 0$ and that

$$f(T)\alpha = b\alpha.$$

Then $\alpha = \sum_j E_j\alpha$ and

$$\begin{aligned} f(T)\alpha &= \sum_j f(T)E_j\alpha \\ &= \sum_j f(c_j)E_j\alpha \\ &= \sum_j bE_j\alpha. \end{aligned}$$

Hence,

$$\begin{aligned} \|\sum_j (f(c_j) - b)E_j\alpha\|^2 &= \sum_j |f(c_j) - b|^2\|E_j\alpha\|^2 \\ &= 0. \end{aligned}$$

Therefore, $f(c_j) = b$ or $E_j\alpha = 0$. By assumption, $\alpha \neq 0$, so there exists an index i such that $E_i\alpha \neq 0$. It follows that $f(c_i) = b$ and hence that $f(S)$ is the spectrum of $f(T)$. Suppose, in fact, that

$$f(S) = \{b_1, \ldots, b_r\}$$

where $b_m \neq b_n$ when $m \neq n$. Let X_m be the set of indices i such that $1 \leq i \leq k$ and $f(c_i) = b_m$. Let $P_m = \sum_i E_i$, the sum being extended over the indices i in X_m. Then P_m is the orthogonal projection of V on the subspace of characteristic vectors belonging to the characteristic value b_m of $f(T)$, and

$$f(T) = \sum_{m=1}^{r} b_m P_m$$

is the spectral resolution of $f(T)$.

Now suppose U is a unitary transformation of V onto V' and that $T' = UTU^{-1}$. Then the equation

$$T\alpha = c\alpha$$

holds if and only if

$$T'U\alpha = cU\alpha.$$

Thus S is the spectrum of T', and U maps each characteristic subspace for T onto the corresponding subspace for T'. In fact, using (9-12), we see that

$$T' = \sum_j c_j E'_j, \qquad E'_j = UE_jU^{-1}$$

is the spectral resolution of T'. Hence

$$\begin{aligned} f(T') &= \sum_j f(c_j)E'_j \\ &= \sum_j f(c_j)UE_jU^{-1} \\ &= U\left(\sum_j f(c_j)E_j\right)U^{-1} \\ &= Uf(T)U^{-1}. \quad \blacksquare \end{aligned}$$

In thinking about the preceding discussion, it is important for one to keep in mind that the spectrum of the normal operator T is the set

$$S = \{c_1, \ldots, c_k\}$$

of distinct characteristic values. When T is represented by a diagonal matrix in a basis of characteristic vectors, it is necessary to repeat each value c_j as many times as the dimension of the corresponding space of characteristic vectors. This is the reason for the change of notation in the following result.

Corollary. *With the assumptions of Theorem* 10, *suppose that* T *is represented in the ordered basis* $\mathcal{B} = \{\alpha_1, \ldots, \alpha_n\}$ *by the diagonal matrix* D *with entries* $d_1, \ldots, d_n$. *Then, in the basis* $\mathcal{B}$, $f(T)$ *is represented by the diagonal matrix* $f(D)$ *with entries* $f(d_1), \ldots, f(d_n)$. *If* $\mathcal{B}' = \{\alpha_1', \ldots, \alpha_n'\}$ *is any other ordered basis and* P *the matrix such that*

$$\alpha_j' = \sum_i P_{ij}\alpha_i$$

then $P^{-1}f(D)P$ *is the matrix of* $f(T)$ *in the basis* $\mathcal{B}'$.

Proof. For each index i, there is a unique j such that $1 \le j \le k$, α_i belongs to $E_j(V)$, and $d_i = c_j$. Hence $f(T)\alpha_i = f(d_i)\alpha_i$ for every i, and

$$\begin{aligned} f(T)\alpha_j' &= \sum_i P_{ij} f(T)\alpha_i \\ &= \sum_i d_i P_{ij}\alpha_i \\ &= \sum_i (DP)_{ij}\alpha_i \\ &= \sum_i (DP)_{ij} \sum_k P_{ki}^{-1}\alpha_k' \\ &= \sum_k (P^{-1}DP)_{kj}\alpha_k'. \quad \blacksquare \end{aligned}$$

It follows from this result that one may form certain functions of a normal matrix. For suppose A is a normal matrix. Then there is an invertible matrix P, in fact a unitary P, such that PAP^{-1} is a diagonal matrix, say D with entries $d_1, \ldots, d_n$. Let f be a complex-valued function which can be applied to $d_1, \ldots, d_n$, and let $f(D)$ be the diagonal matrix with entries $f(d_1), \ldots, f(d_n)$. Then $P^{-1}f(D)P$ is independent of D and just a function of A in the following sense. If Q is another invertible matrix such that QAQ^{-1} is a diagonal matrix D', then f may be applied to the diagonal entries of D' and

$$P^{-1}f(D)P = Q^{-1}f(D')Q.$$

Definition. *Under the above conditions,* $f(A)$ *is defined as* $P^{-1}f(D)P$.

The matrix $f(A)$ may also be characterized in a different way. In doing this, we state without proof some of the results on normal matrices that one obtains by formulating the matrix analogues of the preceding theorems.

Theorem 11. *Let* A *be a normal matrix and* $c_1, \ldots, c_k$ *the distinct complex roots of det* $(xI - A)$. *Let*

$$e_i = \prod_{j \neq i} \left(\frac{x - c_j}{c_i - c_j} \right)$$

and $E_i = e_i(A)$ $(1 \leq i \leq k)$. *Then* $E_iE_j = 0$ *when* $i \neq j$, $E_i^2 = E_i$, $E_i^* = E_i$, *and*

$$I = E_1 + \cdots + E_k.$$

If f *is a complex-valued function whose domain includes* $c_1, \ldots, c_k$, *then*

$$f(A) = f(c_1)E_1 + \cdots + f(c_k)E_k;$$

in particular, $A = c_1E_1 + \cdots + c_kE_k$.

We recall that an operator on an inner product space V is non-negative if T is self-adjoint and $(T\alpha|\alpha) \geq 0$ for every α in V.

Theorem 12. *Let* T *be a diagonalizable normal operator on a finite-dimensional inner product space* V. *Then* T *is self-adjoint, non-negative, or unitary according as each characteristic value of* T *is real, non-negative, or of absolute value* 1.

Proof. Suppose T has the spectral resolution $T = c_1E_1 + \cdots + c_kE_k$, then $T^* = \bar{c}_1E_1 + \cdots + \bar{c}_kE_k$. To say T is self-adjoint is to say $T = T^*$, or

$$(c_1 - \bar{c}_1)E_1 + \cdots + (c_k - \bar{c}_k)E_k = 0.$$

Using the fact that $E_iE_j = 0$ for $i \neq j$, and the fact that no E_j is the zero operator, we see that T is self-adjoint if and only if $c_j = \bar{c}_j$, $j = 1, \ldots, k$. To distinguish the normal operators which are non-negative, let us look at

$$\begin{aligned}(T\alpha|\alpha) &= \left(\sum_{j=1}^{k} c_jE_j\alpha \Big| \sum_{i=1}^{k} E_i\alpha \right) \\ &= \sum_i \sum_j c_j(E_j\alpha|E_i\alpha) \\ &= \sum_j c_j\|E_j\alpha\|^2.\end{aligned}$$

We have used the fact that $(E_j\alpha|E_i\alpha) = 0$ for $i \neq j$. From this it is clear that the condition $(T\alpha|\alpha) \geq 0$ is satisfied if and only if $c_j \geq 0$ for each j. To distinguish the unitary operators, observe that

$$\begin{aligned}TT^* &= c_1\bar{c}_1E_1 + \cdots + c_k\bar{c}_kE_k \\ &= |c_1|^2E_1 + \cdots + |c_k|^2E_k.\end{aligned}$$

If $TT^* = I$, then $I = |c_1|^2E_1 + \cdots + |c_k|^2E_k$, and operating with E_j

$$E_j = |c_j|^2E_j.$$

Since $E_j \neq 0$, we have $|c_j|^2 = 1$ or $|c_j| = 1$. Conversely, if $|c_j|^2 = 1$ for each j, it is clear that $TT^* = I$. ∎

It is important to note that this is a theorem about normal operators. If T is a general linear operator on V which has real characteristic values, it does not follow that T is self-adjoint. The theorem states that if T has real characteristic values, and *if* T is diagonalizable and normal, then T is self-adjoint. A theorem of this type serves to strengthen the analogy between the adjoint operation and the process of forming the conjugate of a complex number. A complex number z is real or of absolute value 1 according as $z = \bar{z}$, or $\bar{z}z = 1$. An operator T is self-adjoint or unitary according as $T = T^*$ or $T^*T = I$.

We are going to prove two theorems now, which are the analogues of these two statements:

(1) Every non-negative number has a unique non-negative square root.

(2) Every complex number is expressible in the form ru, where r is non-negative and $|u| = 1$. This is the polar decomposition $z = re^{i\theta}$ for complex numbers.

Theorem 13. *Let* V *be a finite-dimensional inner product space and* T *a non-negative operator on* V. *Then* T *has a unique non-negative square root, that is, there is one and only one non-negative operator* N *on* V *such that* $N^2 = T$.

Proof. Let $T = c_1E_1 + \cdots + c_kE_k$ be the spectral resolution of T. By Theorem 12, each $c_j \geq 0$. If c is any non-negative real number, let $\sqrt{c}$ denote the non-negative square root of c. Then according to Theorem 11 and (9-12) $N = \sqrt{T}$ is a well-defined diagonalizable normal operator on V. It is non-negative by Theorem 12, and, by an obvious computation, $N^2 = T$.

Now let P be a non-negative operator on V such that $P^2 = T$. We shall prove that $P = N$. Let

$$P = d_1F_1 + \cdots + d_rF_r$$

be the spectral resolution of P. Then $d_j \geq 0$ for each j, since P is non-negative. From $P^2 = T$ we have

$$T = d_1^2F_1 + \cdots + d_r^2F_r.$$

Now $F_1, \ldots, F_r$ satisfy the conditions $I = F_1 + \cdots + F_r$, $F_iF_j = 0$ for $i \neq j$, and no F_j is 0. The numbers $d_1^2, \ldots, d_r^2$ are distinct, because distinct non-negative numbers have distinct squares. By the uniqueness of the spectral resolution of T, we must have $r = k$, and (perhaps reordering) $F_j = E_j$, $d_j^2 = c_j$. Thus $P = N$. ∎

Theorem 14. *Let V be a finite-dimensional inner product space and let T be any linear operator on V. Then there exist a unitary operator U on V and a non-negative operator N on V such that T = UN. The non-negative operator N is unique. If T is invertible, the operator U is also unique.*

Proof. Suppose we have $T = UN$, where U is unitary and N is non-negative. Then $T^* = (UN)^* = N^*U^* = NU^*$. Thus $T^*T = NU^*UN = N^2$. This shows that N is uniquely determined as the non-negative square root of the non-negative operator T^*T.

So, to begin the proof of the existence of U and N, we use Theorem 13 to define N as the unique non-negative square root of T^*T. If T is invertible, then so is N because

$$(N\alpha|N\alpha) = (N^2\alpha|\alpha) = (T^*T\alpha|\alpha) = (T\alpha|T\alpha).$$

In this case, we define $U = TN^{-1}$ and prove that U is unitary. Now $U^* = (TN^{-1})^* = (N^{-1})^*T^* = (N^*)^{-1}T^* = N^{-1}T^*$. Thus

$$\begin{aligned} UU^* &= TN^{-1}N^{-1}T^* \\ &= T(N^{-1})^2T^* \\ &= T(N^2)^{-1}T^* \\ &= T(T^*T)^{-1}T^* \\ &= TT^{-1}(T^*)^{-1}T^* \\ &= I \end{aligned}$$

and U is unitary.

If T is not invertible, we shall have to do a bit more work to define U. We first define U on the range of N. Let α be a vector in the range of N, say $\alpha = N\beta$. We define $U\alpha = T\beta$, motivated by the fact that we want $UN\beta = T\beta$. We must verify that U is well-defined on the range of N; in other words, if $N\beta' = N\beta$ then $T\beta' = T\beta$. We verified above that $||N\gamma||^2 = ||T\gamma||^2$ for every γ in V. Thus, with $\gamma = \beta - \beta'$, we see that $N(\beta - \beta') = 0$ if and only if $T(\beta - \beta') = 0$. So U is well-defined on the range of N and is clearly linear where defined. Now if W is the range of N, we are going to define U on $W^\perp$. To do this, we need the following observation. Since T and N have the same null space, their ranges have the same dimension. Thus $W^\perp$ has the same dimension as the orthogonal complement of the range of T. Therefore, there exists an (inner product space) isomorphism U_0 of $W^\perp$ onto $T(V)^\perp$. Now we have defined U on W, and we define U on $W^\perp$ to be U_0.

Let us repeat the definition of U. Since $V = W \oplus W^\perp$, each α in V is uniquely expressible in the form $\alpha = N\beta + \gamma$, where $N\beta$ is in the range W of N, and γ is in $W^\perp$. We define

$$U\alpha = T\beta + U_0\gamma.$$

This U is clearly linear, and we verified above that it is well-defined. Also

$$\begin{aligned}(U\alpha|U\alpha) &= (T\beta + U_0\gamma|T\beta + U_0\gamma)\\ &= (T\beta|T\beta) + (U_0\gamma|U_0\gamma)\\ &= (N\beta|N\beta) + (\gamma|\gamma)\\ &= (\alpha|\alpha)\end{aligned}$$

and so U is unitary. We also have $UN\beta = T\beta$ for each β. ∎

We call $T = UN$ a **polar decomposition** for T. We certainly cannot call it *the* polar decomposition, since U is not unique. Even when T is invertible, so that U is unique, we have the difficulty that U and N may not commute. Indeed, they commute if and only if T is normal. For example, if $T = UN = NU$, with N non-negative and U unitary, then

$$TT^* = (NU)(NU)^* = NUU^*N = N^2 = T^*T.$$

The general operator T will also have a decomposition $T = N_1U_1$, with N_1 non-negative and U_1 unitary. Here, N_1 will be the non-negative square root of TT^*. We can obtain this result by applying the theorem just proved to the operator T^*, and then taking adjoints.

We turn now to the problem of what can be said about the simultaneous diagonalization of commuting families of normal operators. For this purpose the following terminology is appropriate.

Definitions. *Let $\mathfrak{F}$ be a family of operators on an inner product space* V. *A function* r *on $\mathfrak{F}$ with values in the field* F *of scalars will be called a* **root** *of $\mathfrak{F}$ if there is a non-zero α in* V *such that*

$$\mathrm{T}\alpha = \mathrm{r}(\mathrm{T})\alpha$$

for all T *in $\mathfrak{F}$. For any function* r *from $\mathfrak{F}$ to* F, *let* V(r) *be the set of all α in* V *such that* $\mathrm{T}\alpha = \mathrm{r}(\mathrm{T})\alpha$ *for every* T *in $\mathfrak{F}$.*

Then $V(r)$ is a subspace of V, and r is a root of $\mathfrak{F}$ if and only if $V(r) \neq \{0\}$. Each non-zero α in $V(r)$ is simultaneously a characteristic vector for every T in $\mathfrak{F}$.

Theorem 15. *Let $\mathfrak{F}$ be a commuting family of diagonalizable normal operators on a finite-dimensional inner product space* V. *Then $\mathfrak{F}$ has only a finite number of roots. If* $\mathrm{r}_1, \ldots, \mathrm{r}_k$ *are the distinct roots of $\mathfrak{F}$, then*

(i) $\mathrm{V}(\mathrm{r}_i)$ *is orthogonal to* $\mathrm{V}(\mathrm{r}_j)$ *when* $\mathrm{i} \neq \mathrm{j}$, *and*
(ii) $\mathrm{V} = \mathrm{V}(\mathrm{r}_1) \oplus \cdots \oplus \mathrm{V}(\mathrm{r}_k)$.

Proof. Suppose r and s are distinct roots of F. Then there is an operator T in $\mathfrak{F}$ such that $r(T) \neq s(T)$. Since characteristic vectors belonging to distinct characteristic values of T are necessarily orthogonal, it follows that $V(r)$ is orthogonal to $V(s)$. Because V is finite-dimensional, this implies $\mathfrak{F}$ has at most a finite number of roots. Let $r_1, \ldots, r_k$ be the roots of F. Suppose $\{T_1, \ldots, T_m\}$ is a maximal linearly independent subset of $\mathfrak{F}$, and let

$$\{E_{i1}, E_{i2}, \ldots\}$$

be the resolution of the identity defined by T_i $(1 \leq i \leq m)$. Then the projections E_{ij} form a commutative family. For each E_{ij} is a polynomial in T_i and $T_1, \ldots, T_m$ commute with one another. Since

$$I = (\sum_{j_1} E_{1j_1}) (\sum_{j_2} E_{2j_2}) \cdots (\sum_{j_m} E_{mj_m})$$

each vector α in V may be written in the form

$$\alpha = \sum_{j_1, \ldots, j_m} E_{1j_1}E_{2j_2} \cdots E_{mj_m}\alpha. \tag{9-13}$$

Suppose $j_1, \ldots, j_m$ are indices for which $\beta = E_{1j_1}E_{2j_2} \cdots E_{mj_m}\alpha \neq 0$. Let

$$\beta_i = (\prod_{n \neq i} E_{nj_n}) \alpha.$$

Then $\beta = E_{ij_i}\beta_i$; hence there is a scalar c_i such that

$$T_i\beta = c_i\beta, \qquad 1 \leq i \leq m.$$

For each T in $\mathfrak{F}$, there exist unique scalars b_i such that

$$T = \sum_{i=1}^{m} b_iT_i.$$

Thus

$$\begin{aligned} T\beta &= \sum_i b_iT_i\beta \\ &= (\sum_i b_ic_i) \beta. \end{aligned}$$

The function $T \to \sum_i b_ic_i$ is evidently one of the roots, say r_t of $\mathfrak{F}$, and β lies in $V(r_t)$. Therefore, each non-zero term in (9-13) belongs to one of the spaces $V(r_1), \ldots, V(r_k)$. It follows that V is the orthogonal direct sum of $V(r_1), \ldots, V(r_k)$. ∎

Corollary. *Under the assumptions of the theorem, let* P_j *be the orthogonal projection of* V *on* $V(r_j)$, $(1 \leq j \leq k)$. *Then* $P_iP_j = 0$ *when* $i \neq j$,

$$I = P_1 + \cdots + P_k,$$

and every T *in* $\mathfrak{F}$ *may be written in the form*

$$T = \sum_j r_j(T)P_j. \tag{9-14}$$

Definitions. *The family of orthogonal projections* $\{P_1, \ldots, P_k\}$ *is called the* **resolution of the identity determined by** $\mathfrak{F}$, *and* (9-14) *is the* **spectral resolution of T in terms of this family.**

Although the projections $P_1, \ldots, P_k$ in the preceding corollary are canonically associated with the family $\mathfrak{F}$, they are generally not in $\mathfrak{F}$ nor even linear combinations of operators in $\mathfrak{F}$; however, we shall show that they may be obtained by forming certain products of polynomials in elements of $\mathfrak{F}$.

In the study of any family of linear operators on an inner product space, it is usually profitable to consider the self-adjoint algebra generated by the family.

Definition. *A* **self-adjoint algebra of operators** *on an inner product space V is a linear subalgebra of L(V, V) which contains the adjoint of each of its members.*

An example of a self-adjoint algebra is $L(V, V)$ itself. Since the intersection of any collection of self-adjoint algebras is again a self-adjoint algebra, the following terminology is meaningful.

Definition. *If $\mathfrak{F}$ is a family of linear operators on a finite-dimensional inner product space, the* **self-adjoint algebra generated by** *$\mathfrak{F}$ is the smallest self-adjoint algebra which contains $\mathfrak{F}$.*

Theorem 16. *Let $\mathfrak{F}$ be a commuting family of diagonalizable normal operators on a finite-dimensional inner product space V, and let $\mathfrak{A}$ be the self-adjoint algebra generated by $\mathfrak{F}$ and the identity operator. Let $\{P_1, \ldots, P_k\}$ be the resolution of the identity defined by $\mathfrak{F}$. Then $\mathfrak{A}$ is the set of all operators on V of the form*

$$T = \sum_{j=1}^{k} c_j P_j \tag{9-15}$$

where $c_1, \ldots, c_k$ are arbitrary scalars.

Proof. Let $\mathfrak{C}$ denote the set of all operators on V of the form (9-15). Then $\mathfrak{C}$ contains the identity operator and the adjoint

$$T^* = \sum_j \bar{c}_j P_j$$

of each of its members. If $T = \sum_j c_j P_j$ and $U = \sum_j d_j P_j$, then for every scalar a

$$aT + U = \sum_j (ac + d_j) P_j$$

and

$$\begin{aligned} TU &= \sum_{i,j} c_i d_j P_i P_j \\ &= \sum_j c_j d_j P_j \\ &= UT. \end{aligned}$$

Thus $\mathfrak{C}$ is a self-adjoint commutative algebra containing $\mathfrak{F}$ and the identity operator. Therefore $\mathfrak{C}$ contains $\mathfrak{A}$.

Now let $r_1, \ldots, r_k$ be all the roots of $\mathcal{F}$. Then for each pair of indices (i, n) with $i \neq n$, there is an operator T_{in} in $\mathcal{F}$ such that $r_i(T_{in}) \neq r_n(T_{in})$. Let $a_{in} = r_i(T_{in}) - r_n(T_{in})$ and $b_{in} = r_n(T_{in})$. Then the linear operator

$$Q_i = \prod_{n \neq i} a_{in}^{-1}(T_{in} - b_{in}I)$$

is an element of the algebra $\mathcal{A}$. We will show that $Q_i = P_i$ $(1 \leq i \leq k)$. For this, suppose $j \neq i$ and that α is an arbitrary vector in $V(r_j)$. Then

$$\begin{aligned} T_{ij}\alpha &= r_j(T_{ij})\alpha \\ &= b_{ij}\alpha \end{aligned}$$

so that $(T_{ij} - b_{ij}I)\alpha = 0$. Since the factors in Q_i all commute, it follows that $Q_i\alpha = 0$. Hence Q_i agrees with P_i on $V(r_j)$ whenever $j \neq i$. Now suppose α is a vector in $V(r_i)$. Then $T_{in}\alpha = r_i(T_{in})\alpha$, and

$$a_{in}^{-1}(T_{in} - b_{in}I)\alpha = a_{in}^{-1}[r_i(T_{in}) - r_n(T_{in})]\alpha = \alpha.$$

Thus $Q_i\alpha = \alpha$ and Q_i agrees with P_i on $V(r_i)$; therefore, $Q_i = P_i$ for $i = 1, \ldots, k$. From this it follows that $\mathcal{A} = \mathcal{C}$. ∎

The theorem shows that the algebra $\mathcal{A}$ is commutative and that each element of $\mathcal{A}$ is a diagonalizable normal operator. We show next that $\mathcal{A}$ has a single generator.

Corollary. *Under the assumptions of the theorem, there is an operator* T *in* $\mathcal{A}$ *such that every member of* $\mathcal{A}$ *is a polynomial in* T.

Proof. Let $T = \sum_{j=1}^{k} t_jP_j$ where $t_1, \ldots, t_k$ are distinct scalars. Then

$$T^n = \sum_{j=1}^{k} t_j^nP_j$$

for $n = 1, 2, \ldots$. If

$$f = \sum_{n=1}^{s} a_nx^n$$

it follows that

$$\begin{aligned} f(T) &= \sum_{n=1}^{s} a_nT^n = \sum_{n=1}^{s} \sum_{j=1}^{k} a_nt_j^nP_j \\ &= \sum_{j=1}^{k} \left(\sum_{n=1}^{s} a_nt_j^n \right) P_j \\ &= \sum_{j=1}^{k} f(t_j)P_j. \end{aligned}$$

Given an arbitrary

$$U = \sum_{j=1}^{k} c_jP_j$$

in $\mathcal{A}$, there is a polynomial f such that $f(t_j) = c_j$ $(1 \leq j \leq k)$, and for any such f, $U = f(T)$. ∎

Exercises

1. Give a reasonable definition of a non-negative $n \times n$ matrix, and then prove that such a matrix has a unique non-negative square root.

2. Let A be an $n \times n$ matrix with complex entries such that $A^* = -A$, and let $B = e^A$. Show that

(a) $\det B = e^{\operatorname{tr} A}$;
(b) $B^* = e^{-A}$;
(c) B is unitary.

3. If U and T are normal operators which commute, prove that $U + T$ and UT are normal.

4. Let T be a linear operator on the finite-dimensional complex inner product space V. Prove that the following ten statements about T are equivalent.

(a) T is normal.
(b) $\|T\alpha\| = \|T^*\alpha\|$ for every α in V.
(c) $T = T_1 + iT_2$, where T_1 and T_2 are self-adjoint and $T_1T_2 = T_2T_1$.
(d) If α is a vector and c a scalar such that $T\alpha = c\alpha$, then $T^*\alpha = \bar{c}\alpha$.
(e) There is an orthonormal basis for V consisting of characteristic vectors for T.
(f) There is an orthonormal basis $\mathcal{B}$ such that $[T]_{\mathcal{B}}$ is diagonal.
(g) There is a polynomial g with complex coefficients such that $T^* = g(T)$.
(h) Every subspace which is invariant under T is also invariant under T^*.
(i) $T = NU$, where N is non-negative, U is unitary, and N commutes with U.
(j) $T = c_1E_1 + \cdots + c_kE_k$, where $I = E_1 + \cdots + E_k$, $E_iE_j = 0$ for $i \neq j$, and $E_j^2 = E_j = E_j^*$.

5. Use Exercise 3 to show that any commuting family of normal operators (not necessarily diagonalizable ones) on a finite-dimensional inner product space generates a commutative self-adjoint algebra of normal operators.

6. Let V be a finite-dimensional complex inner product space and U a unitary operator on V such that $U\alpha = \alpha$ implies $\alpha = 0$. Let

$$f(z) = i\frac{(1 + z)}{(1 - z)}, \qquad z \neq 1$$

and show that

(a) $f(U) = i(I + U)(I - U)^{-1}$;
(b) $f(U)$ is self-adjoint;
(c) for every self-adjoint operator T on V, the operator

$$U = (T - iI)(T + iI)^{-1}$$

is unitary and such that $T = f(U)$.

7. Let V be the space of complex $n \times n$ matrices equipped with the inner product

$$(A|B) = \operatorname{tr}(AB^*).$$

If B is an element of V, let L_B, R_B, and T_B denote the linear operators on V defined by

(a) $L_B(A) = BA$.
(b) $R_B(A) = AB$.
(c) $T_B(A) = BA - AB$.

Consider the three families of operators obtained by letting B vary over all diagonal matrices. Show that each of these families is a commutative self-adjoint algebra and find their spectral resolutions.

8. If B is an arbitrary member of the inner product space in Exercise 7, show that L_B is unitarily equivalent to R_{B^t}.

9. Let V be the inner product space in Exercise 7 and G the group of unitary matrices in V. If B is in G, let C_B denote the linear operator on V defined by

$$C_B(A) = BAB^{-1}.$$

Show that

(a) C_B is a unitary operator on V;
(b) $C_{B_1B_2} = C_{B_1}C_{B_2}$;
(c) there is no unitary transformation U on V such that

$$UL_BU^{-1} = C_B$$

for all B in G.

10. Let $\mathfrak{F}$ be any family of linear operators on a finite-dimensional inner product space V and $\mathfrak{A}$ the self-adjoint algebra generated by $\mathfrak{F}$. Show that

(a) each root of $\mathfrak{A}$ defines a root of $\mathfrak{F}$;
(b) each root r of $\mathfrak{A}$ is a multiplicative linear function on A, i.e.,

$$r(TU) = r(T)r(U)$$
$$r(cT + U) = cr(T) + r(U)$$

for all T and U in $\mathfrak{A}$ and all scalars c.

11. Let $\mathfrak{F}$ be a commuting family of diagonalizable normal operators on a finite-dimensional inner product space V; and let $\mathfrak{A}$ be the self-adjoint algebra generated by $\mathfrak{F}$ and the identity operator I. Show that each root of $\mathfrak{A}$ is different from 0, and that for each root r of $\mathfrak{F}$ there is a unique root s of $\mathfrak{A}$ such that $s(T) = r(T)$ for all T in $\mathfrak{F}$.

12. Let $\mathfrak{F}$ be a commuting family of diagonalizable normal operators on a finite-dimensional inner product space V and A_0 the self-adjoint algebra generated by $\mathfrak{F}$. Let $\mathfrak{A}$ be the self-adjoint algebra generated by $\mathfrak{F}$ and the identity operator I. Show that

(a) $\mathfrak{A}$ is the set of all operators on V of the form $cI + T$ where c is a scalar and T an operator in $\mathfrak{A}_0$

(b) There is at most one root r of $\mathfrak{A}$ such that $r(T) = 0$ for all T in $\mathfrak{A}_0$.

(c) If one of the roots of $\mathfrak{A}$ is 0 on $\mathfrak{A}_0$, the projections $P_1, \ldots, P_k$ in the resolution of the identity defined by $\mathfrak{F}$ may be indexed in such a way that $\mathfrak{A}_0$ consists of all operators on V of the form

$$T = \sum_{j=2}^{k} c_j P_j$$

where $c_2, \ldots, c_k$ are arbitrary scalars.

(d) $\mathfrak{A} = \mathfrak{A}_0$ if and only if for each root r of $\mathfrak{A}$ there exists an operator T in $\mathfrak{A}_0$ such that $r(T) \neq 0$.

9.6. Further Properties of Normal Operators

In Section 8.5 we developed the basic properties of self-adjoint and normal operators, using the simplest and most direct methods possible. In Section 9.5 we considered various aspects of spectral theory. Here we prove some results of a more technical nature which are mainly about normal operators on real spaces.

We shall begin by proving a sharper version of the primary decomposition theorem of Chapter 6 for normal operators. It applies to both the real and complex cases.

Theorem 17. *Let* T *be a normal operator on a finite-dimensional inner product space* V. *Let* p *be the minimal polynomial for* T *and* $p_1, \cdots, p_k$ *its distinct monic prime factors. Then each* p_j *occurs with multiplicity* 1 *in the factorization of* p *and has degree* 1 *or* 2. *Suppose* W_j *is the null space of* $p_j(T)$. *Then*

(i) W_j *is orthogonal to* W_i *when* $i \neq j$;

(ii) $V = W_1 \oplus \cdots \oplus W_k$;

(iii) W_j *is invariant under* T, *and* p_j *is the minimal polynomial for the restriction of* T *to* W_j;

(iv) *for every* j, *there is a polynomial* e_j *with coefficients in the scalar field such that* $e_j(T)$ *is the orthogonal projection of* V *on* W_j.

In the proof we use certain basic facts which we state as lemmas.

Lemma 1. *Let* N *be a normal operator on an inner product space* W. *Then the null space of* N *is the orthogonal complement of its range.*

Proof. Suppose $(\alpha|N\beta) = 0$ for all β in W. Then $(N^*\alpha|\beta) = 0$ for all β; hence $N^*\alpha = 0$. By Theorem 19 of Chapter 8, this implies $N\alpha = 0$. Conversely, if $N\alpha = 0$, then $N^*\alpha = 0$, and

$$(N^*\alpha|\beta) = (\alpha|N\beta) = 0$$

for all β in W. ∎

Lemma 2. *If* N *is a normal operator and* α *is a vector such that* $N^2\alpha = 0$, *then* $N\alpha = 0$.

Proof. Suppose N is normal and that $N^2\alpha = 0$. Then $N\alpha$ lies in the range of N and also lies in the null space of N. By Lemma 1, this implies $N\alpha = 0$. ∎

Lemma 3. *Let* T *be a normal operator and* f *any polynomial with coefficients in the scalar field. Then* f(T) *is also normal.*

Proof. Suppose $f = a_0 + a_1x + \cdots + a_nx^n$. Then

$$f(T) = a_0I + a_1T + \cdots + a_nT^n$$

and

$$f(T)^* = \bar{a}_0I + \bar{a}_1T^* + \cdots + \bar{a}_n(T^*)^n.$$

Since $T^*T = TT^*$, it follows that $f(T)$ commutes with $f(T)^*$. ∎

Lemma 4. *Let* T *be a normal operator and* f, g *relatively prime polynomials with coefficients in the scalar field. Suppose α and β are vectors such that* $f(T)\alpha = 0$ *and* $g(T)\beta = 0$. *Then* $(\alpha|\beta) = 0$.

Proof. There are polynomials a and b with coefficients in the scalar field such that $af + bg = 1$. Thus

$$a(T)f(T) + b(T)g(T) = I$$

and $\alpha = g(T)b(T)\alpha$. It follows that

$$(\alpha|\beta) = (g(T)b(T)\alpha|\beta) = (b(T)\alpha|g(T)^*\beta).$$

By assumption $g(T)\beta = 0$. By Lemma 3, $g(T)$ is normal. Therefore, by Theorem 19 of Chapter 8, $g(T)^*\beta = 0$; hence $(\alpha|\beta) = 0$. ∎

Proof of Theorem 17. Recall that the minimal polynomial for T is the monic polynomial of least degree among all polynomials f such that $f(T) = 0$. The existence of such polynomials follows from the assumption that V is finite-dimensional. Suppose some prime factor p_j of p is repeated. Then $p = p_j^2g$ for some polynomial g. Since $p(T) = 0$, it follows that

$$(p_j(T))^2g(T)\alpha = 0$$

for every α in V. By Lemma 3, $p_j(T)$ is normal. Thus Lemma 2 implies

$$p_j(T)g(T)\alpha = 0$$

for every α in V. But this contradicts the assumption that p has least degree among all f such that $f(T) = 0$. Therefore, $p = p_1 \cdots p_k$. If V is a complex inner product space each p_j is necessarily of the form

$$p_j = x - c_j$$

with c_j real or complex. On the other hand, if V is a real inner product space, then $p_j = x_j - c_j$ with c_j in R or

$$p_j = (x - c)(x - \bar{c})$$

where c is a non-real complex number.

Now let $f_j = p/p_j$. Then, since $f_1, \ldots, f_k$ are relatively prime, there exist polynomials g_j with coefficients in the scalar field such that

$$1 = \sum_j f_j g_j. \tag{9-16}$$

We briefly indicate how such g_j may be constructed. If $p_j = x - c_j$, then $f_j(c_j) \neq 0$, and for g_j we take the scalar polynomial $1/f_j(c_j)$. When every p_j is of this form, the $f_j g_j$ are the familiar Lagrange polynomials associated with $c_1, \ldots, c_k$, and (9-16) is clearly valid. Suppose some $p_j = (x - c)(x - \bar{c})$ with c a non-real complex number. Then V is a real inner product space, and we take

$$g_j = \frac{x - \bar{c}}{s} + \frac{x - c}{\bar{s}}$$

where $s = (c - \bar{c})f_j(c)$. Then

$$g_j = \frac{(s + \bar{s})x - (cs + \bar{c}s)}{s\bar{s}}$$

so that g_j is a polynomial with real coefficients. If p has degree n, then

$$1 - \sum_j f_j g_j$$

is a polynomial with real coefficients of degree at most $n - 1$; moreover, it vanishes at each of the n (complex) roots of p, and hence is identically 0.

Now let α be an arbitrary vector in V. Then by (9-16)

$$\alpha = \sum_j f_j(T)g_j(T)\alpha$$

and since $p_j(T)f_j(T) = 0$, it follows that $f_j(T)g_j(T)\alpha$ is in W_j for every j. By Lemma 4, W_j is orthogonal to W_i whenever $i \neq j$. Therefore, V is the orthogonal direct sum of $W_1, \ldots, W_k$. If β is any vector in W_j, then

$$p_j(T)T\beta = Tp_j(T)\beta = 0;$$

thus W_j is invariant under T. Let T_j be the restriction of T to W_j. Then $p_j(T_j) = 0$, so that p_j is divisible by the minimal polynomial for T_j. Since p_j is irreducible over the scalar field, it follows that p_j is the minimal polynomial for T_j.

Next, let $e_j = f_j g_j$ and $E_j = e_j(T)$. Then for every vector α in V, $E_j\alpha$ is in W_j, and

$$\alpha = \sum_j E_j\alpha.$$

Thus $\alpha - E_i\alpha = \sum_{j \neq i} E_j\alpha$; since W_j is orthogonal to W_i when $j \neq i$, this implies that $\alpha - E_i\alpha$ is in $W_i^{\perp}$. It now follows from Theorem 4 of Chapter 8 that E_i is the orthogonal projection of V on W_i. ∎

Definition. *We call the subspaces* W_j $(1 \leq j \leq k)$ *the* **primary components of V under T**.

Corollary. *Let* T *be a normal operator on a finite-dimensional inner product space* V *and* $W_1, \ldots, W_k$ *the primary components of* V *under* T. *Suppose* W *is a subspace of* V *which is invariant under* T. *Then*

$$W = \sum_j W \cap W_j.$$

Proof. Clearly W contains $\sum\limits_j W \cap W_j$. On the other hand, W, being invariant under T, is invariant under every polynomial in T. In particular, W is invariant under the orthogonal projection E_j of V on W_j. If α is in W, it follows that $E_j\alpha$ is in $W \cap W_j$, and, at the same time, $\alpha = \sum\limits_j E_j\alpha$. Therefore W is contained in $\sum\limits_j W \cap W_j$. ∎

Theorem 17 shows that every normal operator T on a finite-dimensional inner product space is canonically specified by a finite number of normal operators T_j, defined on the primary components W_j of V under T, each of whose minimal polynomials is irreducible over the field of scalars. To complete our understanding of normal operators it is necessary to study normal operators of this special type.

A normal operator whose minimal polynomial is of degree 1 is clearly just a scalar multiple of the identity. On the other hand, when the minimal polynomial is irreducible and of degree 2 the situation is more complicated.

Example 1. Suppose $r > 0$ and that θ is a real number which is not an integral multiple of π. Let T be the linear operator on R^2 whose matrix in the standard orthonormal basis is

$$A = r\begin{bmatrix} \cos\theta & -\sin\theta \\ \sin\theta & \cos\theta \end{bmatrix}.$$

Then T is a scalar multiple of an orthogonal transformation and hence normal. Let p be the characteristic polynomial of T. Then

$$\begin{aligned} p &= \det(xI - A) \\ &= (x - r\cos\theta)^2 + r^2\sin^2\theta \\ &= x - 2r\cos\theta x + r^2. \end{aligned}$$

Let $a = r\cos\theta$, $b = r\sin\theta$, and $c = a + ib$. Then $b \neq 0$, $c = re^{i\theta}$

$$A = \begin{bmatrix} a & -b \\ b & a \end{bmatrix}$$

and $p = (x - c)(x - \bar{c})$. Hence p is irreducible over R. Since p is divisible by the minimal polynomial for T, it follows that p is the minimal polynomial.

This example suggests the following converse.

Theorem 18. *Let* T *be a normal operator on a finite-dimensional real inner product space* V *and* p *its minimal polynomial. Suppose*

$$p = (x - a)^2 + b^2$$

where a *and* b *are real and* $b \neq 0$. *Then there is an integer* $s > 0$ *such that* p^s *is the characteristic polynomial for* T, *and there exist subspaces* $V_1, \ldots, V_s$ *of* V *such that*

(i) V_j *is orthogonal to* V_i *when* $i \neq j$;

(ii) $V = V_1 \oplus \cdots \oplus V_s$;

(iii) *each* V_j *has an orthonormal basis* $\{\alpha_j, \beta_j\}$ *with the property that*

$$T\alpha_j = a\alpha_j + b\beta_j$$
$$T\beta_j = -b\alpha_j + a\beta_j.$$

In other words, if $r = \sqrt{a^2 + b^2}$ and θ is chosen so that $a = r \cos \theta$ and $b = r \sin \theta$, then V is an orthogonal direct sum of two-dimensional subspaces V_j on each of which T acts as 'r times rotation through the angle θ'.

The proof of Theorem 18 will be based on the following result.

Lemma. *Let* V *be a real inner product space and* S *a normal operator on* V *such that* $S^2 + I = 0$. *Let* α *be any vector in* V *and* $\beta = S\alpha$. *Then*

$$S^*\alpha = -\beta$$
(9-17)
$$S^*\beta = \alpha$$

$(\alpha|\beta) = 0$, *and* $||\alpha|| = ||\beta||$.

Proof. We have $S\alpha = \beta$ and $S\beta = S^2\alpha = -\alpha$. Therefore

$$0 = ||S\alpha - \beta||^2 + ||S\beta + \alpha||^2 = ||S\alpha||^2 - 2(S\alpha|\beta) + ||\beta||^2 + ||S\beta||^2 + 2(S\beta|\alpha) + ||\alpha||^2.$$

Since S is normal, it follows that

$$0 = ||S^*\alpha||^2 - 2(S^*\beta|\alpha) + ||\beta||^2 + ||S^*\beta||^2 + 2(S^*\alpha|\beta) + ||\alpha||^2 = ||S^*\alpha + \beta||^2 + ||S^*\beta - \alpha||^2.$$

This implies (9-17); hence

$$(\alpha|\beta) = (S^*\beta|\beta) = (\beta|S\beta) = (\beta|-\alpha) = -(\alpha|\beta)$$

and $(\alpha|\beta) = 0$. Similarly

$$||\alpha||^2 = (S^*\beta|\alpha) = (\beta|S\alpha) = ||\beta||^2. \blacksquare$$

Proof of Theorem 18. Let $V_1, \ldots, V_s$ be a maximal collection of two-dimensional subspaces satisfying (i) and (ii), and the additional conditions

$$T^*\alpha_j = a\alpha_j - b\beta_j, \qquad 1 \le j \le s. \tag{9-18}$$
$$T^*\beta_j = b\alpha_j + a\beta_j$$

Let $W = V_1 + \cdots + V_s$. Then W is the orthogonal direct sum of $V_1, \ldots, V_s$. We shall show that $W = V$. Suppose that this is not the case. Then $W^\perp \neq \{0\}$. Moreover, since (iii) and (9-18) imply that W is invariant under T and T^*, it follows that $W^\perp$ is invariant under T^* and $T = T^{**}$. Let $S = b^{-1}(T - aI)$. Then $S^* = b^{-1}(T^* - aI)$, $S^*S = SS^*$, and $W^\perp$ is invariant under S and S^*. Since $(T - aI)^2 + b^2I = 0$, it follows that $S^2 + I = 0$. Let α be any vector of norm 1 in $W^\perp$ and set $\beta = S\alpha$. Then β is in $W^\perp$ and $S\beta = -\alpha$. Since $T = aI + bS$, this implies

$$T\alpha = a\alpha + b\beta$$
$$T\beta = -b\alpha + a\beta.$$

By the lemma, $S^*\alpha = -\beta$, $S^*\beta = \alpha$, $(\alpha|\beta) = 0$, and $\|\beta\| = 1$. Because $T^* = aI + bS^*$, it follows that

$$T^*\alpha = a\alpha - b\beta$$
$$T^*\beta = b\alpha + a\beta.$$

But this contradicts the fact that $V_1, \ldots, V_s$ is a maximal collection of subspaces satisfying (i), (iii), and (9-18). Therefore, $W = V$, and since

$$\det \begin{bmatrix} x - a & b \\ -b & x - a \end{bmatrix} = (x - a)^2 + b^2$$

it follows from (i), (ii) and (iii) that

$$\det (xI - T) = [(x - a)^2 + b^2]^s. \quad \blacksquare$$

Corollary. *Under the conditions of the theorem,* T *is invertible, and*

$$T^* = (a^2 + b^2)T^{-1}.$$

Proof. Since

$$\begin{bmatrix} a & -b \\ b & a \end{bmatrix}\begin{bmatrix} a & b \\ -b & a \end{bmatrix} = \begin{bmatrix} a^2 + b^2 & 0 \\ 0 & a^2 + b^2 \end{bmatrix}$$

it follows from (iii) and (9-18) that $TT^* = (a^2 + b^2)I$. Hence T is invertible and $T^* = (a^2 + b^2)T^{-1}$.

Theorem 19. *Let* T *be a normal operator on a finite-dimensional inner product space* V. *Then any linear operator that commutes with* T *also commutes with* T*. *Moreover, every subspace invariant under* T *is also invariant under* T*.

Proof. Suppose U is a linear operator on V that commutes with T. Let E_j be the orthogonal projection of V on the primary component

W_j $(1 \leq j \leq k)$ of V under T. Then E_j is a polynomial in T and hence commutes with U. Thus

$$E_jUE_j = UE_j^2 = UE_j.$$

Thus $U(W_j)$ is a subset of W_j. Let T_j and U_j denote the restrictions of T and U to W_j. Suppose I_j is the identity operator on W_j. Then U_j commutes with T_j, and if $T_j = c_jI_j$, it is clear that U_j also commutes with $T_j^* = \bar{c}_jI_j$. On the other hand, if T_j is not a scalar multiple of I_j, then T_j is invertible and there exist real numbers a_j and b_j such that

$$T_j^* = (a_j^2 + b_j^2)T_j^{-1}.$$

Since $U_jT_j = T_jU_j$, it follows that $T_j^{-1}U_j = U_jT_j^{-1}$. Therefore U_j commutes with T_j^* in both cases. Now T^* also commutes with E_j, and hence W_j is invariant under T^*. Moreover for every α and β in W_j

$$(T_j\alpha|\beta) = (T\alpha|\beta) = (\alpha|T^*\beta) = (\alpha|T_j^*\beta).$$

Since $T^*(W_j)$ is contained in W_j, this implies T_j^* is the restriction of T^* to W_j. Thus

$$UT^*\alpha_j = T^*U\alpha_j$$

for every α_j in W_j. Since V is the sum of $W_1, \ldots, W_k$, it follows that

$$UT^*\alpha = T^*U\alpha$$

for every α in V and hence that U commutes with T^*.

Now suppose W is a subspace of V that is invariant under T, and let $Z_j = W \cap W_j$. By the corollary to Theorem 17, $W = \sum_j Z_j$. Thus it suffices to show that each Z_j is invariant under T_j^*. This is clear if $T_j = c_jI$. When this is not the case, T_j is invertible and maps Z_j into and hence onto Z_j. Thus $T_j^{-1}(Z_j) = Z_j$, and since

$$T_j^* = (a_j^2 + b_j^2)T_j^{-1}$$

it follows that $T^*(Z_j)$ is contained in Z_j, for every j. ∎

Suppose T is a normal operator on a finite-dimensional inner product space V. Let W be a subspace invariant under T. Then the preceding corollary shows that W is invariant under T^*. From this it follows that $W^\perp$ is invariant under $T^{**} = T$ (and hence under T^* as well). Using this fact one can easily prove the following strengthened version of the cyclic decomposition theorem given in Chapter 7.

Theorem 20. *Let* T *be a normal linear operator on a finite-dimensional inner product space* V *(dim* $\mathrm{V} \geq 1$*). Then there exist* r *non-zero vectors* $\alpha_1, \ldots, \alpha_r$ *in* V *with respective* T*-annihilators* $e_1, \ldots, e_r$ *such that*

(i) $\mathrm{V} = \mathrm{Z}(\alpha_1; \mathrm{T}) \oplus \cdots \oplus \mathrm{Z}(\alpha_r; \mathrm{T})$;
(ii) *if* $1 \leq k \leq r - 1$, *then* e_{k+1} *divides* e_k;

(iii) $Z(\alpha_j; T)$ *is orthogonal to* $Z(\alpha_k; T)$ *when* $j \neq k$. *Furthermore, the integer* r *and the annihilators* $e_1, \ldots, e_r$ *are uniquely determined by conditions* (i) *and* (ii) *and the fact that no* α_k *is* 0.

Corollary. *If* A *is a normal matrix with real (complex) entries, then there is a real orthogonal (unitary) matrix* P *such that* $P^{-1}AP$ *is in rational canonical form.*

It follows that two normal matrices A and B are unitarily equivalent if and only if they have the same rational form; A and B are orthogonally equivalent if they have real entries and the same rational form.

On the other hand, there is a simpler criterion for the unitary equivalence of normal matrices and normal operators.

Definitions. *Let* V *and* V' *be inner product spaces over the same field. A linear transformation*

$$U: V \to V'$$

is called a **unitary transformation** *if it maps* V *onto* V' *and preserves inner products. If* T *is a linear operator on* V *and* T' *a linear operator on* V', *then* T *is* **unitarily equivalent** *to* T' *if there exists a unitary transformation* U *of* V *onto* V' *such that*

$$UTU^{-1} = T'.$$

Lemma. *Let* V *and* V' *be finite-dimensional inner product spaces over the same field. Suppose* T *is a linear operator on* V *and that* T' *is a linear operator on* V'. *Then* T *is unitarily equivalent to* T' *if and only if there is an orthonormal basis* $\mathcal{B}$ *of* V *and an orthonormal basis* $\mathcal{B}'$ *of* V' *such that*

$$[T]_{\mathcal{B}} = [T']_{\mathcal{B}'}.$$

Proof. Suppose there is a unitary transformation U of V onto V' such that $UTU^{-1} = T'$. Let $\mathcal{B} = \{\alpha_1, \ldots, \alpha_n\}$ be any (ordered) orthonormal basis for V. Let $\alpha_j' = U\alpha_j$ $(1 \leq j \leq n)$. Then $\mathcal{B}' = \{\alpha_1', \ldots, \alpha_n'\}$ is an orthonormal basis for V' and setting

$$T\alpha_j = \sum_{k=1}^{n} A_{kj}\alpha_k$$

we see that

$$\begin{aligned} T'\alpha_j' &= UT\alpha_j \\ &= \sum_k A_{kj}U\alpha_k \\ &= \sum_k A_{kj}\alpha_k' \end{aligned}$$

Hence $[T]_{\mathcal{B}} = A = [T']_{\mathcal{B}'}$.

Conversely, suppose there is an orthonormal basis $\mathcal{B}$ of V and an orthonormal basis $\mathcal{B}'$ of V' such that

$$[T]_{\mathcal{B}} = [T']_{\mathcal{B}'}$$

and let $A = [T]_{\mathcal{B}}$. Suppose $\mathcal{B} = \{\alpha_1, \ldots, \alpha_n\}$ and that $\mathcal{B}' = \{\alpha'_1, \ldots, \alpha'_n\}$. Let U be the linear transformation of V into V' such that $U\alpha_j = \alpha'_j$ $(1 \leq j \leq n)$. Then U is a unitary transformation of V onto V', and

$$\begin{aligned} UTU^{-1}\alpha'_j &= UT\alpha_j \\ &= U \sum_k A_{kj}\alpha_k \\ &= \sum_k A_{kj}\alpha'_k. \end{aligned}$$

Therefore, $UTU^{-1}\alpha'_j = T'\alpha'_j$ $(1 \leq j \leq n)$, and this implies $UTU^{-1} = T'$. ∎

It follows immediately from the lemma that unitarily equivalent operators on finite-dimensional spaces have the same characteristic polynomial. For normal operators the converse is valid.

Theorem 21. *Let* V *and* V′ *be finite-dimensional inner product spaces over the same field. Suppose* T *is a normal operator on* V *and that* T′ *is a normal operator on* V′. *Then* T *is unitarily equivalent to* T′ *if and only if* T *and* T′ *have the same characteristic polynomial.*

Proof. Suppose T and T' have the same characteristic polynomial f. Let W_j $(1 \leq j \leq k)$ be the primary components of V under T and T_j the restriction of T to W_j. Suppose I_j is the identity operator on W_j. Then

$$f = \prod_{j=1}^{k} \det (xI_j - T_j).$$

Let p_j be the minimal polynomial for T_j. If $p_j = x - c_j$ it is clear that

$$\det (xI_j - T_j) = (x - c_j)^{s_j}$$

where s_j is the dimension of W_j. On the other hand, if $p_j = (x - a_j)^2 + b_j^2$ with a_j, b_j real and $b_j \neq 0$, then it follows from Theorem 18 that

$$\det (xI_j - T_j) = p_j^{s_j}$$

where in this case $2s_j$ is the dimension of W_j. Therefore $f = \prod_j p_j^{s_j}$. Now we can also compute f by the same method using the primary components of V' under T'. Since $p_1, \ldots, p_k$ are distinct primes, it follows from the uniqueness of the prime factorization of f that there are exactly k primary components W'_j $(1 \leq j \leq k)$ of V' under T' and that these may be indexed in such a way that p_j is the minimal polynomial for the restriction T'_j of T' to W'_j. If $p_j = x - c_j$, then $T_j = c_jI_j$ and $T'_j = c_jI'_j$ where I'_j is the identity operator on W_j'. In this case it is evident that T_j is unitarily equivalent to T_j'. If $p_j = (x - a_j)^2 + b_j^2$, as above, then using the lemma and Theorem 20, we again see that T_j is unitarily equivalent to T_j'. Thus for each j there are orthonormal bases $\mathcal{B}_j$ and $\mathcal{B}_j'$ of W_j and W_j', respectively, such that

$$[T_j]_{\mathcal{B}_j} = [T_j']_{\mathcal{B}_j'}.$$

Now let U be the linear transformation of V into V' that maps each $\mathcal{B}_j$ onto $\mathcal{B}_j'$. Then U is a unitary transformation of V onto V' such that $UTU^{-1} = T'$. ∎

10. Bilinear Forms

10.1. Bilinear Forms

In this chapter, we treat bilinear forms on finite-dimensional vector spaces. The reader will probably observe a similarity between some of the material and the discussion of determinants in Chapter 5 and of inner products and forms in Chapter 8 and in Chapter 9. The relation between bilinear forms and inner products is particularly strong; however, this chapter does not presuppose any of the material in Chapter 8 or Chapter 9. The reader who is not familiar with inner products would probably profit by reading the first part of Chapter 8 as he reads the discussion of bilinear forms.

This first section treats the space of bilinear forms on a vector space of dimension n. The matrix of a bilinear form in an ordered basis is introduced, and the isomorphism between the space of forms and the space of $n \times n$ matrices is established. The rank of a bilinear form is defined, and non-degenerate bilinear forms are introduced. The second section discusses symmetric bilinear forms and their diagonalization. The third section treats skew-symmetric bilinear forms. The fourth section discusses the group preserving a non-degenerate bilinear form, with special attention given to the orthogonal groups, the pseudo-orthogonal groups, and a particular pseudo-orthogonal group—the Lorentz group.

Definition. *Let* V *be a vector space over the field* F. *A* **bilinear form** *on* V *is a function* f, *which assigns to each ordered pair of vectors* α, β *in* V *a scalar* $f(\alpha, \beta)$ *in* F, *and which satisfies*

$$(10\text{-}1)\qquad \begin{aligned} f(c\alpha_1 + \alpha_2, \beta) &= cf(\alpha_1, \beta) + f(\alpha_2, \beta)\\ f(\alpha, c\beta_1 + \beta_2) &= cf(\alpha, \beta_1) + f(\alpha, \beta_2). \end{aligned}$$

If we let $V \times V$ denote the set of all ordered pairs of vectors in V, this definition can be rephrased as follows: A bilinear form on V is a function f from $V \times V$ into F which is linear as a function of either of its arguments when the other is fixed. The zero function from $V \times V$ into F is clearly a bilinear form. It is also true that any linear combination of bilinear forms on V is again a bilinear form. To prove this, it is sufficient to consider linear combinations of the type $cf + g$, where f and g are bilinear forms on V. The proof that $cf + g$ satisfies (10-1) is similar to many others we have given, and we shall thus omit it. All this may be summarized by saying that the set of all bilinear forms on V is a subspace of the space of all functions from $V \times V$ into F (Example 3, Chapter 2). We shall denote the space of bilinear forms on V by $L(V, V, F)$.

EXAMPLE 1. Let V be a vector space over the field F and let L_1 and L_2 be linear functions on V. Define f by

$$f(\alpha, \beta) = L_1(\alpha)L_2(\beta).$$

If we fix β and regard f as a function of α, then we simply have a scalar multiple of the linear functional L_1. With α fixed, f is a scalar multiple of L_2. Thus it is clear that f is a bilinear form on V.

EXAMPLE 2. Let m and n be positive integers and F a field. Let V be the vector space of all $m \times n$ matrices over F. Let A be a fixed $m \times m$ matrix over F. Define

$$f_A(X, Y) = \operatorname{tr}(X^tAY).$$

Then f_A is a bilinear form on V. For, if X, Y, and Z are $m \times n$ matrices over F,

$$\begin{aligned} f_A(cX + Z, Y) &= \operatorname{tr}[(cX + Z)^tAY]\\ &= \operatorname{tr}(cXtAY) + \operatorname{tr}(Z^tAY)\\ &= cf_A(X, Y) + f_A(Z, Y). \end{aligned}$$

Of course, we have used the fact that the transpose operation and the trace function are linear. It is even easier to show that f_A is linear as a function of its second argument. In the special case $n = 1$, the matrix X^tAY is 1×1, i.e., a scalar, and the bilinear form is simply

$$\begin{aligned} f_A(X, Y) &= X^tAY\\ &= \sum_i \sum_j A_{ij}x_iy_j. \end{aligned}$$

We shall presently show that every bilinear form on the space of $m \times 1$ matrices is of this type, i.e., is f_A for some $m \times m$ matrix A.

EXAMPLE 3. Let F be a field. Let us find all bilinear forms on the space F^2. Suppose f is such a bilinear form. If $\alpha = (x_1, x_2)$ and $\beta = (y_1, y_2)$ are vectors in F^2, then

$$\begin{aligned} f(\alpha, \beta) &= f(x_1\epsilon_1 + x_2\epsilon_2, \beta) \\ &= x_1 f(\epsilon_1, \beta) + x_2 f(\epsilon_2, \beta) \\ &= x_1 f(\epsilon_1, y_1\epsilon_1 + y_2\epsilon_2) + x_2 f(\epsilon_2, y_1\epsilon_1 + y_2\epsilon_2) \\ &= x_1 y_1 f(\epsilon_1, \epsilon_1) + x_1 y_2 f(\epsilon_1, \epsilon_2) + x_2 y_1 f(\epsilon_2, \epsilon_1) + x_2 y_2 f(\epsilon_2, \epsilon_2). \end{aligned}$$

Thus f is completely determined by the four scalars $A_{ij} = f(\epsilon_i, \epsilon_j)$ by

$$\begin{aligned} f(\alpha, \beta) &= A_{11}x_1y_1 + A_{12}x_1y_2 + A_{21}x_2y_1 + A_{22}x_2y_2 \\ &= \sum_{i,j} A_{ij}x_iy_j. \end{aligned}$$

If X and Y are the coordinate matrices of α and β, and if A is the 2×2 matrix with entries $A(i, j) = A_{ij} = f(\epsilon_i, \epsilon_j)$, then

$$f(\alpha, \beta) = X^tAY. \tag{10-2}$$

We observed in Example 2 that if A is any 2×2 matrix over F, then (10-2) defines a bilinear form on F^2. We see that the bilinear forms on F^2 are precisely those obtained from a 2×2 matrix as in (10-2).

The discussion in Example 3 can be generalized so as to describe all bilinear forms on a finite-dimensional vector space. Let V be a finite-dimensional vector space over the field F and let $\mathcal{B} = \{\alpha_1, \ldots, \alpha_n\}$ be an ordered basis for V. Suppose f is a bilinear form on V. If

$$\alpha = x_1\alpha_1 + \cdots + x_n\alpha_n \quad \text{and} \quad \beta = y_1\alpha_1 + \cdots + y_n\alpha_n$$

are vectors in V, then

$$\begin{aligned} f(\alpha, \beta) &= f\left(\sum_i x_i\alpha_i, \beta\right) \\ &= \sum_i x_i f(\alpha_i, \beta) \\ &= \sum_i x_i f\left(\alpha_i, \sum_j y_j\alpha_j\right) \\ &= \sum_i \sum_j x_i y_j f(\alpha_i, \alpha_j). \end{aligned}$$

If we let $A_{ij} = f(\alpha_i, \alpha_j)$, then

$$\begin{aligned} f(\alpha, \beta) &= \sum_i \sum_j A_{ij}x_iy_j \\ &= X^tAY \end{aligned}$$

where X and Y are the coordinate matrices of α and β in the ordered basis $\mathcal{B}$. Thus every bilinear form on V is of the type

$$f(\alpha, \beta) = [\alpha]^t_{\mathcal{B}}A[\beta]_{\mathcal{B}} \tag{10-3}$$

for some $n \times n$ matrix A over F. Conversely, if we are given any $n \times n$ matrix A, it is easy to see that (10-3) defines a bilinear form f on V, such that $A_{ij} = f(\alpha_i, \alpha_j)$.

Definition. *Let* V *be a finite-dimensional vector space, and let* $\mathcal{B} = \{\alpha_1, \ldots, \alpha_n\}$ *be an ordered basis for* V. *If* f *is a bilinear form on* V, *the* **matrix of f in the ordered basis** $\mathcal{B}$ *is the* $n \times n$ *matrix* A *with entries* $A_{ij} = f(\alpha_i, \alpha_j)$. *At times, we shall denote this matrix by* $[f]_{\mathcal{B}}$.

Theorem 1. *Let* V *be a finite-dimensional vector space over the field* F. *For each ordered basis* $\mathcal{B}$ *of* V, *the function which associates with each bilinear form on* V *its matrix in the ordered basis* $\mathcal{B}$ *is an isomorphism of the space* L(V, V, F) *onto the space of* $n \times n$ *matrices over the field* F.

Proof. We observed above that $f \to [f]_{\mathcal{B}}$ is a one-one correspondence between the set of bilinear forms on V and the set of all $n \times n$ matrices over F. That this is a linear transformation is easy to see, because

$$(cf + g)(\alpha_i, \alpha_j) = cf(\alpha_i, \alpha_j) + g(\alpha_i, \alpha_j)$$

for each i and j. This simply says that

$$[cf + g]_{\mathcal{B}} = c[f]_{\mathcal{B}} + [g]_{\mathcal{B}}. \quad \blacksquare$$

Corollary. *If* $\mathcal{B} = \{\alpha_1, \ldots, \alpha_n\}$ *is an ordered basis for* V, *and* $\mathcal{B}^* = \{L_1, \ldots, L_n\}$ *is the dual basis for* V*, *then the* n^2 *bilinear forms*

$$f_{ij}(\alpha, \beta) = L_i(\alpha)L_j(\beta), \qquad 1 \leq i \leq n, 1 \leq j \leq n$$

form a basis for the space L(V, V, F). *In particular, the dimension of* L(V, V, F) *is* n^2.

Proof. The dual basis $\{L_1, \ldots, L_n\}$ is essentially defined by the fact that $L_i(\alpha)$ is the ith coordinate of α in the ordered basis $\mathcal{B}$ (for any α in V). Now the functions f_{ij} defined by

$$f_{ij}(\alpha, \beta) = L_i(\alpha)L_j(\beta)$$

are bilinear forms of the type considered in Example 1. If

$$\alpha = x_1\alpha_1 + \cdots + x_n\alpha_n \text{ and } \beta = y_1\alpha_1 + \cdots + y_n\alpha_n,$$

then

$$f_{ij}(\alpha, \beta) = x_iy_j.$$

Let f be any bilinear form on V and let A be the matrix of f in the ordered basis $\mathcal{B}$. Then

$$f(\alpha, \beta) = \sum_{i,j} A_{ij}x_iy_j$$

which simply says that

$$f = \sum_{i,j} A_{ij}f_{ij}.$$

It is now clear that the n^2 forms f_{ij} comprise a basis for $L(V, V, F)$. $\blacksquare$

One can rephrase the proof of the corollary as follows. The bilinear form f_{ij} has as its matrix in the ordered basis $\mathcal{B}$ the matrix 'unit' $E^{i,j}$, whose only non-zero entry is a 1 in row i and column j. Since these matrix units comprise a basis for the space of $n \times n$ matrices, the forms f_{ij} comprise a basis for the space of bilinear forms.

The concept of the matrix of a bilinear form in an ordered basis is similar to that of the matrix of a linear operator in an ordered basis. Just as for linear operators, we shall be interested in what happens to the matrix representing a bilinear form, as we change from one ordered basis to another. So, suppose $\mathcal{B} = \{\alpha_1, \ldots, \alpha_n\}$ and $\mathcal{B}' = \{\alpha'_1, \ldots, \alpha'_n\}$ are two ordered bases for V and that f is a bilinear form on V. How are the matrices $[f]_{\mathcal{B}}$ and $[f]_{\mathcal{B}'}$ related? Well, let P be the (invertible) $n \times n$ matrix such that

$$[\alpha]_{\mathcal{B}} = P[\alpha]_{\mathcal{B}'}$$

for all α in V. In other words, define P by

$$\alpha'_j = \sum_{i=1}^{n} P_{ij}\alpha_i.$$

For any vectors α, β in V

$$\begin{aligned} f(\alpha, \beta) &= [\alpha]^t_{\mathcal{B}}[f]_{\mathcal{B}}[\beta]_{\mathcal{B}} \\ &= (P[\alpha]_{\mathcal{B}'})^t[f]_{\mathcal{B}}P[\beta]_{\mathcal{B}'} \\ &= [\alpha]^t_{\mathcal{B}'}(P^t[f]_{\mathcal{B}}P)[\beta]_{\mathcal{B}'}. \end{aligned}$$

By the definition and uniqueness of the matrix representing f in the ordered basis $\mathcal{B}'$, we must have

$$[f]_{\mathcal{B}'} = P^t[f]_{\mathcal{B}}P. \tag{10-4}$$

EXAMPLE 4. Let V be the vector space R^2. Let f be the bilinear form defined on $\alpha = (x_1, x_2)$ and $\beta = (y_1, y_2)$ by

$$f(\alpha, \beta) = x_1y_1 + x_1y_2 + x_2y_1 + x_2y_2.$$

Now

$$f(\alpha, \beta) = [x_1, x_2]\begin{bmatrix} 1 & 1 \\ 1 & 1 \end{bmatrix}\begin{bmatrix} y_1 \\ y_2 \end{bmatrix}$$

and so the matrix of f in the standard ordered basis $\mathcal{B} = \{\epsilon_1, \epsilon_2\}$ is

$$[f]_{\mathcal{B}} = \begin{bmatrix} 1 & 1 \\ 1 & 1 \end{bmatrix}.$$

Let $\mathcal{B}' = \{\epsilon'_1, \epsilon'_2\}$ be the ordered basis defined by $\epsilon'_1 = (1, -1)$, $\epsilon'_2 = (1, 1)$. In this case, the matrix P which changes coordinates from $\mathcal{B}'$ to $\mathcal{B}$ is

$$P = \begin{bmatrix} 1 & 1 \\ -1 & 1 \end{bmatrix}.$$

Thus

$$\begin{aligned} [f]_{\mathcal{B}'} &= P^t[f]_{\mathcal{B}}P \\ &= \begin{bmatrix} 1 & -1 \\ 1 & 1 \end{bmatrix}\begin{bmatrix} 1 & 1 \\ 1 & 1 \end{bmatrix}\begin{bmatrix} 1 & 1 \\ -1 & 1 \end{bmatrix} \end{aligned}$$

$$= \begin{bmatrix} 1 & -1 \\ 1 & 1 \end{bmatrix} \begin{bmatrix} 0 & 2 \\ 0 & 2 \end{bmatrix}$$

$$= \begin{bmatrix} 0 & 0 \\ 0 & 4 \end{bmatrix}.$$

What this means is that if we express the vectors α and β by means of their coordinates in the basis $\mathcal{B}'$, say

$$\alpha = x_1'\epsilon_1' + x_2'\epsilon_2', \qquad \beta = y_1'\epsilon_1' + y_2'\epsilon_2'$$

then

$$f(\alpha, \beta) = 4x_2'y_2'.$$

One consequence of the change of basis formula (10-4) is the following: If A and B are $n \times n$ matrices which represent the same bilinear form on V in (possibly) different ordered bases, then A and B have the same rank. For, if P is an invertible $n \times n$ matrix and $B = P^tAP$, it is evident that A and B have the same rank. This makes it possible to define the rank of a bilinear form on V as the rank of any matrix which represents the form in an ordered basis for V.

It is desirable to give a more intrinsic definition of the rank of a bilinear form. This can be done as follows: Suppose f is a bilinear form on the vector space V. If we fix a vector α in V, then $f(\alpha, \beta)$ is linear as a function of β. In this way, each fixed α determines a linear functional on V; let us denote this linear functional by $L_f(\alpha)$. To repeat, if α is a vector in V, then $L_f(\alpha)$ is the linear functional on V whose value on any vector β is $f(\alpha, \beta)$. This gives us a transformation $\alpha \rightarrow L_f(\alpha)$ from V into the dual space V^*. Since

$$f(c\alpha_1 + \alpha_2, \beta) = cf(\alpha_1, \beta) + f(\alpha_2, \beta)$$

we see that

$$L_f(c\alpha_1 + \alpha_2) = cL_f(\alpha_1) + L_f(\alpha_2)$$

that is, L_f is a linear transformation from V into V^*.

In a similar manner, f determines a linear transformation R_f from V into V^*. For each fixed β in V, $f(\alpha, \beta)$ is linear as a function of α. We define $R_f(\beta)$ to be the linear functional on V whose value on the vector α is $f(\alpha, \beta)$.

Theorem 2. *Let* f *be a bilinear form on the finite-dimensional vector space* V. *Let* L_f *and* R_f *be the linear transformations from* V *into* V^* *defined by* $(L_f\alpha)(\beta) = f(\alpha, \beta) = (R_f\beta)(\alpha)$. *Then rank* $(L_f) =$ *rank* (R_f).

Proof. One can give a 'coordinate free' proof of this theorem. Such a proof is similar to the proof (in Section 3.7) that the row-rank of a matrix is equal to its column-rank. So, here we shall give a proof which proceeds by choosing a coordinate system (basis) and then using the 'row-rank equals column-rank' theorem.

To prove rank $(L_f) =$ rank (R_f), it will suffice to prove that L_f and

R_f have the same nullity. Let $\mathcal{B}$ be an ordered basis for V, and let $A = [f]_{\mathcal{B}}$. If α and β are vectors in V, with coordinate matrices X and Y in the ordered basis $\mathcal{B}$, then $f(\alpha, \beta) = X^tAY$. Now $R_f(\beta) = 0$ means that $f(\alpha, \beta) = 0$ for every α in V, i.e., that $X^tAY = 0$ for every $n \times 1$ matrix X. The latter condition simply says that $AY = 0$. The nullity of R_f is therefore equal to the dimension of the space of solutions of $AY = 0$.

Similarly, $L_f(\alpha) = 0$ if and only if $X^tAY = 0$ for every $n \times 1$ matrix Y. Thus α is in the null space of L_f if and only if $X^tA = 0$, i.e., $A^tX = 0$. The nullity of L_f is therefore equal to the dimension of the space of solutions of $A^tX = 0$. Since the matrices A and A^t have the same column-rank, we see that

$$\text{nullity } (L_f) = \text{nullity } (R_f). \quad \blacksquare$$

Definition. *If* f *is a bilinear form on the finite-dimensional space* V, *the* **rank** *of* f *is the integer* $r = \text{rank } (L_f) = \text{rank } (R_f)$.

Corollary 1. *The rank of a bilinear form is equal to the rank of the matrix of the form in any ordered basis.*

Corollary 2. *If* f *is a bilinear form on the* n*-dimensional vector space* V, *the following are equivalent:*

(a) *rank* $(f) = n$.
(b) *For each non-zero* α *in* V, *there is a* β *in* V *such that* $f(\alpha, \beta) \neq 0$.
(c) *For each non-zero* β *in* V, *there is an* α *in* V *such that* $f(\alpha, \beta) \neq 0$.

Proof. Statement (b) simply says that the null space of L_f is the zero subspace. Statement (c) says that the null space of R_f is the zero subspace. The linear transformations L_f and R_f have nullity 0 if and only if they have rank n, i.e., if and only if rank $(f) = n$. $\blacksquare$

Definition. *A bilinear form* f *on a vector space* V *is called* **non-degenerate** (*or* **non-singular**) *if it satisfies conditions* (b) *and* (c) *of Corollary* 2.

If V is finite-dimensional, then f is non-degenerate provided f satisfies any one of the three conditions of Corollary 2. In particular, f is non-degenerate (non-singular) if and only if its matrix in some (every) ordered basis for V is a non-singular matrix.

EXAMPLE 5. Let $V = R^n$, and let f be the bilinear form defined on $\alpha = (x_1, \ldots, x_n)$ and $\beta = (y_1, \ldots, y_n)$ by

$$f(\alpha, \beta) = x_1y_1 + \cdots + x_ny_n.$$

Then f is a non-degenerate bilinear form on R^n. The matrix of f in the standard ordered basis is the $n \times n$ identity matrix:

$$f(X, Y) = X^t Y.$$

This f is usually called the dot (or scalar) product. The reader is probably familiar with this bilinear form, at least in the case $n = 3$. Geometrically, the number $f(\alpha, \beta)$ is the product of the length of α, the length of β, and the cosine of the angle between α and β. In particular, $f(\alpha, \beta) = 0$ if and only if the vectors α and β are orthogonal (perpendicular).

Exercises

1. Which of the following functions f, defined on vectors $\alpha = (x_1, x_2)$ and $\beta = (y_1, y_2)$ in R^2, are bilinear forms?

(a) $f(\alpha, \beta) = 1.$
(b) $f(\alpha, \beta) = (x_1 - y_1)^2 + x_2y_2.$
(c) $f(\alpha, \beta) = (x_1 + y_1)^2 - (x_1 - y_1)^2.$
(d) $f(\alpha, \beta) = x_1y_2 - x_2y_1.$

2. Let f be the bilinear form on R^2 defined by

$$f((x_1, y_1), (x_2, y_2)) = x_1y_1 + x_2y_2.$$

Find the matrix of f in each of the following bases:

$$\{(1, 0), (0, 1)\}, \qquad \{(1, -1), (1, 1)\}, \qquad \{(1, 2), (3, 4)\}.$$

3. Let V be the space of all 2×3 matrices over R, and let f be the bilinear form on V defined by $f(X, Y) = \text{trace}\,(X^tAY)$, where

$$A = \begin{bmatrix} 1 & 2 \\ 3 & 4 \end{bmatrix}.$$

Find the matrix of f in the ordered basis

$$\{E^{11}, E^{12}, E^{13}, E^{21}, E^{22}, E^{23}\}$$

where E^{ij} is the matrix whose only non-zero entry is a 1 in row i and column j.

4. Describe explicitly all bilinear forms f on R^3 with the property that $f(\alpha, \beta) = f(\beta, \alpha)$ for all α, β.

5. Describe the bilinear forms on R^3 which satisfy $f(\alpha, \beta) = -f(\beta, \alpha)$ for all α, β.

6. Let n be a positive integer, and let V be the space of all $n \times n$ matrices over the field of complex numbers. Show that the equation

$$f(A, B) = n\,\text{tr}\,(AB) - \text{tr}\,(A)\,\text{tr}\,(B)$$

defines a bilinear form f on V. Is it true that $f(A, B) = f(B, A)$ for all A, B?

7. Let f be the bilinear form defined in Exercise 6. Show that f is degenerate (not non-degenerate). Let V_1 be the subspace of V consisting of the matrices of trace 0, and let f_1 be the restriction of f to V_1. Show that f_1 is non-degenerate.

8. Let f be the bilinear form defined in Exercise 6, and let V_2 be the subspace of V consisting of all matrices A such that trace $(A) = 0$ and $A^* = -A$ (A^* is the conjugate transpose of A). Denote by f_2 the restriction of f to V_2. Show that f_2 is negative definite, i.e., that $f_2(A, A) < 0$ for each non-zero A in V_2.

9. Let f be the bilinear form defined in Exercise 6. Let W be the set of all matrices A in V such that $f(A, B) = 0$ for all B. Show that W is a subspace of V. Describe W explicitly and find its dimension.

10. Let f be any bilinear form on a finite-dimensional vector space V. Let W be the subspace of all β such that $f(\alpha, \beta) = 0$ for every α. Show that

$$\text{rank } f = \dim V - \dim W.$$

Use this result and the result of Exercise 9 to compute the rank of the bilinear form defined in Exercise 6.

11. Let f be a bilinear form on a finite-dimensional vector space V. Suppose V_1 is a subspace of V with the property that the restriction of f to V_1 is non-degenerate. Show that rank $f \geq \dim V_1$.

12. Let f, g be bilinear forms on a finite-dimensional vector space V. Suppose g is non-singular. Show that there exist unique linear operators T_1, T_2 on V such that

$$f(\alpha, \beta) = g(T_1\alpha, \beta) = g(\alpha, T_2\beta)$$

for all α, β.

13. Show that the result given in Exercise 12 need not be true if g is singular.

14. Let f be a bilinear form on a finite-dimensional vector space V. Show that f can be expressed as a product of two linear functionals (i.e., $f(\alpha, \beta) = L_1(\alpha)L_2(\beta)$ for L_1, L_2 in V^*) if and only if f has rank 1.

10.2. Symmetric Bilinear Forms

The main purpose of this section is to answer the following question: If f is a bilinear form on the finite-dimensional vector space V, when is there an ordered basis $\mathcal{B}$ for V in which f is represented by a diagonal matrix? We prove that this is possible if and only if f is a symmetric bilinear form, i.e., $f(\alpha, \beta) = f(\beta, \alpha)$. The theorem is proved only when the scalar field has characteristic zero, that is, that if n is a positive integer the sum $1 + \cdots + 1$ (n times) in F is not 0.

Definition. *Let* f *be a bilinear form on the vector space* V. *We say that* f *is* **symmetric** *if* $f(\alpha, \beta) = f(\beta, \alpha)$ *for all vectors* α, β *in* V.

If V is a finite-dimensional, the bilinear form f is symmetric if and only if its matrix A in some (or every) ordered basis is symmetric, $A^t = A$. To see this, one inquires when the bilinear form

$$f(X, Y) = X^tAY$$

is symmetric. This happens if and only if $X^tAY = Y^tAX$ for all column matrices X and Y. Since X^tAY is a 1×1 matrix, we have $X^tAY = Y^tA^tX$. Thus f is symmetric if and only if $Y^tA^tX = Y^tAX$ for all X, Y. Clearly this just means that $A = A^t$. In particular, one should note that if there is an ordered basis for V in which f is represented by a diagonal matrix, then f is symmetric, for any diagonal matrix is a symmetric matrix.

If f is a symmetric bilinear form, the **quadratic form associated with** f is the function q from V into F defined by

$$q(\alpha) = f(\alpha, \alpha).$$

If F is a subfield of the complex numbers, the symmetric bilinear form f is completely determined by its associated quadratic form, according to the **polarization identity**

$$f(\alpha, \beta) = \tfrac{1}{4}q(\alpha + \beta) - \tfrac{1}{4}q(\alpha - \beta). \tag{10-5}$$

The establishment of (10-5) is a routine computation, which we omit. If f is the bilinear form of Example 5, the dot product, the associated quadratic form is

$$q(x_1, \ldots, x_n) = x_1^2 + \cdots + x_n^2.$$

In other words, $q(\alpha)$ is the square of the length of α. For the bilinear form $f_A(X, Y) = X^tAY$, the associated quadratic form is

$$q_A(X) = X^tAX = \sum_{i,j} A_{ij}x_ix_j.$$

One important class of symmetric bilinear forms consists of the inner products on real vector spaces, discussed in Chapter 8. If V is a *real* vector space, an **inner product** on V is a symmetric bilinear form f on V which satisfies

$$f(\alpha, \alpha) > 0 \quad \text{if} \quad \alpha \neq 0. \tag{10-6}$$

A bilinear form satisfying (10-6) is called **positive definite.** Thus, an inner product on a real vector space is a positive definite, symmetric bilinear form on that space. Note that an inner product is non-degenerate. Two vectors α, β are called **orthogonal** with respect to the inner product f if $f(\alpha, \beta) = 0$. The quadratic form $q(\alpha) = f(\alpha, \alpha)$ takes only non-negative values, and $q(\alpha)$ is usually thought of as the square of the length of α. Of course, these concepts of length and orthogonality stem from the most important example of an inner product—the dot product of Example 5.

If f is any symmetric bilinear form on a vector space V, it is convenient to apply some of the terminology of inner products to f. It is especially convenient to say that α and β are orthogonal with respect to f if $f(\alpha, \beta) = 0$. It is not advisable to think of $f(\alpha, \alpha)$ as the square of the length of α; for example, if V is a complex vector space, we may have $f(\alpha, \alpha) = \sqrt{-1}$, or on a real vector space, $f(\alpha, \alpha) = -2$.

We turn now to the basic theorem of this section. In reading the proof, the reader should find it helpful to think of the special case in which V is a real vector space and f is an inner product on V.

Theorem 3. *Let* V *be a finite-dimensional vector space over a field of characteristic zero, and let* f *be a symmetric bilinear form on* V. *Then there is an ordered basis for* V *in which* f *is represented by a diagonal matrix.*

Proof. What we must find is an ordered basis

$$\mathcal{B} = \{\alpha_1, \ldots, \alpha_n\}$$

such that $f(\alpha_i, \alpha_j) = 0$ for $i \neq j$. If $f = 0$ or $n = 1$, the theorem is obviously true. Thus we may suppose $f \neq 0$ and $n > 1$. If $f(\alpha, \alpha) = 0$ for every α in V, the associated quadratic form q is identically 0, and the polarization identity (10-5) shows that $f = 0$. Thus there is a vector α in V such that $f(\alpha, \alpha) = q(\alpha) \neq 0$. Let W be the one-dimensional subspace of V which is spanned by α, and let $W^\perp$ be the set of all vectors β in V such that $f(\alpha, \beta) = 0$. Now we claim that $V = W \oplus W^\perp$. Certainly the subspaces W and $W^\perp$ are independent. A typical vector in W is $c\alpha$, where c is a scalar. If $c\alpha$ is also in $W^\perp$, then $f(c\alpha, c\alpha) = c^2 f(\alpha, \alpha) = 0$. But $f(\alpha, \alpha) \neq 0$, thus $c = 0$. Also, each vector in V is the sum of a vector in W and a vector in $W^\perp$. For, let γ be any vector in V, and put

$$\beta = \gamma - \frac{f(\gamma, \alpha)}{f(\alpha, \alpha)}\,\alpha.$$

Then

$$f(\alpha, \beta) = f(\alpha, \gamma) - \frac{f(\gamma, \alpha)}{f(\alpha, \alpha)} f(\alpha, \alpha)$$

and since f is symmetric, $f(\alpha, \beta) = 0$. Thus β is in the subspace $W^\perp$. The expression

$$\gamma = \frac{f(\gamma, \alpha)}{f(\alpha, \alpha)}\,\alpha + \beta$$

shows us that $V = W + W^\perp$.

The restriction of f to $W^\perp$ is a symmetric bilinear form on $W^\perp$. Since $W^\perp$ has dimension $(n - 1)$, we may assume by induction that $W^\perp$ has a basis $\{\alpha_2, \ldots, \alpha_n\}$ such that

$$f(\alpha_i, \alpha_j) = 0, \qquad i \neq j \ (i \geq 2, j \geq 2).$$

Putting $\alpha_1 = \alpha$, we obtain a basis $\{\alpha_1, \ldots, \alpha_n\}$ for V such that $f(\alpha_i, \alpha_j) = 0$ for $i \neq j$. ∎

Corollary. *Let* F *be a subfield of the complex numbers, and let* A *be a symmetric* n × n *matrix over* F. *Then there is an invertible* n × n *matrix* P *over* F *such that* P^tAP *is diagonal.*

In case F is the field of real numbers, the invertible matrix P in this corollary can be chosen to be an *orthogonal* matrix, i.e., $P^t = P^{-1}$. In other words, if A is a real symmetric $n \times n$ matrix, there is a real orthogonal matrix P such that P^tAP is diagonal; however, this is not at all apparent from what we did above (see Chapter 8).

Theorem 4. *Let V be a finite-dimensional vector space over the field of complex numbers. Let f be a symmetric bilinear form on V which has rank r. Then there is an ordered basis* $\mathcal{B} = \{\beta_1, \ldots, \beta_n\}$ *for V such that*

(i) *the matrix of f in the ordered basis* $\mathcal{B}$ *is diagonal;*

(ii) $f(\beta_j, \beta_j) = \begin{cases} 1, & j = 1, \ldots, r \\ 0, & j > r. \end{cases}$

Proof. By Theorem 3, there is an ordered basis $\{\alpha_1, \ldots, \alpha_n\}$ for V such that

$$f(\alpha_i, \alpha_j) = 0 \quad \text{for} \quad i \neq j.$$

Since f has rank r, so does its matrix in the ordered basis $\{\alpha_1, \ldots, \alpha_n\}$. Thus we must have $f(\alpha_j, \alpha_j) \neq 0$ for precisely r values of j. By reordering the vectors α_j, we may assume that

$$f(\alpha_j, \alpha_j) \neq 0, \qquad j = 1, \ldots, r.$$

Now we use the fact that the scalar field is the field of complex numbers. If $\sqrt{f(\alpha_j, \alpha_j)}$ denotes any complex square root of $f(\alpha_j, \alpha_j)$, and if we put

$$\beta_j = \begin{cases} \dfrac{1}{\sqrt{f(\alpha_j, \alpha_j)}}\,\alpha_j, & j = 1, \ldots, r \\ \alpha_j, & j > r \end{cases}$$

the basis $\{\beta_1, \ldots, \beta_n\}$ satisfies conditions (i) and (ii). ∎

Of course, Theorem 4 is valid if the scalar field is any subfield of the complex numbers in which each element has a square root. It is not valid, for example, when the scalar field is the field of real numbers. Over the field of real numbers, we have the following substitute for Theorem 4.

Theorem 5. *Let V be an n-dimensional vector space over the field of real numbers, and let f be a symmetric bilinear form on V which has rank r. Then there is an ordered basis* $\{\beta_1, \beta_2, \ldots, \beta_n\}$ *for V in which the matrix of f is diagonal and such that*

$$f(\beta_j, \beta_j) = \pm 1, \qquad j = 1, \ldots, r.$$

Furthermore, the number of basis vectors β_j *for which* $f(\beta_j, \beta_j) = 1$ *is independent of the choice of basis.*

Proof. There is a basis $\{\alpha_1, \ldots, \alpha_n\}$ for V such that

$$\begin{aligned} f(\alpha_i, \alpha_j) &= 0, & i &\neq j \\ f(\alpha_j, \alpha_j) &\neq 0, & 1 &\leq j \leq r \\ f(\alpha_j, \alpha_j) &= 0, & j &> r. \end{aligned}$$

Let

$$\beta_j = |f(\alpha_j, \alpha_j)|^{-1/2}\alpha_j, \qquad 1 \le j \le r$$
$$\beta_j = \alpha_j, \qquad j > r.$$

Then $\{\beta_1, \ldots, \beta_n\}$ is a basis with the stated properties.

Let p be the number of basis vectors β_j for which $f(\beta_j, \beta_j) = 1$; we must show that the number p is independent of the particular basis we have, satisfying the stated conditions. Let V^+ be the subspace of V spanned by the basis vectors β_j for which $f(\beta_j, \beta_j) = 1$, and let V^- be the subspace spanned by the basis vectors β_j for which $f(\beta_j, \beta_j) = -1$. Now $p = \dim V^+$, so it is the uniqueness of the dimension of V^+ which we must demonstrate. It is easy to see that if α is a non-zero vector in V^+, then $f(\alpha, \alpha) > 0$; in other words, f is positive definite on the subspace V^+. Similarly, if α is a non-zero vector in V^-, then $f(\alpha, \alpha) < 0$, i.e., f is negative definite on the subspace V^-. Now let $V^\perp$ be the subspace spanned by the basis vectors β_j for which $f(\beta_j, \beta_j) = 0$. If α is in $V^\perp$, then $f(\alpha, \beta) = 0$ for all β in V.

Since $\{\beta_1, \ldots, \beta_n\}$ is a basis for V, we have

$$V = V^+ \oplus V^- \oplus V^\perp.$$

Furthermore, we claim that if W is any subspace of V on which f is positive definite, then the subspaces W, V^-, and $V^\perp$ are independent. For, suppose α is in W, β is in V^-, γ is in $V^\perp$, and $\alpha + \beta + \gamma = 0$. Then

$$0 = f(\alpha, \alpha + \beta + \gamma) = f(\alpha, \alpha) + f(\alpha, \beta) + f(\alpha, \gamma)$$
$$0 = f(\beta, \alpha + \beta + \gamma) = f(\beta, \alpha) + f(\beta, \beta) + f(\beta, \gamma).$$

Since γ is in $V^\perp$, $f(\alpha, \gamma) = f(\beta, \gamma) = 0$; and since f is symmetric, we obtain

$$0 = f(\alpha, \alpha) + f(\alpha, \beta)$$
$$0 = f(\beta, \beta) + f(\alpha, \beta)$$

hence $f(\alpha, \alpha) = f(\beta, \beta)$. Since $f(\alpha, \alpha) \ge 0$ and $f(\beta, \beta) \le 0$, it follows that

$$f(\alpha, \alpha) = f(\beta, \beta) = 0.$$

But f is positive definite on W and negative definite on V^-. We conclude that $\alpha = \beta = 0$, and hence that $\gamma = 0$ as well.

Since

$$V = V^+ \oplus V^- \oplus V^\perp$$

and W, V^-, $V^\perp$ are independent, we see that $\dim W \le \dim V^+$. That is, if W is any subspace of V on which f is positive definite, the dimension of W cannot exceed the dimension of V^+. If $\mathfrak{B}_1$ is another ordered basis for V which satisfies the conditions of the theorem, we shall have corresponding subspaces V_1^+, V_1^-, and $V_1^\perp$; and, the argument above shows that $\dim V_1^+ \le \dim V^+$. Reversing the argument, we obtain $\dim V^+ \le \dim V_1^+$, and consequently

$$\dim V^+ = \dim V_1^+. \quad \blacksquare$$

There are several comments we should make about the basis $\{\beta_1, \ldots, \beta_n\}$ of Theorem 5 and the associated subspaces V^+, V^-, and $V^\perp$ First, note that $V^\perp$ is exactly the subspace of vectors which are 'orthogonal' to all of V. We noted above that $V^\perp$ is contained in this subspace; but,

$$\dim V^\perp = \dim V - (\dim V^+ + \dim V^-) = \dim V - \operatorname{rank} f$$

so every vector α such that $f(\alpha, \beta) = 0$ for all β must be in $V^\perp$. Thus, the subspace $V^\perp$ is unique. The subspaces V^+ and V^- are not unique; however, their dimensions are unique. The proof of Theorem 5 shows us that dim V^+ is the largest possible dimension of any subspace on which f is positive definite. Similarly, dim V^- is the largest dimension of any subspace on which f is negative definite. Of course

$$\dim V^+ + \dim V^- = \operatorname{rank} f.$$

The number

$$\dim V^+ - \dim V^-$$

is often called the **signature** of f. It is introduced because the dimensions of V^+ and V^- are easily determined from the rank of f and the signature of f.

Perhaps we should make one final comment about the relation of symmetric bilinear forms on real vector spaces to inner products. Suppose V is a finite-dimensional real vector space and that V_1, V_2, V_3 are subspaces of V such that

$$V = V_1 \oplus V_2 \oplus V_3.$$

Suppose that f_1 is an inner product on V_1, and f_2 is an inner product on V_2. We can then define a symmetric bilinear form f on V as follows: If α, β are vectors in V, then we can write

$$\alpha = \alpha_1 + \alpha_2 + \alpha_3 \quad \text{and} \quad \beta = \beta_1 + \beta_2 + \beta_3$$

with α_j and β_j in V_j. Let

$$f(\alpha, \beta) = f_1(\alpha_1, \beta_1) - f_2(\alpha_2, \beta_2).$$

The subspace $V^\perp$ for f will be V_3, V_1 is a suitable V^+ for f, and V_2 is a suitable V^-. One part of the statement of Theorem 5 is that every symmetric bilinear form on V arises in this way. The additional content of the theorem is that an inner product is represented in some ordered basis by the identity matrix.

Exercises

1. The following expressions define quadratic forms q on R^2. Find the symmetric bilinear form f corresponding to each q.

(a) ax_1^2.
(b) bx_1x_2.
(c) cx_2^2.
(d) $2x_1^2 - \frac{1}{3}x_1x_2$.
(e) $x_1^2 + 9x_2^2$.
(f) $3x_1x_2 - x_2^2$.
(g) $4x_1^2 + 6x_1x_2 - 3x_2^2$.

2. Find the matrix, in the standard ordered basis, and the rank of each of the bilinear forms determined in Exercise 1. Indicate which forms are non-degenerate.

3. Let $q(x_1, x_2) = ax_1^2 + bx_1x_2 + cx_2^2$ be the quadratic form associated with a symmetric bilinear form f on R^2. Show that f is non-degenerate if and only if $b^2 - 4ac \neq 0$.

4. Let V be a finite-dimensional vector space over a subfield F of the complex numbers, and let S be the set of all symmetric bilinear forms on V.

(a) Show that S is a subspace of $L(V, V, F)$.
(b) Find dim S.

Let Q be the set of all quadratic forms on V.

(c) Show that Q is a subspace of the space of all functions from V into F.
(d) Describe explicitly an isomorphism T of Q onto S, without reference to a basis.
(e) Let U be a linear operator on V and q an element of Q. Show that the equation $(U^\dagger q)(\alpha) = q(U\alpha)$ defines a quadratic form $U^\dagger q$ on V.
(f) If U is a linear operator on V, show that the function $U^\dagger$ defined in part (e) is a linear operator on Q. Show that $U^\dagger$ is invertible if and only if U is invertible.

5. Let q be the quadratic form on R^2 given by

$$q(x_1, x_2) = ax_1^2 + 2bx_1x_2 + cx_2^2, \qquad a \neq 0.$$

Find an invertible linear operator U on R^2 such that

$$(U^\dagger q)(x_1, x_2) = ax_1^2 + \left(c - \frac{b^2}{a}\right)x_2^2.$$

(*Hint:* To find U^{-1} (and hence U), complete the square. For the definition of $U^\dagger$, see part (e) of Exercise 4.)

6. Let q be the quadratic form on R^2 given by

$$q(x_1, x_2) = 2bx_1x_2.$$

Find an invertible linear operator U on R^2 such that

$$(U^\dagger q)(x_1, x_2) = 2bx_1^2 - 2bx_2^2.$$

7. Let q be the quadratic form on R^3 given by

$$q(x_1, x_2, x_3) = x_1x_2 + 2x_1x_3 + x_3^2.$$

Find an invertible linear operator U on R^3 such that

$$(U^\dagger q)(x_1, x_2, x_3) = x_1^2 - x_2^2 + x_3^2.$$

(*Hint:* Express U as a product of operators similar to those used in Exercises 5 and 6.)

8. Let A be a symmetric $n \times n$ matrix over R, and let q be the quadratic form on R^n given by

$$q(x_1, \ldots, x_n) = \sum_{i,j} A_{ij}x_ix_j.$$

Generalize the method used in Exercise 7 to show that there is an invertible linear operator U on R^n such that

$$(U^\dagger q)(x_1, \ldots, x_n) = \sum_{i=1}^{n} c_i x_i^2$$

where c_i is 1, -1, or 0, $i = 1, \ldots, n$.

9. Let f be a symmetric bilinear form on R^n. Use the result of Exercise 8 to prove the existence of an ordered basis $\mathcal{B}$ such that $[f]_{\mathcal{B}}$ is diagonal.

10. Let V be the real vector space of all 2×2 (complex) Hermitian matrices, that is, 2×2 complex matrices A which satisfy $A_{ij} = \overline{A_{ji}}$.

(a) Show that the equation $q(A) = \det A$ defines a quadratic form q on V.

(b) Let W be the subspace of V of matrices of trace 0. Show that the bilinear form f determined by q is negative definite on the subspace W.

11. Let V be a finite-dimensional vector space and f a non-degenerate symmetric bilinear form on V. Show that for each linear operator T on V there is a unique linear operator T' on V such that $f(T\alpha, \beta) = f(\alpha, T'\beta)$ for all α, β in V. Also show that

$$(T_1T_2)' = T_2'T_1'$$
$$(c_1T_1 + c_2T_2)' = c_1T_1' + c_2T_2'$$
$$(T')' = T.$$

How much of the above is valid without the assumption that T is non-degenerate?

12. Let F be a field and V the space of $n \times 1$ matrices over F. Suppose A is a fixed $n \times n$ matrix over F and f is the bilinear form on V defined by $f(X, Y) = X^tAY$. Suppose f is symmetric and non-degenerate. Let B be an $n \times n$ matrix over F and T the linear operator on V sending X into BX. Find the operator T' of Exercise 11.

13. Let V be a finite-dimensional vector space and f a non-degenerate symmetric bilinear form on V. Associated with f is a 'natural' isomorphism of V onto the dual space V^*, this isomorphism being the transformation L_f of Section 10.1. Using L_f, show that for each basis $\mathcal{B} = \{\alpha_1, \ldots, \alpha_n\}$ of V there exists a unique basis $\mathcal{B}' = \{\alpha_1', \ldots, \alpha_n'\}$ of V such that $f(\alpha_i, \alpha_j') = \delta_{ij}$. Then show that for every vector α in V we have

$$\alpha = \sum_i f(\alpha, \alpha_i')\alpha_i = \sum_i f(\alpha_i, \alpha)\alpha_i'.$$

14. Let V, f, $\mathcal{B}$, and $\mathcal{B}'$ be as in Exercise 13. Suppose T is a linear operator on V and that T' is the operator which f associates with T as in Exercise 11. Show that

(a) $[T']_{\mathcal{B}'} = [T]_{\mathcal{B}}^t$.

(b) $\operatorname{tr}(T) = \operatorname{tr}(T') = \sum_i f(T\alpha_i, \alpha_i')$.

15. Let V, f, $\mathcal{B}$, and $\mathcal{B}'$ be as in Exercise 13. Suppose $[f]_{\mathcal{B}} = A$. Show that

$$\alpha_i' = \sum_j (A^{-1})_{ij}\alpha_j = \sum_j (A^{-1})_{ji}\alpha_j.$$

16. Let F be a field and V the space of $n \times 1$ matrices over F. Suppose A is an invertible, symmetric $n \times n$ matrix over F and that f is the bilinear form on V defined by $f(X, Y) = X^tAY$. Let P be an invertible $n \times n$ matrix over F and $\mathcal{B}$ the basis for V consisting of the columns of P. Show that the basis $\mathcal{B}'$ of Exercise 13 consists of the columns of the matrix $A^{-1}(P^t)^{-1}$.

17. Let V be a finite-dimensional vector space over a field F and f a symmetric bilinear form on V. For each subspace W of V, let $W^\perp$ be the set of all vectors α in V such that $f(\alpha, \beta) = 0$ for every β in W. Show that

(a) $W^\perp$ is a subspace.

(b) $V = \{0\}^\perp$.

(c) $V^\perp = \{0\}$ if and only if f is non-degenerate.

(d) $\text{rank } f = \dim V - \dim V^\perp$.

(e) If $\dim V = n$ and $\dim W = m$, then $\dim W^\perp \geq n - m$. (*Hint:* Let $\{\beta_1, \ldots, \beta_m\}$ be a basis of W and consider the mapping

$$\alpha \to (f(\alpha, \beta_1), \ldots, f(\alpha, \beta_m))$$

of V into F^m.)

(f) The restriction of f to W is non-degenerate if and only if

$$W \cap W^\perp = \{0\}.$$

(g) $V = W \oplus W^\perp$ if and only if the restriction of f to W is non-degenerate.

18. Let V be a finite-dimensional vector space over C and f a non-degenerate symmetric bilinear form on V. Prove that there is a basis $\mathcal{B}$ of V such that $\mathcal{B}' = \mathcal{B}$. (See Exercise 13 for a definition of $\mathcal{B}'$.)

10.3. Skew-Symmetric Bilinear Forms

Throughout this section V will be a vector space over a subfield F of the field of complex numbers. A bilinear form f on V is called **skew-symmetric** if $f(\alpha, \beta) = -f(\beta, \alpha)$ for all vectors α, β in V. We shall prove one theorem concerning the simplification of the matrix of a skew-symmetric bilinear form on a finite-dimensional space V. First, let us make some general observations.

Suppose f is *any* bilinear form on V. If we let

$$\begin{aligned} g(\alpha, \beta) &= \tfrac{1}{2}[f(\alpha, \beta) + f(\beta, \alpha)] \\ h(\alpha, \beta) &= \tfrac{1}{2}[f(\alpha, \beta) - f(\beta, \alpha)] \end{aligned}$$

then it is easy to verify that g is a symmetric bilinear form on V and h is a skew-symmetric bilinear form on V. Also $f = g + h$. Furthermore, this expression for V as the sum of a symmetric and a skew-symmetric form is unique. Thus, the space $L(V, V, F)$ is the direct sum of the subspace of symmetric forms and the subspace of skew-symmetric forms.

If V is finite-dimensional, the bilinear form f is skew-symmetric if and only if its matrix A in some (or every) ordered basis is skew-symmetric, $A^t = -A$. This is proved just as one proves the corresponding fact about symmetric bilinear forms. When f is skew-symmetric, the matrix of f in any ordered basis will have all its diagonal entries 0. This just corresponds to the observation that $f(\alpha, \alpha) = 0$ for every α in V, since $f(\alpha, \alpha) = -f(\alpha, \alpha)$.

Let us suppose f is a non-zero skew-symmetric bilinear form on V. Since $f \neq 0$, there are vectors α, β in V such that $f(\alpha, \beta) \neq 0$. Multiplying α by a suitable scalar, we may assume that $f(\alpha, \beta) = 1$. Let γ be any vector in the subspace spanned by α and β, say $\gamma = c\alpha + d\beta$. Then

$$\begin{aligned} f(\gamma, \alpha) &= f(c\alpha + d\beta, \alpha) = df(\beta, \alpha) = -d \\ f(\gamma, \beta) &= f(c\alpha + d\beta, \beta) = cf(\alpha, \beta) = c \end{aligned}$$

and so

$$\gamma = f(\gamma, \beta)\alpha - f(\gamma, \alpha)\beta. \tag{10-7}$$

In particular, note that α and β are necessarily linearly independent; for, if $\gamma = 0$, then $f(\gamma, \alpha) = f(\gamma, \beta) = 0$.

Let W be the two-dimensional subspace spanned by α and β. Let $W^{\perp}$ be the set of all vectors δ in V such that $f(\delta, \alpha) = f(\delta, \beta) = 0$, that is, the set of all δ such that $f(\delta, \gamma) = 0$ for every γ in the subspace W. We claim that $V = W \oplus W^{\perp}$. For, let ϵ be any vector in V, and

$$\begin{aligned} \gamma &= f(\epsilon, \beta)\alpha - f(\epsilon, \alpha)\beta \\ \delta &= \epsilon - \gamma. \end{aligned}$$

Then γ is in W, and δ is in $W^{\perp}$, for

$$\begin{aligned} f(\delta, \alpha) &= f(\epsilon - f(\epsilon, \beta)\alpha + f(\epsilon, \alpha)\beta, \alpha) \\ &= f(\epsilon, \alpha) + f(\epsilon, \alpha)f(\beta, \alpha) \\ &= 0 \end{aligned}$$

and similarly $f(\delta, \beta) = 0$. Thus every ϵ in V is of the form $\epsilon = \gamma + \delta$, with γ in W and δ in $W^{\perp}$. From (9-7) it is clear that $W \cap W^{\perp} = \{0\}$, and so $V = W \oplus W^{\perp}$.

Now the restriction of f to $W^{\perp}$ is a skew-symmetric bilinear form on $W^{\perp}$. This restriction may be the zero form. If it is not, there are vectors α' and β' in $W^{\perp}$ such that $f(\alpha', \beta') = 1$. If we let W' be the two-dimensional subspace spanned by α' and β', then we shall have

$$V = W \oplus W' \oplus W_0$$

where W_0 is the set of all vectors δ in $W^{\perp}$ such that $f(\alpha', \delta) = f(\beta', \delta) = 0$. If the restriction of f to W_0 is not the zero form, we may select vectors α'', β'' in W_0 such that $f(\alpha'', \beta'') = 1$, and continue.

In the finite-dimensional case it should be clear that we obtain a finite sequence of pairs of vectors,

$$(\alpha_1, \beta_1), (\alpha_2, \beta_2), \ldots, (\alpha_k, \beta_k)$$

with the following properties:

(a) $f(\alpha_j, \beta_j) = 1, j = 1, \ldots, k.$
(b) $f(\alpha_i, \alpha_j) = f(\beta_i, \beta_j) = f(\alpha_i, \beta_j) = 0, i \neq j.$
(c) If W_j is the two-dimensional subspace spanned by α_j and β_j, then

$$V = W_1 \oplus \cdots \oplus W_k \oplus W_0$$

where every vector in W_0 is 'orthogonal' to all α_j and β_j, and the restriction of f to W_0 is the zero form.

Theorem 6. *Let* V *be an* n*-dimensional vector space over a subfield of the complex numbers, and let* f *be a skew-symmetric bilinear form on* V. *Then the rank* r *of* f *is even, and if* r = 2k *there is an ordered basis for* V *in which the matrix of* f *is the direct sum of the* (n − r) × (n − r) *zero matrix and* k *copies of the* 2 × 2 *matrix*

$$\begin{bmatrix} 0 & 1 \\ -1 & 0 \end{bmatrix}.$$

Proof. Let $\alpha_1, \beta_1, \ldots, \alpha_k, \beta_k$ be vectors satisfying conditions (a), (b), and (c) above. Let $\{\gamma_1, \ldots, \gamma_s\}$ be any ordered basis for the subspace W_0. Then

$$\mathcal{B} = \{\alpha_1, \beta_1, \alpha_2, \beta_2, \ldots, \alpha_k, \beta_k, \gamma_1, \ldots, \gamma_s\}$$

is an ordered basis for V. From (a), (b), and (c) it is clear that the matrix of f in the ordered basis $\mathcal{B}$ is the direct sum of the $(n - 2k) \times (n - 2k)$ zero matrix and k copies of the 2×2 matrix

$$\begin{bmatrix} 0 & 1 \\ -1 & 0 \end{bmatrix}. \tag{10-8}$$

Furthermore, it is clear that the rank of this matrix, and hence the rank of f, is $2k$. ∎

One consequence of the above is that if f is a non-degenerate, skew-symmetric bilinear form on V, then the dimension of V must be even. If $\dim V = 2k$, there will be an ordered basis $\{\alpha_1, \beta_1, \ldots, \alpha_k, \beta_k\}$ for V such that

$$f(\alpha_i, \beta_j) = \begin{cases} 0, & i \neq j \\ 1, & i = j \end{cases}$$

$$f(\alpha_i, \alpha_j) = f(\beta_i, \beta_j) = 0.$$

The matrix of f in this ordered basis is the direct sum of k copies of the 2×2 skew-symmetric matrix (10-8). We obtain another standard form for the matrix of a non-degenerate skew-symmetric form if, instead of the ordered basis above, we consider the ordered basis

$$\{\alpha_1, \ldots, \alpha_k, \beta_k, \ldots, \beta_1\}.$$

The reader should find it easy to verify that the matrix of f in the latter ordered basis has the block form

$$\begin{bmatrix} 0 & J \\ -J & 0 \end{bmatrix}$$

where J is the $k \times k$ matrix

$$\begin{bmatrix} 0 & \cdots & 0 & 1 \\ 0 & \cdots & 1 & 0 \\ \vdots & .\cdot^{\cdot} & \vdots & \vdots \\ 1 & \cdots & 0 & 0 \end{bmatrix}.$$

Exercises

1. Let V be a vector space over a field F. Show that the set of all skew-symmetric bilinear forms on V is a subspace of $L(V, V, F)$.

2. Find all skew-symmetric bilinear forms on R^3.

3. Find a basis for the space of all skew-symmetric bilinear forms on R^n.

4. Let f be a symmetric bilinear form on C^n and g a skew-symmetric bilinear form on C^n. Suppose $f + g = 0$. Show that $f = g = 0$.

5. Let V be an n-dimensional vector space over a subfield F of C. Prove the following.

(a) The equation $(Pf)(\alpha, \beta) = \frac{1}{2}f(\alpha, \beta) - \frac{1}{2}f(\beta, \alpha)$ defines a linear operator P on $L(V, V, F)$.

(b) $P^2 = P$, i.e., P is a projection.

(c) rank $P = \dfrac{n(n-1)}{2}$; nullity $P = \dfrac{n(n+1)}{2}$.

(d) If U is a linear operato on V, the equation $(U^\dagger f)(\alpha, \beta) = f(U\alpha, U\beta)$ defines a linear operator $U^\dagger$ on $L(V, V, F)$.

(e) For every linear operator U, the projection P commutes with $U^\dagger$.

6. Prove an analogue of Exercise 11 in Section 10.2 for non-degenerate, skew-symmetric bilinear forms.

7. Let f be a bilinear form on a vector space V. Let L_f and R_f be the mappings of V into V^* associated with f in Section 10.1. Prove that f is skew-symmetric if and only if $L_f = -R_f$.

8. Prove an analogue of Exercise 17 in Section 10.2 for skew-symmetric forms.

9. Let V be a finite-dimensional vector space and L_1, L_2 linear functionals on V. Show that the equation

$$f(\alpha, \beta) = L_1(\alpha)L_2(\beta) - L_1(\beta)L_2(\alpha)$$

defines a skew-symmetric bilinear form on V. Show that $f = 0$ if and only if L_1, L_2 are linearly dependent.

10. Let V be a finite-dimensional vector space over a subfield of the complex numbers and f a skew-symmetric bilinear form on V. Show hat f has rank 2 if

and only if there exist linearly independent linear functionals L_1, L_2 on V such that

$$f(\alpha, \beta) = L_1(\alpha)L_2(\beta) - L_1(\beta)L_2(\alpha).$$

11. Let f be any skew-symmetric bilinear form on R^3. Prove that there are linear functionals L_1, L_2 such that

$$f(\alpha, \beta) = L_1(\alpha)L_2(\beta) - L_1(\beta)L_2(\alpha).$$

12. Let V be a finite-dimensional vector space over a subfield of the complex numbers, and let f, g be skew-symmetric bilinear forms on V. Show that there is an *invertible* linear operator T on V such that $f(T\alpha, T\beta) = g(\alpha, \beta)$ for all α, β if and only if f and g have the same rank.

13. Show that the result of Exercise 12 is valid for symmetric bilinear forms on a complex vector space, but is not valid for symmetric bilinear forms on a real vector space.

10.4. Groups Preserving Bilinear Forms

Let f be a bilinear form on the vector space V, and let T be a linear operator on V. We say that T **preserves** f if $f(T\alpha, T\beta) = f(\alpha, \beta)$ for all α, β in V. For any T and f the function g, defined by $g(\alpha, \beta) = f(T\alpha, T\beta)$, is easily seen to be a bilinear form on V. To say that T preserves f is simply to say $g = f$. The identity operator preserves every bilinear form. If S and T are linear operators which preserve f, the product ST also preserves f; for $f(ST\alpha, ST\beta) = f(T\alpha, T\beta) = f(\alpha, \beta)$. In other words, the collection of linear operators which preserve a given bilinear form is closed under the formation of (operator) products. In general, one cannot say much more about this collection of operators; however, if f is non-degenerate, we have the following.

Theorem 7. *Let* f *be a non-degenerate bilinear form on a finite-dimensional vector space* V. *The set of all linear operators on* V *which preserve* f *is a group under the operation of composition.*

Proof. Let G be the set of linear operators preserving f. We observed that the identity operator is in G and that whenever S and T are in G the composition ST is also in G. From the fact that f is non-degenerate, we shall prove that any operator T in G is invertible, and T^{-1} is also in G. Suppose T preserves f. Let α be a vector in the null space of T. Then for any β in V we have

$$f(\alpha, \beta) = f(T\alpha, T\beta) = f(0, T\beta) = 0.$$

Since f is non-degenerate, $\alpha = 0$. Thus T is invertible. Clearly T^{-1} also preserves f; for

$$f(T^{-1}\alpha, T^{-1}\beta) = f(TT^{-1}\alpha, TT^{-1}\beta) = f(\alpha, \beta). \quad \blacksquare$$

If f is a non-degenerate bilinear form on the finite-dimensional space V, then each ordered basis $\mathcal{B}$ for V determines a group of matrices 'preserving' f. The set of all matrices $[T]_{\mathcal{B}}$, where T is a linear operator preserving f, will be a group under matrix multiplication. There is an alternative description of this group of matrices, as follows. Let $A = [f]_{\mathcal{B}}$, so that if α and β are vectors in V with respective coordinate matrices X and Y relative to $\mathcal{B}$, we shall have

$$f(\alpha, \beta) = X^t A Y.$$

Let T be any linear operator on V and $M = [T]_{\mathcal{B}}$. Then

$$\begin{aligned} f(T\alpha, T\beta) &= (MX)^t A (MY) \\ &= X^t (M^t A M) Y. \end{aligned}$$

Accordingly, T preserves f if and only if $M^t A M = A$. In matrix language then, Theorem 7 says the following: If A is an invertible $n \times n$ matrix, the set of all $n \times n$ matrices M such that $M^t A M = A$ is a group under matrix multiplication. If $A = [f]_{\mathcal{B}}$, then M is in this group of matrices if and only if $M = [T]_{\mathcal{B}}$, where T is a linear operator which preserves f.

Before turning to some examples, let us make one further remark. Suppose f is a bilinear form which is symmetric. A linear operator T preserves f if and only if T preserves the quadratic form

$$q(\alpha) = f(\alpha, \alpha)$$

associated with f. If T preserves f, we certainly have

$$q(T\alpha) = f(T\alpha, T\alpha) = f(\alpha, \alpha) = q(\alpha)$$

for every α in V. Conversely, since f is symmetric, the polarization identity

$$f(\alpha, \beta) = \tfrac{1}{4}q(\alpha + \beta) - \tfrac{1}{4}q(\alpha - \beta)$$

shows us that T preserves f provided that $q(T\gamma) = q(\gamma)$ for each γ in V. (We are assuming here that the scalar field is a subfield of the complex numbers.)

Example 6. Let V be either the space R^n or the space C^n. Let f be the bilinear form

$$f(\alpha, \beta) = \sum_{j=1}^{n} x_j y_j$$

where $\alpha = (x_1, \ldots, x_n)$ and $\beta = (y_1, \ldots, y_n)$. The group preserving f is called the n-dimensional (real or complex) **orthogonal group.** The name 'orthogonal group' is more commonly applied to the associated group of matrices in the standard ordered basis. Since the matrix of f in the standard basis is I, this group consists of the matrices M which satisfy $M^t M = I$. Such a matrix M is called an $n \times n$ (real or complex) **orthogonal matrix.** The two $n \times n$ orthogonal groups are usually denoted $O(n, R)$ and $O(n, C)$. Of course, the orthogonal group is also the group which preserves the quadratic form

$$q(x_1, \ldots, x_n) = x_1^2 + \cdots + x_n^2.$$

EXAMPLE 7. Let f be the symmetric bilinear form on R^n with quadratic form

$$q(x_1, \ldots, x_n) = \sum_{j=1}^{p} x_j^2 - \sum_{j=p+1}^{n} x_j^2.$$

Then f is non-degenerate and has signature $2p - n$. The group of matrices preserving a form of this type is called a **pseudo-orthogonal group.** When $p = n$, we obtain the orthogonal group $O(n, R)$ as a particular type of pseudo-orthogonal group. For each of the $n + 1$ values $p = 0, 1, 2, \ldots, n$, we obtain different bilinear forms f; however, for $p = k$ and $p = n - k$ the forms are negatives of one another and hence have the same associated group. Thus, when n is odd, we have $(n + 1)/2$ pseudo-orthogonal groups of $n \times n$ matrices, and when n is even, we have $(n + 2)/2$ such groups.

Theorem 8. *Let* V *be an* n*-dimensional vector space over the field of complex numbers, and let* f *be a non-degenerate symmetric bilinear form on* V. *Then the group preserving* f *is isomorphic to the complex orthogonal group* O(n, C).

Proof. Of course, by an isomorphism between two groups, we mean a one-one correspondence between their elements which 'preserves' the group operation. Let G be the group of linear operators on V which preserve the bilinear form f. Since f is both symmetric and non-degenerate, Theorem 4 tells us that there is an ordered basis $\mathfrak{B}$ for V in which f is represented by the $n \times n$ identity matrix. Therefore, a linear operator T preserves f if and only if its matrix in the ordered basis $\mathfrak{B}$ is a complex orthogonal matrix. Hence

$$T \rightarrow [T]_{\mathfrak{B}}$$

is an isomorphism of G onto $O(n, C)$. ∎

Theorem 9. *Let* V *be an* n*-dimensional vector space over the field of real numbers, and let* f *be a non-degenerate symmetric bilinear form on* V. *Then the group preserving* f *is isomorphic to an* n × n *pseudo-orthogonal group.*

Proof. Repeat the proof of Theorem 8, using Theorem 5 instead of Theorem 4. ∎

EXAMPLE 8. Let f be the symmetric bilinear form on R^4 with quadratic form

$$q(x, y, z, t) = t^2 - x^2 - y^2 - z^2.$$

A linear operator T on R^4 which preserves this particular bilinear (or quadratic) form is called a **Lorentz transformation,** and the group preserving f is called the **Lorentz group.** We should like to give one method of describing some Lorentz transformations.

Let H be the real vector space of all 2×2 complex matrices A which are Hermitian, $A = A^*$. It is easy to verify that

$$\Phi(x, y, z, t) = \begin{bmatrix} t + x & y + iz \\ y - iz & t - x \end{bmatrix}$$

defines an isomorphism Φ of R^4 onto the space H. Under this isomorphism, the quadratic form q is carried onto the determinant function, that is

$$q(x, y, z, t) = \det \begin{bmatrix} t + x & y + iz \\ y - iz & t - x \end{bmatrix}$$

or

$$q(\alpha) = \det \Phi(\alpha).$$

This suggests that we might study Lorentz transformations on R^4 by studying linear operators on H which preserve determinants.

Let M be any complex 2×2 matrix and for a Hermitian matrix A define

$$U_M(A) = MAM^*.$$

Now MAM^* is also Hermitian. From this it is easy to see that U_M is a (real) linear operator on H. Let us ask when it is true that U_M 'preserves' determinants, i.e., $\det [U_M(A)] = \det A$ for each A in H. Since the determinant of M^* is the complex conjugate of the determinant of M, we see that

$$\det [U_M(A)] = |\det M|^2 \det A.$$

Thus U_M preserves determinants exactly when $\det M$ has absolute value 1.

So now let us select any 2×2 complex matrix M for which $|\det M| = 1$. Then U_M is a linear operator on H which preserves determinants. Define

$$T_M = \Phi^{-1} U_M \Phi.$$

Since Φ is an isomorphism, T_M is a linear operator on R^4. Also, T_M is a Lorentz transformation; for

$$\begin{aligned} q(T_M\alpha) &= q(\Phi^{-1}U_M\Phi\alpha) \\ &= \det (\Phi\Phi^{-1}U_M\Phi\alpha) \\ &= \det (U_M\Phi\alpha) \\ &= \det (\Phi\alpha) \\ &= q(\alpha) \end{aligned}$$

and so T_M preserves the quadratic form q.

By using specific 2×2 matrices M, one can use the method above to compute specific Lorentz transformations. There are two comments which we might make here; they are not difficult to verify.

(1) If M_1 and M_2 are invertible 2×2 matrices with complex entries, then $U_{M_1} = U_{M_2}$ if and only if M_2 is a scalar multiple of M_1. Thus, all of the Lorentz transformations exhibited above are obtainable from unimodular matrices M, that is, from matrices M satisfying $\det M = 1$. If M_1 and M_2 are unimodular matrices such that $M_1 \neq M_2$ and $M_1 \neq -M_2$, then $T_{M_1} \neq T_{M_2}$.

(2) Not every Lorentz transformation is obtainable by the above method.

Exercises

1. Let M be a member of the complex orthogonal group, $O(n, C)$. Show that M^t, $\overline{M}$, and $M^* = \overline{M}^t$ also belong to $O(n, C)$.

2. Suppose M belongs to $O(n, C)$ and that M' is similar to M. Does M' also belong to $O(n, C)$?

3. Let

$$y_j = \sum_{k=1}^{n} M_{jk} x_k$$

where M is a member of $O(n, C)$. Show that

$$\sum_j y_j^2 = \sum_j x_j^2.$$

4. Let M be an $n \times n$ matrix over C with columns $M_1, M_2, \ldots, M_n$. Show that M belongs to $O(n, C)$ if and only if

$$M_j^t M_k = \delta_{jk}.$$

5. Let X be an $n \times 1$ matrix over C. Under what conditions does $O(n, C)$ contain a matrix M whose first column is X?

6. Find a matrix in $O(3, C)$ whose first row is $(2i, 2i, 3)$.

7. Let V be the space of all $n \times 1$ matrices over C and f the bilinear form on V given by $f(X, Y) = X^t Y$. Let M belong to $O(n, C)$. What is the matrix of f in the basis of V consisting of the columns $M_1, M_2, \ldots, M_n$ of M?

8. Let X be an $n \times 1$ matrix over C such that $X^t X = 1$, and I_j be the jth column of the identity matrix. Show there is a matrix M in $O(n, C)$ such that $MX = I_j$. If X has real entries, show there is an M in $O(n, R)$ with the property that $MX = I_j$.

9. Let V be the space of all $n \times 1$ matrices over C, A an $n \times n$ matrix over C, and f the bilinear form on V given by $f(X, Y) = X^t A Y$. Show that f is invariant under $O(n, C)$, i.e., $f(MX, MY) = f(X, Y)$ for all X, Y in V and M in $O(n, C)$, if and only if A commutes with each member of $O(n, C)$.

10. Let S be any set of $n \times n$ matrices over C and S' the set of all $n \times n$ matrices over C which commute with each element of S. Show that S' is an algebra over C.

11. Let F be a subfield of C, V a finite-dimensional vector space over F, and f a non-singular bilinear form on V. If T is a linear operator on V preserving f, prove that $\det T = \pm 1$.

12. Let F be a subfield of C, V the space of $n \times 1$ matrices over F, A an invertible $n \times n$ matrix over F, and f the bilinear form on V given by $f(X, Y) = X^tAY$. If M is an $n \times n$ matrix over F, show that M preserves f if and only if $A^{-1}M^tA = M^{-1}$.

13. Let g be a non-singular bilinear form on a finite-dimensional vector space V. Suppose T is an invertible linear operator on V and that f is the bilinear form on V given by $f(\alpha, \beta) = g(\alpha, T\beta)$. If U is a linear operator on V, find necessary and sufficient conditions for U to preserve f.

14. Let T be a linear operator on C^2 which preserves the quadratic form $x_1^2 - x_2^2$. Show that

(a) $\det(T) = \pm 1$.

(b) If M is the matrix of T in the standard basis, then $M_{22} = \pm M_{11}$, $M_{21} = \pm M_{12}$, $M_{11}^2 - M_{12}^2 = 1$.

(c) If $\det M = 1$, then there is a non-zero complex number c such that

$$M = \frac{1}{2}\begin{bmatrix} c + \frac{1}{c} & c - \frac{1}{c} \\ c - \frac{1}{c} & c + \frac{1}{c} \end{bmatrix}.$$

(d) If $\det M = -1$ then there is a complex number c such that

$$M = \frac{1}{2}\begin{bmatrix} c + \frac{1}{c} & c - \frac{1}{c} \\ -c + \frac{1}{c} & -c - \frac{1}{c} \end{bmatrix}.$$

15. Let f be the bilinear form on C^2 defined by

$$f((x_1, x_2), (y_1, y_2)) = x_1y_2 - x_2y_1.$$

Show that

(a) if T is a linear operator on C^2, then $f(T\alpha, T\beta) = (\det T)f(\alpha, \beta)$ for all α, β in C^2.

(b) T preserves f if and only if $\det T = +1$.

(c) What does (b) say about the group of 2×2 matrices M such that $M^tAM = A$ where

$$A = \begin{bmatrix} 0 & 1 \\ -1 & 0 \end{bmatrix}?$$

16. Let n be a positive integer, I the $n \times n$ identity matrix over C, and J the $2n \times 2n$ matrix given by

$$J = \begin{bmatrix} 0 & I \\ -I & 0 \end{bmatrix}.$$

Let M be a $2n \times 2n$ matrix over C of the form

$$M = \begin{bmatrix} A & B \\ C & D \end{bmatrix}$$

where A, B, C, D are $n \times n$ matrices over C. Find necessary and sufficient conditions on A, B, C, D in order that $M^tJM = J$.

17. Find all bilinear forms on the space of $n \times 1$ matrices over R which are invariant under $O(n, R)$.

18. Find all bilinear forms on the space of $n \times 1$ matrices over C which are invariant under $O(n, C)$.

Appendix

This Appendix separates logically into two parts. The first part, comprising the first three sections, contains certain fundamental concepts which occur throughout the book (indeed, throughout mathematics). It is more in the nature of an introduction for the book than an appendix. The second part is more genuinely an appendix to the text.

Section 1 contains a discussion of sets, their unions and intersections. Section 2 discusses the concept of function, and the related ideas of range, domain, inverse function, and the restriction of a function to a subset of its domain. Section 3 treats equivalence relations. The material in these three sections, especially that in Sections 1 and 2, is presented in a rather concise manner. It is treated more as an agreement upon terminology than as a detailed exposition. In a strict logical sense, this material constitutes a portion of the prerequisites for reading the book; however, the reader should not be discouraged if he does not completely grasp the significance of the ideas on his first reading. These ideas are important, but the reader who is not too familiar with them should find it easier to absorb them if he reviews the discussion from time to time while reading the text proper.

Sections 4 and 5 deal with equivalence relations in the context of linear algebra. Section 4 contains a brief discussion of quotient spaces. It can be read at any time after the first two or three chapters of the book. Section 5 takes a look at some of the equivalence relations which arise in the book, attempting to indicate how some of the results in the book might be interpreted from the point of view of equivalence relations. Section 6 describes the Axiom of choice and its implications for linear algebra.

A.1. Sets

We shall use the words 'set,' 'class,' 'collection,' and 'family' interchangeably, although we give preference to 'set.' If S is a set and x is an object in the set S, we shall say that x is a **member of** S, that x is an **element of** S, that x **belongs to** S, or simply that x is in S. If S has only a finite number of members, $x_1, \ldots, x_n$, we shall often describe S by displaying its members inside braces:

$$S = \{x_1, \ldots, x_n\}.$$

Thus, the set S of positive integers from 1 through 5 would be

$$S = \{1, 2, 3, 4, 5\}.$$

If S and T are sets, we say that S is a **subset of** T, or that S is **contained in** T, if each member of S is a member of T. Each set S is a subset of itself. If S is a subset of T but S and T are not identical, we call S a **proper subset** of T. In other words, S is a proper subset of T provided that S is contained in T but T is not contained in S.

If S and T are sets, the **union of** S **and** T is the set $S \cup T$, consisting of all objects x which are members of either S or T. The **intersection of** S **and** T is the set $S \cap T$, consisting of all x which are members of both S and T. For any two sets, S and T, the intersection $S \cap T$ is a subset of the union $S \cup T$. This should help to clarify the use of the word 'or' which will prevail in this book. When we say that x is either in S or in T, we do not preclude the possibility that x is in both S and T.

In order that the intersection of S and T should always be a set, it is necessary that one introduce the **empty set,** i.e., the set with no members. Then $S \cap T$ is the empty set if and only if S and T have no members in common.

We shall frequently need to discuss the union or intersection of several sets. If $S_1, \ldots, S_n$ are sets, their **union** is the set $\bigcup_{j=1}^{n} S_j$ consisting of all x which are members of at least one of the sets $S_1, \ldots, S_n$. Their **intersection** is the set $\bigcap_{j=1}^{n} S_j$, consisting of all x which are members of each of the sets $S_1, \ldots, S_n$. On a few occasions, we shall discuss the union or intersection of an infinite collection of sets. It should be clear how such unions and intersections are defined. The following example should clarify these definitions and a notation for them.

Example 1. Let R denote the set of all real numbers (the real line). If t is in R, we associate with t a subset S_t of R, defined as follows: S_t consists of all real numbers x which are not less than t.

(a) $S_{t_1} \cup S_{t_2} = S_t$, where t is the smaller of t_1 and t_2.

(b) $S_{t_1} \cap S_{t_2} = S_t$, where t is the larger of t_1 and t_2.

(c) Let I be the unit interval, that is, the set of all t in R satisfying $0 \leq t \leq 1$. Then

$$\bigcup_{t \text{ in } I} S_t = S_0$$

$$\bigcap_{t \text{ in } I} S_t = S_1.$$

A.2. Functions

A **function** consists of the following:

(1) a set X, called the domain of the function;

(2) a set Y, called the co-domain of the function;

(3) a rule (or correspondence) f, which associates with each element x of X a single element $f(x)$ of Y.

If (X, Y, f) is a function, we shall also say f **is a function from** X **into** Y. This is a bit sloppy, since it is not f which is the function; f is the rule of the function. However, this use of the same symbol for the function and its rule provides one with a much more tractable way of speaking about functions. Thus we shall say that f is a function from X into Y, that X is the domain of f, and that Y is the co-domain of f—all this meaning that (X, Y, f) is a function as defined above. There are several other words which are commonly used in place of the word 'function.' Some of these are 'transformation,' 'operator,' and 'mapping.' These are used in contexts where they seem more suggestive in conveying the role played by a particular function.

If f is a function from X into Y, the **range** (or **image**) of f is the set of all $f(x)$, x in X. In other words, the range of f consists of all elements y in Y such that $y = f(x)$ for some x in X. If the range of f is all of Y, we say that f is a **function from** X **onto** Y, or simply that f is **onto.** The range of f is often denoted $f(X)$.

Example 2. (a) Let X be the set of real numbers, and let $Y = X$. Let f be the function from X into Y defined by $f(x) = x^2$. The range of f is the set of all non-negative real numbers. Thus f is not onto.

(b) Let X be the Euclidean plane, and $Y = X$. Let f be defined as follows: If P is a point in the plane, then $f(P)$ is the point obtained by rotating P through 90° (about the origin, in the counterclockwise direction). The range of f is all of Y, i.e., the entire plane, and so f is onto.

(c) Again let X be the Euclidean plane. Coordinatize X as in analytic geometry, using two perpendicular lines to identify the points of X with ordered pairs of real numbers (x_1, x_2). Let Y be the x_1-axis, that is, all points (x_1, x_2) with $x_2 = 0$. If P is a point of X, let $f(P)$ be the point obtained by projecting P onto the x_1-axis, parallel to the x_2-axis. In other words, $f((x_1, x_2)) = (x_1, 0)$. The range of f is all of Y, and so f is onto.

(d) Let X be the set of real numbers, and let Y be the set of positive real numbers. Define a function f from X into Y by $f(x) = e^x$. Then f is a function from X onto Y.

(e) Let X be the set of positive real numbers and Y the set of all real numbers. Let f be the natural logarithm function, that is, the function defined by $f(x) = \log x = \ln x$. Again f is onto, i.e., every real number is the natural logarithm of some positive number.

Suppose that X, Y, and Z are sets, that f is a function from X into Y, and that g is a function from Y into Z. There is associated with f and g a function $g \circ f$ from X into Z, known as the **composition** of g and f. It is defined by

$$(g \circ f)(x) = g(f(x)).$$

For one simple example, let $X = Y = Z$, the set of real numbers; let f, g, h be the functions from X into X defined by

$$f(x) = x^2, \qquad g(x) = e^x, \qquad h(x) = e^{x^2}$$

and then $h = g \circ f$. The composition $g \circ f$ is often denoted simply gf; however, as the above simple example shows, there are times when this may lead to confusion.

One question of interest is the following. Suppose f is a function from X into Y. When is there a function g from Y into X such that $g(f(x)) = x$ for each x in X? If we denote by I the **identity function** on X, that is, the function from X into X defined by $I(x) = x$, we are asking the following: When is there a function g from Y into X such that $g \circ f = I$? Roughly speaking, we want a function g which 'sends each element of Y back where it came from.' In order for such a g to exist, f clearly must be 1:1, that is, f must have the property that if $x_1 \neq x_2$ then $f(x_1) \neq f(x_2)$. If f is 1:1, such a g does exist. It is defined as follows: Let y be an element of Y. If y is in the range of f, then there is an element x in X such that $y = f(x)$; and since f is 1:1, there is exactly one such x. Define $g(y) = x$. If y is not in the range of f, define $g(y)$ to be any element of X. Clearly we then have $g \circ f = I$.

Let f be a function from X into Y. We say that f is **invertible** if there is a function g from Y into X such that

(1) $g \circ f$ is the identity function on X,
(2) $f \circ g$ is the identity function on Y.

We have just seen that if there is a g satisfying (1), then f is 1:1. Similarly, one can see that if there is a g satisfying (2), the range of f is all of Y, i.e., f is onto. Thus, if f is invertible, f is 1:1 and onto. Conversely, if f is 1:1

and onto, there is a function g from Y into X which satisfies (1) and (2). Furthermore, this g is unique. It is the function from Y into X defined by this rule: if y is in Y, then $g(y)$ is the one and only element x in X for which $f(x) = y$.

If f is invertible (1:1 and onto), the **inverse** of f is the unique function f^{-1} from Y into X satisfying

(1′) $f^{-1}(f(x)) = x$, for each x in X,
(2′) $f(f^{-1}(y)) = y$, for each y in Y.

EXAMPLE 3. Let us look at the functions in Example 2.

(a) If $X = Y$, the set of real numbers, and $f(x) = x^2$, then f is not invertible. For f is neither 1:1 nor onto.

(b) If $X = Y$, the Euclidean plane, and f is 'rotation through 90°,' then f is both 1:1 and onto. The inverse function f^{-1} is 'rotation through −90°,' or 'rotation through 270°.'

(c) If X is the plane, Y the x_1-axis, and $f((x_1, x_2)) = (x_1, 0)$, then f is not invertible. For, although f is onto, f is not 1:1.

(d) If X is the set of real numbers, Y the set of positive real numbers, and $f(x) = e^x$, then f is invertible. The function f^{-1} is the natural logarithm function of part (e): $\log e^x = x$, $e^{\log y} = y$.

(e) The inverse of this natural logarithm function is the exponential function of part (d).

Let f be a function from X into Y, and let f_0 be a function from X_0 into Y_0. We call f_0 a **restriction** of f (or a restriction of f to X_0) if

(1) X_0 is a subset of X,
(2) $f_0(x) = f(x)$ for each x in X_0.

Of course, when f_0 is a restriction of f, it follows that Y_0 is a subset of Y. The name 'restriction' comes from the fact that f and f_0 have the same rule, and differ chiefly because we have restricted the domain of definition of the rule to the subset X_0 of X.

If we are given the function f and any subset X_0 of X, there is an obvious way to construct a restriction of f to X_0. We define a function f_0 from X_0 into Y by $f_0(x) = f(x)$ for each x in X_0. One might wonder why we do not call this *the* restriction of f to X_0. The reason is that in discussing restrictions of f we want the freedom to change the co-domain Y, as well as the domain X.

EXAMPLE 4. (a) Let X be the set of real numbers and f the function from X into X defined by $f(x) = x^2$. Then f is not an invertible function, but it is if we restrict its domain to the non-negative real numbers. Let X_0 be the set of non-negative real numbers, and let f_0 be the function from X_0 into X_0 defined by $f_0(x) = x^2$. Then f_0 is a restriction of f to X_0.

Now f is neither 1:1 nor onto, whereas f_0 is both 1:1 and onto. The latter statement simply says that each non-negative number is the square of exactly one non-negative number. The inverse function f_0^{-1} is the function from X_0 into X_0 defined by $f_0^{-1}(x) = \sqrt{x}$.

(b) Let X be the set of real numbers, and let f be the function from X into X defined by $f(x) = x^3 + x^2 + 1$. The range of f is all of X, and so f is onto. The function f is certainly not 1:1, e.g., $f(-1) = f(0)$. But f is 1:1 on X_0, the set of non-negative real numbers, because the derivative of f is positive for $x > 0$. As x ranges over all non-negative numbers, $f(x)$ ranges over all real numbers y such that $y \geq 1$. If we let Y_0 be the set of all $y \geq 1$, and let f_0 be the function from X_0 into Y_0 defined by $f_0(x) = f(x)$, then f_0 is a 1:1 function from X_0 onto Y_0. Accordingly, f_0 has an inverse function f_0^{-1} from Y_0 onto X_0. Any formula for $f_0^{-1}(y)$ is rather complicated.

(c) Again let X be the set of real numbers, and let f be the sine function, that is, the function from X into X defined by $f(x) = \sin x$. The range of f is the set of all y such that $-1 \leq y \leq 1$; hence, f is not onto. Since $f(x + 2\pi) = f(x)$, we see that f is not 1:1. If we let X_0 be the interval $-\pi/2 \leq x \leq \pi/2$, then f is 1:1 on X_0. Let Y_0 be the interval $-1 \leq y \leq 1$, and let f_0 be the function from X_0 into Y_0 defined by $f_0(x) = \sin x$. Then f_0 is a restriction of f to the interval X_0, and f_0 is both 1:1 and onto. This is just another way of saying that, on the interval from $-\pi/2$ to $\pi/2$, the sine function takes each value between -1 and 1 exactly once. The function f_0^{-1} is the inverse sine function:

$$f_0^{-1}(y) = \sin^{-1} y = \text{arc sin } y.$$

(d) This is a general example of a restriction of a function. It is much more typical of the type of restriction we shall use in this book than are the examples in (b) and (c) above. The example in (a) is a special case of this one. Let X be a set and f a function from X into itself. Let X_0 be a subset of X. We say that X_0 is **invariant under** f if for each x in X_0 the element $f(x)$ is in X_0. If X_0 is invariant under f, then f induces a function f_0 from X_0 into itself, by restricting the domain of its definition to X_0. The importance of invariance is that by restricting f to X_0 we can obtain a function from X_0 into itself, rather than simply a function from X_0 into X.

A.3. Equivalence Relations

An equivalence relation is a specific type of relation between pairs of elements in a set. To define an equivalence relation, we must first decide what a 'relation' is.

Certainly a formal definition of 'relation' ought to encompass such familiar relations as '$x = y$,' '$x < y$,' 'x is the mother of y,' and 'x is older than y.' If X is a set, what does it take to determine a relation between pairs of elements of X? What it takes, evidently, is a rule for determining whether, for any two given elements x and y in X, x stands in the given relationship to y or not. Such a rule R, we shall call a (binary) **relation** on X. If we wish to be slightly more precise, we may proceed as follows. Let $X \times X$ denote the set of all ordered pairs (x, y) of elements of X. A binary relation on X is a function R from $X \times X$ into the set $\{0, 1\}$. In other words, R assigns to each ordered pair (x, y) either a 1 or a 0. The idea is that if $R(x, y) = 1$, then x stands in the given relationship to y, and if $R(x, y) = 0$, it does not.

If R is a binary relation on the set X, it is convenient to write xRy when $R(x, y) = 1$. A binary relation R is called

(1) **reflexive,** if xRx for each x in X;
(2) **symmetric,** if yRx whenever xRy;
(3) **transitive,** if xRz whenever xRy and yRz.

An **equivalence relation** on X is a reflexive, symmetric, and transitive binary relation on X.

Example 5. (a) On any set, equality is an equivalence relation. In other words, if xRy means $x = y$, then R is an equivalence relation. For, $x = x$, if $x = y$ then $y = x$, if $x = y$ and $y = z$ then $x = z$. The relation '$x \neq y$' is symmetric, but neither reflexive nor transitive.

(b) Let X be the set of real numbers, and suppose xRy means $x < y$. Then R is not an equivalence relation. It is transitive, but it is neither reflexive nor symmetric. The relation '$x \leq y$' is reflexive and transitive, but not symmetric.

(c) Let E be the Euclidean plane, and let X be the set of all triangles in the plane E. Then congruence is an equivalence relation on X, that is, '$T_1 \cong T_2$' (T_1 is congruent to T_2) is an equivalence relation on the set of all triangles in a plane.

(d) Let X be the set of all integers:

$$\ldots, -2, -1, 0, 1, 2, \ldots.$$

Let n be a fixed positive integer. Define a relation R_n on X by: xR_ny if and only if $(x - y)$ is divisible by n. The relation R_n is called **congruence modulo** n. Instead of xR_ny, one usually writes

$$x \equiv y, \bmod n \qquad (x \text{ is congruent to } y \text{ modulo } n)$$

when $(x - y)$ is divisible by n. For each positive integer n, congruence modulo n is an equivalence relation on the set of integers.

(e) Let X and Y be sets and f a function from X into Y. We define a relation R on X by: x_1Rx_2 if and only if $f(x_1) = f(x_2)$. It is easy to verify that R is an equivalence relation on the set X. As we shall see, this one example actually encompasses all equivalence relations.

Suppose R is an equivalence relation on the set X. If x is an element of X, we let $E(x; R)$ denote the set of all elements y in X such that xRy. This set $E(x; R)$ is called the **equivalence class** of x (for the equivalence relation R). Since R is an equivalence relation, the equivalence classes have the following properties:

(1) Each $E(x; R)$ is non-empty; for, since xRx, the element x belongs to $E(x; R)$.

(2) Let x and y be elements of X. Since R is symmetric, y belongs to $E(x; R)$ if and only if x belongs to $E(y; R)$.

(3) If x and y are elements of X, the equivalence classes $E(x; R)$ and $E(y; R)$ are either identical or they have no members in common. First, suppose xRy. Let z be any element of $E(x; R)$ i.e., an element of X such that xRz. Since R is symmetric, we also have zRx. By assumption xRy, and because R is transitive, we obtain zRy or yRz. This shows that any member of $E(x; R)$ is a member of $E(y; E)$. By the symmetry of R, we likewise see that any member of $E(y; R)$ is a member of $E(x; R)$; hence $E(x; R) = E(y; R)$. Now we argue that if the relation xRy does not hold, then $E(x; R) \cap E(y; R)$ is empty. For, if z is in both these equivalence classes, we have xRz and yRz, thus xRz and zRy, thus xRy.

If we let $\mathfrak{F}$ be the family of equivalence classes for the equivalence relation R, we see that (1) each set in the family $\mathfrak{F}$ is non-empty, (2) each element x of X belongs to one and only one of the sets in the family $\mathfrak{F}$, (3) xRy if and only if x and y belong to the same set in the family $\mathfrak{F}$. Briefly, the equivalence relation R subdivides X into the union of a family of non-overlapping (non-empty) subsets. The argument also goes in the other direction. Suppose $\mathfrak{F}$ is any family of subsets of X which satisfies conditions (1) and (2) immediately above. If we define a relation R by (3), then R is an equivalence relation on X and $\mathfrak{F}$ is the family of equivalence classes for R.

Example 6. Let us see what the equivalence classes are for the equivalence relations in Example 5.

(a) If R is equality on the set X, then the equivalence class of the element x is simply the set $\{x\}$, whose only member is x.

(b) If X is the set of all triangles in a plane, and R is the congruence relation, about all one can say at the outset is that the equivalence class of the triangle T consists of all triangles which are congruent to T. One of the tasks of plane geometry is to give other descriptions of these equivalence classes.

(c) If X is the set of integers and R_n is the relation 'congruence modulo n,' then there are precisely n equivalence classes. Each integer x is uniquely expressible in the form $x = qn + r$, where q and r are integers and $0 \leq r \leq n - 1$. This shows that each x is congruent modulo n to exactly one of the n integers $0, 1, 2, \ldots, n-1$. The equivalence classes are

$$\begin{aligned} E_0 &= \{\ldots, -2n, -n, 0, n, 2n, \ldots\} \\ E_1 &= \{\ldots, 1-2n, 1-n, 1+n, 1+2n, \ldots\} \\ \vdots &= \qquad\qquad \vdots \\ E_{n-1} &= \{\ldots, n-1-2n, n-1-n, n-1, n-1+n, \\ &\qquad\qquad\qquad n-1+2n, \ldots\}. \end{aligned}$$

(d) Suppose X and Y are sets, f is a function from X into Y, and R is the equivalence relation defined by: $x_1 R x_2$ if and only if $f(x_1) = f(x_2)$. The equivalence classes for R are just the largest subsets of X on which f is 'constant.' Another description of the equivalence classes is this. They are in 1:1 correspondence with the members of the range of f. If y is in the range of f, the set of all x in X such that $f(x) = y$ is an equivalence class for R; and this defines a 1:1 correspondence between the members of the range of f and the equivalence classes of R.

Let us make one more comment about equivalence relations. Given an equivalence relation R on X, let $\mathfrak{F}$ be the family of equivalence classes for R. The association of the equivalence class $E(x; R)$ with the element x, defines a function f from X into $\mathfrak{F}$ (indeed, onto $\mathfrak{F}$):

$$f(x) = E(x; R).$$

This shows that R is the equivalence relation associated with a function whose domain is X, as in Example 5(e). What this tells us is that every equivalence relation on the set X is determined as follows. We have a rule (function) f which associates with each element x of X an object $f(x)$, and xRy if and only if $f(x) = f(y)$. Now one should think of $f(x)$ as some property of x, so that what the equivalence relation does (roughly) is to lump together all those elements of X which have this property in common. If the object $f(x)$ is the equivalence class of x, then all one has said is that the common property of the members of an equivalence class is that they belong to the same equivalence class. Obviously this doesn't say much. Generally, there are many different functions f which determine the given equivalence relation as above, and one objective in the study of equivalence relations is to find such an f which gives a meaningful and elementary description of the equivalence relation. In Section A.5 we shall see how this is accomplished for a few special equivalence relations which arise in linear algebra.

A.4. Quotient Spaces

Let V be a vector space over the field F, and let W be a subspace of V. There are, in general, many subspaces W' which are complementary to W, i.e., subspaces with the property that $V = W \oplus W'$. If we have an inner product on V, and W is finite-dimensional, there is a particular subspace which one would probably call the 'natural' complementary subspace for W. This is the orthogonal complement of W. But, if V has no structure in addition to its vector space structure, there is no way of selecting a subspace W' which one could call the natural complementary subspace for W. However, one can construct from V and W a vector space V/W, known as the 'quotient' of V and W, which will play the role of the natural complement to W. This quotient space is not a subspace of V, and so it cannot actually be a subspace complementary to W; but, it is a vector space defined only in terms of V and W, and has the property that it is isomorphic to any subspace W' which is complementary to W.

Let W be a subspace of the vector space V. If α and β are vectors in V, we say that α is **congruent to β modulo W**, if the vector $(\alpha - \beta)$ is in the subspace W. If α is congruent to β modulo W, we write

$$\alpha \equiv \beta, \qquad \text{mod } W.$$

Now congruence modulo W is an equivalence relation on V.

(1) $\alpha \equiv \alpha$, mod W, because $\alpha - \alpha = 0$ is in W.

(2) If $\alpha \equiv \beta$, mod W, then $\beta \equiv \alpha$, mod W. For, since W is a subspace of V, the vector $(\alpha - \beta)$ is in W if and only if $(\beta - \alpha)$ is in W.

(3) If $\alpha \equiv \beta$, mod W, and $\beta \equiv \gamma$, mod W, then $\alpha \equiv \gamma$, mod W. For, if $(\alpha - \beta)$ and $(\beta - \gamma)$ are in W, then $\alpha - \gamma = (\alpha - \beta) + \beta - \gamma)$ is in W.

The equivalence classes for this equivalence relation are known as the **cosets** of W. What is the equivalence class (coset) of a vector α? It consists of all vectors β in V such that $(\beta - \alpha)$ is in W, that is, all vectors β of the form $\beta = \alpha + \gamma$, with γ in W. For this reason, the coset of the vector α is denoted by

$$\alpha + W.$$

It is appropriate to think of the coset of α relative to W as the set of vectors obtained by translating the subspace W by the vector α. To picture these cosets, the reader might think of the following special case. Let V be the space R^2, and let W be a one-dimensional subspace of V. If we picture V as the Euclidean plane, W is a straight line through the origin. If $\alpha = (x_1, x_2)$ is a vector in V, the coset $\alpha + W$ is the straight line which passes through the point (x_1, x_2) and is parallel to W.

The collection of all cosets of W will be denoted by V/W. We now define a vector addition and scalar multiplication on V/W as follows:

$$\begin{aligned}(\alpha + W) + (\beta + W) &= (\alpha + \beta) + W\\ c(\alpha + W) &= (c\alpha) + W.\end{aligned}$$

In other words, the sum of the coset of α and the coset of β is the coset of $(\alpha + \beta)$, and the product of the scalar c and the coset of α is the coset of the vector $c\alpha$. Now many different vectors in V will have the same coset relative to W, and so we must verify that the sum and product above depend only upon the cosets involved. What this means is that we must show the following:

(a) If $\alpha \equiv \alpha'$, mod W, and $\beta \equiv \beta'$, mod W, then

$$\alpha + \beta \rightarrow \alpha' + \beta', \quad \text{mod } W.$$

(2) If $\alpha \equiv \alpha'$, mod W, then $c\alpha \equiv c\alpha'$, mod W.

These facts are easy to verify. (1) If $\alpha - \alpha'$ is in W and $\beta - \beta'$ is in W, then since $(\alpha + \beta) - (\alpha' - \beta') = (\alpha - \alpha') + (\beta - \beta')$, we see that $\alpha + \beta$ is congruent to $\alpha' - \beta'$ modulo W. (2) If $\alpha - \alpha'$ is in W and c is any scalar, then $c\alpha - c\alpha' = c(\alpha - \alpha')$ is in W.

It is now easy to verify that V/W, with the vector addition and scalar multiplication defined above, is a vector space over the field F. One must directly check each of the axioms for a vector space. Each of the properties of vector addition and scalar multiplication follows from the corresponding property of the operations in V. One comment should be made. The zero vector in V/W will be the coset of the zero vector in V. In other words, W is the zero vector in V/W.

The vector space V/W is called the **quotient** (or difference) of V and W. There is a natural linear transformation Q from V onto V/W. It is defined by $Q(\alpha) = \alpha + W$. One should see that we have defined the operations in V/W just so that this transformation Q would be linear. Note that the null space of Q is exactly the subspace W. We call Q the **quotient transformation** (or **quotient mapping**) of V onto V/W.

The relation between the quotient space V/W and subspaces of V which are complementary to W can now be stated as follows.

Theorem. *Let* W *be a subspace of the vector space* V, *and let* Q *be the quotient mapping of* V *onto* V/W. *Suppose* W′ *is a subspace of* V. *Then* $V = W \oplus W'$ *if and only if the restriction of* Q *to* W′ *is an isomorphism of* W′ *onto* V/W.

Proof. Suppose $V = W \oplus W'$. This means that each vector α in V is uniquely expressible in the form $\alpha = \gamma + \gamma'$, with γ in W and γ' in W'. Then $Q\alpha = Q\gamma + Q\gamma' = Q\gamma'$, that is $\alpha + W = \gamma' + W$. This shows that Q maps W' onto V/W, i.e., that $Q(W') = V/W$. Also Q is 1:1 on W'; for suppose γ_1' and γ_2' are vectors in W' and that $Q\gamma_1' = Q\gamma_2'$. Then $Q(\gamma_1' - \gamma_2') = 0$ so that $\gamma_1' - \gamma_2'$ is in W. This vector is also in W', which is disjoint from W; hence $\gamma_1' - \gamma_2' = 0$. The restriction of Q to W' is therefore a one-one linear transformation of W' onto V/W.

Suppose W' is a subspace of V such that Q is one-one on W' and $Q(W') = V/W$. Let α be a vector in V. Then there is a vector γ' in W' such that $Q\gamma' = Q\alpha$, i.e., $\gamma' + W = \alpha + W$. This means that $\alpha = \gamma + \gamma'$ for some vector γ in W. Therefore $V = W + W'$. To see that W and W'

are disjoint, suppose γ is in both W and W'. Since γ is in W, we have $Q\gamma = 0$. But Q is 1:1 on W', and so it must be that $\gamma = 0$. Thus we have $V = W \oplus W'$. ∎

What this theorem really says is that W' is complementary to W if and only if W' is a subspace which contains exactly one element from each coset of W. It shows that when $V = W \oplus W'$, the quotient mapping Q 'identifies' W' with V/W. Briefly $(W \oplus W')/W$ is isomorphic to W' in a 'natural' way.

One rather obvious fact should be noted. If W is a subspace of the finite-dimensional vector space V, then

$$\dim W + \dim (V/W) = \dim V.$$

One can see this from the above theorem. Perhaps it is easier to observe that what this dimension formula says is

$$\text{nullity } (Q) + \text{rank } (Q) = \dim V.$$

It is not our object here to give a detailed treatment of quotient spaces. But there is one fundamental result which we should prove.

Theorem. *Let* V *and* Z *be vector spaces over the field* F. *Suppose* T *is a linear transformation of* V *onto* Z. *If* W *is the null space of* T, *then* Z *is isomorphic to* V/W.

Proof. We define a transformation U from V/W into Z by $U(\alpha + W) = T\alpha$. We must verify that U is well defined, i.e., that if $\alpha + W = \beta + W$ then $T\alpha = T\beta$. This follows from the fact that W is the null space of T; for, $\alpha + W = \beta + W$ means $\alpha - \beta$ is in W, and this happens if and only if $T(\alpha - \beta) = 0$. This shows not only that U is well defined, but also that U is one-one.

It is now easy to verify that U is linear and sends V/W onto Z, because T is a linear transformation of V onto Z. ∎

A.5. Equivalence Relations in Linear Algebra

We shall consider some of the equivalence relations which arise in the text of this book. This is just a sampling of such relations.

(1) Let m and n be positive integers and F a field. Let X be the set of all $m \times n$ matrices over F. Then row-equivalence is an equivalence relation on the set X. The statement 'A is row-equivalent to B' means that A can be obtained from B by a finite succession of elementary row operations. If we write $A \sim B$ for A is row-equivalent to B, then it is not difficult to check the properties (i) $A \sim A$; (ii) if $A \sim B$, then $B \sim A$;

(iii) if $A \sim B$ and $B \sim C$, then $A \sim C$. What do we know about this equivalence relation? Actually, we know a great deal. For example, we know that $A \sim B$ if and only if $A = PB$ for some invertible $m \times m$ matrix P; or, $A \sim B$ if and only if the homogeneous systems of linear equations $AX = 0$ and $BX = 0$ have the same solutions. We also have very explicit information about the equivalence classes for this relation. Each $m \times n$ matrix A is row-equivalent to one and only one row-reduced echelon matrix. What this says is that each equivalence class for this relation contains precisely one row-reduced echelon matrix R; the equivalence class determined by R consists of all matrices $A = PR$, where P is an invertible $m \times m$ matrix. One can also think of this description of the equivalence classes in the following way. Given an $m \times n$ matrix A, we have a rule (function) f which associates with A the row-reduced echelon matrix $f(A)$ which is row-equivalent to A. Row-equivalence is completely determined by f. For, $A \sim B$ if and only if $f(A) = f(B)$, i.e., if and only if A and B have the same row-reduced echelon form.

(2) Let n be a positive integer and F a field. Let X be the set of all $n \times n$ matrices over F. Then similarity is an equivalence relation on X; each $n \times n$ matrix A is similar to itself; if A is similar to B, then B is similar to A; if A is similar to B and B is similar to C, then A is similar to C. We know quite a bit about this equivalence relation too. For example, A is similar to B if and only if A and B represent the same linear operator on F^n in (possibly) different ordered bases. But, we know something much deeper than this. Each $n \times n$ matrix A over F is similar (over F) to one and only one matrix which is in rational form (Chapter 7). In other words, each equivalence class for the relation of similarity contains precisely one matrix which is in rational form. A matrix in rational form is determined by a k-tuple $(p_1, \ldots, p_k)$ of monic polynomials having the property that p_{j+1} divides p_j, $j = 1, \ldots, k - 1$. Thus, we have a function f which associates with each $n \times n$ matrix A a k-tuple $f(A) = (p_1, \ldots, p_k)$ satisfying the divisibility condition p_{j+1} divides p_j. And, A and B are similar if and only if $f(A) = f(B)$.

(3) Here is a special case of Example 2 above. Let X be the set of 3×3 matrices over a field F. We consider the relation of similarity on X. If A and B are 3×3 matrices over F, then A and B are similar if and only if they have the same characteristic polynomial and the same minimal polynomial. Attached to each 3×3 matrix A, we have a pair (f, p) of monic polynomials satisfying

(a) $\deg f = 3$,
(b) p divides f,

f being the characteristic polynomial for A, and p the minimal polynomial for A. Given monic polynomials f and p over F which satisfy (a) and (b), it is easy to exhibit a 3×3 matrix over F, having f and p as its characteristic and minimal polynomials, respectively. What all this tells us is the following. If we consider the relation of similarity on the set of 3×3 matrices over F, the equivalence classes are in one-one correspondence with ordered pairs (f, p) of monic polynomials over F which satisfy (a) and (b).

A.6. The Axiom of Choice

Loosely speaking, the Axiom of Choice is a rule (or principle) of thinking which says that, given a family of non-empty sets, we can choose one element out of each set. To be more precise, suppose that we have an index set A and for each α in A we have an associated set S_α, which is non-empty. To 'choose' one member of each S_α means to give a rule f which associates with each α some element $f(\alpha)$ in the set S_α. The axiom of choice says that this is possible, i.e., given the family of sets $\{S_\alpha\}$, there exists a function f from A into

$$\bigcup_\alpha S_\alpha$$

such that $f(\alpha)$ is in S_α for each α. This principle is accepted by most mathematicians, although many situations arise in which it is far from clear how any explicit function f can be found.

The Axiom of Choice has some startling consequences. Most of them have little or no bearing on the subject matter of this book; however, one consequence is worth mentioning: Every vector space has a basis. For example, the field of real numbers has a basis, as a vector space over the field of rational numbers. In other words, there is a subset S of R which is linearly independent over the field of rationals and has the property that each real number is a rational linear combination of some finite number of elements of S. We shall not stop to derive this vector space result from the Axiom of Choice. For a proof, we refer the reader to the book by Kelley in the bibliography.

Bibliography

Halmos, P., *Finite-Dimensional Vector Spaces,* D. Van Nostrand Co., Princeton, 1958.

Jacobson, N., *Lectures in Abstract Algebra,* II, D. Van Nostrand Co., Princeton, 1953.

Kelley, John L., *General Topology,* D. Van Nostrand Co., Princeton, 1955.

MacLane, S. and Birkhoff, G., *Algebra,* The Macmillan Co., New York, 1967.

Schreier, O. and Sperner, E., *Introduction to Modern Algebra and Matrix Theory,* 2nd Ed., Chelsea Publishing Co., New York, 1955.

van der Waerden, B. L., *Modern Algebra* (two volumes), Rev. Ed., Frederick Ungar Publishing Co., New York, 1969.

Index